WINNER, NATIONAL BOOK CRITICS CIRCLE AWARD FOR BIOGRAPHY
FINALIST, PULITZER PRIZE IN BIOGRAPHY

Praise for *A Beautiful Mind*

"Two paragraphs and I was hooked!" — Oliver Sacks

"A brilliant book." — David Herbert Donald

"Reads like a fine novel." — David Goodstein, *The New York Times*

"Powerfully affecting . . . a three-handkerchief read."
— Charles C. Mann, *The Wall Street Journal*

"A triumph of intellectual biography." — Robert Boynton, *Newsday*

"Might be compared to a Rembrandt portrait, filled with somber shadows and radiant light effects . . . simply a beautiful book." — Marcia Bartusiak, *The Boston Globe*

"A remarkable look into the arcane world of mathematics and the tragedy of madness."
— Simon Singh, *The New York Times Book Review*

"A narrative of compelling power." — John Allen Paulos, *Los Angeles Times*

"A wonderfully absorbing puzzle." — Claire Douglas, *Washington Post Book World*

"A poetical love and coming-of-age story." — Ted Anton, *Chicago Tribune*

"The stuff of classic tragedy." — Robert A. Burton, *San Jose Mercury News*

"A powerful story brilliantly told." — Will St. John, *Detroit Free Press*

"A worthy subject and a fascinating book." — Craig Ryan, *Portland Oregonian*

"A page-turner." — Claiborne Smith, *Austin Chronicle*

"An arresting portrait." — June Kinoshita, *St. Petersburg Times*

"The parabolic arc of an American genius . . . superbly and thrillingly limned."
— Will Blythe, *Mirabella*

"A staggering feat of writing and reporting." — Michael J. Mandel, *BusinessWeek*

"Profoundly sad yet redemptive." — *Worth Magazine*

"Instead of facile theories, the reader enjoys wonder and astonishment."
— Richard Dooling, *Salon*

"Extraordinarily moving." — Jeremy Bernstein, *Commentary*

"Absolutely fascinating." — Jim Holt, *Slate*

"An engrossing, ultimately uplifting book." — Gregg Sapp, *Kirkus Reviews*

"Will touch any reader who understands what it means to hope — or to fear." — *Booklist*

"Unique." — *The Economist*

"A compelling book about a phenomenal figure." — Roy Porter, *The Times*

"Unblinking yet empathic." — Daniel Kevles, *Times Literary Supplement*

"A romantic human story." — Steven McCaffery, *Irish News*

"Genuinely compulsive." — Jon Oberlander, *Sunday Herald*

"An astonishing achievement." — Brian Rotman, *London Review of Books*

"A masterpiece of oral history." — Karl Sigmund, *Nature*

"Be prepared for the birth of a new culture hero." — Peter Wilhelm, *Business Day*

"I defy anyone to read Sylvia Nasar's prologue without being moved."
— Christopher Beauman, *Broadway Ham & High*

"A magnificent biography." — Roy Weintraub, *Journal of the History of Economic Thought*

"High drama." — Wade Roush, *MIT Technology Review*

"Deeply moving." — Paul Trachtman, *Smithsonian Magazine*

"Presented with grace and skill." — Brian Hayes, *The Sciences*

"A must-read with something for everyone." — Keith Devlin, *New Scientist*

"Fascinating, complicated, and studious." — Mark H. Fleisher, *JAMA*

"A deeply moving love story, an account of the centrality of human relationships."
— Richard Wyatt and Kay Jamison, *The New England Journal of Medicine*

"A gripping narrative."
— Kenneth Arrow, Nobel Laureate, *The Times Higher Education Supplement*

A Beautiful Mind

The Life of Mathematical Genius and
Nobel Laureate John Nash

SYLVIA NASAR

Simon & Schuster Paperbacks

NEW YORK   LONDON   TORONTO   SYDNEY

Simon & Schuster Paperbacks
A Division of Simon & Schuster, Inc.
1230 Avenue of the Americas
New York, NY 10020

First Simon & Schuster trade paperback edition July 2011
SIMON & SCHUSTER PAPERBACKS and colophon are registered
trademarks of Simon & Schuster, Inc.
For information about special discounts for bulk purchases,
please contact Simon & Schuster Special Sales:
1-800-456-6798 or business@simonandschuster.com
Designed by Edith Fowler
Manufactured in the United States of America

10   9   8   7   6   5   4   3   2   1

The Library of Congress has cataloged the hardcover edition as follows:
Nasar, Sylvia.
    A beautiful mind : a biography of John Forbes Nash, Jr.,
winner of the Nobel Prize in economics, 1994 /
Sylvia Nasar
       p.   cm.
    Includes bibliographical references and index.
    1. Nash, John F., 1928–   . 2. Mathematicians—
United States—Biography   I. Title
OA29.N25N37   1998
510'.92
[B]—DC21   98-2795

ISBN-13: 978-0-684-81906-8
ISBN-10:   0-684-81906-8
ISBN-13: 978-1-4516-2842-5 (pbk)
ISBN: 978-1-4391-2649-3 (ebook)

The author and publisher gratefully acknowledge permission to reprint material from the
following works:
"The RAND Hymn," words and music by Malvina Reynolds, © copyright 1961 by
Schroeder Music Co. (ASCAP). Used by permission. All rights reserved. "John F. Nash
Jr." (Autobiographical Essay) and "The Work of John Nash in Game Theory" (Nobel
Seminar), in Les Prix Nobel 1994 (Stockholm: Norstedts Tryckeri, 1995). Copyright ©
The Nobel Foundation, 1994. Excerpts from "Waking in the Blue" from Life Studies by
Robert Lowell. Copyright © 1959 by Robert Lowell. Copyright renewed © 1987 by Harriet
Lowell, Sheridan Lowell, and Caroline Lowell. Reprinted by permission of Farrar, Straus
& Giroux, Inc. Excerpts from the letters of Robert Lowell reprinted with the permission of
the Estate of Robert Lowell.

FOR ALICIA ESTHER LARDE NASH

Another race hath been, and other palms are won.
Thanks to the human heart by which we live,
Thanks to its tenderness, its joys, and fears,
To me the meanest flower that blows can give
Thoughts that do often lie too deep for tears.

—WILLIAM WORDSWORTH,
    "Intimations of Immortality"

Contents

Part Three: A Slow Fire Burning

Part Four: The Lost Years

Part Five: The Most Worthy

Foreword

(Adapted from remarks at John Nash's 80th birthday Festschrift)

IN JUNE 2006, I went to St. Petersburg to track down the forty-year-old mathematician who had solved the Poincaré Conjecture. Reputedly a hermit with wild hair and long nails who lived in the woods on mushrooms, he was up for a Fields Medal and a $1 million cash prize but had gone into hiding, not just from the media but from the math community. Meanwhile, some folks in Beijing were claiming that they'd beaten him to the punch. It was a great story—if only we could find him. ❦

After four frustrating days in Russia, my colleague and I hadn't found a soul who had seen or talked to the guy or his family in years. Then, when we had pretty much thrown in the towel, we stumbled on his mother's apartment more or less by accident and, voilà, there was the "hermit," dressed in a sports jacket and Italian loafers, evidently having lunch and watching soccer on TV.

He gestured for us to sit down and explain what we wanted.

"My name is Sylvia Nasar," I began. "I'm a journalist from New York and I'm working on . . ."

He interrupted: "You're a writer?"

I nodded.

"I didn't read the book," he said, "but I saw the movie with Russell Crowe."

The point is that, no matter where in the world you are, you'd have to be a *real* hermit not to know the inspiring story of John Nash.

There are lots of stories about the rise and fall of remarkable individuals. But there are very few stories, much less true stories, with a genuine third act. Nash's story had—has—such a third act. Act III of Nash's life story is his miraculous reawakening.

It is that third act that makes Nash's story resonate with people all over the world—most especially with those who suffer from devastating mental illnesses or love someone who does.

At one point in the movie, when it looks as if things were all over for Nash, his wife, Alicia, takes John's hand, places it over her heart, and says, "I have to believe that something extraordinary is possible."

Something extraordinary *was* possible.

Of all the letters I've received from readers, my favorite came from a homeless man. It arrived in a dirty envelope with no return address, and it was scrawled on neon orange paper. It was signed "Berkeley Baby." It would never have made it past the *New York Times* mailroom after the anthrax scare.

The letter writer turned out to have been the night rewrite editor on the metro desk at the *New York Times* before he was diagnosed with paranoid schizophrenia in the mid-1970s. Since then, he had adopted the name Berkeley Baby and lived on the streets of Berkeley, California, near the university, a forlorn, sad figure not unlike the Phantom of Fine Hall.

He wrote, "John Nash's story gives me hope that one day the world will come back to me too."

The world came back to John Nash after more than thirty years, and it was the third act of his life that drew me to his story in the first place. In the early 1990s, I was an economics reporter at the *New York Times*. I was interviewing a Princeton professor about some trade statistics when he mentioned a rumor that a "crazy mathematician" who hung around the math building might be on the short list for a Nobel prize in economics. "You don't mean the Nash of the Nash equilibrium?" I asked. He told me to call a couple of people in the math department to learn more. By the time I put down the phone, I realized that this was a fairy tale, Greek myth, and Shakespearean tragedy rolled into one.

I didn't write the story immediately. Lots of people wind up on short lists for the Nobel and never win, so writing about him in a newspaper would have been an invasion of privacy. In any case, someone else got the prize in 1993. The next year, however, I saw Nash's name in the Nobel announcement. I ran over to my editor to pitch the story and actually made him cry.

It was a difficult story to get. Nobody who knew any facts was willing to go on the record or even talk to me. Martha Legg, Nash's sister, finally broke the silence about the nature of the illness that had wrecked his life.

Lloyd Shapley, another pioneer of game theory, described Nash as a graduate student in the late 1940s, when he wrote his seminal papers on game theory: "He was immature, he was obnoxious, he was a brat. What redeemed him was a keen, logical, beautiful mind.

So now you know to whom I owe the title of the biography.

Because Nash's story is so familiar, I'd like to share some of the less familiar parts, including how the book came to be and some of the things that happened after the book and movie broke off.

In June 1995, I found myself in Jerusalem. By then, I had written a book proposal, gotten a publisher, and was about to spend a year at the Institute for Advanced Study. Unfortunately, I'd never met my subject or exchanged more than a few words with him on the phone. When I found out that Nash was going to a game theory conference in Jerusalem, I thought I'd go too.

Some will remember what Nash said about John von Neumann, who had given him some of the worst advice ever given to a doctoral student. Fortunately, Nash had ignored von Neumann's advice. Unfortunately for me, he had also decided to ignore the advice of many of his friends and supporters to cooperate with his biographer.

"Dear Mrs. Nasar," a typical note began. "I have decided to take a position of Swiss neutrality . . ."

Everyone knows the phrase "It takes a village." It had taken someone weeks of dogged reporting to put together a six-line CV and a short list of Nash's publications. It took hundreds of sources to piece together his whole story. No single individual, not even Alicia or his sons, knew the whole story.

It turned out to be possible to stitch together thousands of bits and pieces—gathered from hundreds of interviews, dozens of letters, and a smattering of documents—into a narrative. It worked partly because the mathematics community is like a Greek chorus—watching, commenting, remembering, filling in the background, explaining the action.

But ultimately it worked because John Nash was always a star and, all his life, people around him couldn't take their eyes off him, couldn't stop thinking about him. How many of us, years from now, will live as brightly in the memories as he has for so long . . . and long before the fairy tale ending.

And, of course, it worked because Alicia never stopped believing that something extraordinary is possible. She wanted his story told because she thought it would be inspiring for people with mental illnesses.

A friend once asked Nash where Alicia was. John answered "Having dinner with Sylvia." After a pause, he added, "I hope they aren't talking about me."

Actually, Alicia was extremely protective of Nash's privacy and incredibly discreet. There was only one exception: we were in the basement of her bank, sifting through the contents of her safety deposit box looking for photos. She came across these little 2 x 2 snapshots of her and John with Felix Browder at the UC Berkeley swimming pool. That was the beefcake shot that convinced Graydon Carter not to kill the book excerpt in *Vanity Fair*. (Brian Grazer told me he'd bought the rights to the book because Graydon told him to.)

Alicia was holding the picture and chuckling: "Doesn't he have the *greatest* legs?!"

Nash never did agree to give me an interview for the book.

Post-publication meetings between biographers and their living subjects, authorized or not, often take place in lawyers' offices. Ours did not.

Instead, we met at a Broadway play, *Amy's View*, starring Judi Dench. Nash told us that it was his first Broadway play. He and Alicia liked *Proof* better. I was sitting behind them and could see their shoulders shaking with laughter. David Auburn, the playwright of *Proof*, told me that he got the idea of turning the sisters into daughters of a mad mathematician from John's story.

Watching someone get his life back is an incredibly sweet experience—even little things like driving again or having coffee at Starbucks. When I asked Nash, for a *New York Times* story about Nobel prize winners and how they spent their prize money, how the prize had affected his life, he said, well, now he could buy a $2 cup of coffee at Starbucks. "Poor people can't do that," he'd observed.

Watching someone get his life back and in the process touch the lives of millions of people is equally remarkable. People have told me that they'll never again pass someone on the street with matted hair and filthy clothes who's shouting at the air without telling themselves that he or she is someone's child or sibling, someone with a past, and maybe someone, like John Nash, with a future. That's the power of stories.

. . .

The first time I went to the movie set, Ron Howard was filming the wedding scene on the Lower East. All the principals were there because a *New York Times* reporter was going to interview us all.

I met Akiva Goldsman, the screenwriter without whom the movie would never have been made, much less won an Oscar. It was Goldsman's brilliant idea to have the audience see the world through Nash's eyes for the first half of the movie. Stepping into Nash's shoes and then having the rug pulled out from under them, the movie audience was not only drawn into the story but experienced what it's like not to be able to distinguish between reality and delusion.

After Ron Howard screened the movie for John and Alicia, I called them. "So, John, what did you think?"

I don't remember all of his exact words, but I do recall that he mentioned liking three things:

First of all, that it was funny.

Second, John being an action movie fan, that the pacing was fast.

Third . . .

"I think Russell Crowe looks a little like me."

Just in case you think Nash was kidding himself, at a Q&A with Ron Howard at New York University's film school, some mathematicians from the Courant Institute came up to Ron Howard to tell him that John Nash actually had looked like Russell Crowe in the white T-shirt scene.

The movie turned Nash into a celebrity. I was on a flight to Mumbai where I was meeting Amartya Sen, also a Nobel laureate in economics, at a game theory conference. The woman to my left had just asked me why I was going to India when the flight attendant came by with an Indian newspaper, and there was a photograph of John Nash, the keynote speaker, on the front page, right next to one of Sen. All I had to do was to point. In Mumbai, as in Beijing and other places he was invited to speak, he was mobbed by hundreds of reporters and well-wishers.

Nash's story appealed to children and teenagers who were thrilled by the notion that someone really young—and quirky—could accomplish amazing things and outsmart the older generation. And it made math seem cool.

> Dear Mr. Nash,
> Hi! I am 9 years old. My name is Ellie. I am a girl. I really admire you. You are my roll [*sic*] model for a lot of things. I think you are the smartest person who ever lived. I really wish to be like you. I would love to study math. The only problem with that is that I am not very good at math. I can do it. I like it. I am just not good at it. Was that what it was like for you when you were a kid? Please write back. Love, Ellie
>
> P.S. I LOVE your name.

—Sylvia Nasar

Prologue

Where the statue stood
Of Newton with his prism and silent face,
The marble index of a mind for ever
Voyaging through strange seas of Thought, alone.
— WILLIAM WORDSWORTH

JOHN FORBES NASH, JR.—mathematical genius, inventor of a theory of rational behavior, visionary of the thinking machine—had been sitting with his visitor, also a mathematician, for nearly half an hour. It was late on a weekday afternoon in the spring of 1959, and, though it was only May, uncomfortably warm. Nash was slumped in an armchair in one corner of the hospital lounge, carelessly dressed in a nylon shirt that hung limply over his unbelted trousers. His powerful frame was slack as a rag doll's, his finely molded features expressionless. He had been staring dully at a spot immediately in front of the left foot of Harvard professor George Mackey, hardly moving except to brush his long dark hair away from his forehead in a fitful, repetitive motion. His visitor sat upright, oppressed by the silence, acutely conscious that the doors to the room were locked. Mackey finally could contain himself no longer. His voice was slightly querulous, but he strained to be gentle. "How could you," began Mackey, "how could you, a mathematician, a man devoted to reason and logical proof . . . how could you believe that extraterrestrials are sending you messages? How could you believe that you are being recruited by aliens from outer space to save the world? How could you . . . ?"

Nash looked up at last and fixed Mackey with an unblinking stare as cool and dispassionate as that of any bird or snake. "Because," Nash said slowly in his soft, reasonable southern drawl, as if talking to himself, "the ideas I had about supernatural beings came to me the same way that my mathematical ideas did. So I took them seriously." [1]

The young genius from Bluefield, West Virginia—handsome, arrogant, and highly eccentric—burst onto the mathematical scene in 1948. Over the next decade, a decade as notable for its supreme faith in human rationality as for its dark anxieties about mankind's survival,[2] Nash proved himself, in the words of the eminent geometer Mikhail Gromov, "the most remarkable mathematician of the second half of the century."[3] Games of strategy, economic rivalry, computer architecture, the shape of the universe, the geometry of imaginary spaces, the mystery of prime numbers—all engaged his wide-ranging imagination. His ideas were of the deep and wholly unanticipated kind that pushes scientific thinking in new directions.

Geniuses, the mathematician Paul Halmos wrote, "are of two kinds: the ones who are just like all of us, but very much more so, and the ones who, apparently, have an extra human spark. We can all run, and some of us can run the mile in less than 4 minutes; but there is nothing that most of us can do that compares with the creation of the Great G-minor Fugue."[4] Nash's genius was of that mysterious variety more often associated with music and art than with the oldest of all sciences. It wasn't merely that his mind worked faster, that his memory was more retentive, or that his power of concentration was greater. The flashes of intuition were non-rational. Like other great mathematical intuitionists—Georg Friedrich Bernhard Riemann, Jules Henri Poincaré, Srinivasa Ramanujan—Nash saw the vision first, constructing the laborious proofs long afterward. But even after he'd try to explain some astonishing result, the actual route he had taken remained a mystery to others who tried to follow his reasoning. Donald Newman, a mathematician who knew Nash at MIT in the 1950s, used to say about him that "everyone else would climb a peak by looking for a path somewhere on the mountain. Nash would climb another mountain altogether and from that distant peak would shine a searchlight back onto the first peak."[5]

No one was more obsessed with originality, more disdainful of authority, or more jealous of his independence. As a young man he was surrounded by the high priests of twentieth-century science—Albert Einstein, John von Neumann, and Norbert Wiener—but he joined no school, became no one's disciple, got along largely without guides or followers. In almost everything he did—from game theory to geometry—he thumbed his nose at the received wisdom, current fashions, established methods. He almost always worked alone, in his head, usually walking, often whistling Bach. Nash acquired his knowledge of mathematics not mainly from studying what other mathematicians had discovered, but by rediscovering their truths for himself. Eager to astound, he was always on the lookout for the really big problems. When he focused on some new puzzle, he saw dimensions that people who really knew the subject (he never did) initially dismissed as naive or wrongheaded. Even as a student, his indifference to others' skepticism, doubt, and ridicule was awesome.

Nash's faith in rationality and the power of pure thought was extreme, even for a very young mathematician and even for the new age of computers, space travel, and nuclear weapons. Einstein once chided him for wishing to amend relativity theory without studying physics.[6] His heroes were solitary thinkers and supermen like Newton and Nietzsche.[7] Computers and science fiction were his passions. He considered "thinking machines," as he called them, superior in some ways to human beings.[8] At one point, he became fascinated by the possibility that drugs could heighten physical and intellectual performance.[9] He was beguiled by the idea of alien races of hyper-rational beings who had taught themselves to disregard all emotion.[10] Compulsively rational, he wished to turn life's decisions — whether to take the first elevator or wait for the next one, where to bank his money, what job to accept, whether to marry — into calculations of advantage and disadvantage, algorithms or mathematical rules divorced from emotion, convention, and tradition. Even the small act of saying an automatic hello to Nash in a hallway could elicit a furious "Why are you saying hello to me?"[11]

His contemporaries, on the whole, found him immensely strange. They described him as "aloof," "haughty," "without affect," "detached," "spooky," "isolated," and "queer."[12] Nash mingled rather than mixed with his peers. Preoccupied with his own private reality, he seemed not to share their mundane concerns. His manner — slightly cold, a bit superior, somewhat secretive — suggested something "mysterious and unnatural." His remoteness was punctuated by flights of garrulousness about outer space and geopolitical trends, childish pranks, and unpredictable eruptions of anger. But these outbursts were, more often than not, as enigmatic as his silences. "He is not one of us" was a constant refrain. A mathematician at the Institute for Advanced Study remembers meeting Nash for the first time at a crowded student party at Princeton:

> I noticed him very definitely among a lot of other people who were there. He was sitting on the floor in a half-circle discussing something. He made me feel uneasy. He gave me a peculiar feeling. I had a feeling of a certain strangeness. He was different in some way. I was not aware of the extent of his talent. I had no idea he would contribute as much as he really did.[13]

But he did contribute, in a big way. The marvelous paradox was that the ideas themselves were not obscure. In 1958, *Fortune* singled Nash out for his achievements in game theory, algebraic geometry, and nonlinear theory, calling him the most brilliant of the younger generation of new ambidextrous mathematicians who worked in both pure and applied mathematics.[14] Nash's insight into the dynamics of human rivalry — his theory of rational conflict and cooperation — was to become one of the most influential ideas of the twentieth century, transforming the young science of economics the way that Mendel's ideas of genetic transmission, Darwin's model of natural selection, and Newton's celestial mechanics reshaped biology and physics in their day.

It was the great Hungarian-born polymath John von Neumann who first recognized that social behavior could be analyzed as games. Von Neumann's 1928 article on parlor games was the first successful attempt to derive logical and mathematical rules about rivalries.[15] Just as Blake saw the universe in a grain of sand, great scientists have often looked for clues to vast and complex problems in the small, familiar phenomena of daily life. Isaac Newton reached insights about the heavens by juggling wooden balls. Einstein contemplated a boat paddling upriver. Von Neumann pondered the game of poker.

A seemingly trivial and playful pursuit like poker, von Neumann argued, might hold the key to more serious human affairs for two reasons. Both poker and economic competition require a certain type of reasoning, namely the rational calculation of advantage and disadvantage based on some internally consistent system of values ("more is better than less"). And in both, the outcome for any individual actor depends not only on his own actions, but on the independent actions of others.

More than a century earlier, the French economist Antoine-Augustin Cournot had pointed out that problems of economic choice were greatly simplified when either none or a large number of other agents were present.[16] Alone on his island, Robinson Crusoe doesn't have to worry about others whose actions might affect him. Neither, though, do Adam Smith's butchers and bakers. They live in a world with so many actors that their actions, in effect, cancel each other out. But when there is more than one agent but not so many that their influence may be safely ignored, strategic behavior raises a seemingly insoluble problem: "I think that he thinks that I think that he thinks," and so forth.

Von Neumann was able to give a convincing solution to this problem of circular reasoning for games that are two-person, zero-sum games, games in which one player's gain is another's loss. But zero-sum games are the ones least applicable to economics (as one writer put it, the zero-sum game is to game theory "what the twelve-bar blues is to jazz; a polar case, and a point of historical departure"). For situations with many actors and the possibility of mutual gain—the standard economic scenario—von Neumann's superlative instincts failed him. He was convinced that players would have to form coalitions, make explicit agreements, and submit to some higher, centralized authority to enforce those agreements.[17] Quite possibly his conviction reflected his generation's distrust, in the wake of the Depression and in the midst of a world war, of unfettered individualism. Though von Neumann hardly shared the liberal views of Einstein, Bertrand Russell, and the British economist John Maynard Keynes, he shared something of their belief that actions that might be reasonable from the point of view of the individual could produce social chaos. Like them he embraced the then-popular solution to political conflict in the age of nuclear weapons: world government.[18]

The young Nash had wholly different instincts. Where von Neumann's focus was the group, Nash zeroed in on the individual, and by doing so, made game theory relevant to modern economics. In his slender twenty-seven-page doctoral thesis, written when he was twenty-one, Nash created a theory for games in which there was a possibility of mutual gain, inventing a concept that let one cut through the endless chain of reasoning, "I think that you think that I think. . . ."[19] His insight was that the game would be solved when every player independently chose his best response to the other players' best strategies.

Thus, a young man seemingly so out of touch with other people's emotions, not to mention his own, could see clearly that the most human of motives and behavior is as much of a mystery as mathematics itself, that world of ideal platonic forms invented by the human species seemingly by pure introspection (and yet somehow linked to the grossest and most mundane aspects of nature). But Nash had grown up in a boom town in the Appalachian foothills where fortunes were made from the roaring, raw businesses of rails, coal, scrap metal, and electric power. Individual rationality and self-interest, not common agreement on some collective good, seemed sufficient to create a tolerable order. The leap was a short one, from his observations of his hometown to his focus on the logical strategy necessary for the individual to maximize his own advantage and minimize his disadvantages. The Nash equilibrium, once it is explained, sounds obvious, but by formulating the problem of economic competition in the way that he did, Nash showed that a decentralized decision-making process could, in fact, be coherent — giving economics an updated, far more sophisticated version of Adam Smith's great metaphor of the Invisible Hand.

By his late twenties, Nash's insights and discoveries had won him recognition, respect, and autonomy. He had carved out a brilliant career at the apex of the mathematics profession, traveled, lectured, taught, met the most famous mathematicians of his day, and become famous himself. His genius also won him love. He had married a beautiful young physics student who adored him, and fathered a child. It was a brilliant strategy, this genius, this life. A seemingly perfect adaptation.

Many great scientists and philosophers, among them René Descartes, Ludwig Wittgenstein, Immanuel Kant, Thorstein Veblen, Isaac Newton, and Albert Einstein, have had similarly strange and solitary personalities.[20] An emotionally detached, inward-looking temperament can be especially conducive to scientific creativity, psychiatrists and biographers have long observed, just as fiery fluctuations in mood may sometimes be linked to artistic expression. In *The Dynamics of Creation,* Anthony Storr, the British psychiatrist, contends that an individual who "fears love almost as much as he fears hatred" may turn to creative activity not only out of an impulse to experience aesthetic pleasure, or the delight of exercising an active mind, but also to defend himself against anxiety stimulated by conflicting demands for detachment and human contact.[21] In the same vein, Jean-Paul Sartre, the French philosopher and writer, called genius "the brilliant invention of someone who is looking for a way out." Posing the question of why people often are willing to endure frustration and misery in order to create something, even in the absence of large rewards, Storr speculates:

> Some creative people . . . of predominately schizoid or depressive temperaments . . . use their creative capacities in a defensive way. If creative work protects a man from mental illness, it is small wonder that he pursues it with avidity. The schizoid state . . . is characterized by a sense of meaninglessness and futility. For most people, interaction with others provides most of what they require to find meaning and significance in life. For the schizoid person, however, this is not the case. Creative activity is a particularly apt way to express himself . . . the activity is solitary . . . [but] the ability to create and the

productions which result from such ability are generally regarded as possessing value by our society.[22]

Of course, very few people who exhibit "a lifelong pattern of social isolation" and "indifference to the attitudes and feelings of others"—the hallmarks of a so-called schizoid personality—possess great scientific or other creative talent.[23] And the vast majority of people with such strange and solitary temperaments never succumb to severe mental illness.[24] Instead, according to John G. Gunderson, a psychiatrist at Harvard, they tend "to engage in solitary activities which often involve mechanical, scientific, futuristic and other non-human subjects . . . [and] are likely to appear increasingly comfortable over a period of time by forming a stable but distant network of relationships with people around work tasks."[25] Men of scientific genius, however eccentric, rarely become truly insane—the strongest evidence for the potentially protective nature of creativity.[26]

Nash proved a tragic exception. Underneath the brilliant surface of his life, all was chaos and contradiction: his involvements with other men; a secret mistress and a neglected illegitimate son; a deep ambivalence toward the wife who adored him, the university that nurtured him, even his country; and, increasingly, a haunting fear of failure. And the chaos eventually welled up, spilled over, and swept away the fragile edifice of his carefully constructed life.

The first visible signs of Nash's slide from eccentricity into madness appeared when he was thirty and was about to be made a full professor at MIT. The episodes were so cryptic and fleeting that some of Nash's younger colleagues at that institution thought that he was indulging a private joke at their expense. He walked into the common room one winter morning in 1959 carrying *The New York Times* and remarked, to no one in particular, that the story in the upper left-hand corner of the front page contained an encrypted message from inhabitants of another galaxy that only he could decipher.[27] Even months later, after he had stopped teaching, had angrily resigned his professorship, and was incarcerated at a private psychiatric hospital in suburban Boston, one of the nation's leading forensic psychiatrists, an expert who testified in the case of Sacco and Vanzetti, insisted that Nash was perfectly sane. Only a few of those who witnessed the uncanny metamorphosis, Norbert Wiener among them, grasped its true significance.[28]

At thirty years of age, Nash suffered the first shattering episode of paranoid schizophrenia, the most catastrophic, protean, and mysterious of mental illnesses. For the next three decades, Nash suffered from severe delusions, hallucinations, disordered thought and feeling, and a broken will. In the grip of this "cancer of the mind," as the universally dreaded condition is sometimes called, Nash abandoned mathematics, embraced numerology and religious prophecy, and believed himself to be a "messianic figure of great but secret importance." He fled to Europe several times, was hospitalized involuntarily half a dozen times for periods up to a year and a half, was subjected to all sorts of drug and shock treatments, experienced brief remissions and episodes of hope that lasted only a few months, and finally became a sad phantom who haunted the Princeton University campus where he had once been a brilliant graduate student, oddly dressed, muttering to himself, writing mysterious messages on blackboards, year after year.

The origins of schizophrenia are mysterious. The condition was first described in 1806, but no one is certain whether the illness — or, more likely, group of illnesses — existed long before then but had escaped definition or, on the other hand, appeared as an AIDS-like scourge at the start of the industrial age.[29] Roughly 1 percent of the population in all countries succumbs to it.[30] Why it strikes one individual and not another is not known, although the suspicion is that it results from a tangle of inherited vulnerability and life stresses.[31] No element of environment — war, imprisonment, drugs, or upbringing — has ever been proved to cause, by itself, a single instance of the illness.[32] There is now a consensus that schizophrenia has a tendency to run in families, but heredity alone apparently cannot explain why a specific individual develops the full-blown illness.[33]

Eugen Bleuler, who coined the term *schizophrenia* in 1908, describes a "specific type of alteration of thinking, feeling and relation to the external world."[34] The term refers to a splitting of psychic functions, "a peculiar destruction of the inner cohesiveness of the psychic personality."[35] To the person experiencing early symptoms, there is a dislocation of every faculty, of time, space, and body.[36] None of its symptoms — hearing voices, bizarre delusions, extreme apathy or agitation, coldness toward others — is, taken singly, unique to the illness.[37] And symptoms vary so much between individuals and over time for the same individual that the notion of a "typical case" is virtually nonexistent. Even the degree of disability — far more severe, on average, for men — varies wildly. The symptoms can be "slightly, moderately, severely, or absolutely disabling," according to Irving Gottesman, a leading contemporary researcher.[38] Though Nash succumbed at age thirty, the illness can appear at any time from adolescence to advanced middle age.[39] The first episode can last a few weeks or months or several years.[40] The life history of someone with the disease can include only one or two episodes.[41] Isaac Newton, always an eccentric and solitary soul, apparently suffered a psychotic breakdown with paranoid delusions at age fifty-one.[42] The episode, which may have been precipitated by an unhappy attachment to a younger man and the failure of his alchemy experiments, marked the end of Newton's academic career. But, after a year or so, Newton recovered and went on to hold a series of high public positions and to receive many honors. More often, as happened in Nash's case, people with the disease suffer many, progressively more severe episodes that occur at ever shorter intervals. Recovery, almost never complete, runs the gamut from a level tolerable to society to one that may not require permanent hospitalization but in fact does not allow even the semblance of a normal life.[43]

More than any symptom, the defining characteristic of the illness is the profound feeling of incomprehensibility and inaccessibility that sufferers provoke in other people. Psychiatrists describe the person's sense of being separated by a "gulf which defies description" from individuals who seem "totally strange, puzzling, inconceivable, uncanny and incapable of empathy, even to the point of being sinister and frightening."[44] For Nash, the onset of the illness dramatically intensified a pre-existing feeling, on the part of many who knew him, that he was essentially disconnected from them and deeply unknowable. As Storr writes:

> However melancholy a depressive may be, the observer generally feels there is some possibility of emotional contact. The schizoid person, on the other hand, appears withdrawn and inaccessible. His remoteness from human contact makes his state of mind less humanly comprehensible, since his feelings are not communicated. If such a person becomes psychotic (schizophrenic) this lack of connection with people and the external world becomes more obvious; with the result that the sufferer's behavior and utterances appear inconsequential and unpredictable.[45]

Schizophrenia contradicts popular but incorrect views of madness as consisting solely of wild gyrations of mood, or fevered delirium. Someone with schizophrenia is not permanently disoriented or confused, for example, the way that an individual with a brain injury or Alzheimer's might be.[46] He may have, indeed usually does have, a firm grip on certain aspects of present reality. While he was ill, Nash traveled all over Europe and America, got legal help, and learned to write sophisticated computer programs. Schizophrenia is also distinct from manic depressive illness (currently known as bipolar disorder), the illness with which it has most often been confounded in the past.

If anything, schizophrenia can be a ratiocinating illness, particularly in its early phases.[47] From the turn of the century, the great students of schizophrenia noted that its sufferers included people with fine minds and that the delusions which often, though not always, come with the disorder involve subtle, sophisticated, complex flights of thought. Emil Kraepelin, who defined the disorder for the first time in 1896, described "dementia praecox," as he called the illness, not as the shattering of reason but as causing "predominant damage to the emotional life and the will."[48] Louis A. Sass, a psychologist at Rutgers University, calls it "not an escape from reason but an exacerbation of that thoroughgoing illness Dostoevsky imagined . . . at least in some of its forms . . . a heightening rather than a dimming of conscious awareness, and an alienation not from reason but from emotion, instincts and the will."[49]

Nash's mood in the early days of his illness can be described, not as manic or melancholic, but rather as one of heightened awareness, insomniac wakefulness and watchfulness. He began to believe that a great many things that he saw — a telephone number, a red necktie, a dog trotting along the sidewalk, a Hebrew letter, a birthplace, a sentence in *The New York Times* — had a hidden significance, apparent only to him. He found such signs increasingly compelling, so much so that they drove from his consciousness his usual concerns and preoccupations. At the same time, he believed he was on the brink of cosmic insights. He claimed he had found a solution to the greatest unsolved problem in pure mathematics, the so-called Riemann Hypothesis. Later he said he was engaged in an effort to "rewrite the foundations of quantum physics." Still later, he claimed, in a torrent of letters to former colleagues, to have discovered vast conspiracies and the secret meaning of numbers and biblical texts. In a letter to the algebraist Emil Artin, whom he addressed as "a great necromancer and numerologist," Nash wrote:

> I have been considering Algerbiac [sic] questions and have noticed some interesting things that might also interest you . . . I, a while ago, was seized with the concept that numerological calculations dependent on the decimal system might not be sufficiently intrinsic also that language and alphabet structure might contain ancient cultural stereotypes interfering with clear understands [sic] or unbiased thinking. . . . I quickly wrote down a new sequence of symbols. . . . These were associated with (in fact natural, but perhaps not computationally ideal but suited for mystical rituals, incantations and such) system for representing the integers via symbols, based on the products of successive primes.[50]

A predisposition to schizophrenia was probably integral to Nash's exotic style of thought as a mathematician, but the full-blown disease devastated his ability to do creative work. His once-illuminating visions became increasingly obscure, self-contradictory, and full of purely private meanings, accessible only to himself. His longstanding conviction that the universe was rational evolved into a caricature of itself, turning into an unshakable belief that everything had meaning, everything had a reason, nothing was random or coincidental. For much of the time, his grandiose delusions insulated him from the painful reality of all that he had lost. But then would come terrible flashes of awareness. He complained bitterly from time to time of his inability to concentrate and to remember mathematics, which he attributed to shock treatments.[51] He sometimes told others that his enforced idleness made him feel ashamed of himself, worthless.[52] More often, he expressed his suffering wordlessly. On one occasion, sometime during the 1970s, he was sitting at a table in the dining hall at the Institute for Advanced Study—the scholarly haven where he had once discussed his ideas with the likes of Einstein, von Neumann, and Robert Oppenheimer—alone as usual. That morning, an institute staff member recalled, Nash got up, walked over to a wall, and stood there for many minutes, banging his head against the wall, slowly, over and over, eyes tightly shut, fists clenched, his face contorted with anguish.[53]

While Nash the man remained frozen in a dreamlike state, a phantom who haunted Princeton in the 1970s and 1980s scribbling on blackboards and studying religious texts, his name began to surface everywhere—in economics textbooks, articles on evolutionary biology, political science treatises, mathematics journals. It appeared less often in explicit citations of the papers he had written in the 1950s than as an adjective for concepts too universally accepted, too familiar a part of the foundation of many subjects to require a particular reference: "Nash equilibrium," "Nash bargaining solution," "Nash program," "De Giorgi–Nash result," "Nash embedding," "Nash-Moser theorem," "Nash blowing-up."[54] When a massive new encyclopedia of economics, *The New Palgrave,* appeared in 1987, its editors noted that the game theory revolution that had swept through economics "was effected with apparently no new fundamental mathematical theorems beyond those of von Neumann and Nash."[55]

Even as Nash's ideas became more influential — in fields so disparate that almost no one connected the Nash of game theory with Nash the geometer or Nash the analyst — the man himself remained shrouded in obscurity. Most of the young mathematicians and economists who made use of his ideas simply assumed, given the dates of his published articles, that he was dead. Members of the profession who knew otherwise, but were aware of his tragic illness, sometimes treated him as if he were. A 1989 proposal to place Nash on the ballot of the Econometric Society as a potential fellow of the society was treated by society officials as a highly romantic but essentially frivolous gesture — and rejected.[56] No biographical sketch of Nash appeared in *The New Palgrave* alongside sketches of half a dozen other pioneers of game theory.[57]

At around that time, as part of his daily rounds in Princeton, Nash used to turn up at the institute almost every day at breakfast. Sometimes he would cadge cigarettes or spare change, but mostly he kept very much to himself, a silent, furtive figure, gaunt and gray, who sat alone off in a corner, drinking coffee, smoking, spreading out a ragged pile of papers that he carried with him always.[58]

Freeman Dyson, one of the giants of twentieth-century theoretical physics, one-time mathematical prodigy, and author of a dozen metaphorically rich popular books on science, then in his sixties, about five years older than Nash, was one of those who saw Nash every day at the institute.[59] Dyson is a small, lively sprite of a man, father of six children, not at all remote, with an acute interest in people unusual for someone of his profession, and one of those who would greet Nash without expecting any response, but merely as a token of respect.

On one of those gray mornings, sometime in the late 1980s, he said his usual good morning to Nash. "I see your daughter is in the news again today," Nash said to Dyson, whose daughter Esther is a frequently quoted authority on computers. Dyson, who had never heard Nash speak, said later: "I had no idea he was aware of her existence. It was beautiful. I remember the astonishment I felt. What I found most wonderful was this slow awakening. Slowly, he just somehow woke up. Nobody else has ever awakened the way he did."

More signs of recovery followed. Around 1990, Nash began to correspond, via electronic mail, with Enrico Bombieri, for many years a star of the Institute's mathematics faculty.[60] Bombieri, a dashing and erudite Italian, is a winner of the Fields Medal, mathematics' equivalent of the Nobel. He also paints oils, collects wild mushrooms, and polishes gemstones. Bombieri is a number theorist who has been working for a long time on the Riemann Hypothesis. The exchange focused on various conjectures and calculations Nash had begun related to the so-called

ABC conjecture. The letters showed that Nash was once again doing real mathematical research, Bombieri said:

> He was staying very much by himself. But at some point he started talking to people. Then we talked quite a lot about number theory. Sometimes we talked in my office. Sometimes over coffee in the dining hall. Then we began corresponding by e-mail. It's a sharp mind . . . all the suggestions have that toughness . . . there's nothing commonplace about those. . . . Usually when one starts in a field, people remark the obvious, only what is known. In this case, not. He looks at things from a slightly different angle.

A spontaneous recovery from schizophrenia — still widely regarded as a dementing and degenerative disease — is so rare, particularly after so long and severe a course as Nash experienced, that, when it occurs, psychiatrists routinely question the validity of the original diagnosis.[61] But people like Dyson and Bombieri, who had watched Nash around Princeton for years before witnessing the transformation, had no doubt that by the early 1990s he was "a walking miracle."

It is highly unlikely, however, that many people outside this intellectual Olympus would have become privy to these developments, dramatic as they appeared to Princeton insiders, if not for another scene, which also took place on these grounds at the end of the first week of October 1994.

A mathematics seminar was just breaking up. Nash, who now regularly attended such gatherings and sometimes even asked a question or offered some conjecture, was about to duck out. Harold Kuhn, a mathematics professor at the university and Nash's closest friend, caught up with him at the door.[62] Kuhn had telephoned Nash at home earlier that day and suggested that the two of them might go for lunch after the talk. The day was so mild, the outdoors so inviting, the Institute woods so brilliant, that the two men wound up sitting on a bench opposite the mathematics building, at the edge of a vast expanse of lawn, in front of a graceful little Japanese fountain.

Kuhn and Nash had known each other for nearly fifty years. They had both been graduate students at Princeton in the late 1940s, shared the same professors, known the same people, traveled in the same elite mathematical circles. They had not been friends as students, but Kuhn, who spent most of his career in Princeton, had never entirely lost touch with Nash and had, as Nash became more accessible, managed to establish fairly regular contact with him. Kuhn is a shrewd, vigorous, sophisticated man who is not burdened with "the mathematical personality." Not a typical academic, passionate about the arts and liberal political causes, Kuhn is as interested in other people's lives as Nash is remote from them. They were an odd couple, connected not by temperament or experience but by a large fund of common memories and associations.

Kuhn, who had carefully rehearsed what he was going to say, got to the point quickly. "I have something to tell you, John," he began. Nash, as usual, refused to look Kuhn in the face at first, staring instead into the middle distance. Kuhn went on. Nash was to expect an important telephone call at home the following morning, probably around six o'clock. The call would come from Stockholm. It would be made by the Secretary General of the Swedish Academy of Sciences. Kuhn's voice suddenly became hoarse with emotion. Nash now turned his head, concentrating on every word. "He's going to tell you, John," Kuhn concluded, "that you have won a Nobel Prize."

This is the story of John Forbes Nash, Jr. It is a story about the mystery of the human mind, in three acts: genius, madness, reawakening.

PART ONE

A
Beautiful
Mind

1
Bluefield

1928–45

> *I was taught to feel, perhaps too much*
> *The self-sufficing power of solitude.*
> — WILLIAM WORDSWORTH

AMONG JOHN NASH'S EARLIEST MEMORIES is one in which, as a child of about two or three, he is listening to his maternal grandmother play the piano in the front parlor of the old Tazewell Street house, high on a breezy hill overlooking the city of Bluefield, West Virginia.[1]

It was in this parlor that his parents were married on September 6, 1924, a Saturday, at eight in the morning to the chords of a Protestant hymn, amid basketfuls of blue hydrangeas, goldenrod, black-eyed susans, and white and gold marguerites.[2] The thirty-two-year-old groom was tall and gravely handsome. The bride, four years his junior, was a willowy, dark-eyed beauty. Her narrow, brown cut-velvet dress emphasized her slender waist and long, graceful back. She had perhaps chosen its deep shade out of deference to her father's recent death. She carried a bouquet of the same old-fashioned flowers that filled the room, and she wore more of these blooms woven through her thick chestnut hair. The effect was brilliant rather than subdued. The vibrant browns and golds, which would have made a woman with a lighter, more typically southern complexion look wan, embellished her rich coloring and lent her a striking and sophisticated air.

The ceremony, conducted by ministers from Christ Episcopal Church and Bland Street Methodist Church, was simple and brief, witnessed by fewer than a dozen family members and old friends. By eleven o'clock, the newlyweds were standing at the ornate, wrought-iron gate in front of the rambling, white 1890s house waving their goodbyes. Then, according to an account that appeared some weeks later in the Appalachian Power Company's company newsletter, they embarked in the groom's shiny new Dodge for an "extensive tour" through several northern states.[3]

The romantic style of the wedding, and the venturesome honeymoon, hinted at certain qualities in the couple, no longer in the first bloom of youth, that set them somewhat apart from the rest of society in this small American town.

John Forbes Nash, Sr., was "proper, painstaking, and very serious, a very conservative man in every respect," according to his daughter Martha Nash Legg.[4] What saved him from dullness was a sharp, inquiring mind. A Texas native, he came from the rural gentry, teachers and farmers, pious, frugal Puritans and Scottish Baptists who migrated west from New England and the Deep South.[5] He was born in 1892 on his maternal grandparents' plantation on the banks of the Red River in northern Texas, the oldest of three children of Martha Smith and Alexander Quincy Nash. The first few years of his life were spent in Sherman, Texas, where his paternal grandparents, both teachers, had founded the Sherman Institute (later the Mary Nash College for Women), a modest but progressive establishment, where the daughters of Texas's middle class learned deportment, the value of regular physical exercise, and a bit of poetry and botany. His mother had been a student and then a teacher at the college before she married the son of its founders. After his grandparents died, John Sr.'s parents operated the college until a smallpox epidemic forced them to close its doors for good.

His childhood, spent within the precincts of Baptist institutions of higher learning, was unhappy. The unhappiness stemmed largely from his parents' marriage. Martha Nash's obituary refers to "many heavy burdens, responsibilities and disappointments, that made a severe demand on her nervous system and physical force."[6] Her chief burden was Alexander, a strange and unstable individual, a ne'er-do-well and a philanderer who either abandoned his wife and three children soon after the college's demise or, more likely, was thrown out. When precisely Alexander left the family for good or what happened to him after he departed is unclear, but he was in the picture long enough to earn his children's undying enmity and to instill in his youngest son a deep and ever-present hunger for respectability. "He was very concerned with appearances," his daughter Martha later said of her father; "he wanted everything to be very proper."[7]

John Sr.'s mother was a highly intelligent, resourceful woman. After she and her husband separated, Martha Nash supported herself and her two young sons and daughter on her own, working for many years as an administrator at Baylor College, another Baptist institution for girls, in Belton, in central Texas. Obituaries refer to her "fine executive ability" and "remarkable managerial skill." According to the *Baptist Standard,* "She was an unusually capable woman. . . . She had the capacity of managing large enterprises . . . a true daughter of the true Southern gentry." Devout and diligent, Martha was also described as an "efficient and devoted" mother, but her constant struggle against poverty, bad health, and low spirits, along with the shame of growing up in a fatherless household, left its scars on John Sr. and contributed to the emotional reserve he later displayed toward his own children.

Surrounded by unhappiness at home, John Sr. early on found solace and certainty in the realm of science and technology. He studied electrical engineering at Texas Agricultural & Mechanical, graduating around 1912. He enlisted in the army shortly after the United States entered World War I and spent most of his wartime duty as a lieutenant in the 144th Infantry Supply Division in France. When he returned to Texas, he did not go back to his previous job at General Electric, but instead tried his hand at teaching engineering students at Texas A&M. Given his background and interests, he may well have hoped to pursue an academic career. If so, however, those hopes came to nothing. At the end of the academic year, he agreed to take a position in Bluefield with the Appalachian Power Company (now American Electric Power), the utility that would employ him for the next thirty-eight years. By June, he was living in rented rooms in Bluefield.

Photographs of Margaret Virginia Martin — known as Virginia — at the time of her engagement to John Sr. show a smiling, animated woman, stylish and whippet-thin. One account called her "one of the most charming and cultured young ladies of the community."[8] Outgoing and energetic, Virginia was a freer, less rigid spirit than her quiet, reserved husband and a far more active presence in her son's life. Her vitality and forcefulness were such that, years later, her son John, by then in his thirties and seriously ill, would dismiss a report from home that she had been hospitalized for a "nervous breakdown" as simply unbelievable. He would greet the news of her death in 1969 with similar incredulity.[9]

Like her husband, Virginia grew up in a family that valued church and higher education. But there the similarity ended. She was one of four surviving daughters of a popular physician, James Everett Martin, and his wife, Emma, who had moved to Bluefield from North Carolina during the early 1890s. The Martins were a well-to-do, prominent local family. Over time, they acquired a good deal of property in the town, and Dr. Martin eventually gave up his medical practice to manage his real-estate investments and to devote himself to civic affairs. Some accounts refer to him as a one-time postmaster, others as the town's mayor. The Martins' affluence did not protect them from terrible blows — their first child, a boy, died in infancy; Virginia, the second, was left entirely deaf in one ear at age twelve after a bout of scarlet fever; a younger brother was killed in a train wreck; and one of her sisters died in a typhoid epidemic — but on the whole Virginia grew up in a happier atmosphere than her husband. The Martins were also well-educated, and they saw to it that all of their daughters received university educations. Emma Martin was herself unusual in having graduated from a women's college in Tennessee. Virginia studied English, French, German, and Latin first at Martha Washington College and later at West Virginia University. By the time she met her husband-to-be, she had been teaching for six years. She was a born teacher, a talent that she would later lavish on her gifted son. Like her husband, she had seen something beyond the small towns of her home state. Before her marriage, she and another Bluefield teacher, Elizabeth Shelton, spent several summers traveling and attending courses at various universities, including the University of California at Berkeley, Columbia University in New York, and the University of Virginia in Charlottesville.

When the newlyweds returned from their honeymoon, the couple lived at the Tazewell Street house with Virginia's mother and sisters. John Sr. went back to his job at the Appalachian, which in those years consisted largely of driving all over the state inspecting remote power lines. Virginia did not return to teaching. Like most school districts around the country during the 1920s, the Mercer County school system had a marriage bar. Female teachers lost their jobs as soon as they married.[10] But, quite apart from her forced resignation, her new husband had a strong feeling that he ought to provide for his wife and protect her from what he regarded as the shame of having to work, another legacy of his own upbringing.

Bluefield, named for the fields of "azure chicory" in surrounding valleys that grows along every street and alleyway even today, owes its existence to the rolling hills full of coal—"the wildest, most rugged and romantic country to be found in the mountains of Virginia or West Virginia"—that surround the remote little city.[11] Norfolk & Western, in a spirit of "mean force and ignorance," built a line in the 1890s that stretched from Roanoke to Bluefield, which lies in the Appalachians on the easternmost edge of the great Pocahontas coal seam. For a long time, Bluefield was a rough and ready outpost where Jewish merchants, African-American construction workers, and Tazewell County farmers struggled to make a living and where millionaire coal operators, most of whom lived ten miles away in Bramwell, battled Italian, Hungarian, and Polish immigrant laborers, and John L. Lewis and the UMW sat down with the coal operators to negotiate contracts, negotiations that often led to the bloody strikes and lockouts documented in John Sayles's film *Matewan*.

By the 1920s, when the Nashes married, however, Bluefield's character was already changing. Directly on the line between Chicago and Norfolk, the town was becoming an important rail hub and had attracted a prosperous white-collar class of middle managers, lawyers, small businessmen, ministers, and teachers.[12] A real downtown of granite office buildings and stores had sprung up. Handsome churches had also gone up all over town. Snug frame houses with pretty little gardens edged by Rose of Sharon dotted the hills. The town had acquired a daily newspaper, a hospital, and a home for the elderly. Educational institutions, from private kindergartens and dancing schools to two small colleges, one black, one white, were thriving. The radio, telegraph, and telephone, as well as the railroads and, increasingly, the automobile, eased the sense of isolation.

Bluefield was not "a community of scholars," as John Nash later said with more than a hint of irony.[13] Its bustling commercialism, Protestant respectability, and small-town snobbery couldn't have been further removed from the atmosphere of the intellectual hothouses of Budapest and Cambridge which produced John von Neumann and Norbert Wiener. Yet while John Nash was growing up, the town had a sizable group of men with scientific interests and engineering talent, men like John Sr. who were attracted by the railroad, the utility, and the mining companies.[14] Some of those who came to work for the companies wound up as science teachers in the high school or one of the two local colleges. In his autobiographical essay, Nash described "having to learn from the world's knowl-

edge rather than the knowledge of the immediate community" as "a challenge."[15]
But, in fact, Bluefield offered a good deal of stimulation — admittedly, of a down-
to-earth variety — for an inquiring mind; John Nash's subsequent career as a multifaceted mathematician, not to mention a certain pragmatism of character, would
seem to owe something to his Bluefield years.

More than anything, the newly married Nashes were strivers. Solid members of
America's new, upwardly mobile professional middle class, they formed a tight
alliance and devoted themselves to achieving financial security and a respectable
place for themselves in the town's social pyramid.[16] They became Episcopalians,
like many of Bluefield's more prosperous citizens, rather than continuing in the
fundamentalist churches of their youth. Unlike most of Virginia's family, they also
became staunch Republicans, though (so as to be able to vote for a Democratic
cousin in the primaries) not registered party members. They socialized a good deal.
They joined Bluefield's new country club, which was displacing the Protestant
churches as the center of Bluefield's social life. Virginia belonged to various wom-
en's book, bridge, and gardening clubs. John Sr. was a member of the Rotary and
a number of engineering societies. Later on, the only middle-class practice that
they deliberately avoided was sending their son to prep school. Virginia, as her
daughter explained, was "a public-school thinker."

John Sr.'s job with the Appalachian remained secure right through the Depression of the 1930s. The young family fared considerably better in this period than
many of their neighbors and fellow churchgoers, especially the small businessmen.
John Sr.'s paycheck, while hardly munificent, was steady, and frugality did the rest.
All decisions involving the expenditure of money, no matter how modest, were
carefully considered; very often the decision was to avoid, put off, or reduce. There
were no mortgages to be had in those days, no pensions either, even for a rising
young middle manager in one of the nation's largest utilities. Virginia Nash used
to accuse her husband, when they'd had an argument — which they rarely did
within earshot of the children — of being quite likely, in the event that she died
before him, to marry a younger woman and let her squander all the money she,
Virginia, had scraped so hard to save. (Their savings, it turned out, were considerable, however. Even though John Sr. died some thirteen years before Virginia, and
even with the high cost of hospitalizations for John Jr., Virginia barely dipped into
her capital and was able to pass along a trust fund to her children.)

Though they began life as parents in a rental house owned by Emma Martin,
the Nashes were soon able to move to their own modest but comfortable three-
bedroom home in one of the best parts of town, Country Club Hill. Built partly of
cinder blocks that John Sr. was able to buy for a song from a nearby Appalachian
coal-processing plant, the house bore little resemblance to the imposing homes of
the coal families scattered around the hill. But it was within a few hundred yards
of the crest where the club was located, was built to order by a local architect, and
contained all the comforts and conveniences that a small-town, middle-class family

at that time could aspire to: a living room where Virginia's bridge club could be entertained in style, with a fireplace, built-in bookshelves, and graceful wooden trim at the tops of all the doorways, a neat little kitchen with a breakfast nook, a dining room where Sunday dinners of chicken and waffles were served, a real basement that might one day be fitted out with a maid's room, should live-in help be one day possible, and a separate bedroom for each of the two children.

However much they were forced to economize, the Nashes were able to keep up appearances. Virginia had nice clothes, most of which she sewed herself, and allowed herself the weekly luxury of going to a beauty parlor. By the time they moved to their own house, she had a cleaning woman who came once a week. Virginia always had a car to drive, typically a Dodge, which was hardly the norm even among middle-class families at the time. John Sr., of course, had a company car, usually a Buick. The Nashes were a loyal couple, like-minded.

John Forbes Nash, Jr., was born almost exactly four years after his parents' marriage, on June 13, 1928. He first saw the light of day not at home, but in the Bluefield Sanitarium, a small hospital on Ramsey Street that has long since been converted to other uses. Other than that single fact, again suggestive of the Nashes' comfortable circumstances, nothing is now known of his coming into the world. Did Virginia catch influenza during her winter pregnancy? Were there any other complications? Were forceps needed during the delivery? While viral exposure in utero or a subtle birth injury might have played a role in his later mental illness, there is no available record or memory to suggest any such trauma. No anesthesia was required during the delivery, Virginia later told her daughter. The seven-pound baby boy was, as far as anyone still living remembers, apparently healthy, and was soon baptized in the Episcopal Church directly opposite the Martin house on Tazewell Street and given his father's full name. Everyone, however, called him Johnny.

He was a singular little boy, solitary and introverted.[17] The once-dominant view of the origins of the schizoid temperament was that abuse, neglect, or abandonment caused the child to give up the possibility of gratification from human relationships at a very early age.[18] Johnny Nash certainly did not fit this now-discredited paradigm. His parents, especially his mother, were actively loving. In general, one can imagine, on evidence from biographies of many brilliant men who were peculiar and isolated as children, that an inward-looking child might react to intrusive adults by withdrawing further into his own private world or that efforts to make him conform might be met by firm resolve to do things his own way — or perhaps that unsympathetic taunting peers might have a similar effect. But the facts of Nash's childhood, in many ways so typical of the educated classes in small American towns of that era, suggest that his temperament may well have been one that he was born with.

As the vivid memory of his grandmother's piano-playing suggests, Johnny Nash's infancy was spent a good deal in the company not only of his adoring mother, but also of his grandmother, aunts, and young cousins.[19] The Highland

Street house to which the Nashes had moved shortly after his birth was within walking distance of Tazewell Street and Virginia continued to spend a great deal of time there, even after the birth of Johnny's younger sister Martha in 1930. But by the time Johnny was seven or eight, his aunts had come to consider him bookish and slightly odd. While Martha and her cousins rode stick horses, cut paper dolls out of old pattern books, and played house and hide-and-seek in the "almost scary but nice" attic, Johnny could always be found in the parlor with his nose buried in a book or magazine. At home, despite his mother's urgings, he ignored the neighborhood children, preferring to stay indoors alone. His sister spent most of her free time at the pool or playing football and kick ball or taking part in crabapple battles with long, flimsy sticks. But Johnny played by himself with toy airplanes and cars.

Although he was no prodigy, Johnny was a bright and curious child. His mother, with whom he was always closest, responded by making his education a principal focus of her considerable energy. "Mother was a natural teacher," Martha observes. "She liked to read, she liked to teach. She wasn't just a housewife." Virginia, who became actively involved in the PTA, taught Johnny to read by age four, sent him to a private kindergarten, saw to it that he skipped half a grade early in elementary school, tutored him at home and, later on, in high school, had him enroll at Bluefield College to take courses in English, science, and math. John Sr.'s hand in his son's education was less visible. More distant than Virginia, he nonetheless shared his interests with his children — taking Johnny and Martha on Sunday drives to inspect power lines, for example — and, more important, supplied answers to his son's incessant questions about electricity, geology, weather, astronomy, and other technological subjects and the natural world. A neighbor remembers that John Sr. always spoke to his children as if they were adults: "He never gave Johnny a coloring book. He gave him science books." [20]

At school, Johnny's immaturity and social awkwardness were initially more apparent than any special intellectual gifts. His teachers labeled him an under-achiever. He daydreamed or talked incessantly and had trouble following directions, a source of some conflict between him and his mother. His fourth-grade report card, in which music and mathematics were his lowest marks, contained a note to the effect that Johnny needed "improvement in effort, study habits and respect for the rules." He gripped his pencil like a stick, his handwriting was atrocious, and he was somewhat inclined to use his left hand. John Sr. insisted he write only with his right hand. Virginia eventually made him enroll in a penmanship course at a local secretarial college, where he learned a certain style of printing and also how to type. A newspaper clipping from Virginia's scrapbook shows him sitting in a classroom with rows and rows of teenage girls, his eyes rolled up in his head, looking stupefyingly bored. Complaints about his writing, his talking out of turn or even "monopolizing the class discussion," and his sloppiness dogged him right through the end of high school. [21]

His best friends were books, and he was always happiest learning on his own. Nash alludes to his preference obliquely in his autobiographical essay:

> My parents provided an encyclopedia, *Compton's Pictured Encyclopedia,* that I learned a lot from by reading it as a child. And also there were other books available from either our house or the house of the grandparents that were of educational value.[22]

And the best time of day was after dinner every evening when John Sr. would sit at his desk in the small family room off the living room, the size of a sleeping porch, and John Jr. could sprawl in front of the radio, listening to classical music or news reports, or reading either the encyclopedia or the family's stacks of well-worn *Life* and *Time* magazines, and ask his father questions.

His great passion was experimenting. By the time he was twelve or so, he had turned his room into a laboratory. He tinkered with radios, fooled around with electrical gadgets, and did chemistry experiments.[23] A neighbor recalls Johnny rigging the Nash telephone to ring with the receiver off.[24]

Though he had no close companions, he enjoyed performing in front of other children. At one point, he would hold on to a big magnet that was wired with electricity to show how much current he could endure without flinching.[25] Another time, he'd read about an old Indian method for making oneself immune to poison ivy. He wrapped poison ivy leaves in some other leaves and swallowed them whole in front of a couple of other boys.[26]

One afternoon, he went to a carnival that had come to Bluefield.[27] The crowd of children he was with clustered around a sideshow. There was a man sitting in an electric chair holding swords in each of his hands. Sparks flashed and danced between the two tips. He challenged anyone in the crowd to do the same. Johnny Nash, then about twelve, stepped forward and grabbed the swords and repeated the man's trick. "There's nothing to it," he said as he rejoined the others. How did you do that? asked one of the children. "Static electricity," answered Nash before launching into a more detailed explanation.

Johnny's lack of interest in childish pursuits and lack of friends were major sources of worry for his parents. An ongoing effort to make him more "well rounded" became a family obsession.[28] Whether his apparent resolve to march to his own drummer was a question of his temperament or of his parents' concerted efforts to change his nature, the result was his withdrawal into his own private world. Martha, with whom Johnny constantly bickered, recalls:

> Johnny was always different. [My parents] knew he was different. And they knew he was bright. He always wanted to do things his way. Mother insisted I

do things for him, that I include him in my friendships. She wanted me to get him dates. She was right. But I wasn't too keen on showing off my some-what odd brother.

The Nashes pushed Johnny as hard socially as they did academically. At first, it was Boy Scout camp and Sunday Bible classes; later on, lessons at the Floyd Ward dancing school and membership in the John Aldens Society, a youth organi-zation devoted to improving the manners of its members. By high school, the outgoing Martha was always being enlisted to include her older brother when she socialized with friends. And in the summer holidays, the Nashes insisted that Johnny get jobs, including one at the *Bluefield Daily Telegraph.* In order to get him to the paper, "they got up at the wee hours of the night," Martha said. "They thought it was very important in helping make him well rounded. With a brain like John's, it seemed even more important. My mother and father didn't want him to be inside all the time with his hobbies and inventions."[29]

Johnny did not openly rebel — he dutifully trotted off to camp, dancing school, Bible classes, and, later on, blind dates arranged by his sister at Virginia's urging — but he did these things mainly to please his parents, especially his mother, and acquired neither friends nor social graces as a result. He continued to treat sports, going to church, the dances at the country club, visits with his cousins — all the things that so many of his peers found fascinating and enjoyable — as tedious distractions from his books and experiments. Always last to be chosen in softball, Johnny would stand in the right outfield, staring at the clouds above, eating bits of grass. Martha describes one occasion on which Virginia insisted he accompany the family to an Appalachian Power Company dinner. Johnny went, but spent the evening riding up and down in the elevator, which mesmerized him, until it broke — much to his parents' embarrassment. And on his summer jobs he found ways to entertain himself. One of Nash's classmates recalled that Nash, after disappearing for hours from his post at Bluefield Supply, was discovered rigging an elaborate system of mousetraps.[30] At a dance, he pushed a stack of chairs onto the dance floor and danced with them rather than with a girl.[31]

Virginia kept scrapbooks chronicling her children's lives and accomplish-ments. In one of them is a faded and yellowed essay by one Angelo Patri, clipped from a newspaper, covered with her pen marks, underlinings, and circles — poi-gnant hints of her hopes and fears:

Queer little twists and quirks go into the making of an individual. To suppress them all and follow clock and calendar and creed until the individual is lost in the neutral gray of the host is to be less than true to our inheritance. . . . Life, that gorgeous quality of life, is not accomplished by following another man's rules. It is true we have the same hungers and same thirsts, but they are for different things and in different ways and in different seasons. . . . Lay down your own day, follow it to its noon, your own noon, or you will sit in an outer hall listening to the chimes but never reaching high enough to strike your own.[32]

The earliest hint of Johnny's mathematical talent, ironically, was a B-minus in fourth-grade arithmetic. The teacher told Virginia that Johnny couldn't do the work, but it was obvious to his mother that he had merely found his own ways of solving problems. "He was always looking for different ways to do things," his sister commented.[33] More experiences like this followed, especially in high school, when he often succeeded in showing, after a teacher had struggled to produce a laborious, lengthy proof, that the proof could be accomplished in two or three elegant steps.

There is no sign of a mathematical pedigree in Nash's ancestry or any indication that mathematics was much in the air at the Nash household. Virginia Nash was literary. And for all his interest in contemporary developments in science and technology, John Sr. was not well-versed in abstract mathematics. Nash does not recall ever discussing his later research with his father.[34] Martha's recollections of dinner-table discussions were that they revolved around the meaning of words, books the children were reading, and current events.

The first bite of the mathematical apple probably occurred when Nash at around age thirteen or fourteen read E. T. Bell's extraordinary book, *Men of Mathematics* — an experience he alludes to in his autobiographical essay.[35] Bell's book, which was published in 1937, would have given Nash the first glimpse of real mathematics, a heady realm of symbols and mysteries entirely unconnected to the seemingly arbitrary and dull rules of arithmetic and geometry taught in school or even to the entertaining but ultimately trivial calculations that Nash carried out in the course of chemistry and electrical experiments.

Men of Mathematics consists of lively — and, as it turns out, not entirely accurate — biographical sketches.[36] Its flamboyant author, a professor of mathematics at the California Institute of Technology, declared himself disgusted with "the ludicrous untruth of the traditional portrait of the mathematician" as a "slovenly dreamer totally devoid of common sense." He assured his readers that the great mathematicians of history were an exceptionally virile and even adventuresome breed. He sought to prove his point with vivid accounts of infant precocity, monstrously insensitive educational authorities, crushing poverty, jealous rivals, love affairs, royal patronage, and many varieties of early death, including some resulting from duels. He even went so far, in defending mathematicians, as to answer the question "How many of the great mathematicians have been perverts?" None, was his answer. "Some lived celibate lives, usually on account of economic disabilities, but the majority were happily married. . . . The only mathematician discussed here whose life might offer something of interest to a Freudian is Pascal."[37] The book became a bestseller as soon as it appeared.

What makes Bell's account not merely charming, but intellectually seductive, are his lively descriptions of mathematical problems that inspired his subjects when they were young, and his breezy assurance that there were still deep and beautiful problems that could be solved by amateurs, boys of fourteen, to be specific. It was

Bell's essay on Fermat, one of the greatest mathematicians of all time but a perfectly conventional seventeenth-century French magistrate whose life was "quiet, laborious and uneventful," that caught Nash's eye.[38] The main interest of Fermat, who shares the credit for inventing calculus with Newton and analytic geometry with Descartes, was number theory — "the higher arithmetic." Number theory "investigates the mutual relationships of those common whole numbers, 1, 2, 3, 4, 5 . . . which we utter almost as soon as we learn to talk."

For Nash, proving a theorem known as Fermat's Theorem about prime numbers, those mysterious integers that have no divisor besides themselves and one, produced an epiphany of sorts. Other mathematical geniuses, Einstein and Bertrand Russell among them, recount similarly revelatory experiences in early adolescence. Einstein recalled the "wonder" of his first encounter with Euclid at age twelve:

> Here were assertions, as for example the intersection of three altitudes of a triangle at one point which — though by no means evident — could nevertheless be proved with such certainty that any doubt appeared to be out of the question. This lucidity and certainty made an indescribable impression on me.[39]

Nash does not describe his feelings when he succeeded in devising a proof for Fermat's assertion that if n is any whole number and p any prime, then n multiplied by itself p times minus n is divisible by p.[40] But he notes the fact in his autobiographical essay, and his emphasis on this concrete result of his initial encounter with Fermat suggests that the thrill of discovering and exercising his own intellectual powers — as much as any sense of wonder inspired by hitherto unsuspected patterns and meanings — was what made this moment such a memorable one. That thrill has been decisive for many a future mathematician. Bell describes how success in solving a problem posed by Fermat led Carl Friedrich Gauss, the renowned German mathematician, to choose between two careers for which he was similarly talented. "It was this discovery . . . which induced the young man to choose mathematics instead of philology as his life work."[41]

However heady it may have been to prove a theorem of Fermat's, the experience was hardly enough to plant the notion in Nash's mind that he might himself become a mathematician. Although as a high-school student Nash took mathematics at Bluefield College, as late as his senior year, when he already had gone much further into number theory, he still had firmly in mind following in his father's footsteps and becoming an electrical engineer. It was only after he had entered Carnegie Tech, with enough math to skip most entry-level courses, that his professors would convince him mathematics, for a chosen few, was a realistic choice as a profession.

The Japanese attack on the Pearl Harbor naval base in Hawaii, on December 7, 1941, came halfway through Johnny's first year in high school. A few days later,

Johnny and Mop, as he called his younger sister, got a lesson from their father in how to shoot a .22 caliber rifle.[42] He drove them up to a ridge where the power lines cut a wide swath through the scrubby, snow-dusted pine wood. Pointing toward the town below, huddling under a sooty gray cloud, he told them, in the soft, formal way he had of addressing his children, that the Japanese wouldn't rest until they had reached their West Virginia hometown, remote and surrounded by mountains as it was, because blowing up the coal trains was the only way they could cripple the mighty American war machine.

A .22, he said, was only a squirrel gun. You couldn't even kill a deer or a bear with one. But it was easier than a heavier gun for women and children to handle. They had no choice, really. The Japanese wouldn't be satisfied with destroying trains. They'd raze the city, round up all the men, murder all the civilians, even schoolchildren like them. If you could shoot this thing, you might be able to stop someone who was coming after you long enough to run away and hide someplace until the army rescued you. Years later, when Johnny Nash saw secret signs of invaders everywhere and believed that he, and only he, could keep the universe safe, he would be sick with anxiety, shaking and sweating and sleepless for hours and days at a time. But on that bright December afternoon, he was excited and happy as he fingered the rifle.

The war came thundering through Bluefield, West Virginia, in the roaring, rattling shapes of freight car after car heaped high with coal from the great Pocahontas coalfield in the mountains to the west — 40 percent of all the coal fueling the war machine — and troop trains crowded with sailors and soldiers, round-faced farm boys from Iowa and Indiana and edgy factory hands from Pittsburgh and Chicago.[43] The war shook and rattled the city out of its Depression slumber, filling its warehouses and streets, making overnight fortunes for scrap speculators and wheeler-dealers of all kinds. Workers were suddenly in short supply and there were jobs for everybody who wanted them. Bluefield teenagers hung around the train station watching it all, attended war bond rallies (Greer Garson showed up at one), and in school took part in tin can drives and bought war bonds with books of ten-cent stamps they bought in school. The war made a lot of Bluefield boys want to hurry and grow up lest the war be over before they were eligible to join. But Johnny didn't feel that way, his sister recalled. He did become obsessed with inventing secret codes consisting, as one former schoolmate recalled, of weird little animal and people hieroglyphics, sometimes adorned with biblical phrases: *Though the Wealthy Be Great / Roll in splendor and State / I envy them not, / I declare it.*

Adolescence wasn't easy for an intellectually precocious boy with few social skills or athletic interests to help him blend in with his small-town peers. The boys and girls on Country Club Hill let him tag along when they went hiking in the woods, explored caves, and hunted bats.[44] But they found him — his speech, his behavior, the knapsack he insisted on carrying — weird.[45] "He was teased more than average — simply because he was so far out," Donald V. Reynolds, who lived across the street from the Nashes, said. "What he thought of as experimenting, we thought of as crazy. We called him Big Brains."[46] Once some boys in the neighborhood tricked him into a boxing match and he took a beating.[47] But because he was tall, strong, and physically courageous, the teasing only rarely degenerated into outright bullying. He rarely passed up a chance to prove that he was smarter, stronger, braver.

Boredom and simmering adolescent aggression led him to play pranks, occasionally ones with a nasty edge. He caricatured classmates he disliked with weird little cartoons. He later told a fellow mathematician at MIT that, as a youngster, he had sometimes "enjoyed torturing animals."[48] He once constructed a Tinkertoy rocking chair, wired it electrically, and tried to get Martha to sit in it.[49] He played a similar prank on a neighboring child. Nelson Walker, head of Bluefield's Chamber of Commerce, told a newspaper reporter the following story:

> I was a couple of years younger than Johnny. One day I was walking by his house on Country Club Hill and he was sitting on the front steps. He called for me to come over and touch his hands. I walked over to him, and when I touched his hands, I got the biggest shock I'd ever gotten in my life. He had somehow rigged up batteries and wires behind him, so that he wouldn't get shocked but when I touched his hands, I got the living fire shocked out of me. After that he just smiled and I went on my way.[50]

Occasionally the pranks got him into hot water. One incident involving a small explosion in the high school chemistry lab landed him in the principal's office.[51] Another time, he and some other boys were picked up by the police for a curfew violation.[52]

When he was fifteen, Nash and a couple of boys from across the street, Donald Reynolds and Herman Kirchner, began fooling around with homemade explosives.[53] They gathered in Kirchner's basement, which they called their "laboratory," where they made pipe bombs and manufactured their own gunpowder. They constructed cannons out of pipe and shot stuff through them. Once they managed to shoot a candle through a thick wooden board. One day Nash showed up at the lab holding a beaker. "I've just made some nitroglycerin," he announced excitedly. Donald didn't believe him. He told him "to go down to Crystal Rock and throw it over the cliff to see what would happen." Nash did just that. "Luckily," said Reynolds, "it didn't work. He would have blown off the whole side of the mountain." The bombmaking came to a horrifying end one afternoon in January 1944. Herman Kirchner, who was alone at the time, was building yet another pipe bomb when it exploded in his lap, severing an artery. He bled to death in the ambulance that came for him. Donald Reynolds's parents packed him off to boarding school the following fall. For Nash, whose parents may or may not have known the extent of his involvement in the bombmaking, it was a sobering experience that brought home the dangers of his experiments.

He had grown up, essentially, without ever making a close friend. Just as he learned to deflect his parents' criticism of his behavior with his intellectual achievements,

he learned to armor himself against rejection by adopting a hard shell of indifference and using his superior intelligence to strike back. Julia Robinson, the first
woman to become president of the American Mathematical Society, said in her
autobiography that she believed that many mathematicians felt themselves to be
ugly ducklings as children, unlovable and out of kilter with their more conventional, conforming peers.[54] Johnny's apparent sense of superiority, his standoffish-
ness, and his occasional cruelty were ways of coping with uncertainty and
loneliness. What he lost by his lack of genuine interaction with children his own
age was a "lively sense, in reality, of his actual position in the human hierarchy"
that prevents other children with more social contact from feeling either unrealistically weak or unrealistically powerful.[55] If he could not believe he was lovable,
then feeling powerful was a good substitute. As long as he could be successful, his
self-esteem could remain intact.

Johnny chose the time-honored escape route from the confines of small-town
life: He performed well in school. With Virginia's encouragement, he took courses
at Bluefield College. He read voraciously, mostly futuristic fantasy books, popular
science magazines, and real science texts.[56] "He was just an outstanding problem
solver," his high school chemistry teacher later told the *Bluefield Daily Telegraph*.
"When I put a chemistry problem up on the blackboard, all the students would get
out a pencil and a piece of paper. John wouldn't move. He would stare at the
formula on the board, then stand up politely and tell us the answer. He could do
it all in his head. He never even took out a pencil or a piece of paper."[57] This
youthful Gedanken experimentation actually helped shape the way he approached
mathematical problems later on. His peers became more respectful. At a time
when the war was making heroes out of scientists, Johnny's classmates assumed he
was slated to become one.[58]

In high school, Nash became friendly — though not close friends — with a couple
of fellow students, John Williams and John Louthan, both sons of Bluefield College
professors. The three rode a public bus to school together and Johnny helped
Williams with Latin translations. Williams recalled, "We were attracted to him. He
was an interesting guy. That was sort of it. I don't think we ever went over to
John's house. It was pretty much of a school thing."[59] The three also constantly
maneuvered to get out of their classes as much as possible. Before the widespread
use of the SATs, college recruiters routinely came to the high school and would
invite students to take their admissions tests. "We spent many mornings taking
those tests," Williams said.

At the beginning of the year, at Johnny's instigation, they made a bet — no one
remembers for how much — that they could make the honor roll without ever
cracking a book. All three thought they were pretty smart but at the same time
were contemptuous of grinds and teachers' pets. "We kind of got drug into it by
Nash," Williams said. Nash, who was already taking a full load of courses at

Bluefield College, never made the honor society, missing it by a few tenths of a percent. The other two did, though by a hair.

John Sr. suggested that Johnny apply to West Point, a suggestion that, once again, may have reflected the father's anxiety that his son was not growing up well-rounded as much as it did the prospect of free college tuition. But as Martha said, "Even I could see that wouldn't have worked."[60] Whatever fantasies he may have had about becoming a scientist, when asked to describe his career aspirations in an essay, Johnny wrote that he hoped to become an engineer like his father.[61] He and John Sr. wrote an article together describing an improved method for calculating the proper tensions for electric cables and wires — a project that entailed weeks of field measurements — and published the results jointly in an engineering journal.[62] Johnny entered the George Westinghouse competition and won a full scholarship, one of ten that were awarded nationally.[63] The fact that Lloyd Shapley, a son of the famous Harvard astronomer Harlow Shapley, also won a Westinghouse that year made the achievement all the sweeter in the eyes of the Nash family. Johnny was accepted at the Carnegie Institute of Technology. Because of the war all colleges were on accelerated schedules and operated year-round so that students could graduate in three years. Johnny left Bluefield for Pittsburgh, taking a train from nearby Hinton, in mid-June, a few weeks before the VE Day parade celebrating Hitler's defeat.

2

Carnegie Institute
of Technology

June 1945–June 1948

In those days very few people became mathematicians. It was like becoming a concert pianist. — RAOUL BOTT, 1995

NASH WENT TO PITTSBURGH to become a chemical engineer, but his growing interest was in mathematics. It was not long before he abandoned the laboratory and slide rule for Möbius knots and Diophantine equations.[1]

With its smelters, power plants, polluted rivers, and ubiquitous slag heaps, Pittsburgh was a city of violent strikes and frequent floods.[2] So dense was the sulfurous haze that engulfed its downtown that travelers arriving by rail often mistook morning for midnight. The Carnegie Institute of Technology, perched halfway up Squirrel Hill, hardly escaped the inferno. The ivory-colored brick of its buildings — designed, or so students said, to serve as factories should Andrew Carnegie's school fail — were glazed yellow black. Its walkways were gritty with soot particles the size of pebbles. Its students were forced, before a lecture was half over, to brush the cinders from their lecture notes. Even at high noon in midsummer, one could stare directly at the sun without blinking.

In that era, Carnegie was shunned by the local ruling elite, which sent its children east to Harvard and Princeton. Richard Cyert, who joined the Carnegie faculty after the war and would later become its president, recalled, "When I came this place was really very backward."[3] The engineering school, with its two thousand or so students, still resembled the trade school for sons and daughters of electricians and bricklayers that it had been at the turn of the century.

But like so many other colleges right after the war, Carnegie was changing. Robert Doherty, its president, had seized the opportunities created by wartime research to turn the engineering school into a real university. He parlayed defense contracts and the prospect of ballooning enrollments into a big push to recruit brilliant young researchers in math, physics, and economics. "The theoretical sciences were being pushed very hard," recalled Richard Duffin, a mathematician. "Doherty was trying to take CT into the big time."[4]

Corporate giants like Westinghouse, whose headquarters were in Pittsburgh, supplied generous scholarships to lure talented young people to Carnegie. Among the scholarship recipients who entered Carnegie in 1945 were talented youngsters like Andy Warhol, the artist, as well as a group of young men who would eventually, like Nash, shun engineering for science and mathematics.[5]

Nash arrived by train in June 1945; gasoline rationing made car travel impractical.[6] Carnegie Tech was still operating in wartime mode: classes went year-round, most campus activities remained canceled, and most of the fraternity houses were still shut. Within a year the campus would be inundated with veterans and classes would be jammed with these older students. But that June, two months before the war finally ended, it was mostly freshmen and sophomores who were on campus. The scholarship students were housed together in Welch Hall and took most of their classes together — small ones taught by hand-picked instructors, some of whom were first-rate. Nash took his first physics course from Immanuel Estermann, for example, a top-flight physicist who had done much of the experimental work that had netted Otto Stern, a German émigré, the 1943 Nobel Prize for physics.[7]

Nash's engineering aspirations did not survive his first semester, killed off by an unhappy experience in mechanical drawing: "I reacted negatively to the regimentation," he later wrote.[8] But chemistry, his newly chosen major, proved no better suited to his temperament or interests. He worked briefly as a lab assistant for one of his teachers but got into trouble for breaking equipment.[9] He was so bored at his summer job at the Westinghouse Lab that he spent most of his two months there making and polishing a brass egg in the lab's machine shop.[10] The final blow was a C in physical chemistry, which he got after a running dispute with the professor over the lack of rigor of the mathematics in the course. David Lide recalled, "He refused to do the problems the way the professor expected."[11] Of chemistry in general Nash would complain: "It was not a matter of how well one could think . . . but of how well one could handle a pipette and perform titration in the laboratory."[12]

Even as he struggled in the laboratory, Nash was already discovering a brilliant group of newcomers to Carnegie. By his sophomore year, Doherty's program of upgrading the theoretical sciences had brought to Carnegie John Synge, nephew of the Irish playwright John Millington Synge, who became head of the mathematics department. Despite his startling appearance — Synge wore a black patch over one eye and a filter that protruded from one of his nostrils — he was a man of great charm who attracted younger scholars like Richard Duffin, Raoul Bott, and Alexander Weinstein, a European émigré whom Einstein had once invited to become a collaborator.[13] When Albert Tucker, a Princeton topologist who did pathbreaking work in operations research, came to Carnegie to lecture that year, he was so impressed with the depth of mathematical talent at Carnegie that he confessed that he felt as if he were "bringing coals to Newcastle."[14]

From the start, Nash dazzled his mathematics professors; one of them called him "a young Gauss."[15] He took courses in tensor calculus — the mathematical tool used by Einstein to formulate the general theory of relativity — and relativity from Synge.[16] Synge was impressed with Nash's originality and his appetite for difficult problems.[17] He and others began urging Nash to major in mathematics and to consider an academic career. Nash's doubts that one could make a living as a mathematician took some time to overcome. But by the middle of his second year he was concentrating almost exclusively on mathematics. The Westinghouse scholarship administrators were unhappy with Nash's switch to mathematics, but by the time they learned of it, it was a fait accompli.[18]

College is a time when many ugly ducklings discover that they are swans, not just intellectually but socially. Most of the boys in Welch Hall — precocious but immature — found common interests, kindred spirits, and a measure of acceptance painfully lacking in high school. Hans Weinberger recalled, "We were all nerds back in our high schools and here we were able to talk to one another."[19]

Nash was not so lucky. While his professors singled him out as a potential star, his new peers found him weird and socially inept. "He was a country boy, unsophisticated even by our standards," recalled Robert Siegel, a physics major, who remembered that Nash had never attended a symphony performance before.[20] He behaved oddly, playing a single chord on the piano over and over,[21] leaving an ice cream cone melting on top of his castoff clothing in the lounge,[22] walking on his roommate's sleeping body to turn off a light,[23] pouting when he lost a game of bridge.[24]

Nash was rarely invited to go to concerts or restaurants with the group. Paul Zweifel, an avid bridge player, taught Nash how to play bridge, but Nash's pouting and inattention to the details of the game made him a poor partner. "He wanted to talk about the theoretical aspects."[25] Nash roomed with Weinberger for a term, but the two clashed constantly — Nash once pushed Weinberger around to end an argument[26] — and Nash moved into a private room at the end of the hall. "He was extremely lonely," recalled Siegel.[27]

Later in life, as his accomplishments multiplied, his peers would be more apt to be forgiving. But at Carnegie, where he was thrust together with other adolescents around the clock, he became a target. He was not so much bullied — the other boys were afraid of his strength and temper — as ostracized and relentlessly teased. That he was envied for his size and his brains only fueled the teasing. "He was the butt of people's jokes because he was different," recalled George Hinman, a physics student.[28] "Here was a guy who was socially underdeveloped and acting much younger. You do what you can to make his life miserable," Zweifel admitted. "We tormented poor John. We were very unkind. We were obnoxious. We sensed he had a mental problem."[29]

• • •

That first summer, Nash, Paul Zweifel, and a third boy spent an afternoon exploring the subterranean maze of steam tunnels under Carnegie. In the dark, Nash suddenly turned to the others and blurted out, "Gee, if we got trapped down here we'd have to turn homo." Zweifel, who was fifteen, found the remark pretty odd. But during Thanksgiving break, in the deserted dormitory, Nash climbed into Zweifel's bed when the latter was sleeping and made a pass at him.[30]

Away from home, living in close proximity with other adolescents, Nash discovered that he was attracted to other boys. He spoke and acted in ways that seemed natural to him only to find himself exposed to his peers' contempt. Zweifel and other boys in the dormitory started calling Nash "Homo" and "Nash-Mo."[31] "Once the statement was made," George Siegel said, "it stuck. John took a lot."[32] No doubt, he found the label hurtful and humiliating, but his anger is all that anyone witnessed.

The boys made him the butt of various pranks. One time, Weinberger and a couple of others used a footlocker as a battering ram to break down Nash's door.[33] Another time, Zweifel and a few others, knowing of Nash's extreme aversion to cigarette smoke, rigged up a contraption that smoked an entire pack of cigarettes and collected the smoke. "A bunch of us crowded around John's door and blew the smoke under it," Zweifel recalled. "Almost instantaneously, his room filled up with cigarette smoke."[34] Nash exploded in rage. "He came roaring out of his room, picked up Jack [Wachtman], and threw him down on the bed," said Zweifel. "He ripped off Wachtman's shirt and bit him in the back. Then he ran out of the room."

At other times, Nash defended himself the only way he knew how. He wasn't practiced in invective, sarcasm, or ridicule, so he went for childish displays of contempt. " 'You stupid fool,' he'd say," Siegel recalled. "He was openly contemptuous of people who he didn't think were up to his level intellectually. He showed that contempt for all of us: 'You're an ignoramus.' " After a year or so, after he had acquired a reputation for being a genius, he began to hold court in Skibo Hall, the student center.[35] Like the fairground magician with his swords, he would sit in a chair and challenge other students to throw problems at him to solve. A lot of students came to him with their homework. He was a star — but an outcast too.

Nash stared glumly at the announcement tacked to the bulletin board outside the math department office in Administration Hall, which looked, even on the sunniest of days, like the inside of the Lincoln Tunnel. He stood in front of the board for a long time. He hadn't made it into the top five.[36]

Nash's fantasy of instant glory crumbled. The William Lowell Putnam Mathematical Competition was a prestigious national tournament for undergraduates, sponsored by an old-money Boston family known mostly for its Harvard presidents and deans.[37] Today the contest attracts upward of two thousand participants. In

March 1947, it was a decade old and drew about 120. But even then, it was the first chance to establish one's rank in the world of mathematics as well as to seize the limelight.

Then, as now, contestants were given a dozen problems and half an hour each to solve them. The problems were famously difficult. In any given year, the median score out of 120 possible points was zero. That meant that at least half the contestants weren't able to obtain so much as partial credit for even a single problem, and this in spite of the fact that most contestants had been chosen by their departments to compete. To have a prayer of winning—placing in the top five—a young mathematician had to be super-fast or especially ingenious. The prizes involved a nominal amount of money, twenty to forty dollars for each of the top ten contestants, and two hundred to four hundred dollars for each of the top five school teams, but winners became instant mini-celebrities in the mathematics world and were virtually assured a spot in a top graduate program. Different graduate programs pay more or less attention to the Putnam, but at Harvard it is, and always has been, a very, very big deal. That year Harvard pledged a fifteen-hundred-dollar scholarship to one of the winners.

Nash had competed as a freshman and a sophomore. On his second try, he'd managed to get into the top ten, but not the top five. He'd been cocky this time, too. In 1946 a mathematician named Moskovitz tutored the Carnegie Tech team using problems from past exams. Nash was able to solve problems that Moskovitz and the others could not solve. It was a tremendous blow to Nash that George Hinman ranked in the top ten in the 1946 competition and Nash didn't.[38]

Another nineteen-year-old might have shrugged off the disappointment, especially a boy who had been plucked out of a chemical engineering program, welcomed with open arms by the school's mathematicians, and told that he had a brilliant future in mathematics. But for a teenager who had endured a lifetime of rejection by peers, the warm praise of such professors as Richard Duffin and J. L. Synge was too little, too late. Nash craved a more universal form of recognition, recognition based on what he regarded as an objective standard, uncolored by emotion or personal ties. "He always wanted to know where he stood," said Harold Kuhn recently. "It was always important to be in the club."[39] Decades later, after he had acquired a worldwide reputation in pure mathematics and had won a Nobel Prize in economics, Nash hinted in his Nobel autobiography that the Putnam still rankled and implied that the failure played a pivotal role in his graduate career.[40] Today, Nash still tends to identify mathematicians by saying, "Oh, So and So, he won the Putnam three times."

In the fall of 1947, Richard Duffin stood at the board silent and frowning.[41] He was intimately familiar with Hilbert spaces, but he had prepared his lecture too hastily, had wandered down a cul de sac in the course of his proof, and was hopelessly stuck. It happened all the time.

The five students in the advanced graduate class were getting restive. Weinberger, who was Austrian by birth, was often able to explain the fine points of von Neumann's book *Mathematische Grundlagen der Quantenmechanik,* which Duffin was using as a text. But Weinberger was frowning too. After a few moments, everybody turned toward the gawky undergraduate who was squirming in his seat. "Okay, John, you go to the board," said Duffin. "See if you can get me out of trouble." Nash leaped up and strode to the board.[42]

"He was infinitely more sophisticated than the rest of us," said Bott. "He understood the difficult points naturally. When Duffin got stuck, Nash could back him up. The rest of us didn't understand the techniques you needed in this new medium."[43] "He always had good examples and counterexamples," another student recalled.[44]

Afterward, Nash hung around. "I could talk to Nash," Duffin recalled shortly before his death in 1995. "After class one day he started talking about Brouwer's fixed point theorem. He proved it indirectly using the principle of contradiction. That's when you show that if something's not there, something dreadful will happen. Don't know if Nash had ever heard of Brouwer."[45]

Nash took Duffin's course in his third and final year at Carnegie. At nineteen, Nash already had the style of a mature mathematician. Duffin recalled, "He tried to reduce things to something tangible. He tried to relate things to what he knew about. He tried to get a feel for things before he actually tried them. He tried to do little problems with some numbers in them. That's how Ramanujan, who claimed he got his results from spirits, figured things out. Poincaré said he thought of a great theorem getting off a bus."[46]

Nash liked very general problems. He wasn't all that good at solving cute little puzzles. "He was a much more dreamy person," said Bott. "He'd think a long time. Sometimes you could see him thinking. Others would be sitting there with their nose in a book."[47] Weinberger recalled that "Nash knew a lot more than anybody else there. He was working on things we couldn't understand. He had a tremendous body of knowledge. He knew number theory like mad."[48] "Diophantine equations were his love," recalled Siegel. "None of us knew anything about them, but he was working on them then."[49]

It is obvious from these anecdotes that many of Nash's lifelong interests as a mathematician — number theory, Diophantine equations, quantum mechanics, relativity — already fascinated him in his late teens. Memories differ on whether Nash learned about the theory of games at Carnegie.[50] Nash himself does not recall. He did, however, take a course in international trade, his one and only formal course in economics, before graduating.[51] It was in this course that Nash first began to mull over one of the basic insights that eventually led to his Nobel Prize.[52]

By the spring of 1948 — in what would have been his junior year at Carnegie — Nash had been accepted by Harvard, Princeton, Chicago, and Michigan,[53] the four top graduate mathematics programs in the country. Getting into one of these was virtually a prerequisite for eventually landing a good academic appointment.

Harvard was his first choice.[54] Nash told everyone that he believed that Harvard had the best mathematics faculty. Harvard's cachet and social status appealed to him. As a university, Harvard had a national reputation, while Chicago and Princeton, with its largely European faculty, did not. Harvard was, to his mind, simply number one, and the prospect of becoming a Harvard man seemed terribly attractive.

The trouble was that Harvard was offering slightly less money than Princeton. Certain that Harvard's comparative stinginess was the consequence of his less-than-stellar performance in the Putnam competition, Nash decided that Harvard didn't really want him. He responded to the rebuff by refusing to go there. Fifty years later, in his Nobel autobiography Harvard's lukewarm attitude toward him seems still to have stung: "I had been offered fellowships to enter as a graduate student at either Harvard or Princeton. But the Princeton fellowship was somewhat more generous since I had not actually won the Putnam competition."[55]

Princeton was eager. From the 1930s onward, Princeton had a far stronger department and was snaring the lion's share of the best graduate students.[56] Princeton was, as a matter of fact, more selective than Harvard at that point, admitting ten handpicked candidates each year, as opposed to Harvard's twenty-five or so. The Princeton faculty didn't care a hoot about the Putnam, or about tests of any kind, or grades. They paid attention exclusively to the opinions of mathematicians whose views they respected. And once Princeton decided it wanted someone, it pursued him with vigor.

Duffin and Synge were pushing Princeton hard. Princeton was full of purists — topologists, algebraists, number theorists — and Duffin especially regarded Nash as someone obviously suited, by interest and temperament, for a career in the most abstract mathematics. "I thought he would be a completely pure mathematician," Duffin recalled. "Princeton was first in topology. That's why I wanted to send him to Princeton."[57] The only thing Nash really knew about Princeton was that Albert Einstein and John von Neumann were there, along with a bunch of other European émigrés. But the polyglot Princeton mathematical milieu — foreign, Jewish, left-leaning — still seemed to him a distinctly inferior alternative.

Sensing Nash's hesitation, Solomon Lefschetz, the chairman of the Princeton department, had already written to him urging him to choose Princeton.[58] He finally dangled a John S. Kennedy Fellowship.[59] The one-year fellowship was the most prestigious the department had to offer, requiring little or no teaching and guaranteeing a room in Princeton's residential college for graduate students. It was a sign of how much Princeton was panting for Nash. The $1,150 fellowship covered the $450 tuition and was more than ample for the $200 room rent for a year and $14 a week in dining fees, as well as living expenses.[60]

For Nash, that clinched the decision.[61] The difference in the awards could not have been huge in any practical sense. But, then, as so many times later in Nash's life, a relatively trivial amount of money loomed in his decision. It seems clear that Nash calculated Princeton's more generous fellowship as a measure of how Princeton valued him. A personal appeal from Lefschetz, with a flattering reference to his relative youth, also proved decisive. Lefschetz's phrase "We like to catch promising men when they are young and open-minded" struck a chord.[62]

Something else weighed on Nash's mind that last spring at Carnegie. As graduation drew closer, he became more and more worried about being drafted.[63] He thought that the United States might go to war again and was afraid that he might wind up in the infantry. That the army was still shrinking three years after the end of World War II and that the draft had, for all intents and purposes, ground to a standstill, did not make Nash feel safe. The newspapers — of which he was a regular reader — were full of signs, in particular the Russian blockade of Berlin and the subsequent American-British airlift that spring, that the Cold War was heating up. He hated any thought that his personal future might be hostage to forces outside his control and he was obsessed with ways to defend himself against any possible threats to his own autonomy or plans.

So Nash was palpably relieved when Lefschetz offered to help him obtain a summer job with a Navy research project. The project in White Oak, Maryland, was being run by Clifford Ambrose Truesdell, a former student of Lefschetz.[64] Nash wrote to Lefschetz at the beginning of April:

> Should there come a war involving the US I think I should be more useful, and better off, working on some research project than going, say into the infantry. Working on government sponsored research this summer would pave the way toward the more desirable eventuality.[65]

Though Nash did not display outward signs of distress, the disappointments and anxieties of the spring cast a shadow over the summer between his graduation from Carnegie and his arrival at Princeton.

White Oak is a suburb of Washington, D.C. In the summer of 1948, it was a swampy, humid woodland full of raccoons, opossums, and snakes. The mathematicians at White Oak were a hodgepodge of Americans, some of whom had been working for the Navy since the middle of the war, and others, German prisoners of war. Nash found himself a room in downtown Washington, which he rented from a Washington, D.C., police officer. He rode to White Oak in a car pool every day with two of the Germans.[66]

Nash had been looking forward to the summer. Lefschetz had promised that the work would be pure mathematics.[67] Truesdell, quite a good mathematician, was a tolerant supervisor who encouraged the mathematicians in his group to pursue their own research. He essentially gave Nash carte blanche, issuing no instructions and merely saying that he hoped Nash would write something before he left at the end of the summer. But Nash seemed to have trouble working. He made no apparent progress on any of the problems he had mentioned vaguely to Truesdell at the start of the summer, and he never handed in a paper. At the end of the summer, he was forced to apologize to Truesdell for having wasted his time.[68]

Nash spent most of his days, evidently, simply walking around rather aimlessly, lost in thought. Charlotte Truesdell, Truesdell's wife and the project's girl Friday, recalls that Nash seemed terribly young, "like a sixteen-year-old," and almost never spoke to anyone. Once when she asked him what he was thinking, Nash asked whether she, Charlotte, didn't think it would be a good joke if he put live snakes in the chairs of some of the mathematicians. "He didn't do it," she said, "but he thought about it a lot."[69]

3

The Center of the Universe

Princeton, Fall 1948

. . . a quaint ceremonious village. — ALBERT EINSTEIN

. . . the mathematical center of the universe. — HARALD BOHR

NASH ARRIVED in Princeton, New Jersey, on Labor Day 1948, the opening day of Truman's re-election campaign.[1] He was twenty years old. He came by train, directly from Bluefield, via Washington, D.C., and Philadelphia, wearing a new suit and carrying unwieldy suitcases stuffed with bedding and clothes, letters and notes, and a few books. Impatient and eager now, he got off at Princeton Junction, a nondescript little middle-class enclave a few miles from Princeton proper, and hurried onto the Dinky, the small single-track train that shuttles back and forth to the university.

What he saw was a genteel, prerevolutionary village surrounded by gently rolling woodlands, lazy streams, and a patchwork of cornfields.[2] Settled by Quakers toward the end of the seventeenth century, Princeton was the site of a famous Washington victory over the British and, for a brief six-month interlude in 1783, the de facto capital of the new republic. With its college-Gothic buildings nestled among lordly trees, stone churches, and dignified old houses, the town looked every inch the wealthy, manicured exurb of New York and Philadelphia that, in fact, it was. Nassau Street, the town's sleepy main drag, featured a row of "better" men's clothing shops, a couple of taverns, a drugstore, and a bank. It had been paved before the war, but bicycles and pedestrians still accounted for most of the traffic. In *This Side of Paradise,* F. Scott Fitzgerald had described Princeton circa World War I as "the pleasantest country club in America."[3] Einstein called it "a quaint, ceremonious village" in the 1930s.[4] Depression and wars had scarcely changed the place. May Veblen, the wife of a wealthy Princeton mathematician, Oswald Veblen, could still identify by name every single family, white and black, well-to-do and of modest means, in every single house in town.[5] Newcomers invariably felt intimidated by its gentility. One mathematician from the West recalled, "I always felt like my fly was open."[6]

Even the university's mathematics building conjured up images of exclusivity and wealth. "Fine Hall is, I believe, the most luxurious building ever devoted to mathematics," one European émigré wrote enviously.[7] It was a gabled, Neo-Gothic red brick and slate fortress, built in a style reminiscent of the College de France in Paris and Oxford University. Its cornerstone contains a lead box with copies of works by Princeton mathematicians and the tools of the trade — two pencils, one piece of chalk, and, of course, an eraser. Designed by Oswald Veblen, a nephew of the great sociologist Thorstein Veblen, it was meant to be a sanctuary that mathematicians would be "loath to leave."[8] The dim stone corridors that circled the structure were perfect for both solitary pacing and mathematical socializing. The nine "studies" — not offices! — for senior professors had carved paneling, hidden file cabinets, blackboards that opened like altars, oriental carpets, and massive, overstuffed furniture. In a gesture to the urgency of the rapidly advancing mathematical enterprise each office was equipped with a telephone and each lavatory with a reading light. Its well-stocked third-floor library, the richest collection of mathematical journals and books in the world, was open twenty-four hours a day. Mathematicians with a fondness for tennis (the courts were nearby) didn't have to go home before returning to their offices — there was a locker room with showers. When its doors opened in 1921, an undergraduate poet called it "a country club for math, where you could take a bath."

Princeton in 1948 was to mathematicians what Paris once was to painters and novelists, Vienna to psychoanalysts and architects, and ancient Athens to philosophers and playwrights. Harald Bohr, brother of Niels Bohr, the physicist, had declared it "the mathematical center of the universe" in 1936.[9] When the deans of mathematics held their first worldwide meeting after World War II, it was in Princeton.[10] Fine Hall housed the world's most competitive, up-to-the-minute mathematics department. Next door — connected, in fact — was the nation's leading physics department, whose members, including Eugene Wigner, had driven off to Illinois, California, and New Mexico during the war, lugging bits of laboratory equipment, to help build the atomic bomb.[11] A mile or so away, on what had been Olden Farm, was the Institute for Advanced Study, the modern equivalent of Plato's Academy, where Einstein, Gödel, Oppenheimer, and von Neumann scribbled on their blackboards and held their learned discourses.[12] Visitors and students from the four corners of the world streamed to this polyglot mathematical oasis, fifty miles south of New York. What was proposed in a Princeton seminar one week was sure to be debated in Paris and Berkeley the week after, and in Moscow and Tokyo the week after that.

"It is difficult to learn anything about America in Princeton," wrote Einstein's assistant Leopold Infeld in his memoirs, "much more so than to learn about England in Cambridge. In Fine Hall English is spoken with so many different accents that the resultant mixture is termed Fine Hall English. . . . The air is full of mathematical ideas and formulae. You have only to stretch out your hand, close it quickly and you feel that you have caught mathematical air and that a few formulae are stuck to your palm. If one wants to see a famous mathematician one does not need to go to him; it is enough to sit quietly in Princeton, and sooner or later he must come to Fine Hall."[13]

Princeton's unique position in the world of mathematics had been achieved practically overnight, barely a dozen years earlier.[14] The university predated the Republic by a good twenty years. It started out as the College of New Jersey in 1746, founded by Presbyterians. It didn't become Princeton until 1896 and wasn't headed by a layman until 1903 when Woodrow Wilson became its president. Even then, however, Princeton was a university in name only — "a poor place," "an overgrown prep school," particularly when it came to the sciences.[15] In this regard, Princeton merely resembled the rest of the nation, which "admired Yankee ingenuity but saw little use for pure mathematics," as one historian put it. Whereas Europe had three dozen chaired professors who did little except create new mathematics, America had none. Young Americans had to travel to Europe to get training beyond the B.A. The typical American mathematician taught fifteen to twenty hours a week of what amounted to high school mathematics to undergraduates, struggling along on a negligible salary and with very little incentive or opportunity to do research. Forced to drill conic sections into the heads of bored undergraduates, the Princeton professor of mathematics was perhaps not as well off as his forebears of the seventeenth century who practiced law (Fermat), ministered to royalty (Descartes), or occupied professorships with negligible teaching duties (Newton). When Solomon Lefschetz arrived at Princeton in 1924, "There were only seven men there engaged in mathematical research," Lefschetz recalled. "In the beginning we had no quarters. Everyone worked at home."[16] Princeton's physicists were in the same boat, still living in the age of Thomas Edison and Alexander Graham Bell, preoccupied with measuring electricity and supervising endless freshman lab sections.[17] Henry Norris Russell, a distinguished astronomer by the 1920s, fell afoul of the Princeton administration for spending too much time on his own research at the expense of undergraduate teaching. In its disdain for scientific research, Princeton was not very different from Yale or Harvard. Yale refused for seven years to pay a salary to the physicist Willard Gibbs, already famous in Europe, on the grounds that his studies were "irrelevant."[18]

While mathematics and physics at Princeton and other American universities were languishing, a revolution in mathematics and physics was taking place three thousand miles away in such intellectual centers as Göttingen, Berlin, Budapest, Vienna, Paris, and Rome.

John D. Davies, a historian of science, writes of a dramatic revolution in the understanding of the very nature of matter:

> The absolute world of classical Newtonian physics was breaking down and intellectual ferment was everywhere. Then in 1905 an unknown theoretician in the Berne patent office, Albert Einstein, published four epoch-making

papers comparable to Newton's instant leap into fame. The most significant was the so-called Special Theory of Relativity, which proposed that mass was simply congealed energy, energy liberated matter: space and time, previously thought to be absolute, were dependent on relative motion. Ten years later he formulated the General Theory of Relativity, proposing that gravity was a function of matter itself and affected light exactly as it affected material particles. Light, in other words, did not go "straight"; Newton's laws were not the real universe but one seen through the unreal spectacles of gravity. Furthermore, he set forth a set of mathematical laws with which the universe could be described, structural laws and laws of motion.[19]

At around the same time, at the University of Göttingen, a German mathematical genius, David Hilbert, had unleashed a revolution in mathematics. Hilbert set out a famous program in 1900 of which the goal was nothing less than the "axiomatization of all of mathematics so that it could be mechanized and solved in a routine manner." Göttingen became the center of a drive to put existing mathematics on a more secure foundation: "The Hilbert program emerged at the turn of the century as a response to a perceived crisis in mathematics," writes historian Robert Leonard. "The effect was to drive mathematicians to 'clean up' Cantorian set theory, to establish it on a firm axiomatic basis, on the foundation of a limited number of postulates. . . . This marked an important shift in emphasis towards abstraction in mathematics."[20] Mathematics moved further and further away from "intuitive content — in this case, our daily world of surfaces and straight lines — towards a situation in which mathematical terms were leached of their direct empirical content and simply defined axiomatically within the context of the theory. The era of formalism had arrived."

The work of Hilbert and his disciples — among them such future Princeton stars of the 1930s and 1940s as Hermann Weyl and John von Neumann — also triggered a powerful impulse to apply mathematics to problems hitherto considered unamenable to highly formal treatment. Hilbert and others were quite successful in extending the axiomatic approach to a range of topics, the most obvious being physics, in particular the "new physics" of "quantum mechanics," but also to logic and the new theory of games.

But for the first twenty-five years of the century, as Davies writes, Princeton, and indeed the whole American academic community, "stood outside this dramatically swift development."[21] The catalyst for Princeton's transformation into a world capital of mathematics and theoretical physics was an accident — an accident of friendship. Woodrow Wilson, like most other educated Americans of his time, despised mathematics, complaining that "the natural man inevitably rebels against mathematics, a mild form of torture that could only be learned by painful processes of drill."[22] And mathematics played no role whatever in his vision of Princeton as a real university with a graduate college and a system of instruction that emphasized seminars and discussions instead of drills and rote learning. But Wilson's best friend, Henry Burchard Fine, happened to be a mathematician. When Wilson set about hiring literature and history scholars as preceptors, Fine asked him, "Why not a few scientists?" As a gesture of friendship more than anything else, Wilson said yes. After Wilson left the presidency of Princeton for the White House in 1912, Fine became dean of science and proceeded to recruit some top-notch scientists, among them mathematicians G. D. Birkhoff, Oswald Veblen, and Luthor Eisenhart, to teach graduate students. They were known around Princeton as "Fine's research men." The undergraduates, not a single one of whom majored in physics or math, complained bitterly of "brilliant but unintelligible lecturers with foreign accents" and "the European, or demi-God, theory of instruction."

Fine's nucleus of researchers might well have scattered after the dean's premature death in 1928 in a cycling accident on Nassau Street had it not been for several dramatic instances of private philanthropy that turned Princeton into a magnet for the world's biggest mathematical stars. Most people think that America's rise to scientific prominence was a by-product of World War II. But in fact the fortunes accumulated between the gilded eighties and the roaring twenties paved the way.

The Rockefellers made their millions in coal, oil, steel, railroads, and banking — in other words, from the great sweep of industrialization that transformed towns like Bluefield and Pittsburgh in the late nineteenth and early twentieth centuries. When the family and its representatives started to give away some of the money, they were animated by dissatisfaction with the state of higher education in America and a firm belief that "nations that do not cultivate the sciences cannot hold their own."[23] Aware of the scientific revolution sweeping Europe, the Rockefeller Foundation and its offshoots started by sending American graduate students, including Robert Oppenheimer, abroad. By the mid-1920s, the Rockefeller Foundation decided that "instead of sending Mahomet to the Mountain, it would fetch the Mountain here." That is, it decided to import Europeans. To finance the effort, the foundation committed not just its income but $19 million of its capital (close to $150 million in today's dollars). While Wickliffe Rose, a philosopher on Rockefeller's board, scoured such European scientific capitals as Berlin and Budapest to hear about new ideas and meet their authors, the foundation selected three American universities, among them Princeton, to receive the bulk of its largesse. The grants enabled Princeton to establish five European-style research professorships with extravagant salaries, plus a research fund to support graduate and postgraduate students.

Among the first European stars to arrive in Princeton in 1930 were two young geniuses of Hungarian origin, John von Neumann, a brilliant student of Hilbert and Hermann Weyl, and Eugene Wigner, the physicist who went on to win a Nobel Prize in physics in 1963, not for his vital work on the atom bomb but for research on the structure of the atom and its nucleus. The two shared one of the professorships endowed by the Rockefeller Foundation, spending half a year in Princeton and the other half in their home universities of Berlin and Budapest. According to Wigner's autobiography, the men were unhappy at first, homesick for Europe's passionate theoretical discussion and its coffeehouses — the congenial floating seminars of professors and students where the latest research was discussed. Wigner wondered if they were part of the window dressing, like the faux-Gothic buildings. But von Neumann, an enthusiastic admirer of all things American, adapted more quickly.[24] With shrinking opportunities for research in Europe during the Depression, and mounting restrictions on Jews in German universities, they stayed.

A second act of philanthropy, more serendipitous than the Rockefeller enterprise, resulted in the creation of the independent Institute for Advanced Study in Princeton.[25] The Bambergers were department store merchants who opened their first store in Newark and who had gone on to make a huge fortune in the dry-goods business. The owners, a brother and sister, sold out six weeks before the stock market crash of 1929. With a fortune of $25 million between them, they decided to show their gratitude to the state of New Jersey. They had in mind perhaps founding a dental school. An expert on medical education, Abraham Flexner, soon convinced them to drop the idea of a medical school and instead to found a first-rate research institution with no teachers, no students, no classes, but only researchers protected from the vicissitudes and pressures of the outside world. Flexner toyed with the idea of making a school of economics the core of the institute but was soon persuaded that mathematics was a sounder choice since it was more "fundamental." Furthermore, there was infinitely greater consensus among mathematicians on who the best people were. Its location was still up in the air. Newark, with its paint factories and slaughterhouses, offered no attractions for the international band of academic superstars Flexner hoped to recruit. Princeton was more like it. Legend has it that it was Oswald Veblen who convinced the Bambergers that Princeton really could be thought of ("in a topological sense," as he put it) as a suburb of Newark.

With zeal and deep pockets matching those of any impresario, Flexner began a worldwide search for stars, dangling unheard-of salaries, lavish perks, and the promise of complete independence. His undertaking coincided with Hitler's takeover of the German government, the mass expulsion of Jews from German universities, and growing fears of another world war. After three years of delicate negotiation, Einstein, the biggest star of them all, agreed to become the second member of the Institute's School of Mathematics, causing one of his friends in Germany to quip, "The pope of physics has moved and the United States will now become the center for the natural sciences." Kurt Gödel, the Viennese wunderkind of logic, came in 1933 as well, and Hermann Weyl, the reigning star of German mathematics, followed Einstein a year later. Weyl insisted, as a condition of his acceptance, that the Institute appoint a bright light from the next generation. Von Neumann, who had just turned thirty, was lured away from the university to become the Institute's youngest professor. Practically overnight, Princeton had become the new Göttingen.

The Institute professors initially shared the deluxe quarters at Fine Hall with their university colleagues. They moved out in 1939 when the Institute's Fuld Hall, a Neo-Georgian brick building perched in the middle of sweeping English lawns surrounded by woods and a pond just a mile or two from Fine, was built. By the time Einstein and the others moved, the Institute and Princeton professors had become family and the clans continued to mingle like country cousins. They collaborated on research, edited journals jointly, and attended one another's lectures, seminars, and teas. The Institute's proximity made it easier to attract the most brilliant students and faculty to the university, while the university's active mathematics department was a magnet for those visiting or working permanently at the Institute.

By contrast, Harvard, once the jewel of American mathematics, was in "a state of eclipse" by the late 1940s.[26] Its legendary chairman G. D. Birkhoff was dead. Some of its brightest young stars, including Marshall Stone, Marston Morse, and Hassler Whitney, had recently departed, two of them for the Institute for Advanced Study. Einstein had used to complain around the Institute that "Birkhoff is one of the world's great academic anti-Semites." Whether or not this was true, Birkhoff's bias had prevented him from taking advantage of the emigration of the brilliant Jewish mathematicians from Nazi Germany.[27] Indeed, Harvard also had ignored Norbert Wiener, the most brilliant American-born mathematician of his generation, the father of cybernetics and inventor of the rigorous mathematics of Brownian motion. Wiener happened to be a Jew and, like Paul Samuelson, the future Nobel Laureate in economics, he sought refuge at the far end of Cambridge at MIT, then little more than an engineering school on a par with the Carnegie Institute of Technology.[28]

William James, the preeminent American philosopher and older brother of the novelist Henry James, once wrote of a critical mass of geniuses causing a whole civilization to "vibrate and shake."[29] But the man in the street didn't feel the tremors emanating from Princeton until World War II was practically over and these odd men with their funny accents, peculiar dress, and passion for obscure scientific theories became national heroes.

From the start, the European brain drain had an immediate and electrifying effect on American mathematics and theoretical physics. The emigration gathered together a group of geniuses who brought not only broad and deep mathematical know-how, but a set of refreshing new attitudes.[30] In particular, the geographical origin of these mathematicians and physicists positioned them to appreciate the implications of the massive amount of new work that had been done in Europe since the turn of the century and gave them a great affinity for applications of mathematics to physics and engineering. Many of the newcomers were young and at the height of their research careers.

Some historians have called World War II the scientists' war. But because the science required sophisticated mathematics, it was also very much a mathematicians' war, and the war effort tapped the eclectic talents of the Princeton mathematical community.[31] Princeton mathematicians became involved in ciphers and code breaking. A cryptanalytic breakthrough enabled the United States to win a major battle at Midway Island, the turning point in the naval war between the United States and Japan.[32] In Britain, Alan Turing, a Princeton Ph.D., and his group at Bletchley Park broke the Nazi code without the Germans' knowledge, thus turning the tide in the submarine battle for control of the Atlantic.[33]

Oswald Veblen and several of his associates essentially rewrote the science of ballistics at the Aberdeen Proving Ground. Marston Morse, who had recently moved from Harvard to the Institute, headed a related effort in the Office of the Chief of Ordnance.[34] Another mathematician, the Princeton statistician Sam Wilks, made best daily estimates of the position of the German submarine fleet on the basis of the prior day's sighting.[35]

The most dramatic contributions were in the areas of weaponry: radar, infra-red detection devices, bomber aircraft, long-range rockets, and torpedoes with depth charges.[36] The new weapons were extremely costly, and the military needed mathematicians to devise new methods for assessing their effectiveness and the most efficient way to use them. Operations research was a systematic way of coming up with the numbers the military wanted. How many tons of explosive force must a bomb release to do a certain amount of damage? Should airplanes be heavily armored or stripped of defenses to fly faster? Should the Ruhr be bombed, and how many bombs should be used? All these questions required mathematical talent.

The ultimate contribution was, of course, the A-bomb.[37] Wigner at Princeton and Leo Szilard at Columbia composed a letter, which they brought to Einstein to sign, warning President Roosevelt that a German physicist, Otto Hahn, at the Kaiser Friedrich Institute in Berlin had succeeded in splitting the uranium atom. Lise Meitner, an Austrian Jew who was smuggled into Denmark, performed the mathematical calculations on how an atomic bomb could be constructed from these findings. Niels Bohr, the Danish physicist, visited Princeton in 1939 and transmitted the news. "It was they rather than their American born colleagues who sensed the military implications of the new knowledge," wrote Davies. Roosevelt responded by appointing an advisory committee on uranium in October 1939, two months into the war, which eventually became the Manhattan Project.

The war enriched and invigorated American mathematics, vindicated those who had championed the émigrés, and gave the mathematical community a claim on the fruits of the postwar prosperity that was to follow. The war demonstrated not only the power of the new theories but the superiority of sophisticated mathematical analysis over educated guesses. The bomb gave enormous prestige to Einstein's relativity theory, which before then had been seen as a small correction of the still-valuable Newtonian mechanics.

Princeton rode high on the newfound status of mathematics in American society. It found itself on the leading edge not just of topology, algebra, and number theory, but also of computer theory, operations research, and the new theory of games.[38] In 1948, everyone was back and the anxieties and frustrations of the 1930s had been swept away by a feeling of expansiveness and optimism. Science and mathematics were seen as the key to a better postwar world. Suddenly the government, particularly the military, wanted to spend money on pure research. Journals started up. Plans were made for another world mathematical congress, the first since the dark days before the war.

A new generation was crowding in, eager to drink up the wisdom of the older generation, yet full of ideas and attitudes of its own. There were no women yet, of course — with the exception of Oxford's Mary Cartwright, who was in Princeton that year — but Princeton was opening up. Suddenly, being a Jew or a foreigner, having a working-class accent, or graduating from a college that wasn't on the East Coast were no longer automatic bars to a bright young mathematician. The biggest divide on campus was suddenly between "the kids" and the war veterans, who, in their mid-to-late twenties, were starting graduate school alongside twenty-year-olds like Nash. Mathematics was no longer a gentlemen's profession, but a wonderfully dynamic enterprise. "The notion was that the human mind could accomplish anything with mathematical ideas," a Princeton student of that era later recalled. He added: "The postwar years had their threats — the Korean War, the Cold War, China going to the commies — but in fact, in terms of science, there was this tremendous optimism. The sense at Princeton wasn't just that you were close to a great intellectual revolution, but that you were part of it." [39]

4
School of Genius

Princeton, Fall 1948

Conversation enriches the understanding, but solitude is the school of genius.
— *EDWARD GIBBON*

O N NASH'S SECOND AFTERNOON in Princeton, Solomon Lefschetz rounded
up the first-year graduate students in the West Common Room.[1] He was there to
tell them the facts of life, he said, in his French accent, fixing them with his fierce
gaze. And for an hour Lefschetz glared, shouted, and pounded the table with his
gloved, wooden hands, delivering something between a biblical sermon and a drill
sergeant's diatribe.

They were the best, the very best. Each of them had been carefully hand-
picked, like a diamond from a heap of coal. But this was Princeton, where real
mathematicians did real mathematics. Compared to these men, the newcomers
were babies, ignorant, pathetic babies, and Princeton was going to make them grow
up, damn it!

Entrepreneurial and energetic, Lefschetz was the supercharged human loco-
motive that had pulled the Princeton department out of genteel mediocrity right
to the top.[2] He recruited mathematicians with only one criterion in mind: research.
His high-handed and idiosyncratic editorial policies made the *Annals of Mathematics,* Princeton's once-tired quarterly, into the most revered mathematical journal in
the world.[3] He was sometimes accused of caving in to anti-Semitism for refusing to
admit many Jewish students (his rationale being that nobody would hire them
when they completed their degrees),[4] but no one denies that he had brilliant snap
judgment. He exhorted, bossed, and bullied, but with the aim of making the
department great and turning his students into real mathematicians, tough like
himself.

When he came to Princeton in the 1920s, he often said, he was "an
invisible man."[5] He was one of the first Jews on the faculty, loud, rude, and
badly dressed to boot. People pretended not to see him in the hallways and
gave him wide berth at faculty parties. But Lefschetz had overcome far more
formidable obstacles in his life than a bunch of prissy Wasp snobs. He had

been born in Moscow and been educated in France.[6] In love with mathematics, but effectively barred from an academic career in France because he was not a citizen, he studied engineering and emigrated to the United States. At age twenty-three, a terrible accident altered the course of his life. Lefschetz was working for Westinghouse in Pittsburgh when a transformer explosion burned off his hands. His recovery took years, during which he suffered from deep depression, but the accident ultimately became the impetus to pursue his true love, mathematics.[7] He enrolled in a Ph.D. program at Clark University, the university famous for Freud's 1912 lectures on psychoanalysis, soon fell in love with and married another mathematics student, and spent nearly a decade in obscure teaching posts in Nebraska and Kansas. After days of backbreaking teaching, he wrote a series of brilliant, original, and highly influential papers that eventually resulted in a "call" from Princeton. "My years in the west with total hermetic isolation played in my development the role of 'a job in a lighthouse' which Einstein would have every young scientist assume so that he may develop his own ideas in his own way."[8]

Lefschetz valued independent thinking and originality above everything. He was, in fact, contemptuous of elegant or rigorous proofs of what he considered obvious points. He once dismissed a clever new proof of one of his theorems by saying, "Don't come to me with your pretty proofs. We don't bother with that baby stuff around here."[9] Legend had it that he never wrote a correct proof or stated an incorrect theorem.[10] His first comprehensive treatise on topology, a highly influential book in which he coined the term "algebraic topology," "hardly contains one completely correct proof. It was rumored that it had been written during one of Lefschetz' sabbaticals . . . when his students did not have the opportunity to revise it."[11]

He knew most areas of mathematics, but his lectures were usually incoherent. Gian-Carlo Rota, one of his students, describes the start of one lecture on geometry: "Well a Riemann surface is a certain kind of Hausdorff space. You know what a Hausdorff space is, don't you? It's also compact, ok. I guess it is also a manifold. Surely you know what a manifold is. Now let me tell you one non-trivial theorem, the Riemann-Roch theorem."[12]

On this particular afternoon in mid-September 1948, with the new graduate students, Lefschetz was just warming up. "It's important to dress well. Get rid of that thing," he said, pointing to a pen holder. "You look like a workman, not a mathematician," he told one student.[13] "Let a Princeton barber cut your hair," he said to another.[14] They could go to class or not go to class. He didn't give a damn. Grades meant nothing. They were only recorded to please the "goddamn deans." Only the "generals" counted.[15]

There was only one requirement: come to tea.[16] They were absolutely required to come to tea every afternoon. Where else would they meet the finest mathematics faculty in the world? Oh, and if they felt like it, they were free to visit that "embalming parlor," as he liked to call the Institute of Advanced Study, to see if they could catch a glimpse of Einstein, Gödel, or von Neumann.[17] "Remember," he kept repeating, "we're not here to baby you." To Nash, Lefschetz's opening spiel must have sounded as rousing as a Sousa march.

Lefschetz's, hence Princeton's, philosophy of graduate mathematics education had its roots in the great German and French research universities.[18] The main idea was to plunge students, as quickly as possible, into their own research, and to produce an acceptable dissertation quickly. The fact that Princeton's small faculty was, to a man, actively engaged in research itself, was by and large on speaking terms, and was available to supervise students' research, made this a practical approach.[19] Lefschetz wasn't aiming for perfectly polished diamonds and indeed regarded too much polish in a mathematician's youth as antithetical to later creativity. The goal was not erudition, much as erudition might be admired, but turning out men who could make original and important discoveries.

Princeton subjected its students to a maximum of pressure but a wonderful minimum of bureaucracy. Lefschetz was not exaggerating when he said that the department had no course requirements. The department offered courses, true, but enrollment was a fiction, as were grades. Some professors put down all *As*, others all *Cs*, on their grade reports, but both were completely arbitrary.[20] You didn't have to show up a single time to earn them and students' transcripts were, more often than not, works of fiction "to satisfy the Philistines." There were no course examinations. In the language examinations, given by members of the mathematics department, a student was asked to translate a passage of French or German mathematical text. But they were a joke.[21] If you could make neither heads nor tails of the passage — unlikely, since the passages typically contained many mathematical symbols and precious few words — you could get a passing grade merely by promising to learn the passage later. The only test that counted was the general examination, a qualifying examination on five topics, three determined by the department, two by the candidate, at the end of the first, or at latest, second year. However, even the generals were sometimes tailored to the strengths and weaknesses of a student.[22] If, for example, it was known that a student really knew one article well, but only one, the examiners, if they were so moved, might restrict themselves to that paper. The only other hurdle, before beginning the all-important thesis, was to find a senior member of the faculty to sponsor it.

If the faculty, which got to know every student well, decided that so-and-so wasn't going to make it, Lefschetz wasn't shy about not renewing the student's support or simply telling him to leave. You were either succeeding or on your way out. As a result, Princeton students who made it past the generals wound up with doctorates after just two or three years at a time when Harvard students were taking six, seven, or eight years.[23] Harvard, where Nash had yearned to go for the prestige and magic of its name, was at that time a nightmare of bureaucratic red tape, fiefdoms, and faculty with relatively little time to devote to students. Nash could not possibly have realized it fully that first day, but he was lucky to have chosen Princeton over Harvard.

That genius will emerge regardless of circumstance is a widely held belief. The biographer of the great Indian mathematician Ramanujan, for example, claims that the five years that the young Ramanujan spent in complete isolation from other mathematicians, having failed out of school and unable to get as much as a tutoring position, were the key to his stunning discoveries.[24] But when writing Ramanujan's obituary, G. H. Hardy, the Cambridge mathematician who knew him best, called that view, held earlier by himself, "ridiculous sentimentalism." After Ramanujan's death at thirty-three, Hardy wrote that the "the tragedy of Ramanujan was not that he died young, but that, during his five unfortunate years, his genius was misdirected, side-tracked, and to a certain extent distorted."[25]

As was to become increasingly obvious over the months that followed, Princeton's approach to its graduate students, with its combination of complete freedom and relentless pressure to produce, could not have been better suited to someone of Nash's temperament and style as a mathematician, nor more happily designed to elicit the first real proofs of his genius. Nash's great luck, if you want to call it luck, was that he came onto the mathematical scene at a time and to a place tailor-made for his particular needs. He came away with his independence, ambition, and originality intact, having been allowed to acquire a truly first-class training that was to serve him brilliantly.

Like nearly all the other graduate students at Princeton, Nash lived in the Graduate College. The College was a gorgeous, faux-English edifice of dark gray stone surrounding an interior courtyard that sat on a crest overlooking a golf course and lake. It was located about a mile from Fine Hall on the far side of Alexander Road, about halfway between Fine and the Institute for Advanced Study. Especially in winter, when it was dark by the time the afternoon seminar ended, it was a good long walk, and once you were there, you didn't feel like going out again. Its location was the outcome of a fight between Woodrow Wilson and Dean Andrew West.[26] Wilson had wanted the graduate students to mix and mingle with the undergraduates. West wanted to re-create the atmosphere of one of the Oxbridge colleges, far removed from the rowdy, snobbish undergraduate eating clubs on Prospect Street.

In 1948, there were about six hundred graduate students, their ranks swelled by the numbers of returning veterans whose undergraduate or graduate careers had been interrupted by the war.[27] The College, a bit shabbier than before the war and in need of sprucing up, was full, overflowing really, and a good many less lucky first-year students had been turned away and were being forced to lodge in rented rooms in the village. Almost everyone else had to share rooms. Nash, who lived in Pyne Tower, was lucky to get a private room, one of the perks of his fellowship.[28] About fifteen or twenty of the mathematics students, second- and third-year as well as first-year students, and a couple of instructors lived in the college at the time.

Life was masculine, monastic, and scholarly, exactly as Dean West had envisioned.[29] The graduate students ate breakfast, lunch, and dinner together at the cost of fourteen dollars a week. Breakfast and lunch were served in the "breakfast" room, hurried meals that were taken on the run. But dinner, served in Procter

Hall, a refectory very much in the English style, was a more leisurely affair. There were tall windows, long wooden tables, and formal portraits of eminent Princetonians on the walls; the evening prayer was led by Sir Hugh Taylor, the college's dean, or his second in command, the college's master. There were no candles and no wine, but the food was excellent. Gowns were no longer required as before the war (they were reinstated in the early 1950s, and did not disappear for good until the 1970s), but jackets and ties were required.

The atmosphere at dinner was a combination of male debating society, locker room, and seminary. Though historians, English scholars, physicists, and economists all lived cheek by jowl with the mathematicians, the mathematicians segregated themselves as strictly as if they were living under some legal system of apartheid, always occupying a table by themselves.[30] The older, more sophisticated students, namely Harold Kuhn, Leon Henkin, and David Gale, met for sherry in Kuhn's rooms before dinner. Conversation at dinner, sometimes but not always mathematical, was more expansive than at teatime. The talk, one former student recalls, frequently revolved around "politics, music, and girls." Political debate resembled discussions about sports, with more calculation of odds and betting than ideology. In that early fall, the Truman-Dewey race provided a great deal of entertainment. Being a more diverse group, the graduate students were more evenly split between the candidates than the Princeton undergraduates; 98 percent of the undergraduates at Princeton, it turned out, were Dewey supporters. One graduate student even wore a Wallace button for Henry Wallace, the candidate supported by the American Labor party, a Communist front organization.[31]

Girls, or rather the absence of girls, the difficulty of meeting girls, the real or imagined exploits of certain older and more worldly students, were also hot subjects.[32] Very few of the students dated. Women were not allowed in the main dining hall, and, of course, there were no female students. "We are all homosexuals here" was a famous remark made by a resident to fluster the dean's wife.[33] Isolation made the real prospects of meeting a girl remote. A few venturesome souls, organized by a young instructor named John Tukey, went to Thursday night folk dances at the local high school.[34] But most were too shy and self-conscious to do even that. Sir Hugh, a stuffed shirt roundly disliked by the mathematicians, did his best to discourage what little socializing there was. One student was called into the dean's office because a pair of women's panties had been found in his room; it turned out his sister had been visiting and he, to preserve appearances, had moved out for the night. At one point, a seemingly unnecessary rule was handed down that residents of the Graduate College were not allowed to entertain a woman past midnight. The very few students who actually had girlfriends interpreted the rule literally to mean that a woman could be in the room, but couldn't be entertained. Harold Kuhn spent his honeymoon there.[35] The only time and place that women were allowed to join the larger group was Saturday lunch in the Breakfast Room.

In short, social life was rather enveloping—it would be hard to become really lonely—and at the same time limited to other men, in Nash's case specifically to other mathematicians. The parties held in student rooms were thus mostly all-male affairs. Such evenings, as often as not, were devoted to mathematical parties organized by one of the graduate students at Lefschetz's request to entertain some visitor but actually to get his students much-needed job contacts.[36]

The quality, diversity, and sheer volume of mathematics talked about in Princeton every day, by professors, Institute professors, and a steady stream of visitors from all over the world, not to mention the students themselves, were unlike anything Nash had ever imagined, much less experienced. A revolution was taking place in mathematics and Princeton was the center of the action. Topology. Logic. Game theory. There were not only lectures, colloquia, seminars, classes, and weekly meetings at the institute that Einstein and von Neumann occasionally attended, but there were breakfasts, lunches, dinners, and after-dinner parties at the Graduate College, where most of the mathematicians lived, as well as the daily afternoon teas in the common room. Martin Shubik, a young economist studying at Princeton at that time, later wrote that the mathematics department was "electric with ideas and the sheer joy of the hunt. If a stray ten-year-old with bare feet, no tie, torn blue jeans, and an interesting theorem had walked into Fine Hall at tea times, someone would have listened."[37]

Tea was the high point of every day.[38] It was held in Fine Hall between three and four between the last class and the four-thirty seminar that went until five-thirty or six. On Wednesdays it was held in the west common room, or the professor's room as it was also called, and was a far more formal affair, where the self-effacing Mrs. Lefschetz and the other wives of the senior faculty, wearing long gowns and white gloves, poured the tea and passed the cookies. Heavy silver teapots and dainty English bone china were brought out.

On other days, tea was held in the east common room, also known as the students' room, a much-lived-in, funky place full of overstuffed leather armchairs and low tables. The janitor would bring in the tea and cookies a few minutes before three o'clock and the mathematicians, tired from a day of working alone or lecturing or attending seminars, would start drifting in, one by one or in groups. The faculty almost always came, as did most of the graduate students and a sprinkling of more precocious undergraduates. It was very much a family gathering, small and intimate. It is hard to think where a student could get to know as many other mathematicians as well as at Princeton teatime.

The talk was by no means purely formal. Mathematical gossip abounded — who was working on what, who had a nibble from what department, who had run into trouble on his generals. Melvin Hausner, a former Princeton graduate student, later recalled, "You went there to discuss math. To do your own version of gossiping. To meet faculty. To meet friends. We discussed math problems. We shared our readings of recent math papers."[39]

The professors felt it their duty to come, not only to get to know the students but to chat with one another. The great logician Alonzo Church, who looked "like a cross between a panda and an owl," never spoke unless spoken to, and rarely then, would head straight for the cookies, placing one between the fingers of his splayed hand, and munch away.[40] The charismatic algebraist Emil Artin, son of a German opera singer, would fling his gaunt, elegant body into one of the leather armchairs, light a Camel, and opine on Wittgenstein and the like to his disciples, huddled, more or less literally, at his feet.[41] The topologist Ralph Fox, a go master, almost always made a beeline for a game board, motioning some student to join him.[42] Another topologist, Norman Steenrod, a good-looking, friendly midwesterner who had just created a sensation with his now classic exposition of fiber bundles, usually stopped in for a game of chess.[43] Albert Tucker, Lefschetz's right-hand man, was the straitlaced son of a Canadian Methodist minister and Nash's eventual thesis adviser. Tucker always surveyed the room before he came in and would make fussy little adjustments—such as straightening the curtain weights if the drapes happened to be awry, or issuing a word-to-the-wise to a student who was taking too many cookies.[44] More often than not, a few visitors, often from the Institute for Advanced Study, would turn up as well.

The students who gathered at teatime were as remarkable, in a way, as the faculty. Poor Jews, new immigrants, wealthy foreigners, sons of the working classes, veterans in their twenties, and teenagers, the students were a diverse as well as brilliant group, among them John Tate, Serge Lang, Gerard Washnitzer, Harold Kuhn, David Gale, Leon Henkin, and Eugenio Calabi.[45] The teas were heaven for the shy, friendless, and socially awkward, a category in which many of these young men belonged. John Milnor, the most brilliant freshman in the history of the Princeton mathematics department, described it this way: "Everything was new to me. I was awkward socially, shy and isolated. Everything was wonderful. This was a whole new world. Here was a whole community in which I felt very much at home."[46]

The atmosphere was, however, as competitive as it was friendly.[47] Insults and one-upmanship were always major ingredients in teatime banter. The common room was where the young bucks warily sized each other up, bluffed and postured, and locked horns. No culture was more hierarchical than mathematical culture in its precise ranking of individual merit and prestige, yet it was a ranking always in a state of suspense and flux, in which new challenges and scuffles erupted almost daily. Back in their undergraduate colleges, most of these young men had gotten used to being the brightest and best, but now they were bumping up against the brightest and best from other schools. One of the graduate students who entered with Nash admitted, "Competitiveness, it was sort of like breathing. We thrived on it. We were nasty. This guy, he's dumb, we'd say. Therefore he no longer existed."[48]

There were cliques, mostly based on fields. *The* clique at the top of the hierarchy was the topology clique, which clustered around Lefschetz, Fox, and Steenrod. Then came analysis, grouped around Lefschetz's archrival in the department, a civilized and erudite lover of music and art named Bochner. Then came algebra, which consisted of Emil Artin and a handful of anointed followers. Logic, for some reason, was not highly regarded, despite Church's towering reputation among early pioneers of computer theory. The game theory clique around Tucker was considered quite déclassé, an anomaly in this ivory tower of pure mathematics. Each clique had its own thoughts about the importance of its subject and its own way of putting the others down.

Nash had never in his life encountered anything like this exotic little mathematical hothouse. It would soon provide him with the emotional and intellectual context he so much needed to express himself.

5
Genius

Princeton, 1948–49

It is good that I did not let myself be influenced. — LUDWIG WITTGENSTEIN

KAI LAI CHUNG, a mathematics instructor who had survived the horrors of the Japanese conquest of his native China, was surprised to see the door of the Professors' Room standing ajar.[1] It was usually locked. Kai Lai liked to stop by on the rare occasions when it was open and nobody was about. It had the feel of an empty church, no longer imposing and intimidating as it was in the afternoons when it was crowded with mathematical luminaries, but simply a beautiful sanctuary.

The light in the west common room filtered through thick stained-glass windows inlaid with formulae: Newton's law of gravity, Einstein's theory of relativity, Heisenberg's uncertainty principle of quantum mechanics. At the far end, like an altar, was a massive stone fireplace. On one side was a carving of a fly confronting the paradox of the Möbius band. Möbius had given a strip of paper a half twist and connected the ends, creating a seemingly impossible object: a surface with only one side. Kai Lai especially liked to read the whimsical inscription over the fireplace, Einstein's expression of faith in science, "Der Herr Gott ist raffiniert aber Boshaft ist Er nicht," which he took to mean that "the Lord is subtle but not malicious."[2]

On this particular fall morning, as he reached the threshold of the half-open door, Kai Lai stopped abruptly. A few feet away, on the massive table that dominated the room, floating among a sea of papers, sprawled a beautiful dark-haired young man. He lay on his back staring up at the ceiling as if he were outside on a lawn under an elm looking up at the sky through the leaves, perfectly relaxed, motionless, obviously lost in thought, arms folded behind his head. He was whistling softly. Kai Lai recognized the distinctive profile immediately. It was the new graduate student from West Virginia. A trifle shocked and a little embarrassed, Kai Lai backed away from the door and hurried away before Nash could see or hear him.

• • •

The first-year students were an extremely cocky bunch, but Nash immediately struck everyone as a good deal cockier — and odder. His appearance helped create the impression.[3] At twenty, Nash looked young, perhaps younger than he was, but he was no longer a gawky youngster who looked as if he'd just climbed off a tractor. Six foot one, he weighed nearly 170 pounds. He had broad shoulders, a heavily muscled chest, and a tapered waist. He had the build, if not the bearing, of an athlete, "a very strong, very masculine body," one fellow graduate student recalled. He was, moreover, "handsome as a god," according to another student. His high forehead, somewhat protruding ears, distinctive nose, fleshy lips, and small chin gave him the look of an English aristocrat. His hair flopped over his forehead; he was constantly brushing it away. He wore his fingernails very long, which drew attention to his rather limp and beautiful hands and long, delicate fingers. His voice, on the high, reedy side, was cool and southern and had a slightly ironic edge. His speech had an Olympian and ornamental quality that struck others as a bit stilted. Moreover, his expression was somewhat haughty and he smiled to himself in a superior way.

From the start, he was quite visible at teatime. He seemed eager to be noticed and seemed to want to establish that he was smarter than anyone else in the place. A fellow student, who had come to Princeton from the City College of New York, recalled, "He had a way of saying 'trivial' to anything you might have regarded as nontrivial. That could be taken as a put-down." Nash would accuse people of burbling. If somebody was talking on and on, he was just burbling. "ALGEBRA IS BURBLE," Nash once scrawled on a blackboard that another student, an algebraist, would pull down in the midst of a talk. "Hackers" was another favorite Nash term. A hacker was somebody who plodded along, somebody who was doing things not worth doing.[4] As another student put it: "Nash was very interested that everyone would recognize how smart he was, not because he needed this admiration, but anybody who didn't recognize it wasn't on top of things. If anyone wasn't aware, he would take a little trouble to make sure he found out." Another student recalls, "He wanted to be noticed more than anything."[5]

He seized opportunities to boast about his accomplishments. He would mention, out of the blue, that he'd discovered, as an undergraduate, an original proof of Gauss's proof of the fundamental theorem of algebra, one of the great achievements of eighteenth-century mathematics, nowadays taught in advanced courses on the theory of complex variables.[6]

He was a self-declared free thinker. On his Princeton application, in answer to the question "What is your religion?" he wrote "Shinto."[7] He implied that his lineage was superior to that of his fellow students, especially Jewish students. Martin Davis, a fellow student who grew up in a poor family in the Bronx, recalled catching up with Nash when he was ruminating about blood lines and natural aristocracies one day as they were walking from the Graduate College to Fine Hall. "He definitely had a set of beliefs about the aristocracy," said Davis. "He was opposed to racial mixing. He said that miscegenation would result in the deterioration of the racial line. Nash implied that his own blood lines were pretty good." [8] He once asked Davis whether Davis had grown up in a slum.

Nash appeared to be interested in almost everything mathematical — topology, algebraic geometry, logic, and game theory — and he seemed to absorb a tremendous amount about each of these during his first year. [9] He himself recalled, without elaborating, having "studied mathematics fairly broadly" at Princeton. [10] Yet he avoided attending classes. No one recalls sitting in a regular class with him. [11] He did, he later said, begin a course in algebraic topology offered by Steenrod, who essentially founded the field. [12] Steenrod and Samuel Eilenberg had just invented the axioms that were the foundation of homology theory. The stuff was very trendy and the course attracted many students, but Nash decided it was too formal for him and not geometric enough for his taste, so he stopped going.

Nobody remembers seeing Nash with a book during his graduate career either. [13] In fact, he read astonishingly little. "Both Nash and I were dyslexic to some degree," said Eugenio Calabi, a young Italian immigrant who entered Princeton the year before Nash. "I had great difficulty keeping my attention on reading that required great concentration. Then, I just thought of it as laziness. Nash, on the other hand, defended not reading, taking the attitude that learning too much secondhand would stifle creativity and originality. It was a dislike of passivity and giving up control." [14]

Nash's main mode of picking up information he deemed necessary consisted of quizzing various faculty members and fellow students. [15] He carried around a clipboard and constantly made notes to himself. They were little hints to himself, ideas, facts, things he wanted to do, Calabi recalled. His handwriting was almost unreadable. He once explained to Lefschetz that he had to use ruled notebook paper even when writing a letter because without the lines his script formed a "very irregular wavy line." As it was, his notes were full of crossouts and misspellings of even simple words like "InteresEted." [16]

He compensated by learning through conversation in the common room and by attending lectures given by visiting mathematicians. According to Calabi, Nash "was quite systematic in asking shrewd questions and developing his own ideas from the answers. I've seen some of his results in the making." Some of his best ideas came "from things learned only halfway, sometimes even wrongly, and trying to reconstruct them — even if he could not do so completely." [17]

He was always asking probing questions. The questions, not only about game theory, but also about topology and geometry, often contained a kernel of speculation. John Milnor, who entered as a freshman that year, recalls one such question, posed in the common room: Let V_0 be a singular algebraic variety of dimension k, embedded in some smooth variety M_0 and let $M_1 = G_k(M_0)$ be the Grassmann variety of tangent k-planes to M_0. Then V_0 lifts naturally to a k-dimensional variety $V_1 \subset M_1$. Continuing inductively, we obtain a sequence of k-dimensional varieties.

. . . Do we eventually reach a variety V_q which is nonsingular? (As it turns out, Milnor adds, the conjecture has since been proven only in special cases.)[18]

Nash spent most of his time, it appears, simply thinking. He rode bicycles borrowed from the racks in front of the Graduate College in tight little figure eights or ever-smaller concentric circles.[19] He paced around the interior quadrangle of the college. He glided along the gloomy second-floor hallway of Fine, his shoulder pressed firmly against the wall, like a trolley never losing contact with the dark paneled walls.[20] He would lie on a desk or table in the empty common room, or more frequently, in the third-floor library.[21] Almost always, he whistled Bach, most often the Little Fugue.[22] The whistling prompted the mathematics secretaries to complain about Nash to Lefschetz and Tucker.[23]

Melvin Hausner recalled: "He was always buried in thought. He'd sit in the common room by himself. He could easily walk by you and not see you. He was always muttering to himself. Always whistling. Nash was always thinking. . . . If he was lying on a table, it was because he was thinking. Just thinking. You could see he was thinking."[24]

He seemed to be enjoying himself immensely. A profound dislike for merely absorbing knowledge and a strong compulsion to learn by doing is one of the most reliable signs of genius. In Princeton, Nash's thinking began to take on an urgent, focused quality. He was obsessed with learning from scratch. Milnor recalled: "It was as if he wanted to rediscover, for himself, three hundred years of mathematics."[25] Steenrod, who was to become Nash's sounding board as the year wore on, wrote several years later, "More than any other student I have known, Nash believes in learning a subject by doing research in it."[26]

Like the nineteenth-century German mathematician Carl Friedrich Gauss, who complained that "such an overwhelming horde of ideas stormed my mind before I was twenty that I could hardly control them and had time but for a small fraction,"[27] Nash seemed to overflow with ideas. According to Steenrod, "During his first year of graduate work, he presented me with a characterization of a simple closed curve in the plane. This was essentially same as one given by Wilder in 1932. Some time later he devised a system of axioms for topology based on the primitive concept of connectedness. I was able to refer him to papers by Wallace. During his second year, he showed me a definition of a new kind of homology group which proved to be the same as the Reidemeister group based on homotopy chains."[28] What is striking about the ideas that Steenrod attributes to Nash as a first-year student is that they are not merely clever exercises designed to show off the brilliance of a precocious student, but mathematically interesting and important ideas.[29]

Nash was always on the lookout for problems. "He was very much aware of unsolved problems," said Milnor. "He really cross-examined people on what were the important problems. It showed a tremendous amount of ambition."[30] In this search, as in so much else, Nash displayed an uncommon measure of self-confidence and self-importance. On one occasion, not long after his arrival at Princeton, he went to see Einstein and sketched some ideas he had for amending quantum theory.

That first fall in Princeton, Nash sometimes took a slight detour down busy Mercer Street in order to catch a glimpse of Princeton's most remarkable resident.[31] Most mornings between nine and ten, Einstein walked the mile or so from his white clapboard house at 112 Mercer Street to his office at the Institute. On several occasions, Nash managed to brush past the saintly scientist — wearing a baggy sweater, drooping trousers, sandals without socks, and an impassive expression — on the street.[32] He imagined how he might strike up a conversation, stopping Einstein in his tracks with some startling observation.[33] But once when he passed him walking with Kurt Gödel, Nash caught snatches of German and sadly wondered whether his own lack of that language might constitute an insuperable barrier to communicating with the great man.[34]

In 1948, Einstein had been a world cult figure for more than a quarter of a century.[35] His special theory of relativity was published in 1905, as was his assertion that light was propagated in space not as waves but as discrete particles. The general theory of relativity appeared in 1916. Astronomers' confirmation in 1919 that light rays were bent by the sun's gravity — as Einstein had predicted — brought him fame unrivaled by any scientist before or since. Einstein's political activities — on behalf of the A-bomb and then for nuclear disarmament, world government, the state of Israel — added a saintly aura.

For decades, Einstein's main scientific preoccupations had been two, one in which he achieved a measure of success, the other a complete failure.[36] He succeeded in casting doubt on some of the basic tenets of one of the most successful and widely accepted theories in physics — quantum theory — a theory first proposed by himself when he demonstrated the existence of light quanta in 1905, and subsequently developed by Niels Bohr and Werner Heisenberg, who insisted the act of observation changes the object being measured. Einstein's 1935 attack on quantum theory produced a front-page headline in *The New York Times* and has never been satisfactorily refuted; indeed, as of the mid-1990s, the latest experimental evidence has breathed new life into his critique.

His greater preoccupation was the ultimate task of uniting the phenomena of light and gravity into a single theory. Einstein never was able, as one biographer put it, to "accept that the universe was fragmented into relativity on one side and quantum mechanics on the other."[37] On the eve of his seventieth birthday, he was still searching for a single, consistent set of principles that applied to all of the universe's diverse forces and particles and was, in fact, preparing what proved to be his final paper on so-called "unified field theory."[38]

It was a measure of Nash's bravura and the power of his fantasy that he was not content merely to see Einstein but soon requested an audience with him. Just a few weeks into his first term at Princeton, Nash made an appointment to see

Einstein in his office in Fuld Hall. He told Einstein's assistant that he had an idea that he wished to discuss with Professor Einstein.[39]

Einstein's office, a large airy room with a bay window that let in plenty of light, was messy. Einstein's twenty-two-year-old Hungarian assistant—an intense, chain-smoking logician named John Kemeny, who would later invent the computer language BASIC, become president of Dartmouth College, and head a commission to investigate Three Mile Island—ushered Nash in. Einstein's handshake, which ended with a twist, was remarkably firm, and he showed Nash to a large wooden meeting table on the far side of the office.

The late-morning light streaming through the bay window produced a sort of aura around Einstein. Nash, however, quickly got into the substance of his idea while Einstein listened politely, twirled the curls on the back of his head with his finger, sucked on his tobaccoless pipe, and occasionally muttered a remark or asked a question. As he spoke, Nash became aware of a mild form of echolalia: deep, deep, interesting, interesting.[40]

Nash had an idea about "gravity, friction, and radiation," as he later recalled. The friction he was thinking of was the friction that a particle, say a photon, might encounter as it moved through space due to its fluctuating gravitational field interacting with other gravitational fields.[41] Nash had given his hunch enough thought to spend much of the meeting at the blackboard scribbling equations. Soon, Einstein and Kemeny were standing at the blackboard as well.[42] The discussion lasted the better part of an hour. But in the end all that Einstein said, with a kindly smile, was "You had better study some more physics, young man." Nash did not immediately take Einstein's advice and he never wrote a paper on his idea. His youthful foray into physics would become a lifetime interest—though, like Einstein's search for the unified field, it would not be especially fruitful.[43] Many decades later, however, a German physicist published a similar idea.[44]

Nash conspicuously avoided attaching himself to any particular faculty member, either in the department or at the institute. It was not a matter of shyness, his fellow students thought, but rather that he wished to preserve his independence. One mathematician who knew Nash at the time observed: "Nash was determined to keep his intellectual independence. He didn't want to be unduly influenced. He'd talk freely with other students, but he was always worried about getting too close to other professors for fear that he'd be overwhelmed. He didn't want to become dominated. He disliked the whole idea of being intellectually beholden."[45]

He did, however, use at least one faculty member, Steenrod, as a kind of sounding board. Temperamentally, Steenrod was an entirely different character from flamboyant, domineering types like Lefschetz and Bochner, whose lectures, it was said, were "exciting but 90 percent wrong." Steenrod was a careful, methodical man who chose his suits and sports coats according to a mathematical formula and had a mania for thinking up highly logical, if impractical, solutions to social problems like crime.[46] Steenrod also happened to be friendly, helpful, and patient.

He was immensely impressed by Nash, found him more charming than not, and treated the young man's brashness and eccentricity with amused tolerance.[47]

Surrounded for the first time in his life by young men whom he regarded, if not exactly as his equals, at least as worth talking to, Nash preferred picking other students' brains. "Some mathematicians work very much by themselves," said one fellow student. "He liked to exchange ideas."[48] One of the students he sought out was John Milnor, the first of a number of brilliant younger mathematicians to whom Nash was drawn. Tall, lithe, with a baby face and the body of a gymnast, Milnor was only a freshman but he was already the department's golden boy.[49] In his freshman year, in a differential geometry course taught by Albert Tucker, he learned about an unproved conjecture of a Polish topologist, Karol Borsuk, concerning the total curvature of a knotted curve in space. The story goes that Milnor mistook the conjecture for a homework assignment.[50] Whatever the case, he arrived at Tucker's door a few days after with a written proof and the request: "Would you be good enough to point out the flaw in this attempt. I'm sure there is one, but can't find it." Tucker studied it, showed it to Fox and to Shiing-shen Chern. No one could find anything wrong. Tucker encouraged Milnor to submit the proof as a Note to the *Annals of Mathematics*. A few months later Milnor turned in an exquisitely crafted paper with a full theory of the curvature of knotted curves in which the proof of the Borsuk conjecture was a mere by-product. The paper, more substantial than most doctoral dissertations, was published in the *Annals* in 1950. Milnor also dazzled the department — and Nash — by winning the Putnam competition in his second semester at Princeton (in fact, he went on to win it two more times and was offered a Harvard scholarship).[51]

Nash was choosy about whom he would talk mathematics with. Melvin Peisakoff, another student who would later overlap with Nash at the RAND Corporation, recalled: "You couldn't engage him in a long conversation. He'd just walk off in the middle. Or he wouldn't respond at all. I don't remember Nash having a conversation that came to a nice soft landing. I also don't remember him ever having a conversation about mathematics. Even the full professors would discuss problems they were working on with other people."[52]

On one occasion in the common room, however, Nash was sketching an idea when another graduate student got very interested in what he was saying and started to elaborate on the idea.[53] Nash said, "Well, maybe I ought to write a Note for the *Proceedings of the National Academy* on this." The other student said, "Well, Nash, be sure to give me a credit." Nash's reply was, "All right, I'll put in a footnote that So and So was in the room when I had the idea."

Nash was respected but not well liked. He wasn't invited to Kuhn's room for sherry or out with the others when they went to Nassau Street to drink beer. "He wasn't somebody you'd want as a close friend," Calabi recalled. "I don't know many people who felt any warmth for him."[54] Most of the graduate students were slightly odd ducks themselves, beset by shyness, awkwardness, strange mannerisms, and all kinds of physical and psychological tics, but they collectively felt that Nash was even odder. "Nash was out of the ordinary," said a former graduate student from his time. "If he was in a room with twenty people, and they were talking, if you asked an observer who struck you as odd, it would have been Nash. It wasn't anything he consciously did. It was his bearing. His aloofness."[55]

Another recalled, "Nash was totally spooky. He wouldn't look at you. He'd take a lot of time answering a question. If he thought the question was foolish, he wouldn't answer at all. He had no affect. It was a mixture of pride and something else. He was so isolated but there really was underneath it all a warmth and appreciation [for other people]."[56]

When Nash did engage in one of his flights of garrulity, he often seemed to be simply thinking out loud. Hausner remembered, "A lot of us would discount a lot of what Nash said. A lot of the things he said were so far out, you didn't want to engage him. 'What was happening on earth when the Martians took over and there was a period of violence and why such and such.' You wouldn't know what he was talking about. Nash came out with things. They were unfinished and we weren't ready to hear them. I wouldn't want to listen. You didn't feel comfortable with the person."[57]

His sense of humor was not only childish but odd. One former student recalled that Nash was personally responsible for getting the much-despised gown requirement at meals temporarily restored. "First," recounted Felix Browder, who left Princeton in the fall of 1948, "he wrote a letter to Hugh Taylor, a pompous ass who was looking for an excuse, demanding that the custom be restored. After it was, nobody ate in the hall. It didn't make John popular."[58]

He was also capable of frightening people when provoked. Occasionally, the teasing and needling would spill over into a sudden eruption of violence. On one occasion, Nash was baiting one of Artin's students by telling him that the best way into Artin's graces was to catch his beautiful daughter Karin.[59] The student, Serge Lang, who everyone knew was painfully obsessed by his shyness around girls, threw a cup of hot tea in Nash's face. Nash chased him around the table, threw him to the ground, and stuffed ice cubes down the back of his shirt. Another time, Nash picked up a metal ashtray stand—the kind that supports a heavy glass ashtray— and brought it down on Melvin Peisakoff's shins, hard enough to cause considerable pain for a number of weeks.[60]

In the spring of 1949, Nash ran into some trouble.[61] He had acquired some strong supporters on the faculty, namely Steenrod, Lefschetz, and Tucker. Tucker was among those who believed that Nash was "very brilliant and original but rather eccentric," arguing that "his creative ability . . . should make one tolerate his queerness."[62] But not everyone in the department felt that way. Some felt that Nash didn't belong at Princeton at all. Among them was Artin.

Slender, handsome, with ice-blue eyes and a spellbinding voice, Artin looked like a 1920s German matinee idol.[63] He wore a black leather trench coat and sandals throughout the academic year, wore his hair long, and smoked incessantly. *The* representative of "modern" algebra, Artin, who had been recommended by Weyl for the appointment at the institute that von Neumann eventually got, was a wonderful lecturer who admired polish and scholarship, but was famously intolerant of those who did not meet his rather fastidious standards. He was well known for screaming and throwing chalk at students who asked obtuse questions in his classes.

Artin and Nash had clashed a number of times in the common room. Artin was always interested in talking with talented students. Yet he apparently found Nash not only irritatingly brash but also shockingly ignorant.[64] At a faculty meeting in the spring, Artin commented that he could see no way for Nash to pass his generals, which the better students were expected to take at the end of their first year. When Lefschetz proposed an Atomic Energy Commission fellowship for Nash for the following year, Artin opposed it and made it clear he thought it would be better if Nash left Princeton.

Lefschetz and Tucker overruled Artin on the subject of the fellowship.[65] But they dissuaded Nash from sitting for the generals that spring and suggested that he take them in the fall instead. He was safe for the time being, but his unpopularity among some faculty members was to crop up again when he sought, two years later, to join the department as an assistant professor.

6
Games

Princeton, Spring 1949

JOHN VON NEUMANN, aka the Great Man behind his back, was threading his way through the crowd, nattily dressed as always and daintily holding a cup in one hand, a saucer in the other.[1] The students' common room was unusually crowded on this late afternoon in spring. A large audience, from the Institute and physics as well as math, had turned out for So and So's lecture and was lingering over tea. Von Neumann hovered for a moment by two rather sloppily dressed graduate students who hunched over a peculiar-looking piece of cardboard. It was a rhombus covered with hexagons. It looked like a bathroom floor. The two young men were taking turns putting down black and white go stones and had very nearly covered the entire board.

Von Neumann did not ask the students or anyone near him what game they were playing and when Tucker caught his eye momentarily, he averted his glance and quickly moved away. Later that evening, at a faculty dinner, however, he buttonholed Tucker and asked, with studied casualness, "Oh, by the way, what was it that they were playing?" "Nash," answered Tucker, allowing the corners of his mouth to turn up ever so slightly, "Nash."

Games were one of the charming European customs that the émigrés brought with them to Fine Hall in the 1930s. Since then one game or another has always dominated the students' common room. Today it's backgammon, but in the late 1940s it was Kriegspiel, go, and, after it was invented by its namesake, "Nash" or "John."[2]

In Nash's first year, there was a small clique of go players led by Ralph Fox, the genial topologist who had imported it after the war.[3] Fox, who was a passionate Ping-Pong player, had achieved master status in go, not altogether surprising given his mathematical specialty. He was sufficiently expert to have been invited to Japan to play go and to have once invited a well-known Japanese master named Fukuda to play with him at Fine Hall. Fukuda, who also played against Einstein and won, obliterated Fox — to the delight of Nash and some of the other denizens of Fine.[+]

Kriegspiel, however, was the favorite game. A cousin of chess, Kriegspiel was a century-long fad in Prussia. William Poundstone, the author of *Prisoner's Dilemma,* reports that Kriegspiel was devised as an educational game for German military schools in the eighteenth century, originally played on a board consisting of a map of the French-Belgian frontier divided into a grid of thirty-six hundred squares.[5] Von Neumann, growing up in Budapest, played a version of Kriegspiel with his brothers. They drew castles, highways, and coastlines on graph paper, then advanced and retreated armies according to a set of rules. Kriegspiel turned up in the United States after the Civil War, but Poundstone quotes an army officer complaining that the game "cannot readily and intelligently be pursued by anyone who is not a mathematician." Poundstone compared it to learning a foreign language.[6] The version of Kriegspiel that surfaced in the common room in the 1930s was played with three chessboards, of which one — the only one that accurately showed the moves of both players — was visible only to the umpire. The players sat back to back and were ignorant of each other's moves. The umpire told them only whether the moves they made were legal or illegal and also when a piece was taken.

A number of his fellow students remember thinking that Nash spent all of his time at Princeton in the common room playing board games.[7] Nash, who had played chess in high school,[8] played both go and Kriegspiel, the latter frequently with Steenrod or Tukey.[9] He was by no means a brilliant player, but he was unusually aggressive.[10] Games brought out Nash's natural competitiveness and one-upmanship. He would stride into the common room, one former student recalls, where people were playing Kriegspiel, glance at the boards, and say offhandedly but loudly enough for the players to hear, "Oh, white really missed his opportunity when he didn't take castle three moves ago."[11]

One time, a new graduate student was playing go. "He managed not just to overwhelm me but to destroy me by pretending to have made a mistake and letting me think I was catching him in an oversight," Hartley Rogers recalled. "This is regarded by the Japanese as a very invidious way of cheating — *hamate* — poker-type bluffing. That was a lesson both in how much better he was and how much better an actor."[12]

That spring, Nash astounded everyone by inventing an extremely clever game that quickly took over the common room.[13] Piet Hein, a Dane, had invented the game a few years before Nash, and it would be marketed by Parker Brothers in the mid-1950s as Hex. But Nash's invention of the game appears to have been entirely independent.[14]

One can imagine that von Neumann felt a twinge of envy on hearing Tucker tell him that the game he was watching had been dreamed up by a first-year graduate student from West Virginia. Many great mathematicians have amused themselves by thinking up games and puzzles, of course, but it is hard to think of a single one who has invented a game that other mathematicians find intellectually intriguing and esthetically appealing yet that nonmathematical people could enjoy playing.[15] The inventors of games that people do play—whether chess, Kriegspiel, or go—are, of course, lost in the mists of time. Nash's game was his first bona fide invention and the first hard evidence of genius.

The game would likely not have appeared in a physical manifestation, in the Princeton common room or anywhere else, had it not been for another graduate student named David Gale. Gale, a New Yorker who had spent the war in the MIT Radiation Lab, was one of the first men Nash met at the Graduate College.[16] Gale, Kuhn, and Tucker ran the weekly game theory seminar. Now a professor at Berkeley and the editor of a column on games and puzzles in *The Mathematical Intelligencer*, Gale is an aficionado of mathematical puzzles and games. Nash knew of Gale's interest in such games since Gale was in the habit, during mealtimes at the Graduate College, of silently laying down a handful of coins in a pattern or drawing a grid and then abruptly challenging whoever was dining across the table to solve some puzzle. (This is exactly what Gale did when he saw Nash for the first time after a fifty-year hiatus at a small dinner in San Francisco to celebrate Nash's Nobel.)[17]

One morning in late winter 1949, Nash literally ran into the much shorter, wiry Gale on the quadrangle inside the Graduate College. "Gale! I have an example of a game with perfect information," he blurted out. "There's no luck, just pure strategy. I can prove that the first player always wins, but I have no idea what his strategy will be. If the first player loses at this game, it's because he's made a mistake, but nobody knows what the perfect strategy is." [18]

Nash's description was somewhat elliptical, as most of his explanations were. He described the game not in terms of a rhombus with hexagonal tiles, but as a checkerboard. "Assume that two squares are adjacent if they are next to each other in a horizontal or vertical row, but also on the positive diagonal," he said.[19] Then he described what the two players were trying to do.

When Gale finally understood what Nash was trying to tell him, he was captivated. He immediately started to think about how to design an actual game board, something that had apparently never occurred to Nash, who had been toying with the idea of the game since his final year at Carnegie. "You could make it pretty, I thought." Gale, who came from a well-to-do business family, was artistic and a bit of a tinkerer. He also thought, and said as much to Nash, that the game might have some commercial potential.

"So I made a board," said Gale. "People played it using go stones. I left it in Fine Hall. It was the mathematical idea that counted. What I did was just design. I acted as his agent."

"Nash" or "John" is a beautiful example of a zero-sum two-person game with perfect information in which one player always has a winning strategy.[20] Chess and tic-tac-toe are also zero-sum two-person games with perfect information but they can end in draws. "Nash" is really a topological game. As Milnor describes it, an "*n* by *n*" Nash board consists of a rhombus tiled with *n* hexagons on each side.[21] The ideal size is fourteen by fourteen. Two opposite edges of the board are colored black, the other two white. The players use black and white go stones. They take turns placing stones on the hexagons, and once played the pieces are never moved. The black player tries to construct a connected chain of black stones from the black to black boundary. The white player tries to do the same with white stones from the white to white boundary. The game continues until one or the other player succeeds. The game is entertaining because it is challenging and appealing because it involves no complex set of rules as does chess.

Nash proved that, on a symmetrical board, the first player can always win. His proof is extremely deft, "marvelously nonconstructive" in the words of Milnor, who plays it very well.[22] If the board is covered by black and white pieces, there's always a chain that connects black to black or white to white, but never both. As Gale put it, "You can walk from Mexico to Canada or swim from California to New York, but you can't do both."[23] That explains why there can never be a draw as in tic-tac-toe. But as opposed to tic-tac-toe, even if both players try to lose, one will win, like it or not.

The game quickly swept the common room.[24] It brought Nash many admirers, including the young John Milnor, who was beguiled by its ingenuity and beauty. Gale tried to sell the game. He said, "I even went to New York and showed it to several manufacturers. John and I had some agreement that I'd get a share if it sold. But they all said no, a thinking game would never sell. It was a marvelous game though. I then sent it off to Parker Brothers, but I never got a response."[25] Gale is the one who suggested the name Hex in his letter to Parker Brothers, which Parker used for the Dane's game. (Kuhn remembers Nash describing the game to him, very likely over a meal at the college, in terms of points with six arrows emanating from each point, proof, in Kuhn's mind, that his invention was independent of Hein's.)[26] Kuhn made a board for his children, who played it with great delight and saw to it that their children learned it too.[27] Milnor still has a board that he made for his children.[28] His poignant essay on Nash's mathematical contributions for the *Mathematical Intelligencer,* written after Nash's Nobel Prize, begins with a loving and detailed description of the game.

7

John von Neumann

Princeton, 1948–49

JOHN VON NEUMANN was the very brightest star in Princeton's mathematical firmament and the apostle of the new mathematical era. At forty-five, he was universally considered the most cosmopolitan, multifaceted, and *intelligent* mathematician the twentieth century had produced.[1] No one was more responsible for the newly found importance of mathematics in America's intellectual elite. Less of a celebrity than Oppenheimer, not as remote as Einstein, as one biographer put it, von Neumann was the role model for Nash's generation.[2] He held a dozen consultancies, but his presence in Princeton was much felt.[3] "We were all drawn by von Neumann," Harold Kuhn recalled.[4] Nash was to come under his spell.[5]

Possibly the last true polymath, von Neumann made a brilliant career — half a dozen brilliant careers — by plunging fearlessly and frequently into any area where highly abstract mathematical thought could provide fresh insights. His ideas ranged from the first rigorous proof of the ergodic theorem to ways of controlling the weather, from the implosion device for the A-bomb to the theory of games, from a new algebra [of rings of operators] for studying quantum physics to the notion of outfitting computers with stored programs.[6] A giant among pure mathematicians by the time he was thirty years old, he had become in turn physicist, economist, weapons expert, and computer visionary. Of his 150 published papers, 60 are in pure mathematics, 20 in physics, and 60 in applied mathematics, including statistics and game theory.[7] When he died in 1957 of cancer at fifty-three, he was developing a theory of the structure of the human brain.[8]

Unlike the austere and otherworldly G. H. Hardy, the Cambridge number theorist idolized by the previous generation of American mathematicians, von Neumann was worldly and engaged. Hardy abhorred politics, considered applied mathematics repellent, and saw pure mathematics as an esthetic pursuit best practiced for its own sake, like poetry or music.[9] Von Neumann saw no contradiction between the purest mathematics and the grittiest engineering problems or between the role of the detached thinker and the political activist.

He was one of the first of those academic consultants who were always on a train or plane bound for New York, Washington, or Los Angeles, and whose names frequently appeared in the news. He gave up teaching when he went to the Institute in 1933 and gave up full-time research in 1955 to become a powerful member of the Atomic Energy Commission.[10] He was one of the people who told Americans how to think about the bomb and the Russians, as well as how to think about the peaceful uses of atomic energy.[11] An alleged model for Dr. Strangelove in the 1963 Stanley Kubrick film,[12] he was a passionate Cold Warrior, advocating a first strike against Russia[13] and defending nuclear testing.[14] Twice married and wealthy, he loved expensive clothes, hard liquor, fast cars, and dirty jokes.[15] He was a workaholic, blunt and even cold at times.[16] Ultimately he was hard to know; the standing joke around Princeton was that von Neumann was really an extraterrestrial who had learned how to imitate a human perfectly.[17] In public, though, von Neumann exuded Hungarian charm and wit. The parties he gave in his brick mansion on Princeton's fashionable Library Place were "frequent and famous and long," according to Paul Halmos, a mathematician who knew von Neumann.[18] His rapid-fire repartee in any of four languages was packed with references to history, politics, and the stock market.[19]

His memory was astounding and so was the speed with which his mind worked. He could instantly memorize a column of phone numbers and virtually anything else. Stories of von Neumann's beating computers in mammoth feats of calculation abound. Paul Halmos tells the story in an obituary of the first test of von Neumann's electronic computer. Someone suggested a question like "What is the smallest power of 2 with the property that its decimal digit fourth from the right is 7?" As Halmos recounts, "The machine and Johnny started at the same time, and Johnny finished first."[20]

Another time somebody asked him to solve the famous fly puzzle:[21]

Two bicyclists start twenty miles apart and head toward each other, each going at a steady rate of 10 m.p.h. At the same time, a fly that travels at a steady 15 m.p.h. starts from the front wheel of the southbound bicycle and flies to the front wheel of the northbound one, then turns around and flies to the front wheel of the southbound one again, and continues in this manner till he is crushed between the two front wheels. Question: what total distance did the fly cover?

There are two ways to answer the problem. One is to calculate the distance the fly covers on each leg of its trips between the two bicycles and finally sum the infinite series so obtained. The quick way is to observe that the bicycles meet exactly an hour after they start so that the fly had just an hour for his travels; the answer must therefore be 15 miles. When the question was put to von Neumann, he solved it in an instant, and thereby disappointed the questioner: "Oh, you must have heard the trick before!" "What trick," asked von Neumann, "all I did was sum the infinite series."

This seems astounding until one learns that at six, von Neumann could divide two eight-digit numbers in his head.[22]

Born in Budapest to a family of Jewish bankers, von Neumann was undeniably precocious.[23] At age eight, he had mastered calculus. At age twelve, he was reading works aimed at professional mathematicians, such as Emile Borel's *Theorie des Fonctions*. But he also loved to invent mechanical toys and became a child expert on Byzantine history, the Civil War, and the trial of Joan of Arc. When it was time to go off to university, he agreed to study chemical engineering as a compromise with his father, who feared that his son couldn't make a living as a mathematician. Von Neumann kept his bargain by enrolling at the University of Budapest and promptly leaving for Berlin, where he spent his time doing mathematics, including visiting lectures by Einstein, and returning to Budapest at the end of every semester to take examinations. He published his second mathematics paper, in which he gave the modern definition of ordinal numbers which superseded Cantor's, at age nineteen.[24] By age twenty-five he had published ten major papers; by age thirty, nearly three dozen.[25]

As a student in Berlin, von Neumann frequently took the three-hour train trip to Göttingen, where he got to know Hilbert. The relationship led to von Neumann's famous 1928 paper on the axiomatization of set theory. Later he found the first mathematically rigorous proof of the ergodic theorem, solved Hilbert's so-called Fifth Problem for compact groups, invented a new algebra and a new field called "continuous geometry," which is the geometry of dimensions that vary continuously (instead of a fourth dimension, one could now speak of three and three-quarters dimension). He was also a leader in the drive among mathematicians to colonize other disciplines by inventing new approaches.[26] Von Neumann was still in his twenties when he wrote his famous paper on the theory of parlor games and his groundbreaking book on the mathematics of the new quantum physics, *Mathematische Grundlagen der Quantenmechanik*—the one Nash studied in the original German at Carnegie.[27]

Von Neumann was a *privatdozent* first at Berlin and then at Hamburg. He became a half-time professor at Princeton in 1931 and joined the Institute for Advanced Study in 1933 at age thirty. When the war came, his interests shifted once again. Halmos says that "till then he was a top-flight pure mathematician who understood physics; after that he was an applied mathematician who remembered his pure work."[28] During the war, he collaborated with Morgenstern on a twelve-hundred-page manuscript that became *The Theory of Games and Economic Behavior*. He was also the top mathematician in Oppenheimer's Manhattan Project from 1943 onward. His contribution to the A-bomb was his proposal for an implosion method for triggering an explosion with nuclear fuel, an idea credited with shortening the time needed to develop the bomb by as much as a year.[29]

In 1948, he was back at the Institute and very much a presence in Princeton. He did not teach any courses, but he edited and held court at the IAS.[30] He dropped in at Fine Hall teas from time to time. He and Oppenheimer were already deep into their great debate over whether the H-bomb, or the Super, as it was known, could and should be built.[31] He was fascinated by meteorological prediction and control, suggesting once that the north and south poles be dyed blue in order to raise the earth's temperature. He not only showed the physicists, economists, and electrical engineers that formal mathematics could yield fresh breakthroughs in their fields but made the enterprise of applying mathematics to real-world disciplines seem glamorous to the purest of young mathematicians.

By the end of the war, von Neumann's real passion had become computers, though he called his interest in them "obscene."[32] While he did not build the first computer, his ideas about computer architecture were accepted, and he invented mathematical techniques needed for computers. He and his collaborators, who included the future scientific director of IBM, Hermann Goldstine, invented stored rather than hardwired programs, a prototype digital computer, and a system for weather prediction. The theoretically oriented Institute had no interest in building a computer, so von Neumann sold the idea to the Navy, arguing that the Normandy invasion had almost failed because of poor weather predictions. He promoted the MANIAC, as the machine was eventually named, as a device for improving meteorological prediction. More than anything, though, von Neumann was the one who saw the potential of these "thinking machines" most clearly, arguing in a speech in Montreal in 1945 that "many branches of both pure and applied mathematics are in great need of computing instruments to break the present stalemate created by the failure of the purely analytical approach to nonlinear problems."[33]

Everything von Neumann touched was imbued with his glamour. By wading fearlessly into fields far beyond mathematics, he inspired other young geniuses, Nash among them, to do the same. His success in applying similar approaches to dissimilar problems was a green light for younger men who were problem solvers rather than specialists.

8
The Theory of Games

*The invention of deliberately oversimplified theories is one of the major
techniques of science, particularly of the "exact" sciences, which make
extensive use of mathematical analysis. If a biophysicist can usefully employ
simplified models of the cell and the cosmologist simplified models of the
universe then we can reasonably expect that simplified games may prove to be
useful models for more complicated conflicts.* — JOHN WILLIAMS, The
Compleat Strategyst

NASH BECAME AWARE of a new branch of mathematics that was in the air of
Fine Hall. It was an attempt, invented by von Neumann in the 1920s, to construct
a systematic theory of rational human behavior by focusing on games as simple
settings for the exercise of human rationality.

The first edition of *The Theory of Games and Economic Behavior* by von
Neumann and Oskar Morgenstern came out in 1944.[1] Tucker was running a
popular new seminar in Fine on game theory.[2] The Navy, which had made use of
the theory during the war in antisubmarine warfare, was pouring money into game
theory research at Princeton.[3] The pure mathematicians around the department
and at the Institute were inclined to view the new branch of mathematics, with
its social science and military orientation, as "trivial," "just the latest fad," and
"déclassé,"[4] but to many of the students at Princeton at the time it was glamorous,
heady stuff, like everything associated with von Neumann.[5]

Kuhn and Gale were always talking about von Neumann and Morgenstern's
book.[6] Nash attended a lecture by von Neumann, one of the first speakers in
Tucker's seminar.[7] Nash was intrigued by the apparent wealth of interesting, unsolved problems. He soon became one of the regulars at the seminar that met
Thursdays at five o'clock; before long he was identified as a member of "Tucker's
clique."[8]

Mathematicians have always found games intriguing. Just as games of chance led
to probability theory, poker and chess began to interest mathematicians around
Göttingen, the Princeton of its time, in the 1920s.[9] Von Neumann was the first to
provide a complete mathematical description of a game and to prove a fundamental result, the min-max theorem.[10]

Von Neumann's 1928 paper, *Zur Theorie der Gesellschaftspiele,* suggests that the theory of games might have applications to economics: "Any event — given the external conditions and the participants in the situation (provided that the latter are acting of their own free will) — may be regarded as a game of strategy if one looks at the effect it has on the participants," adding, in a footnote, "[this] is the principal problem of classical economics: how is the absolutely selfish 'homo economicus' going to act under given external circumstances."[11] But the focal point of the theory — in von Neumann's lectures and in discussions in mathematical circles during the 1930s — basically remained the exploration of parlor games like chess and poker.[12] It was not until von Neumann met Morgenstern, a fellow émigré, in Princeton in 1938 that the link to economics was forged.[13]

Morgenstern, a tall, imposing expatriate from Vienna who was given to Napoleonic airs, claimed to be the grandson of the Kaiser's father, Friedrich III of Germany.[14] Tall, darkly handsome, "with cool gray eyes and a sensuous mouth," Morgie cut an elegant figure on horseback, and caused a sensation among his students by abruptly marrying a beautiful redhead named Dorothy, a volunteer for the World Federalists many years his junior.[15] Born in Silesia, Germany, in 1902, Morgenstern grew up and was educated in Vienna in a period of great intellectual and artistic ferment.[16] After a three-year fellowship abroad financed by the Rockefeller Foundation, he became a professor and, until the Anschluss, was head of an institute for business cycle research. When Hitler marched into Vienna, Morgenstern happened to be visiting Princeton, and he decided it made sense to stay. He joined the university's economics faculty, but disliked most of his American colleagues. He gravitated to the Institute, where Einstein, von Neumann, and Gödel were working at the time, angling for, but never receiving, an appointment there. "There is a spark missing," he wrote disdainfully to a friend, referring to the University. "It is too provincial."[17]

Morgenstern was, by temperament, a critic. His first book, *Wirtschaftsprognose (Economic Prediction),* was an attempt to prove that forecasting the ups and downs of the economy was a futile endeavor.[18] One reviewer called it as "remarkable for its pessimism as it is for any . . . theoretical innovation."[19] Unlike those in astronomy, economic predictions have the peculiar ability to change outcomes.[20] Predict a shortage, and businesses and consumers will react; the result is a glut.

His larger theme was the failure of economic theory to take proper account of interdependence among economic actors. He saw interdependence as the salient feature of all economic decisions, and he was always criticizing other economists for ignoring it.[21] Robert Leonard, the historian, writes: "To some extent, his increasingly harsh views of economic theory were the product of mathematicians' critical stance on the subject."[22] Von Neumann, he found, "focused on the black hole in the middle of economic theory."[23] According to one of von Neumann's biographers, Morgenstern "interested him in aspects of economic situations, specifically in problems of exchange of goods between two or more persons, in problems of monopoly, oligopoly and free competition. It was in a discussion of attempts to schematize mathematically such processes that the present shape of this theory began to take form." [24]

Morgenstern yearned to do "something in the truly scientific spirit." [25] He convinced von Neumann to write a treatise with him arguing that the theory of games was the correct foundation for all economic theory. Morgenstern, who had studied philosophy, not mathematics, could not contribute to the elaboration of the theory, but played muse and producer. [26] Von Neumann wrote almost the whole twelve-hundred-page treatise, but it was Morgenstern who crafted the book's provocative introduction and framed the issues in such a way that the book captured the attention of the mathematical and economic community. [27]

The Theory of Games and Economic Behavior was in every way a revolutionary book. In line with Morgenstern's agenda, the book was "a blistering attack" on the prevailing paradigm in economics and the Olympian Keynesian perspective, in which individual incentives and individual behavior were often subsumed, as well as an attempt to ground the theory in individual psychology. It was also an effort to reform social theory by applying mathematics as the language of scientific logic, in particular set theory and combinatorial methods. The authors wrapped the new theory in the mantle of past scientific revolutions, implicitly comparing their treatise to Newton's *Principia* and the effort to put economics on a rigorous mathematical footing to Newton's mathematization, using his invention of the calculus, of physics. [28] One reviewer, Leo Hurwicz, wrote, "Ten more such books and the future of economics is assured." [29]

The essence of von Neumann and Morgenstern's message was that economics was a hopelessly unscientific discipline whose leading members were busily peddling solutions to pressing problems of the day—such as stabilizing employment—without the benefit of any scientific basis for their proposals. [30] The fact that much of economic theory had been dressed up in the language of calculus struck them as "exaggerated" and a failure. [31] This was not, they said, because of the "human element" or because of poor measurement of economic variables. [32] Rather, they claimed, "Economic problems are not formulated clearly and are often stated in such vague terms as to make mathematical treatment *a priori* appear hopeless because it is quite uncertain what the problems really are." [33]

Instead of pretending that they had the expertise to solve urgent social problems, economists should devote themselves to "the gradual development of a theory." [34] The authors argued that a new theory of games was "the proper instrument with which to develop a theory of economic behavior." [35] The authors claimed that "the typical problems of economic behavior become strictly identical with the mathematical notions of suitable games of strategy." [36] Under the heading "necessary limitations of the objectives," von Neumann and Morgenstern admitted that their efforts to apply the new theory to economic problems had led them to "results that are already fairly well known," but defended themselves by contending that exact proofs for many well-known economic propositions had been lacking.[37]

> Before they have been given the respective proofs, theory simply does not exist as a scientific theory. The movements of the planets were known long before their courses had been calculated and explained by Newton's theory. . . .
>
> We believe that it is necessary to know as much as possible about the behavior of the individual and about the simplest forms of exchange. This standpoint was actually adopted with remarkable success by the founders of the marginal utility school, but nevertheless it is not generally accepted. Economists frequently point to much larger, more burning questions and brush everything aside which prevents them from making statements about them. The experience of more advanced sciences, for example, physics, indicates this impatience merely delays progress, including the treatment of the burning questions.

When the book appeared in 1944, von Neumann's reputation was at its peak. It got the kind of public attention — including a breathless front-page story in *The New York Times* — that no other densely mathematical work had ever received, with the exception of Einstein's papers on the special and general theories of relativity.[38] Within two or three years, a dozen reviews appeared by top mathematicians and economists.[39]

The timing, as Morgenstern had sensed, was perfect. The war had unleashed a search for systematic attacks on all sorts of problems in a wide variety of fields, especially economics, previously thought to be institutional and historical in character. Quite apart from the new theory of games, a major transformation was under way — led by Samuelson's *Foundations of Economic Theory* — making economic theory more rigorous through the use of calculus and advanced statistical methods.[40] Von Neumann was critical of these efforts, but they surely prepared the ground for the reception of game theory.[41]

Economists were actually somewhat standoffish, at least compared to mathematicians, but Morgenstern's antagonism to the economics profession no doubt contributed to that reaction. Samuelson later complained to Leonard, the historian, that although Morgenstern made "great claims, he himself lacked the mathematical wherewithal to substantiate them. Moreover [Morgenstern] had the irksome habit of always invoking the authority of some physical scientist or another."[42] In Princeton, Jacob Viner, the chairman of the economics department, heaped scorn on the unpopular Morgenstern by saying that if game theory couldn't even solve a game like chess, what good was it, since economics was far more complicated than chess?[43]

It must have become obvious to Nash fairly early on that "the bible," as *The Theory of Games and Economic Behavior* was known to students, though mathematically innovative, contained no fundamental new theorems beyond von Neumann's stunning min-max theorem.[44] He reasoned that von Neumann had succeeded neither in solving a major outstanding problem in economics using the new theory nor in making any major advance in the theory itself.[45] Not a single one of its applications to economics did more than restate problems that economists had already grappled with.[46] More important, the best-developed part of the theory —which took up one-third of the book—concerned zero-sum two-person games, which, because they are games of total conflict, appeared to have little applicability in social science.[47] Von Neumann's theory of games of more than two players, another large chunk of the book, was incomplete.[48] He couldn't prove that a solution existed for all such games.[49] The last eighty pages of *The Theory of Games and Economic Behavior* dealt with non-zero-sum games, but von Neumann's theory reduced such games formally to zero-sum games by introducing a fictitious player who consumes the excess or makes up the deficit.[50] As one commentator was later to write, "This artifice helped but did not suffice for a completely adequate treatment of the non-zero-sum case. This is unfortunate because such games are the most likely to be found useful in practice."[51]

To an ambitious young mathematician like Nash, the gaps and flaws in von Neumann's theory were as alluring as the puzzling absence of ether through which light waves were supposed to travel was to the young Einstein. Nash immediately began thinking about the problem that von Neumann and Morgenstern described as *the* most important test of the new theory.

9

The Bargaining Problem

Princeton, Spring 1949

We hope however to obtain a real understanding of the problem of exchange by studying it from an altogether different angle; that is, from the perspective of a "game of strategy." — VON NEUMANN AND MORGENSTERN, The Theory of Games and Economic Behavior, second edition, 1947

NASH WROTE HIS FIRST PAPER, one of the great classics of modern economics, during his second term at Princeton.[1] "The Bargaining Problem" is a remarkably down-to-earth work for a mathematician, especially a young mathematician. Yet no one but a brilliant mathematician could have conceived the idea. In the paper, Nash, whose economics training consisted of a single undergraduate course taken at Carnegie, adopted "an altogether different angle" on one of the oldest problems in economics and proposed a completely surprising solution.[2] By so doing, he showed that behavior that economists had long considered part of human psychology, and therefore beyond the reach of economic reasoning, was, in fact, amenable to systematic analysis.

The idea of exchange, the basis of economics, is nearly as old as man, and deal-making has been the stuff of legend since the Levantine kings and the pharaohs traded gold and chariots for weapons and slaves.[3] Despite the rise of the great impersonal capitalist marketplace, with its millions of buyers and sellers who never meet face-to-face, the one-on-one bargain — involving wealthy individuals, powerful governments, labor unions, or giant corporations — dominates the headlines. But two centuries after the publication of Adam Smith's *The Wealth of Nations,* there were still no principles of economics that could tell one how the parties to a potential bargain would interact, or how they would split up the pie.[4]

The economist who first posed the problem of the bargain was a reclusive Oxford don, Francis Ysidro Edgeworth, in 1881.[5] Edgeworth and several of his Victorian contemporaries were the first to abandon the historical and philosophical tradition of Smith, Ricardo, and Marx and to attempt to replace it with the mathematical tradition of physics, writes Robert Heilbroner in *The Worldly Philoso-phers.*[6]

> Edgeworth was not fascinated with economics because it justified or explained or condemned the world, or because it opened new vistas, bright or gloomy, into the future. This odd soul was fascinated by economics because economics dealt with *quantities* and because anything that dealt with quantities could be translated into *mathematics.*[7]

Edgeworth thought of people as so many profit-and-loss calculators and recognized that the world of perfect competition had "certain properties peculiarly favorable to mathematical calculation; namely a certain indefinite multiplicity and dividedness, analogous to that infinity and infinitesimality which facilitate so large a portion of Mathematical Physics . . . (consider the theory of Atoms, and all applications of the Differential Calculus)."[8]

The weak link in his creation, as Edgeworth was uncomfortably aware, was that people simply did not behave in a purely competitive fashion. Rather, they did not behave this way all the time. True, they acted on their own. But, equally often, they collaborated, cooperated, struck deals, evidently also out of self-interest. They joined trade unions, they formed governments, they established large enterprises and cartels. His mathematical models captured the results of competition, but the consequences of cooperation proved elusive.[9]

> Is it peace or war? asks the lover of "Maud" of economic competition. It is both, pax or pact between contractors during contract, war, when some of the contractors without consent of others contract.
>
> The first principle of Economics is that every agent is actuated only by self-interest. The workings of this principle may be viewed under two aspects, according as the agent acts without, or with, the consent of others affected by his actions. In a wide sense, the first species of action may be called war; the second contract.

Obviously, parties to a bargain were acting on the expectation that cooperation would yield more than acting alone. Somehow, the parties reached an agreement to share the pie. How they would split it depended on bargaining power, but on that score economic theory had nothing to say and there was no way of finding one solution in the haystack of possible solutions that met this rather broad criterion. Edgeworth admitted defeat: "The general answer is — (a) Contract without competition is indeterminate."[10]

Over the next century, a half-dozen great economists, including the Englishmen John Hicks and Alfred Marshall and the Dane F. Zeuthen, took up Edgeworth's problem, but they, too, ended up throwing up their hands.[11] Von

Neumann and Morgenstern suggested that the answer lay in reformulating the problem as a game of strategy, but they themselves did not succeed in solving it.[12]

Nash took a completely novel approach to the problem of predicting how two rational bargainers will interact. Instead of defining a solution directly, he started by writing down a set of reasonable conditions that any plausible solution would have to satisfy and then looked at where they took him.

This is called the axiomatic approach — a method that had swept mathematics in the 1920s, was used by von Neumann in his book on quantum theory and his papers on set theory, and was in its heyday at Princeton in the late 1940s.[13] Nash's paper is one of the first to apply the axiomatic method to a problem in the social sciences.[14]

Recall that Edgeworth had called the problem of the bargain "indeterminate." In other words, if all one knew about the bargainers were their preferences, one couldn't predict how they would interact or how they would divide the pie. The reason for the indeterminacy would have been obvious to Nash. There wasn't enough information so one had to make additional assumptions.

Nash's theory assumes that both sides' expectations about each other's behavior are based on the intrinsic features of the bargaining situation itself. The essence of a situation that results in a deal is "two individuals who have the opportunity to collaborate for mutual benefit in more than one way."[15] How they will split the gain, he reasoned, reflects how much the deal is worth to each individual.

He started by asking the question, What reasonable conditions would any solution — any split — have to satisfy? He then posed four conditions and, using an ingenious mathematical argument, showed that, if his axioms held, a unique solution existed that maximized the product of the players' utilities. In a sense, his contribution was not so much to "solve" the problem as to state it in a simple and precise way so as to show that unique solutions were possible.

The striking feature of Nash's paper is not its difficulty, or its depth, or even its elegance and generality, but rather that it provides an answer to an important problem. Reading Nash's paper today, one is struck most by its originality. The ideas seem to come out of the blue. There is some basis for this impression. Nash arrived at his essential idea — the notion that the bargain depended on a combination of the negotiators' back-up alternatives and the potential benefits of striking a deal — as an undergraduate at Carnegie Tech before he came to Princeton, before he started attending Tucker's game theory seminar, and before he had read von Neumann and Morgenstern's book. It occurred to him while he was sitting in the only economics course he would ever attend.[16]

The course, on international trade, was taught by a clever and young Viennese émigré in his thirties named Bert Hoselitz. Hoselitz, who emphasized theory in his course, had degrees in law and economics, the latter from the University of Chicago.[17] International agreements between governments and between monopolies had dominated trade, especially in commodities, between the wars, and Hoselitz was an expert on the subject of international cartels and trade.[18] Nash took the course in his final semester, in the spring of 1948, simply to fulfill degree requirements.[19] As always, though, the big, unsolved problem was the bait.

That problem concerned trade deals between countries with separate currencies, as he told Roger Myerson, a game theorist at Northwestern University, in 1996.[20] One of Nash's axioms, if applied in an international trade context, asserts that the outcome of the bargain shouldn't change if one country revalued its currency. Once at Princeton, Nash would have quickly learned about von Neumann and Morgenstern's theory and recognized that the arguments that he'd thought of in Hoselitz's class had a much wider applicability.[21] Very likely Nash sketched his ideas for a bargaining solution in Tucker's seminar and was urged by Oskar Morgenstern — whom Nash invariably referred to as Oskar La Morgue — to write a paper.[22]

Legend, possibly encouraged by Nash himself, soon had it that he'd written the whole paper in Hoselitz's class — much as Milnor solved the Borsuk problem in knot theory as a homework assignment — and that he had arrived at Princeton with the bargaining paper tucked into his briefcase.[23] Nash has since corrected the record.[24] But when the paper was published in 1950, in *Econometrica,* the leading journal of mathematical economics, Nash was careful to retain full credit for the ideas: "The author wishes to acknowledge the assistance of Professors von Neumann and Morgenstern who read the original form of the paper and gave helpful advice as to the presentation."[25] And in his Nobel autobiography, Nash makes it clear that it was his interest in the bargaining problem that brought him into contact with the game theory group at Princeton, not the other way around: "as a result of that exposure to economic ideas and problems I arrived at the idea that led to the paper 'The Bargaining Problem' which was later published in *Econometrica.* And it was this idea which in turn, when I was a graduate student at Princeton, *led to my interest in the game theory studies there.*"[26]

10

Nash's Rival Idea

Princeton, 1949–50

I was playing a non-cooperative game in relation to von Neumann rather than simply seeking to join his coalition. —JOHN F. NASH, JR., 1993

IN THE SUMMER OF 1949, Albert Tucker caught the mumps from one of his children.[1] He had planned to be in Palo Alto, California, where he was to spend his sabbatical year, by the end of August. Instead, he was in his office at Fine, gathering up some books and papers, when Nash walked in to ask whether Tucker would be willing to supervise his thesis.

Nash's request caught him by surprise.[2] Tucker had little direct contact with Nash during the latter's first year and had been under the impression that he would probably write a thesis with Steenrod. But Nash, who offered no real explanation, told Tucker only that he thought he had found some "good results related to game theory." Tucker, who was still feeling out of sorts and eager to get home, agreed to become his adviser only because he was sure that Nash would still be in the early stages of his research by the time he returned to Princeton the following summer.

Six weeks later, Nash and another student were buying beers for a crowd of graduate students and professors in the bar in the basement of the Nassau Inn — as tradition demanded of men who had just passed their generals.[3] The mathematicians were growing more boisterous and drunken by the minute. A limerick competition was in full swing. The object was to invent the cleverest, dirtiest rhyme about a member of the Princeton mathematics department, preferably about one of the ones present, and shout it out at the top of one's lungs.[4] At one point, a shaggy Scot aptly named Macbeath jumped to his feet, beer bottle in hand, and began to belt out stanza after stanza of a popular and salacious drinking song, with the others chiming in for the chorus: "I put my hand upon her breast/She said, 'Young man, I like that best'/(Chorus) Gosh, gore, blimey, how ashamed I was."[5]

That night, with its quaint, masculine rite of passage, marked the effective end of Nash's years as a student. He had been trapped in Princeton for an entire hot and sticky summer, forced to put aside the interesting problems he had been thinking about, to cram for the general examination.[6] Luckily, Lefschetz had appointed a friendly trio of examiners: Church, Steenrod, and a visiting professor from Stanford, Donald Spencer.[7] The whole nerve-racking event had gone rather well.

Many mathematicians, most famously the French genius Henri Poincaré, have testified to the value of leaving a partially solved problem alone for a while and letting the unconscious work behind the scenes. In an oft-quoted passage from a 1908 essay about the genesis of mathematical discovery, Poincaré writes:[8]

> For fifteen days I struggled to prove that no functions analogous to those I have since called Fuchsian functions could exist. I was then very ignorant. Every day I sat down at my work table where I spent an hour or two; I tried a great number of combinations and arrived at no result. . . .
>
> I then left Caen where I was living at the time, to participate in a geological trip sponsored by the School of Mines. The exigencies of travel made me forget my mathematical labors; reaching Coutances we took a bus for some excursion or another. The instant I put my foot on the step the idea came to me, apparently with nothing whatever in my previous thoughts having prepared me for it.

Nash's "wasted" summer, with its enforced break from his research, proved unexpectedly fruitful, allowing several vague hunches from the spring to crystallize and mature. That October, he started to experience a virtual storm of ideas. Among them was his brilliant insight into human behavior: the Nash equilibrium.

Nash went to see von Neumann a few days after he passed his generals.[9] He wanted, he had told the secretary cockily, to discuss an idea that might be of interest to Professor von Neumann. It was a rather audacious thing for a graduate student to do.[10] Von Neumann was a public figure, had very little contact with Princeton graduate students outside of occasional lectures, and generally discouraged them from seeking him out with their research problems. But it was typical of Nash, who had gone to see Einstein the year before with the germ of an idea.

Von Neumann was sitting at an enormous desk, looking more like a prosperous bank president than an academic in his expensive three-piece suit, silk tie, and jaunty pocket handkerchief.[11] He had the preoccupied air of a busy executive. At the time, he was holding a dozen consultancies, "arguing the ear off Robert Oppenheimer" over the development of the H-bomb, and overseeing the construction and programming of two prototype computers.[12] He gestured Nash to sit down. He knew who Nash was, of course, but seemed a bit puzzled by his visit.

He listened carefully, with his head cocked slightly to one side and his fingers tapping. Nash started to describe the proof he had in mind for an equilibrium in games of more than two players. But before he had gotten out more than a few disjointed sentences, von Neumann interrupted, jumped ahead to the yet unstated conclusion of Nash's argument, and said abruptly, "That's trivial, you know. That's just a fixed point theorem."[13]

It is not altogether surprising that the two geniuses should clash. They came at game theory from two opposing views of the way people interact. Von Neumann, who had come of age in European café discussions and collaborated on the bomb and computers, thought of people as social beings who were always communicating. It was quite natural for him to emphasize the central importance of coalitions and joint action in society. Nash tended to think of people as out of touch with one another and acting on their own. For him, a perspective founded on the ways that people react to individual incentives seemed far more natural.

Von Neumann's rejection of Nash's bid for attention and approval must have hurt, however, and one guesses that it was even more painful than Einstein's earlier but kindlier dismissal. He never approached von Neumann again. Nash later rationalized von Neumann's reaction as the naturally defensive posture of an established thinker to a younger rival's idea, a view that may say more about what was in Nash's mind when he approached von Neumann than about the older man. Nash was certainly conscious that he was implicitly challenging von Neumann. Nash noted in his Nobel autobiography that his ideas *"deviated somewhat from the 'line' (as if of 'political party lines') of von Neumann and Morgenstern's book."*[14]

Valleius, the Roman philosopher, was the first to offer a theory for why geniuses often appeared, not as lonely giants, but in clusters in particular fields in particular cities. He was thinking of Plato and Aristotle, Pythagoras and Archimedes, and Aeschylus, Euripides, Sophocles, and Aristophanes, but there are many later examples as well, including Newton and Locke, or Freud, Jung, and Adler. He speculated that creative geniuses inspired envy as well as emulation and attracted younger men who were motivated to complete and recast the original contribution.[15]

In a letter to Robert Leonard, Nash wrote a further twist: "I was playing a non-cooperative game in relation to von Neumann rather than simply seeking to join his coalition. And of course, it was psychologically natural for him not to be entirely pleased by a rival theoretical approach."[16] In his opinion, von Neumann never behaved unfairly. Nash compares himself to a young physicist who challenged Einstein, noting that Einstein was initially critical of Kaluza's five-dimensional unified theory of gravitational and electric fields but later supported its publication.[17] Nash, so often oblivious to the feelings and motivations of other people, was quick, in this case, to pick up on certain emotional undercurrents, especially envy and jealousy. In a way, he saw rejection as the price genius must pay.

A few days after the disastrous meeting with von Neumann, Nash accosted

David Gale. "I think I've found a way to generalize von Neumann's min-max theorem," he blurted out. "The fundamental idea is that in a two-person zero-sum solution, the best strategy for both is . . . The whole theory is built on it. And it works with any number of people and doesn't have to be a zero-sum game."[18] Gale recalls Nash's saying, "I'd call this an equilibrium point." The idea of equilibrium is that it is a natural resting point that tends to persist. Unlike von Neumann, Gale saw Nash's point. "Hmm," he said, "that's quite a thesis." Gale realized that Nash's idea applied to a far broader class of real-world situations than von Neumann's notion of zero-sum games. "He had a concept that generalized to disarmament," Gale said later. But Gale was less entranced by the possible applications of Nash's idea than its elegance and generality. "The mathematics was so beautiful. It was so right mathematically."

Once again, Gale acted as Nash's agent. "I said this is a great result," Gale recalled. "This should get priority." He told Nash that he was sure that Nash had a brilliant thesis in hand. But he also urged Nash to take credit for the result right away before someone else came up with a similar idea. Gale suggested asking a member of the National Academy of Sciences to submit the proof to the academy's monthly proceedings. "He was spacey. He would never have thought of doing that," Gale said recently, "so he gave me his proof and I drafted the NAS note." Lefschetz submitted the note immediately and it appeared in the November proceedings.[19] Gale added later, "I certainly knew right away that it was a thesis. I didn't know it was a Nobel."[20]

Almost fifty years later, two months before his death, Tucker could not recall getting Nash's first draft of the thesis, which Nash mailed to him at Stanford, or his own reaction on reading it, other than being surprised that Nash had produced a result so quickly. He was certain, however, that he had not been bowled over. He said: "Whether or not this was of any interest to economists wasn't known."[21]

Nash used to say that Tucker was "a machine," implying that Tucker was methodical but unimaginative.[22] But, in fact, Nash was quite astute to have chosen him as an adviser. Tucker, a Canadian, Methodist rigidity notwithstanding, possessed a rare willingness to defend unconventional ideas and individuals. A truly fine teacher, he firmly believed that students should choose research topics they felt passionate about, not ones they merely believed would appeal to their professors. A few years later, it was Tucker who convinced another young, offbeat genius who would go on to become one of the fathers of artificial intelligence, Marvin L. Minsky, to drop the mainstream but boring mathematics problem he had chosen as a thesis topic and instead to write on his real passion, the structure of the brain.[23] Tucker always claimed that he did little more than sign off on Nash's slender, twenty-seven-page dissertation—"There was no essential role played by me," Tucker said—but he encouraged Nash to get it out quickly and defended its merits within the department.[24] Kuhn, who was close to Tucker at the time, later recalled: "The thesis itself was completed and submitted after the persistent urging and counsel of Professor Tucker. John always wanted to add more material, and Tucker had the wisdom to say, 'Get the results out early.' "[25]

Tucker responded to Nash's first draft by demanding that Nash include a concrete example of his equilibrium idea. He also suggested a number of changes in Nash's presentation. "I urged him to deal with a particular case rather than only a general case,"[26] Tucker said. The recommendation, to his mind, was largely esthetic. "When you deal with the general case you have to deal with sophisticated notation that is very hard to read," he said.[27] Nash responded with a prolonged silence that was in fact a measure of his fury. "He reacted unfavorably, largely by expressing nothing. I didn't hear from him again for a long time," Tucker recalled.[28]

Nash was actually considering dropping the thesis with Tucker and pursuing another topic, an ambitious problem in algebraic geometry, with Steenrod instead.[29] He chose to interpret Tucker's demands for revisions — along with von Neumann's coldly dismissive reaction — as signs that the department would not accept his work on game theory for a dissertation. However, Tucker, who could be surprisingly forceful, eventually convinced Nash to stick with his original conception — and to make the requested changes. "Nash had an answer for everything," he said. "You couldn't catch him out in a mathematical fault."[30] A May 10 letter to Lefschetz reads: "It is not necessary that I see the revised draft, for he has kept me informed (almost daily) of the progress of the revision."[31] Tucker adds, "I was delighted to notice a pleasant change of attitude in Nash during the course of our long correspondence on his work. He became much more cooperative and appreciative towards the end. I wrote to him like a Dutch uncle, but I suspect you or someone else at the Princeton end had some influence in effecting the change."[32]

The entire edifice of game theory rests on two theorems: von Neumann's min-max theorem of 1928 and Nash's equilibrium theorem of 1950.[33] One can think of Nash's theorem as a generalization of von Neumann's, as Nash did, but also as a radical departure. Von Neumann's theorem was the cornerstone of his theory of games of pure opposition, so-called two-person zero-sum games. But two-person zero-sum games have virtually no relevance to the real world.[34] Even in war there is almost always something to be gained from cooperation. Nash introduced the distinction between cooperative and noncooperative games.[35] Cooperative games are games in which players can make enforceable agreements with other players. In other words, as a group they can fully commit themselves to specific strategies. In contrast, in a noncooperative game, such collective commitment is impossible. There are no enforceable agreements. By broadening the theory to include games that involved a mix of cooperation and competition, Nash succeeded in opening the door to applications of game theory to economics, political science, sociology, and, ultimately, evolutionary biology.[36]

Although Nash used the same strategic form as von Neumann had proposed, his approach is radically different. More than half of the von Neumann and

Morgenstern book deals with cooperative theory. In addition, von Neumann and Morgenstern's solution concept — something called a stable set — does not exist for every game. By contrast, Nash proved on page six of his thesis that every noncooperative game with any number of players has at least one Nash equilibrium point.

To understand the beauty of Nash's result, write Avinash Dixit and Barry Nalebuff in *Thinking Strategically,* one begins with the notion that interdependence is the distinguishing feature of games of strategy.[37] The outcome of a game for one player depends on what all the other players choose to do and vice versa. Games like tic-tac-toe and chess involve one kind of interdependence. The players move in sequence, each aware of the other's moves. The principle for a player in a sequential-move game is to look ahead and reason back. Each player tries to figure out how the other players will respond to his current move, how he will respond in turn, and so forth. The player anticipates where his initial decision will ultimately lead and uses the information to make his current best choice. In principle, any game that ends after a finite sequence of moves can be solved completely. The player's best strategy can be determined by looking ahead to every possible outcome. For chess, in contrast to tic-tac-toe, the calculations are too complex for the human brain — or even for computer programs written by humans. Players look a few moves ahead and try to evaluate the resultant positions on the basis of experience.

Games like poker, on the other hand, involve simultaneous moves. "In contrast to the linear chain of reasoning for sequential games, a game with simultaneous moves involves a logical circle," write Dixit and Nalebuff. "Although players act at the same time, in ignorance of other players' current actions, each is forced to think about the fact that there are other players who in turn are similarly aware.[38] Poker is an example of, 'I think he thinks that I think that he thinks that I think . . .' Each must figuratively put himself in the shoes of all and try to calculate the outcome. His own best action is an integral part of the calculation."

Such circular reasoning would seem to have no conclusion. Nash squared the circle using a concept of equilibrium whereby each player picks his best response to what the others do. Players look for a set of choices such that each person's strategy is best for him when all others are playing their best strategies.

Sometimes one person's best choice is the same no matter what the others do. That is called a dominant strategy for that player. At other times, one player has a uniformly bad choice — a dominated strategy — in the sense that some other choice is best for him irrespective of what the others do. The search for equilibrium should begin by looking for dominant strategies and eliminating dominated ones. But these are special and relatively rare cases. In most games each player's best choice does depend on what the others do, and one must turn to Nash's construct.

Nash defined equilibrium as a situation in which no player could improve his or her position by choosing an alternative available strategy, without implying that each person's privately held best choice will lead to a collectively optimal result. He proved that for a certain very broad class of games of any number of players, at least one equilibrium exists — so long as one allows mixed strategies. But some games have many equilibria and others, relatively rare ones that fall outside the class he defined, may have none.

Today, Nash's concept of equilibrium from strategic games is one of the basic paradigms in social sciences and biology.[39] It is largely the success of his vision that has been responsible for the acceptance of game theory as, in the words of *The New Palgrave,* "a powerful and elegant method of tackling a subject that had become increasingly baroque, much as Newtonian methods of celestial mechanics had displaced the primitive and increasingly *ad hoc* methods of the ancients."[40] Like many great scientific ideas, from Newton's theory of gravitation to Darwin's theory of natural selection, Nash's idea seemed initially too simple to be truly interesting, too narrow to be widely applicable, and, later on, so obvious that its discovery by *someone* was deemed all but inevitable.[41] As Reinhard Selten, the German economist who shared the 1994 Nobel with Nash and John C. Harsanyi, said: "Nobody would have foretold the great impact of the Nash equilibrium on economics and social science in general. It was even less expected that Nash's equilibrium point concept would ever have any significance for biological theory."[42] Its significance was not immediately recognized, not even by the brash twenty-one-year-old author himself, and certainly not by the genius who inspired Nash, von Neumann.[43]

11

Lloyd

Princeton, 1950

*All mathematicians live in two different worlds. They live in a crystalline
world of perfect platonic forms. An ice palace. But they also live in the
common world where things are transient, ambiguous, subject to vicissitudes.
Mathematicians go backward and forward from one world to another. They're
adults in the crystalline world, infants in the real one. — S. CAPPELL, Courant
Institute of Mathematics, 1996*

AT TWENTY-ONE, Nash the mathematical genius had emerged and connected with the larger community of mathematicians around him, but Nash the
man remained largely hidden behind a wall of detached eccentricity. He was quite
popular with his professors, but utterly out of touch with his peers. His interactions
with most of the men his own age seemed motivated by an aggressive competitiveness and the most cold considerations of self-interest. His fellow students believed
that Nash had felt nothing remotely resembling love, friendship, or real sympathy,
but as far as they were able to judge, Nash was perfectly at home in this arid state
of emotional isolation.

This was not the case, however. Nash, like all human beings, wanted to be
close to someone, and at the beginning of his second year at Princeton he had
finally found what he was looking for. The friendship with Lloyd Shapley, an older
student, was the first of a series of emotional attachments Nash formed to other
men, mostly brilliant mathematical rivals, usually younger. These relationships,
which usually began with mutual admiration and intense intellectual exchange,
soon became one-sided and typically ended in rejection. The relationship with
Shapley foundered within a year, although Nash never completely lost touch with
him over the decades to follow — all through his long illness and after he began to
recover — when he and Shapley became direct competitors for the Nobel Prize.

When he first moved into the Graduate College a few doors down from Nash in
the fall of 1949, Lloyd Shapley had just turned twenty-six, five years and eleven

days older than Nash.[1] No one could have presented a stronger contrast with the childish, boorish, handsome, and uninhibited boy wonder from West Virginia.

Born and bred in Cambridge, Massachusetts, Shapley was one of five children of one of the most famous and revered scientists in America, the Harvard astronomer Harlow Shapley. The senior Shapley was a public figure known to every educated household, and also one of the most politically active.[2] In 1950, he was accorded the dubious honor of being the first prominent scientist to appear on the earliest of Senator Joseph McCarthy's famous lists of crypto-communists.[3]

Lloyd Shapley was a war hero.[4] He was drafted in 1943. He refused an offer to become an officer. That same year, as a sergeant in the Army Air Corps in Sheng-Du, China, Shapley got a Bronze Star for breaking the Japanese weather code. In 1945, he went back to Harvard, where he had begun to study mathematics before he was drafted, and finished his B.A. in mathematics in 1948.

When Shapley showed up at Princeton, von Neumann already considered him the brightest young star in game theory research.[5] Shapley had spent the year after graduating from Harvard at the RAND Corporation, a think tank in Santa Monica that was attempting to use game theory applications to solve military problems, and came to Princeton while technically on leave from RAND. He was immediately recognized as brilliant and quite sophisticated in his thinking. One contemporary remembers that he "talked good math, knew a lot."[6] He did extraordinarily hard double crostics from *The New York Times* without using a pencil.[7] He was a fiercely competitive and highly accomplished player of Kriegspiel[8] and go. "Everybody knew that his game was strictly his own," said another fellow student. "He went out of his way to find nonstandard moves. No one was going to anticipate them."[9] He was also well read. He played the piano beautifully.[10] His manner suggested an acute awareness of pedigree and prospects. When Lefschetz wrote him a letter telling him of a very generous grant if he came to Princeton, for example, Shapley replied loftily and with a hint of disdain, "Dear Lefschetz, The arrangements are satisfactory. Go ahead with the formalities. Shapley."[11]

Shapley was by no means as self-confident as his imperious note to Lefschetz implied. His appearance can only be described as rather strange. Tall, dark, and so thin that his clothing hung from him like a scarecrow's, Shapley reminded one young woman of a giant insect; another contemporary says he looked like a horse.[12] His normally gentle demeanor and ironic banter hid a violent temper and a harshly self-critical streak.[13] When challenged in some unexpected fashion, he could become hysterical, literally vibrating and shaking with fury.[14] His perfectionism, which would later prevent him from publishing a large portion of his research, was extreme.[15] He was, moreover, acutely self-conscious about being a few years older than some of the brilliant young men around the Princeton mathematics department.[16]

Nash was one of the first students Shapley met at the Graduate College. For a time, they shared a bathroom. Both of them attended Tucker's game theory seminar every Thursday, now run by Kuhn and Gale while Tucker was at Stanford. The best way to describe the impression Nash made on Shapley when the two first talked about mathematics is to say that Nash took Shapley's breath away. Shapley could, of course, see what the others saw — the childishness, brattiness, obnoxiousness — but he saw a great deal more. He was dazzled by what he would later describe as Nash's "keen, beautiful, logical mind."[17] Instead of being alienated like the others by the younger man's odd manner and weird behavior, he interpreted these simply as signs of immaturity. "Nash was spiteful, a child with a social IQ of 12, but Lloyd did appreciate talent," recalled Martin Shubik.[18]

As for Nash, starved for affection, how could he not be drawn to Shapley? In Nash's eyes, Shapley had it all. A brilliant mathematician. War hero. Harvard man. A son of Harlow. Favorite of von Neumann and, soon, of Tucker as well. Shapley, who was popular with faculty and students alike, was one of the very few around Princeton, other than Milnor, who could really hold Nash's attention in a mathematical conversation, challenge him, and help him to pursue the implications of his own reasoning. And, for that reason — along with his open admiration and obvious sympathy — he was one who could engage Nash's emotions.

Nash acted like a thirteen-year-old having his first crush. He pestered Shapley mercilessly.[19] He made a point of disrupting his beloved Kriegspiel games, sometimes by sweeping the pieces to the ground. He rifled through his mail. He read the papers on his desk. He left notes for Shapley: "Nash was here!" He played all kinds of pranks on him.

Shapley's greatest eccentricity at the time was his claim that he was on a twenty-five-hour sleep cycle.[20] He worked and slept at extremely odd hours, often transposing night and day. "Every once in a while he'd disappear from sight," another student recalled. "That's what he said. We accepted anything."[21] Waking Shapley when he was lost to the world became an ongoing prank. "A group of us was attending a regular seminar at the institute given by de Rham and Kodaira. We were always very anxious to go but only three or four of us had cars. Lloyd Shapley was one but there was one difficulty. Lloyd liked to sleep late and was often asleep at two o'clock in the afternoon. So we had to devise all sorts of ways to wake him. We dropped hot candle wax on him. I devised another method. We played 45-rpm records of Lloyd's favorite Chinese music without the little insert so that it oscillated all over the place (and made excruciating noise)."[22] Nash once tried to wake Shapley by climbing on his bed, straddling him and dropping water in his ear with an eyedropper.[23]

Sometimes the jokes, also aimed at other friends of Shapley's, got totally out of hand. Shapley shared his room at the college with a graduate student in economics, Martin Shubik, who became interested in game theory and also developed a lifelong friendship with Shapley. Shubik recalled: "Nash's idea of a joke was to unscrew the electric light bulb in the bathroom. There was a glass shade under the bulb, which he filled full of water. We could easily have gotten electrocuted. Did he intend to electrocute me? I'm not sure he didn't intend to."[24]

Shubik, whom Nash insisted on calling Shoobie-Woobie, was a frequent target of Nash's digs. A typical putdown, from a postscript to a note ostensibly commiserating with Shubik after the latter was injured in a car accident: "Oscar le Morgue would like for someone . . . to blast Baumol [William Baumol, then the rising young star of the Princeton economics department] for his impudence in publishing a paper attacking confusedly the only true utility. It's beneath his dignity, but he doesn't really think you're the best man for the job because . . . 'Shubik does not write very clearly.' "[25]

John McCarthy, one of the inventors of artificial intelligence, also befriended Shapley and apparently aroused Nash's jealousy. One day McCarthy got an inquiry from a Philadelphia haberdashery about a massive shirt order he had placed.[26] How good was his credit, the company wanted to know? McCarthy, who hadn't placed any such order, immediately suspected Nash and asked Shapley if Nash was the culprit. Shapley confirmed that this was highly likely. McCarthy asked the company for the original order. Sure enough, a postcard came back with Nash's unmistakable scrawl in green ink, the color Nash always used. Shubik and McCarthy cornered Nash and confronted him. "There was no denying what he had done. We threatened him with postal inspectors. The post office refused to merely bawl him out. 'If we do anything, we'll prosecute him,' they said." Concluding that Nash had learned his lesson, Shubik and McCarthy dropped the matter. Another time, he rigged up McCarthy's bed so that it would collapse when McCarthy tried to crawl under the covers.[27]

It was Shapley who reacted to Nash's absurd behavior with amused tolerance, who proposed that they might channel his mischievous impulses in a more intellectually constructive way. So Nash, Shapley, Shubik, and McCarthy, along with another student named Mel Hausner, invented a game involving coalitions and double-crosses. Nash called the game — which was later published under the name "So Long, Sucker" — Fuck Your Buddy.[28] The game is played with a pile of different-colored poker chips. Nash and the others crafted a complicated set of rules designed to force players to join forces with one another to advance, but ultimately to double-cross one another in order to win. The point of the game was to produce psychological mayhem, and, apparently it often did. McCarthy remembers losing his temper after Nash cold-bloodedly dumped him on the second-to-last round, and Nash was absolutely astonished that McCarthy could get so emotional. "But I didn't need you anymore," Nash kept saying, over and over.[29]

By and large, Shapley tried to play the role of mentor. He came to Nash's aid, for example, when Tucker demanded that Nash include a concrete example of an equilibrium point in his thesis and Nash couldn't think of a good one. Shapley spent weeks working out an elaborate but convincing example of Nash's equilibrium concept involving three-handed poker, another Shapley specialty.[30]

The friendship between the men always had a competitive edge.[31] Shapley, who started out as the slightly older and wiser half of the relationship, may have resented Nash's reputation as a genius. He kept remarking on "running starts," and he made it clear that he felt he was being left behind.[32] Nash's stubborn independence in the face of well-meant advice, instead of delighting, began to irk. Nash's real sin, though, may have been to publish three important papers in the space of one year, long before Shapley had even come close to finding a thesis topic for himself.[33] In one of them, Nash beat Shapley to the punch on a problem they were both working on and had spent many hours discussing.[34]

But Shapley actually had good reason to feel secure. Despite Nash's brilliant dissertation, the consensus at Princeton at the time was that it was Shapley who was the real star of the next generation and inheritor of the von Neumann mantle. Tucker wrote in 1953: Shapley is "the best young American mathematician working in the subject."[35] As a person, Tucker added, Shapley is "agreeable, cooperative and well-liked by faculty and students."[36] A letter from Frederic Bohnenblust, Shapley's mentor at RAND, dated 1953, says Shapley "perhaps lacked the wherewithal to develop a theory and depended on others for ideas," but added that he thought him "second only to the creator of the theory of games, John von Neumann."[37] A letter from von Neumann dated January 1954 said: "I know Shapley very well and I think he is VERY good. I would put him above Bohnenblust and I would bracket him with Segal and Birkhoff."[38]

But something other than graduate-student rivalry caused a sudden break. By the middle of the next year, by which time Nash had already completed his thesis and was on the job market, Shapley told a fellow student that he would not return to RAND if Nash, who had been offered a permanent post there, were to accept it.[39] Fifty years later, Shapley made a point of correcting anyone who suggested that he and Nash had ever been close friends.[40]

12

The War of Wits

RAND, Summer 1950

Oh, the RAND Corporation is the boon of the world;
They think all day for a fee.
They sit and play games about going up in flames,
For counters they use you and me, Honey Bee,
For counters they use you and me.
— MALVINA REYNOLDS, *"The RAND Hymn," 1961*

THE DC-3 SHOOK as it droned past the desert and mountains toward the opaque Pacific and water-colored sky. Los Angeles lay thousands of feet below, resembling some science-fiction vision of a space colony under its sulfurous blanket of haze. Nash had boarded the TWA flight in New York almost twenty-four hours earlier. He had not slept at all. He was rumpled, sweaty, cramped, and exhausted, but as the plane descended, he hardly registered these discomforts. His attention was wholly absorbed by the exotic panorama and his own intense excitement.

Flying was still a highly novel experience in 1950, no more so than for a twenty-two-year-old West Virginian whose travels had mostly been limited to the Norfolk & Western runs between Roanoke and Princeton. Nash's first flight marked the beginning of his career as a consultant for the secretive RAND Corporation. RAND is a civilian think tank in Santa Monica, described by *Fortune* in 1951 as "the Air Force's big-brain-buying venture,"[1] where brilliant academics pondered nuclear war and the new theory of games. Nash's on-and-off encounter with RAND over the next four years was a transforming experience in his life. His association with RAND, at the height of the Cold War, started promisingly in the summer of 1950, just as the Korean War began, and ended traumatically in the summer of 1954, when McCarthyism reached its peak.

On a purely personal level, Nash's view of the world and himself was permanently and subtly colored by the RAND Zeitgeist — its worship of the rational life and quantification, its geopolitical obsessions, and its weirdly compelling mix of Olympian detachment, paranoia, and megalomania. Intellectually, it was another story. From the moment of his arrival, Nash began actively disengaging himself from the interests and individuals that brought him to RAND in the first place, retreating from game theory and moving rapidly into pure mathematics, a process of disengagement that would repeat itself several times over the rest of the decade.

Nothing like the RAND of the early 1950s has existed before or since.[2] It was the original think tank, a strange hybrid of which the unique mission was to apply rational analysis and the latest quantitative methods to the problem of how to use the terrifying new nuclear weaponry to forestall war with Russia — or to win a war if deterrence failed. The people of RAND were there to think the unthinkable, in Herman Kahn's famous phrase.[3] It attracted some of the best minds in mathematics, physics, political science, and economics. RAND may well have been the model for Isaac Asimov's *Foundation* series, about a RANDlike organization full of hyper-rational social scientists — psychohistorians — who are supposed to save the galaxy from chaos.[4] And Kahn and von Neumann, RAND's most celebrated thinkers, were among the alleged models for Dr. Strangelove.[5] Although its heyday lasted a decade or less, RAND's way of looking at human conflict not only shaped America's defense in the second half of the century but also made a deep and lasting impression on American social science. RAND had its roots in World War II, when the American military, for the first time in its history, had recruited legions of scientists, mathematicians, and economists and used them to help win the war. As Fred Kaplan writes of RAND's role in nuclear strategy,[6]

> [World War II was] a war in which the talents of scientists were exploited to an unprecedented, almost extravagant degree. First, there were all the new inventions of warfare — radar, infrared detection devices, bomber aircraft, long-range rockets, torpedoes with depth charges, as well as the atomic bomb. Second, the military had only the vaguest of ideas about how to use these inventions. . . . Someone had to devise new techniques for these new weapons, new methods of assessing their effectiveness and the most efficient way to use them. It was a task that fell to the scientists.

Initially, the scientists worked on narrow technical problems — for example, how to build the bomb, how deep to set the charges, the choice of targets. But when it became clear that people didn't know the best way to use this incredibly expensive and destructive weaponry, they were increasingly drawn into discussions of strategy.

The advent of the bomb turned the temporary wartime partnership between the military and the scientific establishment into a continuing relationship. The Air Force, which controlled the new weaponry, emerged after the war as the linchpin of the national defense. "Whole conceptions of modern warfare, the nature of international relations, the question of world order, the function of weaponry, had to be thought through again. Nobody knew the answers," Kaplan writes.[7] Again the military turned to the academic community. As Oskar Morgenstern, also a RAND consultant during the 1950s, put it in his book on defense issues: "Military matters have become so complex and so involved that the ordinary experience and training of the generals and admirals were no longer sufficient to master the problems. . . . More often than not their attitude is, 'here is a big problem. Can you help us?' And this is not restricted to the making of new bombs, better fuel, a new guidance system or what have you. It often comprises tactical and strategic use of the things on hand and the things only planned."[8] *Fortune* magazine put it more succinctly: "If World War II was a war of weapons, another conflict would include on both sides a war of wits at the highest level of knowledge."[9]

In the final days of the war, the Air Force generals began to worry about the brain drain of top scientists.[10] How to keep the best and brightest thinking about military problems was far from obvious. Men of the caliber of John von Neumann would hardly sign up for the civil service. But scientists would have to have access to secrets so one couldn't just rely on contracts with universities. The solution was a private nonprofit organization outside the military but with close ties to the Air Force. In the fall of 1945, General Henry "Hap" Arnold promised to give Douglas Aircraft $10 million of leftover wartime procurement funds for a research venture to be called Project RAND (for "research and development," though wits later insisted the acronym stood for "research and nondevelopment"). The project was housed on the third floor of Douglas's Santa Monica plant. Friction between Douglas and the new entity led to a spinoff as a private nonprofit corporation in 1946, which was when RAND moved to its downtown offices.

RAND's Air Force contract gave it an amazingly free hand, according to William Poundstone's history of RAND. The contract called for research on intercontinental warfare, which, given the dominant role of nuclear weaponry, effectively gave RAND an unrestricted license to roam over the front lines of the U.S. defense strategy. Within these guidelines, RAND scientists could study anything that interested them. RAND could also refuse specific studies requested by the Air Force.

From the beginning, RAND's work was a curious mix of narrowly focused engineering, cost-benefit studies, and blue-sky conjecture. A now-famous 1946 study, completed more than a decade before the launch of *Sputnik* in 1957, proved remarkably prescient. In "Preliminary Design of an Experimental World-Circling Spaceship," RAND scientists argued that "the nation which first makes significant achievements in space travel will be acknowledged as the world leader in both military and scientific techniques. To visualize the impact on the world, one can imagine the consternation and admiration that would be felt here if the US were to discover suddenly that some other nation had already put up a successful satellite."[11]

RAND's civilian scientists soon made a mark on American defense policy. Poundstone reports that RAND played a leading role in the development of the ICBM; RAND convinced the Air Force to adopt in-flight refueling of jet bombers; it was responsible for the fail-safe protocol whereby bombers are kept in the air at all times and during a crisis head for targets in an enemy nation. Its worry that a psychotic individual in a position of power could trigger a nuclear war convinced the Air Force to adopt a safer button that required cooperation of several individuals to arm and detonate a nuclear warhead.

To be plucked from academe and initiated into the secret world of the military had become something of a rite of passage for the mathematical elite. In World War II, the very best had traveled into the New Mexico desert to Los Alamos to work on the A-bomb alongside von Neumann, and to Bletchley Park north of London to help Turing and his team break the Nazi code.[12] Many others, less well known or simply younger, wound up at dozens of less famous sites working on weapon design, encryption, bomb targeting, and submarine chases.[13]

The recruitment of scientists by the military hadn't stopped when the war ended, much to everyone's surprise. Many of the mathematicians and scientists did not return to their quiet prewar routines but instead took on military research contracts, made frequent visits to the Pentagon and the Atomic Energy Commission, and, in a few cases, stayed on at Los Alamos and the other government weapons labs. For an elite cadre of applied mathematicians, computer engineers, political scientists, and economists RAND was the equivalent of Los Alamos.[14]

The problems the military asked the scientists to solve called for new theories and new techniques, which in turn attracted the top scientific talent on which RAND's credibility depended. "We had so many practical problems that involved mathematicians and we didn't have the right tools," said Bruno Augenstein, a former RAND vice-president, years later. "So we had to invent or perfect the tools."[15] Mostly, according to Duncan Luce, a psychologist who was a consultant at RAND, "RAND capitalized on ideas that surfaced during the war."[16] These were scientific, or at least systematic, approaches to problems that had been previously considered the exclusive province of men of "experience." They included such topics as logistics, submarine research, and air defense. Operations research, linear programming, dynamic programming, and systems analysis were all techniques that RAND brought to bear on the problem of "thinking the unthinkable." Of all the new tools, game theory was far and away the most sophisticated.

The spirit of quantification, however, was contagious, and it was at RAND, more than anywhere else, that game theory in particular and mathematical modeling in general entered the mainstream of postwar thinking in economics. At that point, the military was the only government sponsor of pure research in the social sciences — a role later taken over by the National Science Foundation — and it bankrolled a great many ideas that turned out to have little true relevance for the military but a great deal for other endeavors. RAND attracted a younger generation of mathematically sophisticated economists who embraced the new methods and tools, including the computer, and attempted to turn economics from a branch of political philosophy into a precise, predictive science.

Take Kenneth Arrow, one of the early Nobel Laureates in economics. When Arrow came to RAND in 1948, he was an unknown youngster.[17] His famous thesis, written in the as-yet-unfamiliar language of symbolic logic, was a product of a RAND assignment. The assignment was to demonstrate that it was okay to apply game theory, which is formulated in terms of individuals, to aggregations of many individuals, namely nations. Arrow was asked to write a memorandum showing how it could be done. As it turned out, the memorandum became Arrow's dissertation, an attempt to restate the theories of British economist John Hicks in modern mathematical language. "That was it! It took about five days to write in September 1948," he recalled. "When every attempt failed I thought of the impossibility theorem."[18] Arrow showed that it is logically impossible to add up the choices of individuals into an unambiguous social choice not just under a constitution based on the principle of majority rule, but under every conceivable constitution except dictatorship. Arrow's theorem, along with his proof of the existence of a competitive equilibrium, which also owes something to Nash, earned him the Nobel Prize in 1972 and ushered in the use of sophisticated mathematics in economic theory.

Other giants of modern economics who did seminal work at RAND in the early 1950s included Paul A. Samuelson, probably the most influential economist of the twentieth century, and Herbert Simon, who pioneered the study of decision-making inside organizations.

RAND's location was part of its allure. The corporation's headquarters, in a once-sleepy beach colony, lies five miles to the south of the Santa Monica Mountains at the far end of the Malibu Crescent, just west of Los Angeles. In the early 1950s, Santa Monica looked the way Nash imagined that certain towns in Italy or France might look. Wide avenues were lined with pencil-thin palm trees. Cream-colored houses were topped with tiled roofs and encircled by shoulder-high walls. Seaside hotels and rest homes were across from a seaside promenade. The magentas and reds of the bougainvillea and hibiscus were improbably intense. The breeze, surprisingly cool, smelled of oleander and seawater. Some of the best work was done in beach chairs.

RAND itself was tucked out of sight of the ocean on Fourth and Broadway at the edge of Santa Monica's slightly rundown business district. The 1920s bank building was a white stucco affair ornamented with Victorian flourishes. The building had recently housed the presses of the *Santa Monica Evening Outlook;* the newspaper had moved catty-corner to a former Chevy dealership when RAND moved in. By 1950, RAND was already spilling over into several annexes located over storefronts, including ones occupied by the *Outlook* and a bicycle shop. A year later, when *Fortune* magazine discreetly introduced RAND to the wider public, it described "bright walls shining through fog-sunny days and its wide, white-lighted windows shining on uninterruptedly through the night. The building is never closed, nor is it ever really open."[19]

It was one of the most difficult buildings in the United States to get into, *Fortune* said. On Nash's first day, members of RAND's uniformed, armed police force stood guard in front of the building and in its lobby, scrutinizing him closely and memorizing his face.[20] After that, for the rest of the summer and in subsequent years, the guards always greeted him with a cool, respectful "Hello, Dr. Nash." There were no ID cards in those days. Inside were a series of locked doors, with offices clustered by types of security clearance needed to gain access to them. The math division occupied a group of small private offices in the middle of the first floor, upstairs from the electronics shop where von Neumann's new computer, the Johnniac, stood.[21] Nash got an office to himself, a small windowless cubicle whose walls didn't quite extend to the ceiling, with a desk, blackboard, fan, and, of course, a safe.

RAND bristled with self-confidence, a sense of mission, an esprit de corps.[22] Military uniforms signaled visitors from Washington. Executives from defense firms came for meetings. The consultants, mostly under thirty, carried briefcases, smoked pipes, and walked around looking self-important. Big shots like von Neumann and Herman Kahn had shouting matches in the hallways.[23] There was a feeling around the place of "wanting to outrun the enemy," as a former RAND vice-president later put it.[24] Arrow, who was an army veteran from the Bronx, said, "We were all convinced that the mission was important though there was lots of room for intellectual vision." [25]

RAND's sense of mission was propelled largely by a single fact: Russia had the A-bomb. That shocking news had been delivered by President Truman the previous fall, a mere four years after Nagasaki and Hiroshima, and many years before Washington had expected it. The military had hard evidence, the president said in a speech on September 13, 1949, of a nuclear explosion deep inside the Soviet Union.[26] Nobody in the scientific community, especially around Princeton, where von Neumann and Oppenheimer were engaged in an almost daily debate over the wisdom of pushing ahead with the Super, doubted that the Soviets were capable of developing nuclear weapons.[27] The shock was that they had succeeded so quickly. Physicists and mathematicians, who were less convinced of Russia's scientific and technological backwardness, had been warning the administration all along that predictions by senior government officials that America's nuclear monopoly would persist another ten, fifteen, or twenty years were hopelessly naive, but the sense of being caught off guard was still very great.[28] The news effectively ended the debate over the hydrogen bomb more or less immediately. By the time the president delivered the news of the Soviet explosion to the public, he had authorized a crash program at Los Alamos to design and manufacture an H-bomb.[29]

It was unthinkable that such destructive power would be unleashed. Therefore RAND insisted that it was necessary to ponder the possibility.[30] The rational life was worshiped to an almost absurd degree. RAND was full of men and women committed to the idea that systematic thought and quantification were the key to the most complex problems. Facts, preferably detached from emotion, convention, and preconception, reigned supreme. If reducing complex political and military choices, including the problem of nuclear war, to mathematical formulae could produce light, why then the same approach must be good for more mundane matters. RAND scientists tried to tell their wives that the decision whether to buy or not to buy a washing machine was an "optimization problem."[31]

RAND was privy to the military's most highly guarded secrets at a time when the nation was growing increasingly nervous about the safeguarding of those secrets to the point of paranoia. From the summer of 1950 on, RAND would be increasingly affected by the growing alarm over Russian access to American military secrets.[32] It began with the Fuchs trial in the winter of 1950.[33] Fuchs was a German émigré scientist who had fled to Britain during the war and eventually wound up working with von Neumann and Edward Teller at Los Alamos. A clandestine member of the British Communist Party, Fuchs subsequently confessed in January 1950 to passing atomic secrets to the Russians and was tried and convicted in London that February. Senator Joseph McCarthy had embarked that same month on his anticommunist campaign, accusing the federal government of security breaches.[34] Four years later, in April of 1954, Robert Oppenheimer, the former head of the Manhattan Project, the director of the Institute for Advanced Study, and the most famous scientist in America, was declared a security risk by Eisenhower and stripped of his security clearances in the full glare of national publicity.[35] The ostensible reason was Oppenheimer's youthful left-wing associations, but the real reason, as von Neumann and most scientists testified at the time, was Oppenheimer's refusal to support the development of the H-bomb.

The fact that McCarthy himself ultimately became a target of censure would do little to dispel the atmosphere of paranoia and intimidation at RAND, which lived on Air Force and AEC money and had projects on the H-bomb and ICBMs.[36] Most of what the mathematicians worked on was not in fact classified, but that didn't matter. RAND, which harbored a collection of oddballs like Richard Bellman (a former Princeton mathematician who had all kinds of communist associations, mostly accidental, including a chance encounter with a cousin of Julius and Ethel Rosenberg), would become particularly careful about minding its Ps and Qs.[37]

Everybody needed a top-secret clearance. People who arrived without a temporary security clearance were banished to "quarantine" or "preclearance" and weren't permitted to sit with everybody else. Nash's secret clearance was granted on October 25, 1950.[38] His recollection that he had a top-secret clearance — which a large contingent in the math division did have — is probably faulty. Nash also recalls that he applied for a Q clearance in 1952.[39] Any consultant to the math division who worked on Atomic Energy Commission contracts was required to have a Q clearance because of access to documents related to the construction and use of nuclear weapons. But despite a November 10, 1952, postcard to his parents telling them that he had applied for a higher clearance at RAND, Nash now says it was never approved — meaning that his role at RAND was largely confined to highly theoretical excercises as opposed to applications of game theory concepts to actual questions of nuclear strategy — the province of men like von Neumann, Herman Kahn, and Thomas Schelling.[40]

Everyone had a safe in his office for storing classified documents, and everyone was warned about taking documents out of the building or talking out of school.[41] Papers had to be put in the safes at the end of every day. There were spot checks. There was a public address system and there were parts of the building that were off-limits to people who didn't have a Q clearance.

By 1953, soon after Eisenhower issued a new set of security guidelines, security consciousness, in the sense of not overlooking anyone who might be thought remotely unreliable, grew.[42] The Eisenhower guidelines broadened the grounds for denying a clearance or stripping someone of an existing clearance. Without a doubt, fear about potential leaks brought to a boil many simmering antagonisms against individuals and groups who posed little or no actual threat to security. Almost any sign of nonconformity, political or personal, came to be considered a potential security breach. The notion, for example, that homosexuals were unreliable, because of either poor judgment or vulnerability to blackmail, was first codified in the Eisenhower guidelines.

Like the decade itself, RAND had a split personality. Its style was informal. It tolerated quirky people. It was in some ways more democratic than a university. Almost everyone, including von Neumann, was called by his or her first name, except by the guards, never Doctor or Professor or Sir. Graduate students rubbed shoulders with full professors in a way unimaginable in most academic departments. RAND's president, a former Douglas Aircraft executive, was a spit-and-polish man who was almost never seen in a suit and tie. All but one or two of the mathematicians, including Nash, came to work in short-sleeved shirts. Appearances were so casual that one mathematician, who found it all very déclassé, felt obliged to rebel by wearing a three-piece suit and a tie to the office every day.[43]

Practical jokes were as much a part of the RAND culture as pipes and crew-cuts. Mathematicians and physicists mixed rubber bands into the pipe tobacco, substituted dog biscuits for cookies, and tilted desks so pencils rolled onto the floor.[44] Wit was greatly appreciated. When John Williams, the head of RAND's mathematics department, wrote a primer on game theory, published as a RAND study, it was illustrated with funny little cartoon figures and full of jokey examples starring John Nash, Alex Mood, Lloyd Shapley, John Milnor, and other members of the math department.[45]

The mathematicians were, as usual, the freest spirits.[46] They had no set hours. If they wanted to come into their offices at 3:00 A.M., fine. Shapley, who had come back from Princeton for the summer and continued to insist on the sanctity of his sleep cycle, was rarely seen before midafternoon. Another man, an electrical engineer named Hastings, typically slept in the "shop" next to his beloved computer.

Lunches were long, much to the annoyance of RAND's engineers, who prided themselves on sticking to a more respectable routine. The mathematicians mostly took their bag lunches to a conference room and pulled out chessboards. They invariably played Kriegspiel, usually in total silence, occasionally punctuated by a wrathful outburst from Shapley, who frequently lost his temper over an umpire's or opponent's error. Even though the games typically lasted well into the afternoon, they were rarely finished and finally reluctantly abandoned midgame. Poker and bridge groups met after hours.

There were no afternoon teas, formal seminars, or faculty meetings at RAND. Unlike the physicists and engineers, the mathematicians usually worked alone. The idea was that they would work on their own ideas but would help solve the myriad problems encountered by researchers, picking up problems to solve as the spirit moved them.[47] People would drift into each other's offices or, more frequently, simply stop to chat in the corridors near the coffee stations. The grids and courtyards of RAND's permanent headquarters — to which the mathematics group moved in 1953, the year before Nash's final summer at RAND — were designed, by John Williams, as it happens, "to maximize chance meetings."[48] Through such encounters new research was "announced" and mathematicians got hooked on problems that colleagues in other departments wanted solved. Most of the work wasn't reported formally, and even when it was published as RAND memoranda, there was no formal approval process. A consultant would simply go to the math department secretaries, hand over a handwritten paper, and a day or two later a RAND memorandum would appear.[49] Published reports for outside circulation didn't go through a much more rigorous vetting process.

This copacetic atmosphere was mostly Williams's doing.[50] Witty and charming, weighing close to three hundred pounds, expensively suited, Williams looked like a businessman always about to reach into his pocket to pull out a wad of twenties. An astronomer from Arizona who had spent a couple of years in Princeton attending lectures in Fine Hall, playing poker, and developing an enthusiasm for the theory of games, Williams had been a dollar-a-year man in Washington during the war and became RAND's fifth employee afterward. Williams hated flying. He loved fast cars. At one point, he spent an entire year outfitting his chocolate-brown Jaguar with a powerful Cadillac engine. It had taken substantial RAND resources (RAND had a repair shop) and considerable bravado to install the thing. Cadillac and Jaguar mechanics had both dismissed the idea as impractical, but Williams had prevailed. He disproved the mechanics' conventional wisdom in late-night, 125-mile-an-hour drives along the Pacific Coast Highway.

Williams's approach to management would have made him very much at home in Silicon Valley today: "Williams had a theory," recalled his deputy, Alexander Mood, also a former Princetonian. "He believed people should be left alone. He was a great believer in basic research. He was a very relaxed administrator. That's why people thought the math division was pretty weird."[51] Williams's letter to von Neumann offering the mathematician a two-hundred-dollar-a-month retainer conveys the man's style. The letter said, "The only part of your thinking we'd like to bid for systematically is that which you spend shaving: we'd like you to pass on to us any ideas that come to you while so engaged."[52] When Williams first arrived, RAND was a tiny annex inside a mammoth Douglas Aircraft factory where thirty thousand workers punched time cards every day. Williams was the one who freed the mathematicians from the clock and then proceeded to demand coffee and blackboards for his mathematicians, explaining that not providing these would guarantee that none of them would produce anything worthwhile. After RAND and Douglas Aircraft parted company, Williams went further. He insisted that the building be open twenty-four hours a day instead of just between eight and five. He got private offices. He set up coffee stations that had their own special full-time maintenance crew. He mollified the engineers and the Air Force generals, who wondered why the hell the mathematicians had to be allowed to be themselves.

Everyone soon knew Nash by sight. He roamed the halls incessantly.[53] He was usually chewing an empty paper coffee cup that was clamped firmly between his teeth. He would glide through the corridors for hours at a time, frowning, lost in thought, shirt untucked, his powerfully built shoulders hunched forward, his sharp Nixonian nose leading the way. Sometimes he wore a small, ironic smile that suggested some secret amusement not likely to be shared with anyone he might encounter. When he did meet someone he knew, he rarely greeted him by name or even acknowledged his presence unless spoken to first, and then not always. When he wasn't chewing a coffee cup, he whistled, often the same tune, from Bach's *The Art of the Fugue,* over and over again.[54]

His legend had preceded him. In the eyes of his new colleagues, Arrow recalled, Nash was "a young genius who could do anything, a guy who liked solving problems."[55] Mathematicians who were struggling with tricky problems quickly learned to collar him by planting themselves squarely in his path. Nash's curiosity was easily piqued, they discovered, provided that the problem struck him as interesting and the speaker mathematically competent. He was usually more than willing to step around to their offices to look at masses of messy equations on their blackboards.

Williams's deputy, Alex Mood, was one of the first to try.[56] A gentle giant of a man with a dry wit and easy manner, Mood happened to be oppressed by a problem left over from a first, ill-fated thesis attempt at Princeton before the war. He had found a better derivation of a famous solution, he felt, but his proof was overly long, too complicated, and distressingly inelegant. Could Nash come up with something "shorter, simpler"? Nash listened and stared, frowned and walked away. But the very next day, he was back at Mood's door with a clever and entirely unanticipated solution. Nash had "sidestepped the whole induction by regarding integers as variables and sending them to revealing limits." As much as anything else, Mood was charmed by Nash's style. "When he found a problem," Mood recalled, "he sat down and started attacking it immediately. He didn't, like some of his colleagues, browse through the library to see what related stuff had already been done."

Williams too was immediately taken with Nash and took him under his wing. He frequently told others that Nash had greater insight into mathematical structure than any mathematician he had ever known, an extraordinary remark from a man who spent the late 1930s in Fine Hall and was an intimate of von Neumann's. "He knew which factors of a hundred thousand were the most important," Williams used to say.[57] He liked to describe how Nash would come into an office, stare at a blackboard dense with equations, and stand there silently, meditating. "Then," Williams would say, "he'd solve the whole thing. He could *see* the structure."

However, Nash mostly kept to himself. He talked about his own research rarely and then only with a select few. When he did, it was not usually because he was looking for help. "It wasn't so much that he sought advice," another consultant recalled. "You were a reflecting mirror. He was his own creative object."[58] The only person he regularly sought out at RAND was Shapley, and fairly soon people around the mathematics division started to think of the two as a pair, RAND's Wunderkinder.

Still, Nash's eccentricity soon became fodder for RAND's gossip mill. "He reinforced RAND's idea that mathematicians were a bit crazy," Mood said.[59] His office, in which he could rarely be found, was a godawful mess. When he left at the end of that summer he did so without bothering to clean out his desk. The staffer who was saddled with the chore found, among other things, "banana peels. Bank statements for Swiss bank accounts with thousands of dollars in them. One or two hundred dollars in cash. Classified documents. The C-1 isometric embedding paper."[60]

Some people found Nash absurdly childish. He was fond of playing adolescent jokes on his colleagues. Knowing that his whistling irritated one particular music-loving mathematician, who frequently asked him to stop, he once left behind a recording of his whistling on the man's Dictaphone.[61] RAND's blue-collar police force and maintenance crew found Nash an entertaining subject. They would watch him as he left the building walking north on Fourth Avenue. On several occasions some of them complained to a RAND manager that they had seen Nash tiptoing exaggeratedly along the avenue, stalking flocks of pigeons, and then suddenly rushing forward, "trying to kick 'em."[62]

13

Game Theory at RAND

We hope [the theory of games] will work, just as we hoped in 1942 that the atomic bomb would work. — ANONYMOUS PENTAGON SCIENTIST *to* Fortune, *1949*

NASH'S NOVEL IDEA about games with many players had preceded him at RAND by several months. The first version of his elegant proof of the existence of equilibrium for games with many players—two skimpy pages in the November 1949 issue of the National Academy of Sciences proceedings—swept through the white stucco building at Fourth and Broadway like a California brushfire.[1]

The biggest appeal of the Nash equilibrium concept was its promise of liberation from the two-person zero-sum game. The mathematicians, military strategists, and economists at RAND had focused almost exclusively on games of total conflict—my win is your loss or vice versa—between two players. Shapley and Dresher's 1949 review of game theory research at RAND refers to the organization's "preoccupation with the zero-sum two person game."[2] That preoccupation was natural, given that these were games for which the von Neumann theory was both sound and reasonably complete. Zero-sum games also seemed to fit the problem—nuclear conflict between two superpowers—which absorbed most of RAND's attention.

Only it really didn't. At least some of the researchers at RAND were already chafing at the central assumption of a fixed payoff in such games, Arrow recalled.[3] As weapons got ever more destructive, even all-out war had ceased to be a situation of pure conflict in which opponents had no common interest whatever. Inflicting the greatest amount of damage on an enemy—bombing him back to the Stone Age—no longer made any sense, as American strategists realized during the final phase of the campaign against Germany when they decided not to destroy the coal mines and industrial complexes of the Ruhr.[4] As Thomas C. Schelling, one of RAND's nuclear strategists, would put it a decade later,[5]

In international affairs, there is mutual dependence as well as opposition. Pure conflict, in which the interests of two antagonists are completely opposed, is a special case; it would arise in a war of complete extermination, otherwise not even in war. The possibility of mutual accommodation is as important and dramatic as the element of conflict. Concepts like deterrence, limited war, and disarmament, as well as negotiation, are concerned with the common interest and mutual dependence that can exist between participants in a conflict.

Schelling goes on to say why this is so: "These are games in which, though the element of conflict provides the dramatic interest, mutual dependence is part of the logical structure and demands some kind of collaboration or mutual accommodation—tacit, if not explicit—even if only in the avoidance of mutual disaster."[6]

In 1950, at least the economists at RAND were aware that if game theory were to evolve into a descriptive theory that could be usefully applied to real-life military and economic conflicts, one had to focus on games that allowed for cooperation as well as conflict. "Everybody was already bothered by the zero-sum game," Arrow recalled."You're trying to decide whether to go to war or not. You couldn't say that the losses to the losers were gains to the winner. It was a troublesome thing."[7]

Military strategists were the first to seize on the ideas of game theory. Most economists ignored *The Theory of Games and Economic Behavior* and the few that didn't, like John Kenneth Galbraith writing in *Fortune* and Carl Kaysen, later director of the Institute for Advanced Study, turn out to have had significant contact with military strategists during the war.[8] An article in *Fortune* in 1949 by John McDonald made it clear that the military hoped to use von Neumann's theory of games to work out intelligence missions, bombing patterns, and nuclear defense strategy.[9] On the lookout for new ideas and with plenty of money to spend, the Air Force embraced game theory with the same enthusiasm with which the Prussian military had embraced probability theory a couple hundred years earlier.[10]

Game theory had already made its debut in military planning rooms. It had been used during the war to develop antisubmarine tactics when German submarines were destroying American military transports. As McDonald reported in *Fortune:*[11]

The military application of "Games" was begun early in the last war, some time in fact before the publication of the complete theory, by ASWOEG (Anti-Submarine Warfare Operations Evaluation Group). Mathematicians in the group had got hold of von Neumann's first paper on poker, published in 1928.

But von Neumann actually spent his frenetic visits to Santa Monica almost exclusively with the computer engineers and the nuclear scientists.[12] His enormous prestige and Williams's deft salesmanship led to a major concentration on game theory at RAND from 1947 into the 1950s. The hope was that game theory would provide the mathematical underpinning for a theory of human conflict and spread to disciplines other than mathematics. Williams convinced the Air Force to let

RAND create two new divisions, economics and social science. By the time Nash arrived, a "trust" of game theory research had grown up at RAND including such game theorists as Lloyd S. Shapley, J. C. McKinsey, N. Dalkey, F. B. Thompson, and H. F. Bohnenblust, such pure mathematicians as John Milnor, statisticians David Blackwell, Sam Karlin, and Abraham Girschick, and economists Paul Samuelson, Kenneth Arrow, and Herbert Simon.[13]

Most of the RAND military applications of game theory concerned tactics. Air battles between fighters and bombers were modeled as duels.[14] The strategic problem in a duel is one of timing. For each opponent, having the first shot maximizes the chance of a miss. But having the better shot also maximizes the chance for being hit. The question is when to fire. There's a tradeoff. By waiting a little longer each opponent improves his own chance of scoring a hit, but also increases the risk of being shot down. Such duels can be both noisy and silent. With "silent guns," the duelist doesn't know the other has fired unless he is hit. Therefore, neither participant knows whether the other still has a bullet or has fired and missed and is now defenseless.

A report by Dresher and Shapley summarizing RAND's game theory research between the fall of 1947 and the spring of 1949 gives the flavor.[15] The mathematicians describe a problem of staggered attacks in a bombing mission:

> Problem
> A single intercepter base, having I fighters, is located on a base line. Each fighter has a given endurance. If a fighter, vectored out against a bomber attack, has not yet engaged his original target, then at the option of the ground controller he may be vectored back to engage a second attack.
> The attacker has a stock of N bombers and A bombs. The attacker chooses two points to attack and sends N_1 bombers including A_1 bomb carriers on the first attack and t minutes later he sends $N_2 = N - N_1$ bombers including $A_2 = A - A_1$ carriers on the second attack.
> The payoff to the attacker is the number of bomb carriers that are not destroyed by the fighters.
> Solution
> Both players have pure optimal strategies. An optimal strategy of the attacker is to attack both targets simultaneously and distribute the A bomb carriers in proportion to the number of bombers in each attack. An optimum strategy of the defender is to dispatch interceptors in proportion to the number of attacking bombers and not to revector fighters. The value of the game to the attacker will be
> $$V = \max (0, A(1 - 1/Nk))$$
> where k is the kill probability of the fighter

The game Nash had in mind could be solved without communication or collaboration. Von Neumann had long believed that the RAND researchers ought to focus on cooperative games, conflicts in which players have the opportunity to communicate and collaborate and are able "to discuss the situation and agree on a rational joint plan of action, an agreement that is assumed to be enforceable." [16] In cooperative games, players form coalitions and reach agreements. The key assumption is that there's an umpire around to enforce the agreement. The mathematics of cooperative games, like the mathematics of zero-sum games, is rich and elegant. But most economists, like Arrow, were cool to the idea.[17] It was like saying, they thought, that the only hope for preventing a dangerous and wasteful nuclear arms race lay in appointing a world government with the power to enforce simultaneous disarmament. World government, as it happens, was a popular idea among mathematicians and scientists at the time. Albert Einstein, Bertrand Russell, and indeed much of the world's intellectual elite subscribed to some version of "one worldism." [18] Even von Neumann tipped his hat to the notion, conservative hawk that he was. But most social scientists were dubious that any nation, much less the Soviets, would cede sovereignty to such an extent. Cooperative game theory also seemed to have little relevance to most economic, political, and military problems. As Arrow jokingly put it, "You did have cooperative game theory. But I couldn't force the other side to cooperate." [19]

By demonstrating that noncooperative games, games that did not involve joint actions, had stable solutions, said Arrow, "Nash suddenly provided a framework to ask the right questions." At RAND, he added, it immediately led "a lot of people to calculate equilibrium points."

News of Nash's equilibrium result also inspired the most famous game of strategy in all of social science: the Prisoner's Dilemma. The Prisoner's Dilemma was partly invented at RAND, some months before Nash arrived, by two RAND mathematicians who responded to Nash's idea with more skepticism than appreciation of the revolution that Nash's concept of a game would inspire.[20] The actual tale of prisoners used to illustrate the game's significance was invented by Nash's Princeton mentor, Al Tucker, who used it to explain what game theory was all about to an audience of psychologists at Stanford.[21]

As Tucker told the story, the police arrest two suspects and question them in separate rooms.[22] Each one is given the choice of confessing, implicating the other, or keeping silent. The central feature of the game is that no matter what the other suspect does, each (considered alone) would be better off if he confessed. If the other confesses, the suspect in question ought to do the same and thereby avoid an especially harsh penalty for holding out. If the other remains silent, he can get especially lenient treatment for turning state's witness. Confession is the dominant strategy. The irony is that both prisoners (considered together) would be better off if neither confessed — that is, if they cooperated — but since each is aware of the other's incentive to confess, it is "rational" for both to confess.

Since 1950, the Prisoner's Dilemma has spawned an enormous psychology literature on determinants of cooperation and defection.[23] On a conceptual level, the game highlights the fact that Nash equilibria — defined as each player's following his best strategy assuming that the other players will follow their best strategy — aren't necessarily the best solution from the vantage point of the group of players.[24] Thus, the Prisoner's Dilemma contradicts Adam Smith's metaphor of the Invisible Hand in economics. When each person in the game pursues his private interest, he does not necessarily promote the best interest of the collective.

The arms race between the Soviet Union and the United States could be thought of as a Prisoner's Dilemma. Both nations might be better off if they cooperated and avoided the race. Yet the dominant strategy is for each to arm itself to the teeth. However, it doesn't appear that Dresher and Flood, Tucker, or, for that matter, von Neumann, thought of the Prisoner's Dilemma in the context of superpower rivalry.[25] For them, the game was simply an interesting challenge to Nash's idea.

The very afternoon that Dresher and Flood learned of Nash's equilibrium idea, they ran an experiment using Williams and a UCLA economist, Armen Alchian, as guinea pigs.[26] Poundstone says that Flood and Dresher "wondered if real people playing the game — especially people who had never heard of Nash or equilibrium points — would be drawn mysteriously to the equilibrium strategy. Flood and Dresher doubted it. The mathematicians ran their experiment one hundred times."

Nash's theory predicted that both players would play their dominant strategies, even though playing their dominated strategies would have left both better off. Though Williams and Alchian didn't always cooperate, the results hardly resembled a Nash equilibrium. Dresher and Flood argued, and von Neumann apparently agreed, that their experiment showed that players tended not to choose Nash equilibrium strategies and instead were likely to "split the difference."

As it turns out, Williams and Alchian chose to cooperate more often than they chose to cheat. Comments recorded after each player decided on strategy but before he learned the other player's strategy show that Williams realized that players ought to cooperate to maximize their winnings. When Alchian didn't cooperate, Williams punished him, then went back to cooperating next round.

Nash, who learned of the experiment from Tucker, sent Dresher and Flood a note — later published as a footnote in their report — disagreeing with their interpretation:[27]

> The flaw in the experiment as a test of equilibrium point theory is that the experiment really amounts to having the players play one large multi-move game. One cannot just as well think of the thing as a sequence of independent games as one can in zero-sum cases. There is too much interaction. . . . It is really striking however how inefficient [Player One] and [Player Two] were in obtaining the rewards. *One would have thought them more rational.*

Nash managed to solve a problem at RAND that he and Shapley had both been working on the previous year. The problem was to devise a model of negotiation between two parties — whose interests neither coincided nor were diametrically opposed — that the players could use to determine what threats they should use in the process of negotiating. Nash beat Shapley to the punch. "We all worked on this problem," Martin Shubik later wrote in a memoir of his Princeton experiences, "but Nash managed to formulate a good model of the two-person bargain utilizing threat moves to start with."[28]

Instead of deriving the solution axiomatically — that is, listing desirable properties of a "reasonable" solution and then proving that these properties actually point to a unique outcome — as he had in formulating his original model of bargaining, Nash laid out a four-step negotiation.[29] Stage One: Each player chooses a threat. This is what I'll be forced to do if we can't make a deal, that is, if our demands are incompatible. Stage Two: The players inform each other of the threats. Stage Three: Each player chooses a demand, that is, an outcome worth a certain amount to him. If the bargain doesn't guarantee him that amount, he won't agree to a deal. Stage Four: If it turns out that a deal exists that satisfies both players' demands, the players get what they ask for. Otherwise, the threats have to be executed. It turns out that the game has an infinite number of Nash equilibria, but Nash gave an ingenious argument for selecting a unique stable equilibrium that coincides with the bargaining solution he previously derived axiomatically. He showed that each player had an "optimal" threat, that is, a threat that ensures that a deal is struck no matter what strategy the other player chooses.

Nash initially wrote up his results in a RAND memorandum dated August 31, 1950, suggesting that he managed to finish the paper just before leaving RAND for Bluefield.[30] A longer and more descriptive version of the paper was eventually accepted by *Econometrica,* which had published "The Bargaining Problem" that April. Accepted for publication sometime during the following academic year, "Two Person Cooperative Games" did not in fact appear until January 1953.[31] It was Nash's last significant contribution to the theory of games.

Nobody at RAND solved any big new problems in the theory of noncooperative games. For all intents and purposes, Nash stopped working in the field in 1950. The dominant thrust of game theory at RAND came from the mathematicians, particularly Shapley, and they were guided less by applications than by the mathematics themselves. During the 1950s Shapley focused on cooperative games, which were necessarily of limited interest not only to economists but also to military strategists.

The justification of all mathematical models is that, oversimplified, unrealistic, and even false as they may be in some respect, they force analysts to confront possibilities that would not have occurred to them otherwise. The history of physics and medicine abounds with wrong or incomplete theories that throw just enough light to allow some other big breakthroughs. The atom bomb, for example, was built before physicists understood the structure of particles.

The most significant application of game theory to a military problem grew straight out of the theory of duels and helped shape what was probably RAND's single most influential strategic study. The study was the brainchild of Al Wohlstetter, a mathematician who joined RAND's economics group in early 1951, about six months after Nash joined the mathematics group.

According to Kaplan, the SAC operational plan in the early 1950s was to fly bombers from the United States to overseas bases and then to mobilize and launch an attack against the Soviet Union from there.[32] The Air Force's whole deterrence strategy was based on the idea of the power of the H-bomb and America's ability to respond in kind to any attack. Apparently, no one before Wohlstetter had focused on vulnerability to a first strike aimed, not at American cities, but at wiping out the SAC force, then concentrated in a small number of foreign bases within striking distance of the Soviet Union. Kaplan writes:

> Up to that point, most military applications of game theory had focused on tactics — the best way to plan a fighter-bomber duel, how to design bomber formations or execute anti-submarine warfare campaigns. But Wohlstetter would carry it further. It was this insistence on figuring out one's own best moves in light of the enemy's best moves that provoked Wohlstetter to look at a map and to conclude that the closer we are to them, the closer they are to us — the easier it is for us to hit them, the easier it is for them to hit us. Wohlstetter and his team estimated that a mere 120 bombs . . . could destroy 75 to 85 percent of the B-47 bombers while they casually sat on overseas bases. The SAC, seemingly the most powerful strike force in the world, was appearing to be so vulnerable in so many ways that merely putting the plan into action . . . created a target so concentrated that it invited a pre-emptive attack from the Soviet Union.[33]

Wohlstetter's study had an electrifying effect on the Air Force establishment. With its focus on American vulnerability and the temptation of a Soviet surprise attack, the study also rationalized a paranoia in the military establishment that seeped into the body politic and wound up as national hysteria over the supposed "missile gap" in the second half of the 1950s. The RAND report, Fred Kaplan writes, "legitimized a basic fear of the enemy and the unknown through mathematical calculation and rational analysis, providing the techniques and the general perspective through which the new and rather scary situation — the Soviet Union's acquisition of long range nuclear weapons — could be discussed and acted upon."[34]

The golden age at RAND, from the point of view of the mathematicians, strategic thinkers, and economists, was already coming to a close.[35] After a time, RAND's sponsors grew less enthusiastic about pure research, less tolerant of idiosyncrasies, and more demanding. Mathematicians got bored and frustrated with game theory. Consultants stopped coming and permanent staffers drifted to universities. Nash never returned after the summer of 1954. Flood left for Columbia University in

1953. Von Neumann, who in any case had played a very small role in the group after inspiring it, dropped his RAND consultancy in 1954 when he accepted an appointment as a member of the Atomic Energy Commission.

Game theory, in any case, was going out of vogue at RAND. R. Duncan Luce and Howard Raiffa concluded in their 1957 book, *Games and Decisions:* "We have the historical fact that many social scientists have become disillusioned with game theory. Initially there was a naive band-wagon feeling that game theory solved innumerable problems of sociology and economics, or that, at least it made their solution a practical matter of a few years' work. This has not turned out to be the case." [36] The military strategists were of the same mind. "Whenever we speak of deterrence, atomic blackmail, the balance of terror ... we are evidently deep in game theory," Thomas Schelling wrote in 1960, "yet formal game theory has contributed little to the clarification of these ideas." [37]

14

The Draft

Princeton, 1950–51

NEITHER THE PROSPECT of playing military strategist, nor living in Santa Monica, nor earning a handsome salary tempted Nash to accept Williams's offer of a permanent post at the think tank. Nash shared little of RAND's camaraderie or sense of mission. He wanted to work on his own and to have the freedom to roam all over mathematics. To do that, he would have to obtain a faculty position at a leading university.

For the moment, he planned to spend the upcoming academic year in Princeton. Tucker had arranged for his support by assigning him to teach a section of undergraduate calculus[1] and making him a research assistant on his Office of Naval Research grant.[2] In fact, Nash intended to devote most of his energy to his own research and to looking for an academic opening for the following fall. But before he could turn to these matters, he was forced to confront an immediate threat to his career plans, namely, the Korean War.

North Korea had invaded the South on June 25, 1950, about the time that Nash was flying to Santa Monica.[3] A week later Truman promised to send American troops to repel the invasion. The first reinforcements landed July 19. By July 31, Truman had issued an order to the Selective Service to call up one hundred thousand young men right away, twenty thousand immediately. A week or two later, John Sr. and Virginia wrote that Nash might be in imminent danger of being drafted. Like most Republicans, they disliked Truman and had their doubts about the war. They urged Nash to come to Bluefield as soon as practical to talk with members of the local draft board personally to sound them out about a II-A. Surely, they said, Nash was more valuable at RAND or at Princeton than in uniform.

When Nash left RAND at the very end of August, he flew from Los Angeles to Boston and spent a day at the world mathematical congress, which was meeting in Cambridge.[4] He presented his algebraic manifolds result to a small audience there, a nice distinction for a young mathematician. But he was anxious to get back to Bluefield and didn't stay for most of the meetings.

He was determined to do all he could to avoid the draft. With a war on, even an unpopular and undeclared war, who knew how long he would have to serve? Any interruption of his research could jeopardize his dream of joining a top-ranked mathematics department. Returning World War II veterans had flooded the job market and enrollments were falling because of the draft. In two years there would be another crop of brilliant youngsters clamoring for the handful of instructorships. His game theory thesis had been greeted with a mix of indifference and derision by the pure mathematicians, so his only hope of a good offer, he felt, was to finish his paper on algebraic manifolds.

Besides, he had no wish to become part of someone else's larger design and dreaded the thought of military life — his hawkish instincts and southern background notwithstanding. He had been one of the few boys at Beaver High who hadn't prayed for World War II to last long enough so that he would have a chance to serve. Life in the army, with its mindless regimentation, stultifying routines, and lack of privacy, revolted him, and he had heard enough stories from other mathematicians to dread being herded together with the kind of rude, uneducated young men whose company he had been only too happy to escape when he left Bluefield for Carnegie Tech.

Nash proceeded methodically. Once back in Bluefield, he called on two members of the board, including its chairman, a retired attorney named T. H. Scott, whom he later described as "a rock-ribbed Republican (Truman = moron = Roosevelt)," and a Dr. H. L. Dickason, the president of Bluefield State, a black junior college on the far side of the town.[5] He made it his business to find out as much as he could about the men who would be deciding his fate. As it turned out, the board had only a fuzzy sense of what Nash was doing. Until he showed up at the Peery building, they had no idea that he had already received his doctorate and had assumed he was returning to Princeton that fall as a student. His student deferment had not yet been canceled.

His meeting with Scott did nothing to ease his anxiety. The board was already working through its list of twenty-two-year-olds. Now that the board knew that he was no longer a graduate student, he might very well be in the next call, which was scheduled for the twentieth of the month, less than two weeks away. Nash mentioned that he was doing classified research for the military, and described both his affiliation with RAND and the ONR project at Princeton. Scott did not rule out the possibility of granting an occupational deferment, but he expressed some skepticism that a young mathematician could be indispensable, except in uniform, in a national emergency. Nash felt slightly better about his meeting with Dickason, who had taught math and physics before the war and appeared to be impressed by Nash's Princeton degree and associates. It was probably Dickason who tipped Nash off to the fact that merely filing an application for a II-A, an occupational deferment, would temporarily halt the wheels of the draft machinery and take him out of the pool of potential draftees at least until the board had time to consider his II-A application.

Nash wasted no time. In Bluefield, he went to the library and read the Selective Service law. He thought about the board's psychology. He wrote to Tucker, to the Office of Naval Research in Washington, and no doubt also to Williams at RAND, though there is no record of such a letter.[6] (A letter from the

Office of Naval Research in Washington, received by Al Tucker on September 15, begins, "John Nash has written me asking if ONR can help get him a draft deferment.") Nash asked them to request a II-A deferment, but urged them to state only the bare facts, promising more information later—so that "heavier guns may be later rolled out without the appearance" of merely repeating the initial statements.[7] He was intent on buying as much time as possible. Later on, in other circumstances, Nash would repeatedly express his dislike and resentment of "politics" and "politicking." But, impractical, childish, and detached from everyday concerns as he was in some ways, he was quite capable of plotting strategy, ferreting out necessary facts, making use of his father's connections, and most of all, marshaling allies and supporters.

Tucker, the university, the Navy, and RAND responded sympathetically and promptly, claiming in unison that he was irreplaceable, it would take years to train a substitute, and his work was "essential to the welfare and security of this nation."[8] Fred D. Rigby at the Office of Naval Research in Washington advised Tucker that the best route to take was for a university officer to ask the New York branch of the ONR to write to the Bluefield draft board. "This process is said to work well. Normally, it takes place after the man is put in 1-A, but there is no rule against its use in advance of that event."[9] Rigby also noted that "this kind of question is coming up frequently these days," suggesting that Nash was hardly alone among young academics with Defense Department affiliations seeking to avoid the draft. Rigby also promised that, should the branch office action fail, "we will then make a second try directly with the national selective service organization," adding, however, that in all likelihood "this will not be necessary."[10]

The concerted effort to save Nash from the draft was not much different from similar efforts made for a great many other young scientists at the time. The Korean War did not inspire the same patriotic fervor as World War II.[11] Many academics regarded defense research as a kind of alternative service and the notion of exempting especially accomplished and valuable individuals had antecedents even in World War II.[12] Kuhn remembers trying but failing to join the Navy's V-12 program, which would have allowed him to spend the war attending the same classes at Caltech that he would have attended as a civilian, only in uniform. He wound up in the infantry only because he failed the Navy's tougher physical.[13] Korea did not prompt the massive draft evasion of the Vietnam era, de facto a working-class war, but among a certain elite in Nash's generation there was a sense of entitlement and a lack of embarrassment about obtaining special treatment.

The urgency of Nash's efforts to avoid the draft suggests deeper fears than those related to career ambitions or personal convenience. His was a personality for which regimentation, loss of autonomy, and close contact with strangers were not merely unpleasant, but highly threatening. With some justification, Nash would later blame the onset of his illness partly on the stress of teaching, a far milder form of regimentation than military life. His fear of being drafted remained acute long after the Korean War ended and after he turned twenty-six (the age cut-off for draft eligibility). It eventually reached delusional proportions and helped drive him to attempt to abandon his American citizenship and seek political asylum abroad.

Interestingly, Nash's gut instinct has since been validated by schizophrenia researchers.[14] None of the life events known to produce mental disorders such as depression or anxiety neurosis — combat, death of a loved one, divorce, loss of a job — have ever been convincingly implicated in the onset of schizophrenia. But several studies have since shown that basic military training during peacetime can precipitate schizophrenia in men with a hitherto unsuspected vulnerability to the illness.[15] Although the study subjects were all carefully screened for mental illnesses, hospitalization rates for schizophrenia turned out to be abnormally high, especially for draftees.

Rigby's prediction was soon borne out. A handwritten note dated September 15 from the files of Princeton's dean of faculty, Douglas Brown, records a telephone call from Agnes Henry, the mathematics department secretary, who informed the dean's secretary that John Nash had telephoned her asking the dean to write to the Office of Naval Research.[16] A few days later Nash filled out a university form, "Information Needed in a National Emergency," in which he stated that he was registered at Local Board 12 in Bluefield, that his current classification was I-A, and that he had a "chance for 2-A, application pending."[17] The form noted that Nash was engaged in project 727, Tucker's ONR logistics grant. In response to the question "Are you engaged in any other research work or consultation of possible national interest?" Nash responded yes and listed "consultant for RAND corporation." A note, added perhaps by the head of Princeton's grants office, mentioned that Nash had spent "3 years or more on the theory of games and related fields. Wrote paper in this field when at Carnegie Tech as undergraduate. Two years to get Ph.D. at Princeton. Dr. Rigby has already told NY to support."

The university immediately wrote to ONR stating that "this project is considered by the Logistics Branch of ONR, Washington as a very important contribution in the present national emergency. Dr. Nash is a key member of our staff in this project and is one of the very few individuals in the country who have been trained in this field."[18] The ONR followed, on September 28, with a letter to the draft board saying that Nash was "a key research assistant" and "this contract is an essential part of the Navy Department's research and development program and is in the interest of national safety."[19]

RAND protected Nash as well. RAND's former manager of security, Richard Best, recalls writing letters for Nash and another mathematician from Princeton, Mel Peisakoff, to "save" them from the draft.[20] (Peisakoff's recollection differs from Best's, however; he says he wanted to enlist but that his superiors at RAND wouldn't let him.)[21] "We had a lot of reservists and a great many young people," said Best. "In 1948, the average age was 28.35 years. The personnel office wasn't well [equipped to handle the situation]. I wrote some form letters to the draft board for Nash," he recalled.[22]

Nash's lobbying campaign worked, though he was not immediately granted the desired II-A. By October 6, the university informed Nash that "you seem to be safe until June 30."[23] Apparently, the board had simply postponed the designation for active service until June 30, 1951. The university advised Nash, "I would suggest that we defer any further action until next spring, at which time, we can again apply for a II-A classification and can consider an appeal if this should be rejected."[24] But, at least for now, he had prevented the military from wrecking his plans. More important, by protecting his personal freedom, Nash may have protected the integrity of his personality and won the ability to function well for longer than he might otherwise have.

15

A Beautiful Theorem

Princeton, 1950–51

S TRANGE AS IT MAY NOW SEEM, the dissertation that would one day win Nash a Nobel wasn't highly regarded enough to assure him an offer from a top academic department. Game theory did not inspire much interest or respect among the mathematical elite, von Neumann's prestige notwithstanding. Indeed, Nash's mentors at Carnegie and Princeton were vaguely disappointed in him; they had expected the youngster who had re-proved theorems of Brouwer and Gauss to tackle a really deep problem in an abstract field like topology.[1] Even his biggest fan, Tucker, had concluded that while Nash could "hold his own in pure mathematics," it was not "his real strength."[2]

Having successfully sidestepped the threat of the draft, Nash now began working on a paper that he hoped would win him recognition as a pure mathematician.[3] The problem concerned geometric objects called manifolds, which were of great interest to mathematicians at that time. Manifolds were a new way of looking at the world, so much so that even defining them sometimes tripped up eminent mathematicians. At Princeton, Salomon Bochner, one of the leading analysts of his day and a fine lecturer, used to walk into his graduate classes, start to give a definition of a manifold, get hopelessly bogged down, and finally give up, saying with an exasperated air, before moving on, "Well, you all know what a manifold is."[4]

In one dimension, a manifold may be a straight line, in two dimensions a plane, or the surface of a cube, a balloon, or a doughnut. The defining feature of a manifold is that, from the vantage point of any spot on such an object, the immediate vicinity looks like perfectly regular and normal Euclidean space. Think of yourself shrunk to the size of a pinpoint, sitting on the surface of a doughnut. Look around you, and it seems that you're sitting on a flat disk. Go down one dimension and sit on a curve, and the stretch nearby looks like a straight line. Should you be perched on a three-dimensional manifold, however esoteric, your immediate neighborhood would look like the interior of a ball. In other words, how the object appears from afar may be quite different from the way it appears to your nearsighted eye.

By 1950, topologists were having a field day with manifolds, redefining every object in sight topologically. The diversity and sheer number of manifolds is such that today, although all two-dimensional objects have been defined topologically, not all three- and four-dimensional objects — of which there is literally an infinite assortment — have been so precisely described. Manifolds turn up in a wide variety of physical problems, including some in cosmology, where they are often very hard to cope with. The notoriously difficult three-body problem proposed by King Oskar II of Sweden and Norway in 1885 for a mathematical competition in which Poincaré took part, which entails predicting the orbits of any three heavenly bodies — such as the sun, moon, and earth — is one in which manifolds figure largely.[5]

Nash became fascinated with the subject of manifolds at Carnegie.[6] But it is likely that his ideas did not crystallize until after he came to Princeton and began having regular conversations with Steenrod. In his Nobel autobiography, Nash says that, right around the time that he got his equilibrium result for *n*-person games, that is, in the fall of 1949, he also made "a nice discovery relating to manifolds and real algebraic varieties."[7] This is the result that he had considered writing up as a dissertation after von Neumann's cool reaction to his ideas about equilibrium for games with many players.

The discovery came long before Nash had worked out the laborious steps of the actual proof. Nash always worked backward in his head. He would mull over a problem and, at some point, have a flash of insight, an intuition, a vision of the solution he was seeking. These insights typically came early on, as was the case, for example, with the bargaining problem, sometimes years before he was able, through prolonged effort, to work out a series of logical steps that would lead one to his conclusion. Other great mathematicians — Riemann, Poincaré, Wiener — have also worked in this way.[8] One mathematician, describing the way he thought Nash's mind worked, said: "He was the kind of mathematician for whom the geometric, visual insight was the strongest part of his talent. He would see a mathematical situation as a picture in his mind. Whatever a mathematician does has to be justified by a rigorous proof. But that's not how the solution presents itself to him. Instead, it's a bunch of intuitive threads that have to be woven together. And some of the early ones present themselves visually."[9]

With Steenrod's encouragement,[10] Nash gave a short talk on his theorem at the International Congress of Mathematicians in Cambridge in September 1950.[11] Judging from the published abstract, however, Nash was still missing essential elements of his proof. Nash planned to complete it at Princeton. Unfortunately for Nash, Steenrod was on leave in France.[12] Lefschetz, who undoubtedly was pressing Nash to have the paper ready before the annual job market got under way in February, urged Nash to go to Donald Spencer, the visiting professor who had been on Nash's generals committee and had just been hired away from Stanford, and to use Spencer as a sounding board for completing the paper.[13]

As a visiting professor, Spencer occupied a tiny office squeezed between Artin's huge corner office and an equally grand study belonging to William Feller. Spencer, as Lefschetz wrote to the dean of faculty, was "probably the most attractive mathematician in America at that moment," as well as "one of the most versatile American born mathematicians."[14] A doctor's son, Spencer grew up in Colorado and was admitted to Harvard, where he intended to study medicine. Instead, he wound up at MIT studying theoretical aerodynamics and then at Cambridge, England, where he became a student of J. E. Littlewood, Hardy's great coauthor.[15] Spencer did brilliant work in complex analysis, a branch of pure mathematics that has widespread engineering applications.[16] He was a much sought-after collaborator, his most celebrated collaboration being with the Japanese mathematician Kunihiko Kodaira, a Fields medalist.[17] Spencer himself won the Bôcher Prize.[18] Although he primarily worked in highly theoretical fields, he nonetheless had some applied interests, namely hydrodynamics.[19]

A lively, voluble man, Spencer was "sometimes daunting in his reckless energy."[20] His appetite for difficult problems was boundless, his powers of concentration impressive. He could drink enormous quantities of alcohol — five martinis out of "bird bath" glasses — and still talk circles around other mathematicians.[21] A man whose natural exuberance hid a darker tendency toward depression and introspection, Spencer's appetite for abstraction was accompanied by an extraordinary empathy for colleagues who were in trouble.[22]

He did not, however, suffer fools gladly. The first draft of Nash's paper gave Spencer little confidence that the younger mathematician was up to the task he'd set for himself. "I didn't know what he was going to do, really. But I didn't think he was going to get anywhere."[23] But for months, Nash showed up at Spencer's door once or twice a week. Each time he would lecture Spencer on his problem for an hour or two. Nash would stand at the blackboard, writing down equations and expounding his points. Spencer would sit and listen and then shoot holes in Nash's arguments.

Spencer's initial skepticism slowly gave way to respect. He was impressed by the calm, professional way that Nash responded to his most outrageous challenges and his fussiest objections. "He wasn't defensive. He was absorbed in his work. He responded thoughtfully." He also liked Nash for not being a whiner. Nash never talked about himself, Spencer recalled. "Unlike other students who felt underappreciated," he said, "Nash never complained." The more he listened to Nash, moreover, the more Spencer appreciated the sheer originality of the problem. "It was *not* a problem that somebody gave Nash. People didn't *give* Nash problems. He was highly original. Nobody else could have thought of this problem."

Many breakthroughs in mathematics come from seeing unsuspected relationships between objects that seem intractable and ones that mathematicians have already got their arms around.

Nash had in mind a very broad category of manifolds, all manifolds that are compact (meaning that they are bounded and do not run off into infinity the way a plane does, but are self-enclosed like a sphere) and smooth (meaning that they have no sharp bends or corners, as there are, for example, on the surface of a cube). His "nice discovery," essentially, was that these objects were more manageable than they appeared at first glance because they were in fact closely related to a simpler class of objects called real algebraic varieties, something previously unsuspected.

Algebraic varieties are, like manifolds, also geometric objects, but they are objects defined by a locus of points described by one or more algebraic equations. Thus $x^2 + y^2 = 1$ represents a circle in the plane, while $xy = 1$ represents a hyperbola. Nash's theorem states the following: Given any smooth compact k-dimensional manifold M, there exists a real algebraic variety V in R^{2k+1} and a connected component W of V so that W is a smooth manifold diffeomorphic to M.[24] In plain English, Nash is asserting that for any manifold it is possible to find an algebraic variety one of whose parts corresponds in some essential way to the original object. To do this, he goes on to say, one has to go to higher dimensions.

Nash's result was a big surprise, as the mathematicians who nominated Nash for membership in the National Academy of Sciences in 1996 were to write: "It had been assumed that smooth manifolds were much more general objects than varieties."[25] Today, Nash's result still impresses mathematicians as "beautiful" and "striking"—quite apart from any applicability. "Just to conceive of the theorem was remarkable," said Michael Artin, professor of mathematics at MIT.[26] Artin and Barry Mazur, a mathematician at Harvard, used Nash's result in a 1965 paper to estimate periodic points of a dynamical system.[27]

Just as biologists want to find many species distinguished by only minor differences to trace evolutionary patterns, mathematicians seek to fill in the gaps in the continuum between bare topological spaces at one end and very elaborate structures like algebraic varieties at the other. Finding a missing link in this great chain —as Nash did with this result—opened up new avenues for solving problems. "If you wanted to solve a problem in topology, as Mike and I did," said Mazur recently, "you could climb one rung of the ladder and use techniques from algebraic geometry."[28]

What impressed Steenrod and Spencer, and later on, mathematicians of Artin and Mazur's generation, was Nash's audacity. First, the notion that every manifold could be described by a polynomial equation is a larger-than-life thought, if only because the immense number and sheer variety of manifolds would seem to make it inherently unlikely that all could be described in so relatively simple a fashion. Second, believing that one could prove such a thing also involves daring, even hubris. The result Nash was aiming for would have seemed "too strong" and therefore improbable and unprovable. Other mathematicians before Nash had spotted relationships between some manifolds and some algebraic varieties, but had treated these correspondences very narrowly, as highly special and unusual cases.[29]

By early winter, Spencer and Nash were satisfied that the result was solid and that the various parts of the lengthy proof were correct. Although Nash did not get around to submitting a final draft of his paper to the *Annals of Mathematics* until October 1951,[30] Steenrod, in any case, vouched for the results that February, referring to "a piece of research which he has nearly completed, and with which I am well acquainted since he used me as a sounding board."[31] Spencer thought game theory was so boring that he never bothered to ask Nash in the course of that whole year what it was that he had proved in his thesis.[32]

Nash's paper on algebraic manifolds — the only one he was ever truly satisfied with, though it was not his deepest work[33] — established Nash as a pure mathematician of the first rank. It did not, however, save him from a blow that fell that winter.

Nash hoped for an offer from the Princeton mathematics department. Although the department's stated policy was not to hire its own students, it did not, as a matter of practice, pass up ones of exceptional promise. Lefschetz and Tucker very likely dropped hints that an offer was a real possibility. Although most of the faculty other than Tucker neither understood nor displayed any interest in his thesis topic, they were aware that it had been greeted with respect by economists.[34]

In January, Tucker and Lefschetz made a formal proposal that Nash be offered an assistant professorship.[35] Bochner and Steenrod were strongly in favor, although Steenrod, of course, was not present at the discussion. The proposal, however, was doomed to failure. No appointment could be made without unanimous support in a department as small as Princeton's, and at least three members of the faculty, including Emil Artin, voiced strong opposition. Artin simply did not feel that he could live with Nash, whom he regarded as aggressive, abrasive, and arrogant, in such a small department.[36] Artin, who supervised the honors calculus program in which Nash taught for a term, also complained that Nash couldn't teach or get along with students.[37]

So the appointment wasn't offered. It was a bitter moment. The thought must have occurred to Nash that he was being rejected less on the basis of his work than on the basis of his personality. It was an even greater blow because the same faculty made it clear that it hoped that John Milnor, only a junior by this time, would one day become part of the Princeton faculty.[38]

The job market, while not as bad as in the Depression, was nonetheless rather bleak, the Korean War having cut into university enrollments. Having been turned down by Princeton, Nash knew he would be lucky to get a temporary instructorship in a respectable department.

Both MIT and Chicago, it turns out, were interested in hiring Nash as an instructor.[39] Bochner had the ear of William Ted Martin, the new chairman of the MIT mathematics department, and strongly urged Martin to offer Nash an instructorship.[40] Bochner urged Martin to ignore the gossip about Nash's supposedly difficult personality. Tucker, meanwhile, was pushing Chicago to do the same.[41] When MIT offered Nash a C. L. E. Moore instructorship, Nash, who liked the idea of living in Cambridge, accepted.[42]

16

MIT

BY THE END OF JUNE, Nash was in Boston living in a cheap room on the Boston side of the Charles.[1] Every morning he walked across the Harvard Bridge, over the yellow-gray river to east Cambridge where MIT's modern, aggressively utilitarian campus lay sprawled between the river and a swath of factories and warehouses. Even before he reached the far side, he could smell the factory smells, including the distinct odors of chocolate and soap mingling together from a Necco candy factory and a P&G detergent plant.[2] As he turned right onto Memorial Drive, he could see Building Two looming ahead, a featureless block of cement painted an "alarming brown," just to the right of the new library, then under construction.[3] His office was on the third floor next to the stairwell in a corner suite assigned to several instructors, a spare, narrow room with a high ceiling, overlooking the river and the low Boston skyline beyond.[4]

In 1951, before *Sputnik* and Vietnam, MIT was not exactly an intellectual backwater, but it was nothing like what it is today. The Lincoln Laboratory was famous for its wartime research, but its future academic superstars were still relatively unknown youngsters, and powerhouse departments for which it has since become known—economics, linguistics, computer science, mathematics—were either infants or gleams in some academic's eye. It was, in spirit and in fact, still very much the nation's leading engineering school, not a great research university.[5]

An environment more antithetical to the hothouse atmosphere of Princeton is hard to imagine. MIT's large scale and modern contours made it feel like the behemoth state universities of the Midwest. The military, as well as industry, loomed awfully large, so large that MIT's armed, plainclothes campus security force existed solely for the purpose of guarding the half-dozen "classified" sites scattered around the campus and preventing those without proper security clearances and identification from wandering in. ROTC and courses in military science were required of all MIT's two-thousand-plus undergraduate men.[6] The academic departments like mathematics and economics existed pretty much to cater to the engineering student—in Paul Samuelson's words, "a pretty crude animal."[7] All counted as "service departments," gas stations where engineers pulled up to get their tanks filled with obligatory doses of fairly elementary mathematics, physics, and chemistry.[8] Economics, for example, had no graduate program at all until the war.[9] Physics had no Nobel Laureates on its faculty at the time.[10] Teaching loads were heavy—sixteen hours a week was not uncommon for senior faculty—and were weighted toward large introductory courses like calculus, statistics, and linear algebra.[11] Its faculty were younger, less well known, and less credentialed than Harvard's, Yale's, or Princeton's.

"There were advantages," said Samuelson. "A lot of the MIT faculty didn't have Ph.D.'s. I came without a formal degree. Solow came before he had a formal degree. We were treated magnificently. It was more of a meritocracy." He added, "People would say, doesn't everybody do that? Not up the river, we'd answer. How do you explain that? We're Avis, we try harder." [12]

Socially, MIT was dominated by an old guard not of high-society intellectuals, but of middle-class Republicans and engineers. "It certainly was not a faculty club populated by cultivated Brahmins," said Samuelson, who was then twenty-five years old: "When I came [in 1940] it was 85 percent engineering, 15 percent science." [13]

MIT also had a less exclusionary tradition than Harvard or even Princeton. By the 1950s, perhaps 40 percent of the mathematics faculty and students at MIT were Jewish.[14] Bright youngsters from New York City public schools, effectively barred even then from attending Princeton as undergraduates, went there. Princeton was "out of the question for a Jew," recalls Joseph Kohn, who enrolled as a freshman at MIT in 1950. "At Brooklyn Tech the greatest thing in the world was sending a student to MIT." [15]

Still smarting from his rejection by Princeton, Nash arrived at Building Two with something of a chip on his shoulder, a feeling that he was a swan among ducks. MIT was already changing, however. Indeed, bringing a brilliant young researcher like Nash on board in the mathematics department was itself a sign of that shift.

There was money all of a sudden, not just for teaching the exploding numbers of students, but for research.[16] The amounts were small by post-*Sputnik* standards or even those of today, but huge by prewar standards. Support for science, initially fueled by the successes during World War II, was now growing because of the Cold War. It came not just from the Army, Navy, and Air Force but from the Atomic Energy Commission and the Central Intelligence Agency. MIT wasn't unique. Other institutions, from the big state universities in the upper Midwest to Stanford, grew up the same way. There was also the talent. Physics got many of the Los Alamos people. Electrical engineering was becoming a magnet for the first generation of computer scientists, an eclectic group of neurobiologists, applied mathematicians, and assorted visionaries like Jerome Lettvin and Walter Pitts, who saw the computer as a model for studying the architecture and functioning of the human brain.[17] "It was very much a growing environment and science was a growing sphere," said Samuelson, adding that after the war, the 85 percent–15 percent split between engineering and science had shifted to 50 percent–50 percent. He added: "It was the upswing in money . . . that made this possible. That was part of the whole postwar pattern." [18]

Mathematics was on the verge of becoming an important department, although that was not obvious to everyone at the time. The department had one famous name, Norbert Wiener (who wound up at MIT largely thanks to Harvard's anti-Semitism), and two or three first-rate younger men, including the topologist George Whitehead and the analyst Norman Levinson. But otherwise, mathematics consisted largely of competent teachers rather than great researchers — "a few giants but a lot of mediocrities."[19]

The man who changed all that was appointed chairman of the department in 1947. William Ted Martin, called Ted by everyone who knew him, was the tall, skinny, loquacious son of an Arkansas country doctor. Blond and blue-eyed with a sunny disposition and a ready grin, Martin was married to the granddaughter of a president of Smith College and revved up with ambition. A man whose innate decency would turn him into one of Nash's protectors after Nash became ill, Martin would soon endure his own trial by fire. At the height of the McCarthy witch hunt, Martin's secret past as an underground member of the Communist Party in the late 1930s and early 1940s would be exposed, threatening both his career and his vision for the department.[20] But in 1951 the past was still safely buried. A "sparkplug of a chairman," his real talent was for making things happen, wheedling money out of the MIT administration, the Navy, and the Air Force, and using it to great, indeed astounding, effect.[21]

One of Martin's strokes of genius was figuring out that the cheapest and quickest way to upgrade the department was not to reel in a few more big names, but to lure young hotshots there for a year or two and handle them, as much as possible, with kid gloves. Copying Harvard's Benjamin Pierce Fellows, Martin created C. L. E. Moore Instructorships, so called in honor of MIT's most distinguished mathematician in the 1920s.[22] Moore Instructors weren't expected to join the permanent faculty. The idea was to get a stream of talent that would act as a catalyst, firing up MIT's humdrum atmosphere and attracting better students, the best of whom now automatically went to the Ivies and Chicago.

Since he wouldn't have to live with them for long, or so he thought, Martin wasn't scared of difficult personalities. "Bochner said Nash was worth appointing. 'Don't worry about anything!' " Martin recalled.[23] And Martin didn't. He came to value Nash, not just as "a brilliant and creative young man," but as an ally in his quest to make the department great. He would come to particularly rely on Nash's absolute intellectual honesty: "When Nash mentioned somebody [as a potential hire], you didn't wonder if he was a crony or a relative. If Nash said he was top flight, you didn't need much in the way of outside references."

The most attractive figure at MIT from Nash's point of view was Norbert Wiener. Wiener was, in some ways, an American John von Neumann, a polymath of great originality who made stunning contributions in pure mathematics up until the beginning of World War II and then embarked on a second and equally astounding career in applied mathematics.[24] Like von Neumann, Wiener is known to the public for his later work. He was, among other things, the father of cybernetics, the application of mathematics and engineering to communications and control problems.

Wiener was also famously eccentric. His appearance alone was remarkable. His beard, Samuelson recalled after Wiener's death in 1964, was like "the Ancient Mariner's."[25] He puffed on fat cigars. He waddled like a duck, a myopic parody of an absentminded professor. His extraordinary upbringing at the hands of his father, Leo, was the subject of two popular books, *I Am a Genius* and *I Am a Mathematician,* the first of which became a bestseller in the early 1950s. Prolific as he was, Wiener generated as many anecdotes about himself as theorems. He hardly seemed to know where he was. He would ask, for example, "When we met, was I walking to the faculty club or away from it? For in the latter case I've already had my lunch."[26] He was notoriously insecure. If he encountered someone he knew carrying a book under his arm, he would, as likely as not, ask anxiously whether his name was in the book.[27] Friends and admirers traced this feature of his personality to his obsessive and overbearing father, who once bragged that he could turn a broomstick into a mathematician, and to Harvard's anti-Semitism, which cost Wiener an appointment in Birkhoff's department. As Samuelson said in a eulogy after Wiener's death: "The exodus from Harvard dealt a lasting psychic trauma to Norbert Wiener. It did not help that his father was a Harvard professor . . . or that Norbert's mother regarded his move as a cruel comedown in life."[28]

Wiener's colleagues at MIT knew that he suffered from periods of manic excitability followed by severe depressions, constantly threatened to resign, and sometimes spoke of suicide. "When he was high he'd run all over MIT telling people his latest theorem," Zipporah "Fagi" Levinson, the wife of Norman Levinson, recalled. "You couldn't stop him."[29] At times, he would come to the Levinsons' house, weeping, and say that he wished to kill himself.[30] One of Wiener's ever-present fears was that he would go mad; his brother Theo, as well as two nephews, suffered from schizophrenia.[31]

Perhaps because of his own psychological struggles, Wiener had an acute empathy for other people's trials. "He was egotistical and childish, but also very sensitive to the real needs of others," Mrs. Levinson recalled.[32] When a younger colleague was writing a book but couldn't afford a typewriter, Wiener showed up at his door unannounced with a Royal portable under his arm.

When Nash arrived at MIT in 1951, Wiener embraced him enthusiastically and encouraged Nash's growing interest in the subject of fluid dynamics — an interest that eventually led Nash to his most important work. For example, Nash sent Wiener a note in November 1952, inviting him to a seminar Nash was to give on "turbulence via statistical mechanics, collision functions, etc."[33] His postscript, saying, "I've found the smoothing effect in definite form now," suggests that Nash talked about his research with Wiener, something he did with almost no one else in the department. Nash saw Wiener, a genius who was at once adulated and isolated, as a kindred spirit and fellow exile.[34] He copied some of Wiener's more extreme mannerisms, his own form of homage to the older man.[35]

•   •   •

But Nash was to become far closer to Norman Levinson, a first-rate mathematician and a man of extraordinary character, who would play a role in Nash's career similar to those of Steenrod and Tucker at Princeton — a combination of sounding board and father substitute. Levinson, then in his early forties, was more enigmatic than Martin but far more accessible than Wiener.[36] Wiry, of medium height, with craggy features, Levinson was a fine teacher who rarely displayed the slightest facial expression and never referred to his own accomplishments. He suffered from hypochondria and from wide mood swings, long manic periods of intense creative activity followed by months, sometimes years, of depression in which nothing interested him. A former Communist like Martin, Levinson would suffer doubly during the McCarthy years when he endured not only notoriety and threats to his career as a mathematician, but his teenage daughter's slide into mental illness.[37] Despite these burdens, Levinson was, and would long remain, by far the most respected member of the department. Thoughtful, decisive, and attuned to the personal as well as intellectual needs of those around him, Levinson was father confessor and wise elder, the one whose judgments were constantly sought and carried most weight, on everything from research to appointments.

His personal history was one of individual triumph over bleak beginnings. Born in Lynn, Massachusetts, just before World War I, Levinson was the son of a shoe factory worker who earned eight dollars a week and whose education consisted of attending a yeshiva for a few years. His mother was illiterate. Despite a childhood of desperate poverty and an education that consisted of attending rundown vocational schools, Levinson's brilliance was undeniable. He managed, with the help of Wiener, who spotted his talent, to attend MIT and, later, Cambridge. At Cambridge, he became a protégé of G. H. Hardy and embarked on a series of brilliant papers on ordinary differential equations. "He was very uncouth, very provincial," his wife, Zipporah, who met Levinson soon after he returned from England, recalled in 1995. "He was highly opinionated and too ignorant to know that he didn't know everything. But he'd plunge in and make a good paper, despite the fact that he didn't know the literature. Wiener ignored his rough edges."

Like many promising young Jewish mathematicians of his generation, Levinson had difficulty getting an academic post when he returned to the States, and it was Hardy who, while visiting Harvard in 1937, was ultimately responsible for Levinson's appointment that year at MIT. The university's provost, Vannevar Bush, had turned down Wiener's recommendation that Levinson be offered an assistant professorship when Hardy, who at that time was both an outspoken opponent of Nazi anti-Semitism and the most prominent member of the German mathematical society, went with Wiener to the provost's office to protest. "Tell me, Mr. Bush, do you think you're running an engineering school or a theological seminary?" he is supposed to have said. When the provost gave a puzzled frown, Hardy went on: "If it isn't, why not hire Levinson?"

Nash was attracted by Levinson's strong personality and by a quality that he both shared and admired, namely Levinson's uncommon willingness to tackle new and difficult problems. Levinson was an early pioneer in the theory of partial differential equations, recognized by a Bôcher Prize, and the author of an important theorem in the quantum theory of scattering of particles. Most remarkably, when he was in his early sixties and already suffering from the brain tumor that would eventually kill him, Levinson achieved the most important result of his career, the solution to a part of the famous Riemann Hypothesis.[38] In many ways, Levinson was a role model for Nash.

17

Bad Boys

People considered him a bad boy — but a great one. — DONALD J. NEWMAN, *1995*

The Great Man . . . is colder, harder, less hesitating, and without fear of "opinion"; he lacks the virtues that accompany respect and "respectability," and altogether everything that is the "virtue of the herd." If he cannot lead, he goes alone. . . . He knows he is incommunicable: he finds it tasteless to be familiar. . . . When not speaking to himself, he wears a mask. There is a solitude within him that is inaccessible to praise or blame. — FRIEDRICH NIETZSCHE, The Will to Power

NASH WAS just twenty-three years old when he became an MIT instructor. He was not only the youngest member of the faculty, but younger than many of the graduate students. His boyish looks and adolescent behavior won him nicknames like Li'l Abner and the Kid Professor.[1]

By MIT standards of that time, the teaching duties of C. L. E. Moore instructors were light. But Nash found them irksome nonetheless — as he did everything that interfered with his research or smacked of routine. Later, he would be one of the few active researchers on the faculty who avoided giving courses in his own research area. Partly, it was a matter of temperament, partly a matter of calculation. He shrewdly realized that his advancement did not depend on how well or poorly he performed in front of students. He'd advise other instructors, "If you're at MIT, forget about teaching. Just do research."[2]

Perhaps for this reason, Nash was mostly assigned required courses for undergraduates. In the seven years of his teaching career at MIT, he seems to have taught only three graduate courses, all introductory, one in logic in his second year, one in probability, and a third, in the fall of 1958, in game theory.[3] Mostly, it seems, he taught different sections of undergraduate calculus.

His lectures were closer to free association than exposition. Once, he described how he planned to teach complex numbers to freshmen: "Let's see . . . I'd tell them i equals square root of minus one. But I'd also tell them that it could be minus the square root of minus one. Then so how would you decide which one. . . ." He started to wander. Just what freshmen needed, the listener said, in disgusted tones, in 1995. "He didn't care whether the students learned or not, made outrageous demands, and talked about subjects that were either irrelevant or far too advanced."[4] He was a tough grader too.

At times his ideas about the classroom had more to do with playing mind games than pedagogy. Robert Aumann, who later became a distinguished game theoretician and was then a freshman at MIT, described Nash's escapades in the classroom as "flamboyant" and "mischievous."[5] Joseph Kohn, later the chairman of the Princeton mathematics department, called him "a bit of a gamester."[6] During the 1952 Stevenson-Eisenhower race, Nash was convinced, quite rightly as it turned out, that Eisenhower would win. Most of the students supported Stevenson. He made elaborate bets with the students that were constructed so that he would win regardless of who won the election. The very brightest students were amused, but most were frightened away and soon the better-informed students started to avoid his courses altogether.

In his first year at MIT, Nash taught an analysis course for advanced undergraduates. The course was supposed to be an introductory look at calculus in which students weren't just learning manipulations but rather absolutely solid proofs of statements and how to construct such proofs. Between the first and second semesters of the yearlong course, the number of students dwindled from about thirty to five.

Kohn recalled: "He gave a one-hour test. He handed out blue books where you filled in your name and the course number on the cover. When the bell rang, you were supposed to turn over the exam sheet and start working on the test. There were four problems. Problem number one was 'What is your name?' The other three problems were fairly hard. Since I knew by then how his mind worked, I made sure to write next to number one, 'My name is Joseph Kohn.' People who assumed that writing their name on the cover was enough got twenty-five points taken off."[7]

Putting classic unsolved problems on exams was another of Nash's favorite tricks, Aumann recalled: "The students were supposed to show that pi is an irrational number. Later, when Nash was upbraided by the chairman of the department for putting the equivalent of Fermat's Last Theorem on a final, he responded by saying that people have a mental picture that this is a difficult problem. Maybe that's the stumbling block. Maybe, if people didn't realize that the problem was 'hard,' they could solve it."[8]

On another occasion, one of Nash's graders actually confronted him after he put the following question on a test:

If you make up a bunch of fractions of pi 3.141592. . . . If you start from the decimal point, take the first digit, and place decimal point to the left, you get .1
Then take the next 2 digits .41
Then take the next 3 digits .592
And so on and so on.
You get a sequence of fractions between 0 and 1.

What are the limit points of this set of numbers?
(A limit point is a point such that in any open interval containing it, however small, there are an infinite number of numbers from the sequence.)[9]

The grader immediately realized that it was a question that nobody had ever answered. The decimal expansion of pi isn't a famous outstanding problem, but it's the kind of thing mathematicians ask each other, not undergraduates. Only one fact has been proved, namely, that it has to have at least one limit point. It was clear that the students should know that there was at least one limit. But Nash thought that he knew, intuitively, that every number between 0 and 1 should be a limit point. He felt strongly that he knew the answer intuitively, which is of course quite different from having a solid proof. "It was a sort of strange thing to do," said the grader, in 1996.

Nash's propensity for tricks of this kind was so well known that it became the occasion of a small joke on him, George Whitehead, a topologist in the department at the time, recalled in a conversation in 1995.[10] Nash was teaching a large section of the same freshman calculus course that several graduate students were also teaching. All the sections had a prescribed and identical final and all the tests were graded together. A test, signed J. Forbes Hacker, Jr., with all wrong answers, came back, "hacker" being a double-entendre referring both to Nash's favorite putdown, which was "hack," and MIT slang for jokester. (It was hackers, for example, who one night removed a car belonging to Donald Spencer, who was briefly an instructor at MIT before the war, from its parking space on Massachusetts Avenue, deconstructed it, and left it for him to find when he walked into his classroom the next morning, once again fully assembled.) On another occasion, messages appeared on several blackboards around Building Two: THIS IS HATE JOHN NASH DAY![11]

Still, Nash could be charming to students he regarded as mathematically talented, and such students found much to admire. To a select few, often undergraduates, Nash made himself "very, very available for chatting about mathematics," Barry Mazur, a number theorist at Harvard who first encountered Nash during his freshman year at MIT, recalled. "It was amazing what he was willing to talk about. There was a sense of infinite time in every conversation."

Once Mazur and Nash were chatting in the common room. Someone mentioned a classical theorem by a disciple of Gauss, Peter Gustave Lejeune Dirichlet, that states that there are an infinite number of prime numbers in certain arithmetic progressions. "It's the kind of thing that one just accepts or perhaps goes off and looks up afterwards," Mazur said. Nash, however, jumped up, went to the board, and "for hours and hours elegantly thought through the proof from first principles" for Mazur's benefit.[12]

Outside the classroom, Nash alternated between the sort of behavior for which he was famous at Princeton — pacing in Building Two's cavernous hallways whistling Bach — and bouts of sociability. By day, he spent very little time in the office suite that he shared with the other Moore Instructors. Mostly, he spent his time in the mathematics common room—a far cry from the one in Fine Hall, a ratty and nondescript lounge directly below the instructors' offices, at the bottom of a flight of stairs.

The social atmosphere of the MIT common room resembled some of the more raucous scenes from the cult movie *If,* about a British public school that is taken over by its "boys." Nash imported the Princeton practice of a regular tea hour to MIT, but not any of its more genteel customs.[13] "He wanted to be the quickest," Isadore M. Singer, a fellow Moore Instructor, recalled in 1994. "He was a real competitor."[14] Just as he had at Princeton, Nash liked jumping into a conversation, throwing out challenges and being challenged. He liked solving problems.

Students and an occasional professor played games, including go, chess, a great favorite of Wiener's despite lack of skill at the game,[15] and bridge. (Nash, Singer recalled, was hopeless at bridge. "It was absurd," Singer said. "He had no sense of the laws of probability in cards.")[16] Many of the games, however, were made up on the spur of the moment. One day a group made up an index of eccentricity by which various department members were ranked. Wiener, not Nash, drew the highest score.[17] Another time, everyone played a version of charades that involved drawing abstract pictures representing people around the department. A graduate student drew a highly elaborate picture of what appeared to be a taxi. Nobody could guess who it was supposed to be. The picture, it turned out, was meant to be a Nash, the car manufactured in the 1940s and 1950s, and was supposed to signify Nash the Hack, again, a reference to Nash's favorite putdown of those he regarded as plodders.[18]

The crowd in the common room was dominated by a handful of fast-talking, wisecracking veterans of Stuyvesant High School and the Bronx High School of Science math teams and the City College "Math Table"—a once-famous table in City's cafeteria at which an entire generation of math students, mostly working-class Jews and immigrants, honed their skills in problem solving and repartee.[19]

It was a brasher, rougher crowd, less uptight and more tolerant than the one in Fine Hall, and an audience more to Nash's liking. Showing off wasn't regarded as a crime if you knew your stuff. Lack of social graces was considered part and parcel of being real mathematicians. "Their attitudes were famously nonbourgeois, exhibitionistic, dissolute," Felix Browder recalled.[20] If anything, all of them placed a certain premium on eccentricity and outrageousness, although by today's standards what went for unconventional behavior and manners was, by and large, mild —depending on certain turns of phrase, brands of humor, and little deviations in dress. One fellow insisted on wearing pants with fly buttons with a button or two undone.[21] One graduate student recalled: "At that time we thought of eccentricity and being good in math as going together. We were all enjoying ourselves by being a little bit wild. We thought of ourselves as taking advantage of being bright by ignoring conventions we didn't like. We turned ourselves a little bit into characters."[22]

In this circle, Nash learned to make a virtue of necessity, styling himself self-consciously as a "free thinker." He announced that he was an atheist.[23] He created his own vocabulary.[24] He began conversations in midstream with "Let's take this aspect." He referred to people as "humanoids."

Nash picked up the mannerisms of other eccentric geniuses. For example, Wiener, who was terribly nearsighted, would keep one of his fingers in the groove in the walls between the wall tiles and the plaster, as he navigated his way hesitantly through the corridors. Nash did the same thing.[25] D. J. Newman condemned all music after Beethoven. Nash would stalk into the music library and tell anyone who was listening to anything more modern, "That's junk."[26] Levinson, whose daughter suffered from manic depression, hated psychiatrists. Nash adopted a similarly vehement stance against the profession.[27] Warren Ambrose detested conventional greetings like "How are you?" Nash followed suit.[28]

Marvin Minsky, whom Nash had known during his final year in Princeton and whom he regarded as the most intelligent "humanoid" of all, recalled: "We shared a similarly cynical view of the world. We'd think of a mathematical reason for why something was the way it was. We thought of radical, mathematical solutions to social problems. At one point, Nash suggested a complete transfusion for something. If there was a problem, we were good at finding a really ridiculously extreme solution."[29] One time he said that parents should "self-destruct," that is, commit suicide, and hand over all their holdings to their children. It would be not only convenient but principled, Nash said, according to Herta Newman, the wife of Nash's friend Donald Newman.[30] Another time he told a class of undergraduates that American citizens' voting rights should be made proportional to their income (or perhaps it was wealth).[31] In many ways Nash's views were more suited to nineteenth-century England's elitist political landscape than to the predominantly left-wing counterculture of the MIT math department of the 1950s.

Nevertheless, he adopted a touch of flamboyance about his dress. He wore translucent white Dacron shirts sans undershirt, others thought, to show off his powerful physique.[32] He bought a camera and spent much of his time browsing through photography books.[33] For a time, he read and talked a great deal about experimenting with mind-altering drugs like heroin — although there is no evidence that he ever tried any.[34] His growing heterogeneity of interests and heterodoxy could, with hindsight, be seen as the first overt signs of a growing alienation from convention and society that would later evolve into a radical sense of separateness and disconnection.

But, at the moment, these postures enhanced rather than detracted from Nash's social appeal. Nash's status as an instructor and his growing reputation as a mathematician brought him newfound respect. He was now considered interesting company. His arrogance was seen as evidence of his genius, and so was his eccentricity, a source of both amusement and grudging respect, the other side of the genius coin, as it were. Fagi Levinson, the department's den mother, said in 1996: "For Nash to deviate from convention is not as shocking as you might think. They were all prima donnas. If a mathematician was mediocre he had to toe the line and be conventional. If he was good, anything went." [35]

Jerome Neuwirth, a graduate student at MIT, said, "When your solution turns out to be right, we give you your due. We give you a lot of leeway. Had Nash been less of a mathematician, he wouldn't have gotten away with his nastiness." [36] Donald Newman added, "People were annoyed with him because he was flippant, but not really annoyed. They considered him a bad boy, but a great one, a great golden boy." [37]

The gang around Nash included Newman, aka D.J., a Harvard graduate student who spent most of his time at MIT hanging out with his old friends from City College and with Nash, because "Harvard was too snooty." [38] Other members of the group included Walter Weissblum, a brilliant sad sack, drunk, and hunchback with a heart of gold, who never finished his degree; [39] Harry Gonshor, who later became a professor at Rutgers, an oddball who wore Coke-bottle glasses, looked as if he were floating on air, and once proved a theorem so that it could be stated as "AFL = CIO"; [40] Gustave Solomon, the most humane of the group, later a coinventor of the Reed-Solomon code; [41] Leopold "Poldy" Flatto, an inveterate people-watcher and storyteller; [42] and, after 1952, Jacob Leon Bricker, the group's Woody Allen. [43]

Neuwirth, a latecomer to the group, said, "Who were we? What were we trying to do? Every group has its own currency. Our only currency is what we were thinking. Who's smart? Who's doing what? What can you solve? How far did you get? It doesn't sound nice but it was exciting." [44]

Nash's closest equal, in brains, competitiveness, and general superciliousness, was Newman. Newman was considered a genius and the best problem solver of the group. [45] A big, brash, blond swaggerer, Newman had the distinction, very impressive to Nash, of being a three-time Putnam winner. He was already a husband and father, with responsibilities that, however, did little to cramp his flamboyant style. He drove a flashy white Thunderbird with red leather seats that he liked to drag race along Memorial Drive in the middle of the night. As an undergraduate at City College, he'd been famous for stunts like turning up in the class of some unfortunate mathematics professor bearing an enormous tree branch, leaves and all, that he claimed was for a biology class.

Nash and Newman immediately recognized each other as kindred spirits. "They loved to spark each other," Arthur Singer recalled. [46] "They admired each other's sarcasm," said Mattuck. "It was all good-natured. But D.J. could make cracks much faster. He had instant recall when it came to mathematics. People used to say that D.J. could solve any problem that could be done in twenty-four hours.

Newman didn't have the power of Nash's sustained concentration. Nash could think about a problem for half a year."[47]

Newman went to a seminar given by Nash. "I sat in on some of Nash's lectures," said Newman, who was intrigued rather than put off. "It was different, kind of exciting. He wandered, unlike most lecturers, because he liked to explore a lot of things at once. It was kind of nice. . . . We chewed each other out," Newman recalled. "Nash and I were friendly friends."[48]

Thanks to the acceptance of Newman and his friends, Nash acquired a real social life. The crowd often ate lunch together in Walker Memorial, but it also gathered after hours at various cheap restaurants, coffee shops, and beer halls that were as plentiful in 1950s Cambridge and Boston as they are today, places that didn't mind if you nursed a beer all night and were willing to write separate checks.[49] They included famous Boston restaurants like Durgin Park, which served generous helpings of traditional New England dishes, including a sinfully delicious roast beef and Indian pudding; Jake Wirth, an old-style German establishment with a mammoth oak bar; and the Wursthaus in Harvard Square. Other favorites were Cronin's, Chez Dreyfus, and the Newbury Steakhouse. The Hayes-Bickford and the Waldorf, which were both Horn & Hardart–style coffee shops, open most of the night, were also frequent gathering places. At other times, everybody would hang out at some graduate student's apartment, or go to parties given by the Martins, Levinsons, and in the mid-1950s, the Minskys.

Within his new circle, Nash strove to constantly underscore his own uniqueness, superiority, and self-sufficiency. "I'm Nash with a capital N!" his whole manner shouted.[50] He was always saying that only one or two people in the department— Wiener was always one of these—were up to his standard. His putdowns were legendary. "You're a child," was a favorite expression. "You don't know crap. How trivial! How stupid! You'll never do anything!" he would say.[51]

He loved to perform. At parties, he acted rather than conversed. Once, at the Minskys', Nash demanded that his listeners challenge him with a difficult mathematical problem. He said, "I've had a few drinks. Are my thinking powers stronger or weaker on drink?"[52]

He was not above dissembling slightly to wow an audience.[53] He would pout if he was bested in an argument.[54] And he hated being challenged by someone he considered to be an inferior. One day in the common room, a group of students was talking about a famous World War II logistics puzzle, the "Jeep" problem.[55] The essence of the Jeep problem is that you want to cross the two-thousand-mile-wide Sahara desert but the Jeep's gas tank holds only enough gas to travel two hundred miles. The only way to cross the desert is to follow a two-steps-forward, one-step-back strategy: to load up the Jeep with cans of gasoline, drive, say, one hundred miles, drop off the cans, and go back to the starting point. Then you get more cans of gas, go one hundred miles, unload some and use some to top off the gas in the tank, go another one hundred miles, and go back, picking up some more gasoline. The question is, how many gallons would be needed?

There is no optimal solution to the problem, as it turns out. Everybody was proposing solutions. Nash threw out a number. Nash's grader that term, Seymour Haber, proposed a number half as big. Nash contemptuously dismissed Haber's solution. When Haber insisted that he prove it, Nash said, "My solution's much better."

Haber recounted: "I didn't see it. I insisted that he prove it. He didn't want to. He said it was obvious. I still wouldn't accept his assertion. So he did the calculation. He turned out to be mostly right, but he was extremely annoyed with me. He was angry for my having forced him to do this grungy work when it was perfectly clear all along what the answer was. He was angry with me for some period afterward."

Nor was he above putting the audience down. A typical example: at lunch one day, a graduate student was describing an axiomatic approach to a problem outlined by one of his professors. Nash fairly exploded, "Don't give me all that crap! Tell me how you'd solve the problem. You haven't learned anything. All these concepts don't mean a thing." [56]

Nash's putdowns of other mathematicians earned him the sobriquet "Gnash." Nash responded, "G obviously stands for genius. In fact, there are few geniuses these days here at MIT. Me, of course, and also Norbert Wiener. Even Norbert may no longer be a genius, but there is evidence that he once was." After that, he referred to Gnu (Newman) and G-squared (Andrew Gleason, a young Harvard professor who had just solved Hilbert's fifth problem). [57]

When John McCarthy, whom Nash knew from Princeton, gave a seminar in the department, Nash pulled him aside afterward and said, "There are too many journals. There are too many trashy papers being published. There are too many guys doing research. Only a few of us should be in research. The rest of them should be in sin x" — a snide reference to the tables at the back of high-school trigonometry books. [58]

Nash flaunted his social snobbery, a legacy of his Bluefield upbringing. He implied that he came from old money. [59] He would sniff wine at a party and say, "This is an adequate Chianti." [60] Nowhere was his snobbery more evident than in his reaction to being "a non-Jew in a definitely Jewish atmosphere." [61] Later, when Nash became paranoid and embraced all sorts of strange delusions, he wrote letters to Newman and others addressed to "Jewboy," became obsessed with the state of Israel, and talked about "Krypto-Zionist conspiracies." [62] But in the early 1950s, his attitude was merely one of social superiority. He frequently told Newman that he looked "too Jewish." [63] Like Groucho Marx, he was inclined not to admire any club that accepted him. Nash displayed a contempt for people and things he considered beneath him. As Fred Brauer, another instructor at MIT, put it forty years later, "That covered a lot of territory." [64]

18

Experiments

RAND, Summer 1952

O NE AFTERNOON during Nash's second summer in Santa Monica, he and Harold N. Shapiro, another mathematician from RAND, were swimming in the surf off Santa Monica Beach just south of the pier.[1] The ocean was fairly rough. Below the breakwater, Santa Monica Beach was a narrow and steep strip of sand with breakers that were usually six to ten feet high. It was a favorite of body surfers.

Nash and Shapiro were far from shore when they were caught in a powerful current that swept them farther out. Both men were strong swimmers. Nash was "built like a Greek god," Shapiro recalled, and he, too, was sturdy and muscular. But Shapiro remembers being dragged under the waves, briefly overpowered by the current, and very frightened. Nash seemed to be struggling as well. "It was hard work getting back to shore," Shapiro said. When the two young men finally reached the beach, they threw themselves on the sand, exhausted and breathing heavily. Shapiro recalled lying there, thinking how lucky they were not to have drowned. To his amazement, however, Nash jumped to his feet after a moment or two and announced he was going back into the water. "I wonder if that was an accident," Nash said in a calm and detached tone. "I think I'll go back in and see."

At the beginning of that second summer, Nash had driven cross-country from Bluefield to Santa Monica in a rusty old Dodge. He and John Milnor, who was by now a graduate student at Princeton, made the trip together, though Milnor drove his own car.[2] Traveling with them were Nash's younger sister Martha and Ruth Hincks, a journalism major at the University of North Carolina in Chapel Hill, who joined them at the last minute.[3] They met in Chapel Hill, then drove on to Bluefield. Hincks remembers being warned not to let slip that Martha would be sharing the apartment with Milnor as well as Nash. She recalled in 1997 that this secretiveness struck her as strange. As they started out, Ruth drove with Nash, Martha with Milnor. Ruth was struck by Nash's complete indifference to her. "I was slim, attractive, intelligent," she recalled in 1997. Nash "never even noticed that I was there," she said. She was also struck by the seemingly distant relationship between Nash and Milnor. "They just sort of stood around. They could have met the day before. They never referred to shared experiences. They didn't seem to really know each other." Even the relationship between brother and sister seemed "a little standoffish, not affectionate at all," said Ruth. "I don't think I saw any affection from anybody on that trip."

They traveled on U.S. 40, which took them through Kansas and Nebraska.[4] They stopped once for a day in Grand Lakes, Colorado, where they all went horseback riding, and also in Salt Lake City, where they visited the Mormon Temple. The men put the young women in charge of divvying up all the motel, restaurant, and gas bills. All should have been fine for these young people, privileged as few were, in 1952, to be traveling cross-country on their own. Yet before the trip was over, Nash and Ruth had quarreled, and Martha, who had been riding with Milnor, was forced, reluctantly, to ride with her older brother for the remainder of the journey.[5]

It started as a fine adventure. Martha had just graduated from Chapel Hill, and had traveled very little before.[6] Tall and striking like her brother, Martha was extremely intelligent. In spite of a fierce determination not to be regarded as an egghead and an oddball, Martha had won a Pepsi-Cola scholarship by beating every boy at Beaver High on the SATs and had received invitations to apply to Radcliffe, Smith, and other top women's schools. Her father, however, had turned down the scholarship on her behalf, saying that the family could afford tuition at a nearby school, and Martha wound up at St. Mary's, a junior college attended mostly by well-to-do southern girls who brought fur coats with them, rode horses, and were themselves being groomed not for the job but for the marriage market. After graduating from St. Mary's, she went on to the University of North Carolina, where she completed a teaching degree.

John had persuaded his parents that it would be good for Martha to spend a summer in Santa Monica, suggesting that he could get more work done if Martha kept house for him.[7] Martha, who had never been away from home except at college, was eager to go. Once the plans were made, John also made no secret of his hope that his sister and John Milnor would take an interest in each other.

It was Nash who had proposed that they all travel together. Milnor and Nash, of course, had known each other since Milnor was a freshman at Princeton four years earlier. Though he had not yet completed his dissertation, Milnor had already been asked by Princeton to join its faculty. Nash confessed to Martha that he was jealous of Milnor's abilities, but he was clearly also charmed by Milnor's self-effacing personality, his brilliantly lucid mind, and the younger man's lanky good looks.

Ruth said her good-byes as soon as the quartet arrived in Santa Monica. Martha, Nash, and Milnor rented a small furnished apartment at the top of a rambling Spanish-style villa on Georgina Avenue, a stately street in the old section of Santa Monica and ten minutes' walk via Palisades Park from RAND.[8] Nobody did much cooking or housekeeping. A guest who had been invited for lunch said: "The place hadn't been cleaned — ever. There were dust balls and dirty dishes. After looking around — they obviously hadn't prepared a meal — I decided to ask for eggs. John pushed the remnants of a previously fried egg aside in the frying pan. 'Very nice people,' I thought to myself."[9] Martha got a job in a bakery. She hardly saw her two roommates, who seemed to spend most of their waking hours inside the RAND headquarters. Martha tried to visit their offices one day but was barred by the guards because she had no security clearance.[10] She and Milnor went out to dinner once in the first week or two, but despite their many hours together in the car, Milnor was uneasy and painfully tongue-tied, and it became clear to Martha that no romance was in the offing.[11]

The two men worked mostly on their own. Milnor wrote a lovely paper called "Games Against Nature."[12] Nash dabbled with games that could be played using a computer.[13] He was, by this time, chiefly concerned with mathematical problems that arise in the study of fluid dynamics. A paper on war games was merely a half-hearted effort, designed to justify his employment at RAND and to be hastily drafted before he returned to Cambridge at the beginning of September.[14]

But Nash and Milnor did collaborate on one project, an experiment on bargaining involving hired subjects, that was to become, unexpectedly, a much-cited classic.[15] The experiment, designed with two researchers from the University of Michigan who were also at RAND for the summer, anticipated by several decades the now-thriving field of experimental economics.

The RAND experiments grew more or less directly out of the habit of playing games that the mathematicians indulged in their spare time. Inventing new games and trying them out, always with the inventors as subjects, had been a popular pastime at Princeton. Many of the players had, like Nash, only recently outgrown boyhood passions for chemistry and electricity experiments. The idea of recording the play to see whether people played the way the theory predicted was already a bit of a tradition at RAND, inaugurated by the famous Prisoner's Dilemma experiment. Martha was astonished to learn that the volunteers were earning fifty dollars a day "to play games."[16]

The experiment, which was conducted over a two-day period, was designed to test how well different theories of coalitions and bargaining held up when real people were making the decisions.[17] Von Neumann and Morgenstern, with their interest in games with many players, focused on coalitions, groups of people who act in unison. They argued that rational players would calculate the benefits of joining every possible coalition and choose the best one — that is, the one that was most advantageous to them — whether they were business executives intent on collusion or workers who wanted to join a union.

Nash, Milnor, and the other researchers hired eight subjects, college students and housewives. They devised different games, mostly with four rotating players, one with as many as seven. The game mimicked the general, "*n*-person" game of von Neumann's theory. Subjects were told they could win cash by forming coalitions, and the specific amounts that would be awarded to each possible coalition. To be eligible to win, however, the coalition partners had to commit in advance to a given division of the winnings.

According to Al Roth, a leading experimental economist, the experiment yielded two insights that proved highly influential.[18] For one thing, it drew attention to information possessed by participants: If the same players play the game repeatedly, the authors concluded, players tend to "regard a run of plays as a single play of a more complicated game." Second, like the Prisoner's Dilemma experiment devised by Melvin Dresher and Merrill Flood in 1950, it showed that players' decisions were often motivated by concerns about fairness. In particular, in situations in which neither player had a privileged position, players typically opted to "split the difference."

For the designers of the experiment, however, the results merely cast doubt on the predictive power of game theory and undermined whatever confidence they still had in the subject. Milnor was particularly disillusioned.[19] Though he continued at RAND as a consultant for another decade, he lost interest in mathematical models of social interaction, concluding that they were not likely to evolve to a useful or intellectually satisfying stage in the foreseeable future. The strong assumptions of rationality on which both the work of von Neumann and Nash were constructed struck him as particularly fatal. After Nash won the Nobel Prize in 1994, Milnor wrote an essay on Nash's mathematical work in which he essentially adopted the widespread view among pure mathematicians that Nash's work on game theory was trivial compared with his subsequent work in pure mathematics. In the essay, Milnor writes:

> As with any theory which constructs a mathematical model for some real-life problem, we must ask how realistic the model is. Does it help us to understand the real world? Does it make predictions which can be tested? . . .
>
> First let us ask about the realism of the underlying model. The hypothesis is that all of the players are rational, that they understand the precise rules of the game, and that they have complete information about the objectives of all of the other players. Clearly, this is seldom completely true.
>
> One point which should particularly be noticed is the linearity hypothesis in Nash's theorem. This is a direct application of the von Neumann–Morgenstern theory of numerical utility; the claim that it is possible to measure the relative desirability of different possible outcomes by a real-valued function which is linear with respect to probabilities. . . . My own belief is that this is quite reasonable as a normative theory, but that it may not be realistic as a descriptive theory.
>
> Evidently, Nash's theory was not a finished answer to the problem of understanding competitive situations. In fact, it should be emphasized that no simple mathematical theory can provide a complete answer, since the psychology of the players and the mechanism of their interaction may be crucial to a more precise understanding.[20]

Nevertheless, decades later, economists, differing with Milnor, came to regard this "failure" of an experiment as a very worthwhile one. Casual as the experiment was in one sense, it became a model for a new method of economic research, one that had never before been tried in the two hundred years since Adam Smith dreamed up the Invisible Hand. The feeling was that even if the experiments weren't sophisticated enough to show how people's brains work, watching the way people played games could draw researchers' attention to elements of interaction — such as signaling or implicit threats — that couldn't be derived axiomatically.[21]

By the time the experiment was run the relationship between Nash and Milnor had become strained, and Milnor had moved out of the Georgina Avenue apartment.

Milnor says now that Nash made a sexual overture toward him. "I was very naive and very homophobic," said Milnor. "It wasn't the kind of thing people talked about then."[22] But what Nash felt toward Milnor may have been something close to love. A dozen years later, in a letter to Milnor, Nash wrote: "Concerning love, I know a conjugation: amo, amas, amat, amamus, amatis, amant. Perhaps amas is also the imperative, love! Perhaps one must be very masculine to use the imperative."[23]

19

Reds

Spring 1953

Now, the thing I think would interest the committee very greatly, if you could possibly explain to them . . . Doctor . . . how you can account for what would seem to be an abnormally large percentage of communists at MIT? — ROBERT L. KUNZIG, Counsel, HUAC, April 22, 1953

T HE COLD WAR promised to be the sugar daddy of the MIT mathematics department, but McCarthyism — which blamed the setbacks in that war on sinister conspiracies and domestic subversion — threatened to devour it.

While Nash and his graduate student friends were shooting each other down and playing games in the mathematics common room, FBI investigators were fanning out around Cambridge, rifling through trash cans, placing individuals under surveillance, and questioning neighbors, colleagues, students, and even children.[1] Their targets, as Nash and everyone else at MIT would learn in early 1953, included the chairman and the deputy chairman of the MIT mathematics department, as well as a tenured full professor of mathematics, Dirk Struik — all three one-time members, indeed, leading members, of the Cambridge cell of the Communist Party. All three were subpoenaed by the House Un-American Activities Committee.[2] It was a state of siege and everyone in the mathematics department felt the threat.

At the time, Nash was no doubt far more preoccupied with the draft — not to mention growing complications of his personal life — than with the possible repercussions for himself of the persecution of his benefactors. Nevertheless, the whole episode was a warning that the world he and other mathematicians inhabited was an extremely fragile one. A congressional committee could destroy your career, just as your draft board could send you halfway around the world.

The whole thing had begun as a farce.[3] McCarthy's original list of communists, announced in February 1950, was studded with academics, including the father of Nash's friend Lloyd Shapley, Harvard astronomy professor Harlow Shapley, whom McCarthy incorrectly identified to reporters as "Howard Shipley, astrologer." But as the red hunt gathered momentum, the entire scientific community felt vulnerable. Princeton's Solomon Lefschetz would be identified as a possible communist sympathizer by an investigative body.[4] Within a year, Robert Oppenheimer, head of the Manhattan Project, one of the most revered scientists in America and the director of the Institute for Advanced Study, would be humiliated by the McCarthyites.

When the subpoenas were issued, nobody knew how MIT would handle the matter. Other universities had responded with immediate firings and suspensions.[5] "McCarthyism was a big threat to these schools," Zipporah Levinson, Norman Levinson's widow, recalled. "During the war the government had started pouring money into them. The threat was that the research money would dry up. It was a bread-and-butter issue."[6] Martin and Levinson were certain that they were about to lose their jobs and wind up blacklisted for good, like so many others. Levinson talked about becoming a plumber and specializing in the repair of furnaces. The investigators had their eye on the three Browder boys — sons of former Communist Party head Earl Browder, who had all studied or were studying mathematics at MIT and were scholarship recipients, as well.[7]

"MIT was turned topsy-turvy," Mrs. Levinson recalled. "The faculty debated and debated how to prove that MIT was patriotic. There was strong pressure to name names."[8] As it turned out, Karl Compton, the president of the university and an outspoken liberal who was a supporter of the Chinese revolution and a critic of Chiang Kai-shek, may have felt that he himself would soon be subpoenaed. He hired a white-shoe Boston law firm, Choate, Hall & Steward, to defend Martin, Levinson, and the others for a minimal fee.[9] By April, when Martin and Levinson were forced to testify, *The Tech* was running daily stories and anti-McCarthy sentiment was running high on campus.[10]

There is no evidence that the FBI ever questioned Nash or any other students or faculty in the department, or asked for depositions, in an effort to establish a link between Levinson's and Martin's Communist Party membership and classified defense research — a link that probably never existed, given that both left the party soon after the end of the war. The graduate students and junior faculty in the department stood on the sidelines and watched lives and careers ruined and homes, even car insurance, lost. "By that time, young people had prospects, jobs, optimism," Mrs. Levinson recalled. "The younger people — Nash's group — didn't want to be too friendly. They were scared. They distanced themselves."[11]

Martin and several others named their former associates. Norman Levinson refused to name anyone who had not been previously named. "Ted and Izzy Amadur hemmed and hawed. Norman knew that Ted Martin and Izzy would cooperate. They spilled all the names. Norman said he'd talk freely about the party but that he wouldn't name names. The lawyer told Norman, no you don't have to say any names. He'd cooperate, but he wouldn't give any names."[12] Martin gave a pathetic, frightened performance. Levinson's testimony, by contrast, demonstrated the qualities of intellect and character that made him such a force in the mathematics community. In a series of forceful and eloquent answers to direct questioning, he managed at one and the same time to defend the youthful idealism that led him into the party, attack the intellectual poverty of communism, and, implicitly, call into question the committee's assumption that communism was a threat to the nation. He spoke out against the hounding of former party members and asked the committee to take a stand against the blacklisting of Browder's oldest son, Felix, who had finished his Ph.D. and was unable to obtain an academic post.

Thanks to MIT's support and the compromises they struck, Levinson and the others kept their jobs. But the whole dispiriting affair, which had been preceded by months of harassment and threats, left deep scars on everyone involved. Martin, in particular, was shattered and deeply depressed, and was unable, nearly forty-five years later, to talk about it.[13] Levinson's younger daughter, a student in junior high school, suffered a breakdown and was diagnosed with manic depression. Levinson and his wife blamed it partly on her being harassed by the FBI.[14] And those on the periphery, ostensibly unaffected, learned a lesson, namely that the world they so very much took for granted was dangerously fragile and vulnerable to forces beyond its control.

Nash took no part in the heated discussions among some of the graduate students over the morality of the mathematicians' decision to cooperate with the government.[15] Any discussion of morality raised for him the specter of hypocrisy. But the angry, frightening, turbulent time would supply him with some of the prosecutory demons that came to haunt him later.[16]

20

Geometry

There are two kinds of mathematical contributions: work that's important to the history of mathematics and work that's simply a triumph of the human spirit. — PAUL J. COHEN, 1996

IN THE SPRING OF 1953, Paul Halmos, a mathematician at the University of Chicago, received the following letter from his old friend Warren Ambrose, a colleague of Nash's:

There's no significant news from here, as always. Martin is appointing John Nash to an Assistant Professorship (not the Nash at Illinois, the one out of Princeton by Steenrod) and I'm pretty annoyed at that. Nash is a childish bright guy who wants to be "basically original," which I suppose is fine for those who have some basic originality in them. He also makes a damned fool of himself in various ways contrary to this philosophy. He recently heard of the unsolved problem about imbedding a Riemannian manifold isometrically in Euclidean space, felt that this was his sort of thing, provided the problem were sufficiently worthwhile to justify his efforts; so he proceeded to write to everyone in the math society to check on that, was told that it probably was, and proceeded to announce that he had solved it, modulo details, and told Mackey he would like to talk about it at the Harvard colloquium. Meanwhile he went to Levinson to inquire about a differential equation that intervened and Levinson says it is a system of partial differential equations and if he could only [get] to the essentially simpler analog of a single ordinary differential equation it would be a damned good paper — and Nash had only the vaguest notions about the whole thing. So it is generally conceded he is getting nowhere and making an even bigger ass of himself than he has been previously supposed by those with less insight than myself. But we've got him and saved ourselves the possibility of having gotten a real mathematician. He's a bright guy but conceited as Hell, childish as Wiener, hasty as X, obstreperous as Y, for arbitrary X and Y.[1]

• • •

Ambrose had every reason to be both skeptical and annoyed.

Ambrose was a moody, intense, somewhat frustrated mathematician in his late thirties, full, as his letter indicates, of black humor.[2] He was a radical and nonconformist. He married three times. He gave a lecture on "Why I am an atheist." He once tried to defend some left-wing demonstrators against police in Argentina — and got himself beaten up and jailed for his efforts. He was also a jazz fanatic, a personal friend of Charlie Parker, and a fine trumpet player.[3] Handsome, solidly built, with a boxer's broken nose — the consequence of an accident in an elevator! — he was one of the most popular members of the department. He and Nash clashed from the start.

Ambrose's manner was calculated to give an impression of stupidity: "I'm a simple man, I can't understand this." Robert Aumann recalled: "Ambrose came to class one day with one shoelace tied and the other untied. 'Did you know your right shoelace is untied?' we asked. 'Oh, my God,' he said, 'I tied the left one and thought the other must be tied by considerations of symmetry.' "[4]

The older faculty in the department mostly ignored Nash's putdowns and jibes. Ambrose did not. Soon a tit-for-tat rivalry was under way. Ambrose was famous, among other things, for detail. His blackboard notes were so dense that rather than attempt the impossible task of copying them, one of his assistants used to photograph them.[5] Nash, who disliked laborious, step-by-step expositions, found much to mock. When Ambrose wrote what Nash considered an ugly argument on the blackboard during a seminar, Nash would mutter, "Hack, Hack," from the back of the room.[6]

Nash made Ambrose the target of several pranks. "Seminar on the REAL mathematics!" read a sign that Nash posted one day. "The seminar will meet weekly Thursdays at 2 P.M. in the Common Room." Thursday at 2:00 P.M. was the hour that Ambrose taught his graduate course in analysis.[7] On another occasion, after Ambrose delivered a lecture at the Harvard mathematics colloquium, Nash arranged to have a large bouquet of red roses delivered to the podium as if Ambrose were a ballerina taking her bows.[8]

Ambrose needled back. He wrote "Fuck Myself" on the "To Do" list that Nash kept hanging over his desk on a clipboard.[9] It was he who nicknamed Nash "Gnash" for constantly making belittling remarks about other mathematicians.[10] And, during a discussion in the common room, after one of Nash's diatribes about hacks and drones, Ambrose said disgustedly, "If you're so good, why don't you solve the embedding problem for manifolds?" — a notoriously difficult problem that had been around since it was posed by Riemann.[11]

So Nash did.

Two years later at the University of Chicago, Nash began a lecture describing his first really big theorem by saying, "I did this because of a bet."[12] Nash's opening statement spoke volumes about who he was. He was a mathematician who viewed mathematics not as a grand scheme, but as a collection of challenging problems. In the taxonomy of mathematicians, there are problem solvers and theoreticians, and, by temperament, Nash belonged to the first group. He was not a game theorist, analyst, algebraist, geometer, topologist, or mathematical physicist. But he zeroed in on areas in these fields where essentially nobody had achieved anything. The thing was to find an interesting question that he could say something about.

Before taking on Ambrose's challenge, Nash wanted to be certain that solving the problem would cover him with glory. He not only quizzed various experts on the problem's importance, but, according to Felix Browder, another Moore Instructor, claimed to have proved the result long before he actually had.[13] When a mathematician at Harvard confronted Nash, recalled Browder, "Nash explained that he wanted to find out whether it was worth working on."[14]

"The discussion of manifolds was everywhere," said Joseph Kohn in 1995, gesturing to the air around him. "The precise question that Ambrose asked Nash in the common room one day was the following: Is it possible to embed any Riemannian manifold in a Euclidean space?"[15]

It's a "deep philosophical question" concerning the foundations of geometry that virtually every mathematician — from Riemann and Hilbert to Elie-Joseph Cartan and Hermann Weyl — working in the field of differential geometry for the past century had asked himself.[16] The question, first posed explicitly by Ludwig Schläfli in the 1870s, had evolved naturally from a progression of other questions that had been posed and partly answered beginning in the mid-nineteenth century.[17] First mathematicians studied ordinary curves, then surfaces, and finally, thanks to Riemann, a sickly German genius and one of the great figures of nineteenth-century mathematics, geometric objects in higher dimensions. Riemann discovered examples of manifolds inside Euclidean spaces. But in the early 1950s interest shifted to manifolds partly because of the large role that distorted space and time relationships had in Einstein's theory of relativity.

Nash's own description of the embedding problem in his 1995 Nobel autobiography hints at the reason he wished to make sure that solving the problem would be worth the effort: "This problem, although classical, was not much talked about as an outstanding problem. It was not like, for example, the four-color conjecture."[18]

Embedding involves portraying a geometric object as — or, a bit more precisely, making it a subset of — some space in some dimension. Take the surface of a balloon. You can't put it on a blackboard, which is a two-dimensional space. But you can make it a subset of spaces of three or more dimensions. Now take a slightly more complicated object, say a Klein bottle. A Klein bottle looks like a tin can whose lid and bottom have been removed and whose top has been stretched around and reconnected through the side to the bottom. If you think about it, it's obvious that if you try that in three-dimensional space, the thing intersects itself. That's bad from a mathematical point of view because the neighborhood in the immediate vicinity of the intersection looks weird and irregular, and attempts to calculate various attributes like distance or rates of change in that part of the object tend to blow up. But put the same Klein bottle into a space of four dimensions and the thing no longer intersects itself. Like a ball embedded in three-space, a Klein bottle in four-space becomes a perfectly well-behaved manifold.

Nash's theorem stated that any kind of surface that embodied a special notion of smoothness can actually be embedded in Euclidean space. He showed that you could fold the manifold like a silk handkerchief, without distorting it. Nobody would have expected Nash's theorem to be true. In fact, everyone would have expected it to be false. "It showed incredible originality," said Mikhail Gromov, the geometer whose book *Partial Differential Relations* builds on Nash's work. He went on:

> Many of us have the power to develop existing ideas. We follow paths prepared by others. But most of us could never produce anything comparable to what Nash produced. It's like lightning striking. Psychologically the barrier he broke is absolutely fantastic. He has completely changed the perspective on partial differential equations. There has been some tendency in recent decades to move from harmony to chaos. Nash says chaos is just around the corner.[19]

John Conway, the Princeton mathematician who discovered surreal numbers and invented the game of Life, called Nash's result "one of the most important pieces of mathematical analysis in this century."[20]

It was also, one must add, a deliberate jab at then-fashionable approaches to Riemannian manifolds, just as Nash's approach to the theory of games was a direct challenge to von Neumann's. Ambrose, for example, was himself involved in a highly abstract and conceptual description of such manifolds at the time. As Jürgen Moser, a young German mathematician who came to know Nash well in the mid-1950s, put it, "Nash didn't like that style of mathematics at all. He was out to show that this, to his mind, exotic approach was completely unnecessary since any such manifold was simply a submanifold of a high dimensional Euclidean space."[21]

Nash's more important achievement may have been the powerful technique he invented to obtain his result. In order to prove his theorem, Nash had to confront a seemingly insurmountable obstacle, solving a certain set of partial differential equations that were impossible to solve with existing methods.

That obstacle cropped up in many mathematical and physical problems. It was the difficulty that Levinson, according to Ambrose's letter, pointed out to Nash, and it is a difficulty that crops up in many, many problems — in particular, nonlinear problems.[22] Typically, in solving an equation, the thing that is given is some function, and one finds estimates of derivatives of a solution in terms of derivatives of the given function. Nash's solution was remarkable in that the *a priori* estimates lost derivatives. Nobody knew how to deal with such equations. Nash invented a novel iterative method — a procedure for making a series of educated guesses — for finding roots of equations, and combined it with a technique for smoothing to counteract the loss of derivatives.[23]

1. Virginia Nash with her children, Johnny and Martha, Bluefield, West Virginia, April 1935.

2. Martha and Johnny on a family holiday in Texas, circa 1939.

3. John Nash, Sr., napping in the company car, Bluefield, 1940s.

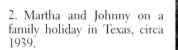

4, 5. John Nash standing tall—*above left*, at age six in Bluefield, and, *above right*, at his graduation at age twenty-one in Princeton, May 1950.

6. *Below left*, John Nash and his sister, Martha, Bluefield, fall 1948.

7. *Below right*, Martha, John Sr., John Jr., and Virginia Nash, Roanoke, summer 1954.

8. *Above left,* John Nash, Cambridge, Massachusetts, in the early 1950s.

9. *Above,* the common-room crowd at MIT, Cambridge: *left to right,* John Nash, Walter Weissblum, Israel Young, Donald Newman, Jacob Bricker.

10, 11. *Left,* Eleanor Stier in Boston in 1956, and, *below left,* in 1955 with her and John Nash's son, John David Stier.

12. *Below,* John Nash and John David.

13. Alicia Lopez-Harrison de Larde and Carlos Larde with their children, Rolando and Alicia, San Salvador, circa 1937.

14. Alicia Larde, John Nash's future wife, San Salvador, circa 1940.

15. John and Alicia Nash after their wedding, Washington, D.C., February 1957.

16. Berkeley, California, summer 1957: *left to right*, an unidentified person, John (standing) and Alicia Nash, and Felix and Eva Browder.

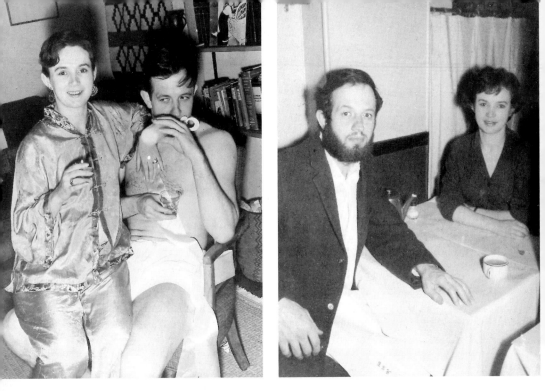

17. *Above left*, Alicia and John Nash (drinking out of a baby's bottle) at a New Year's Eve costume party, Needham, Massachusetts, 1958.

18. *Above right*, John and Alicia Nash in a Chinese restaurant, Paris, winter 1960.

19. *Below left*, Alicia Nash holding their son, John Charles Martin Nash, Washington, D.C., 1960.

20. *Below right*, John Nash with his niece Karla Nash, San Francisco, winter 1967.

21, 22. John Nash with his sons, John David Stier, *above left*, and John Charles Nash, *above right*, Princeton Junction, circa 1977.

23. John Charles Nash on the day he received his Ph.D. from Rutgers University, May 1985.

24, 25, 26. *Above left*, John and Alicia Nash at the Nobel ceremony in Stockholm, December 1994. *Above right*, John Nash bowing to the audience after receiving the Nobel medal from the King of Sweden, and, *below*, lecturing at the University of Uppsala a few days later.

Newman described Nash as a "very poetic, different kind of thinker."[24] In this instance, Nash used differential calculus, not geometric pictures or algebraic manipulations, methods that were classical outgrowths of nineteenth-century calculus. The technique is now referred to as the Nash–Moser theorem, although there is no dispute that Nash was its originator.[25] Jürgen Moser was to show how Nash's technique could be modified and applied to celestial mechanics — the movement of planets — especially for establishing the stability of periodic orbits.[26]

Nash solved the problem in two steps. He discovered that one could embed a Riemannian manifold in a three-dimensional space if one ignored smoothness.[27] One had, so to speak, to crumple it up. It was a remarkable result, a strange and interesting result, but a mathematical curiosity, or so it seemed.[28] Mathematicians were interested in embedding without wrinkles, embedding in which the smoothness of the manifold could be preserved.

In his autobiographical essay, Nash wrote:

So as it happened, as soon as I heard in conversation at MIT about the question of embeddability being open I began to study it. The first break led to a curious result about the embeddability being realizable in surprisingly low-dimensional ambient spaces provided that one would accept that the embedding would have only limited smoothness. And later, with "heavy analysis," the problem was solved in terms of embedding with a more proper degree of smoothness.[29]

Nash presented his initial, "curious" result at a seminar in Princeton, most likely in the spring of 1953, at around the same time that Ambrose wrote his scathing letter to Halmos. Emil Artin was in the audience. He made no secret of his doubts.

"Well, that's all well and good, but what about the embedding theorem?" said Artin. "You'll never get it."

"I'll get it next week," Nash shot back.[30]

One night, possibly en route to this very talk, Nash was hurtling down the Merritt Parkway.[31] Poldy Flatto was riding with him as far as the Bronx. Flatto, like all the other graduate students, knew that Nash was working on the embedding problem. Most likely to get Nash's goat and have the pleasure of watching his reaction, he mentioned that Jacob Schwartz, a brilliant young mathematician at Yale whom Nash knew slightly, was also working on the problem.

Nash became quite agitated. He gripped the steering wheel and almost shouted at Flatto, asking whether he had meant to say that Schwartz had solved the problem. "I didn't say that," Flatto corrected. "I said I heard he was working on it."

"Working on it?" Nash replied, his whole body now the picture of relaxation. "Well, then there's nothing to worry about. He doesn't have the insights I have."

Schwartz was indeed working on the same problem. Later, after Nash had produced his solution, Schwartz wrote a book on the subject of implicit-function theorems. He recalled in 1996:

I got half the idea independently, but I couldn't get the other half. It's easy to see an approximate statement to the effect that not every surface can be exactly embedded, but that you can come arbitrarily close. I got that idea and I was able to produce the proof of the easy half in a day. But then I realized that there was a technical problem. I worked on it for a month and couldn't see any way to make headway. I ran into an absolute stone wall. I didn't know what to do. Nash worked on that problem for two years with a sort of ferocious, fantastic tenacity until he broke through it.[32]

Week after week, Nash would turn up in Levinson's office, much as he had in Spencer's at Princeton. He would describe to Levinson what he had done and Levinson would show him why it didn't work. Isadore Singer, a fellow Moore instructor, recalled:

He'd show the solutions to Levinson. The first few times he was dead wrong. But he didn't give up. As he saw the problem get harder and harder, he applied himself more, and more and more. He was motivated just to show everybody how good he was, sure, but on the other hand he didn't give up even when the problem turned out to be much harder than expected. He put more and more of himself into it.[33]

There is no way of knowing what enables one man to crack a big problem while another man, also brilliant, fails. Some geniuses have been sprinters who have solved problems quickly. Nash was a long-distance runner. If Nash defied von Neumann in his approach to the theory of games, he now took on the received wisdom of nearly a century. He went into a classical domain where everybody believed that they understood what was possible and not possible. "It took enormous courage to attack these problems," said Paul Cohen, a mathematician at Stanford University and a Fields medalist.[34] His tolerance for solitude, great confidence in his own intuition, indifference to criticism — all detectable at a young age but now prominent and impermeable features of his personality — served him well. He was a hard worker by habit. He worked mostly at night in his MIT office — from ten in the evening until 3:00 A.M. — and on weekends as well, with, as one observer said, "no references but his own mind" and his "supreme self-confidence." Schwartz called it "the ability to continue punching the wall until the stone breaks."

The most eloquent description of Nash's single-minded attack on the problem comes from Moser:

> The difficulty [that Levinson had pointed out], to anyone in his right mind, would have stopped them cold and caused them to abandon the problem. But Nash was different. If he had a hunch, conventional criticisms didn't stop him. He had no background knowledge. It was totally uncanny. Nobody could understand how somebody like that could do it. He was the only person I ever saw with that kind of power, just brute mental power.[35]

The editors of the *Annals of Mathematics* hardly knew what to make of Nash's manuscript when it landed on their desks at the end of October 1954. It hardly had the look of a mathematics paper. It was as thick as a book, printed by hand rather than typed, and chaotic. It made use of concepts and terminology more familiar to engineers than to mathematicians. So they sent it to a mathematician at Brown University, Herbert Federer, an Austrian-born refugee from Nazism and a pioneer in surface area theory, who, although only thirty-four, already had a reputation for high standards, superb taste, and an unusual willingness to tackle difficult manuscripts.[36]

Mathematics is often described, quite rightly, as the most solitary of endeavors. But when a serious mathematician announces that he has found the solution to an important problem, at least one other serious mathematician, and sometimes several, as a matter of longstanding tradition that goes back hundreds of years, will set aside his own work for weeks and months at a time, as one former collaborator of Federer's put it, "to make a go of it and straighten everything out."[37] Nash's manuscript presented Federer with a sensationally complicated puzzle and he attacked the task with relish.

The collaboration between author and referee took months. A large correspondence, many telephone conversations, and numerous drafts ensued. Nash did not submit the revised version of the paper until nearly the end of the following summer. His acknowledgment to Federer was, by Nash's standards, effusive: "I am profoundly indebted to H. Federer, to whom may be traced most of the improvement over the first chaotic formulation of this work."[38]

Armand Borel, who was a visiting professor at Chicago when Nash gave a lecture on his embedding theorem, remembers the audience's shocked reaction. "Nobody believed his proof at first," he recalled in 1995. "People were very skeptical. It looked like a [beguiling] idea. But when there's no technique, you are skeptical. You dream about a vision. Usually you're missing something. People did not challenge him publicly, but they talked privately."[39] (Characteristically, Nash's report to his parents merely said "talks went well.")[40]

Gian-Carlo Rota, professor of mathematics and philosophy at MIT, confirmed Borel's account. "One of the great experts on the subject told me that if one of his graduate students had proposed such an outlandish idea he'd throw him out of his office."[41]

The result was so unexpected, and Nash's methods so novel, that even the experts had tremendous difficulty understanding what he had done. Nash used to leave drafts lying around the MIT common room.[42] A former MIT graduate student recalls a long and confused discussion between Ambrose, Singer, and Masatake Kuranishi (a mathematician at Columbia University who later applied Nash's result) in which each one tried to explain Nash's result to the other, without much success.[43]

Jack Schwartz recalled:

Nash's solution was not just novel, but very mysterious, a mysterious set of weird inequalities that all came together. In my explication of it I sort of looked at what happened and could generalize and give an abstract form and realize it was applicable to situations other than the specific one he treated. But I didn't quite get to the bottom of it either.[44]

Later, Heinz Hopf, professor of mathematics in Zurich and a past president of the International Mathematical Union, "a great man with a small build, friendly, radiating a warm glow, who knew everything about differential geometry," gave a talk on Nash's embedding theorem in New York.[45] Usually Hopf's lectures were models of crystalline clarity. Moser, who was in the audience, recalled: "So we thought, 'NOW we'll understand what Nash did.' He was naturally skeptical. He would have been an important validator of Nash's work. But as the lecture went on, my God, Hopf was befuddled himself. He couldn't convey a complete picture. He was completely overwhelmed."[46]

Several years later, Jürgen Moser tried to get Nash to explain how he had overcome the difficulties that Levinson had originally pointed out. "I did not learn so much from him. When he talked, he was vague, hand waving, 'You have to control this. You have to watch out for that.' You couldn't follow him. But his written paper was complete and correct."[47] Federer not only edited Nash's paper to make it more accessible, but also was the first to convince the mathematical community that Nash's theorem was indeed correct.

Martin's surprise proposal, in the early part of 1953, to offer Nash a permanent faculty position set off a storm of controversy among the eighteen-member mathematics faculty.[48] Levinson and Wiener were among Nash's strongest supporters. But others, like Warren Ambrose and George Whitehead, the distinguished topologist, were opposed. Moore Instructorships weren't meant to lead to tenure-track positions. More to the point, Nash had made plenty of enemies and few friends in his first year and a half. His disdainful manner toward his colleagues and his poor record as a teacher rubbed many the wrong way.

Mostly, however, Nash's opponents were of the opinion that he hadn't proved he could produce. Whitehead recalled, "He talked big. Some of us were not sure he could live up to his claims."[49] Ambrose, not surprisingly, felt similarly. Even Nash's champions could not have been completely certain. Flatto remembered one occasion on which Nash came to Levinson's office to ask Levinson whether he'd read a draft of his embedding paper. Levinson said, "To tell you the truth I don't have enough background in this area to pass judgment."[50]

When Nash finally succeeded, Ambrose did what a fine mathematician and sterling human being would do. His applause was as loud as or louder than anyone else's. The bantering became friendlier and, among other things, Ambrose took to telling his musical friends that Nash's whistling was the purest, most beautiful tone he had ever heard.[51]

PART TWO

Separate Lives

21

Singularity

Nash was leading all these separate lives. Completely separate lives. —ARTHUR
MATTUCK, 1997

ALL THROUGH HIS CHILDHOOD, adolescence, and brilliant student career,
Nash had seemed largely to live inside his own head, immune to the emotional
forces that bind people together. His overriding interest was in patterns, not people,
and his greatest need was making sense of the chaos within and without by em-
ploying, to the largest possible extent, the resources of his own powerful, fearless,
fertile mind. His apparent lack of ordinary human needs was, if anything, a matter
of pride and satisfaction to him, confirming his own uniqueness. He thought of
himself as a rationalist, a free thinker, a sort of Spock of the starship *Enterprise*.
But now, as he entered early adulthood, this unfettered persona was shown to be
partly a fiction or at least partly superseded. In those first years at MIT, he discovered that he had some of the same wishes as others. The cerebral, playful, calculating, and episodic connections that had once sufficed no longer served. In five short
years, between the ages of twenty-four and twenty-nine, Nash became emotionally
involved with at least three other men. He acquired and then abandoned a secret
mistress who bore his child. And he courted — or rather was courted by — a woman
who became his wife.

As these initial intimate connections multiplied and became ever-present elements in his consciousness, Nash's formerly solitary but coherent existence became
at once richer and more discontinuous, separate and parallel existences that reflected an emerging adult but a fragmented and contradictory self. The others on
whom he now depended occupied different compartments of his life and often, for
long periods, knew nothing of one another or of the nature of the others' relation
to Nash. Only Nash was in the know. His life resembled a play in which successive
scenes are acted by only two characters. One character is in all of them while the
second changes from scene to scene. The second character seems no longer to
exist when he disappears from the boards.

More than a decade later, when he was already ill, Nash himself provided a
metaphor for his life during the MIT years, a metaphor that he couched in his first
language, the language of mathematics: B squared $+$ RTF $= 0$, a "very personal"
equation Nash included in a 1968 postcard that begins, "Dear Mattuck, Thinking

that you will understand this concept better than most I wish to explain . . ." The equation represents a three-dimensional hyperspace, which has a singularity at the origin, in four-dimensional space. Nash is the singularity, the special point, and the other variables are people who affected him — in this instance, men with whom he had friendships or relationships.[1]

Inevitably, the accretion of significant relationships with others brings with it demands for integration — the necessity of having to choose. Nash had little desire to choose one emotional connection over another. By not choosing, he could avoid, or at least minimize, both dependence and demands. To satisfy his own emotional needs for connectedness meant he inevitably made others look to him to satisfy theirs. Yet while he was preoccupied with the effect of others on him, he mostly ignored — indeed, seemed unable to grasp — his effect on others. He had in fact no more sense of "the Other" than does a very young child. He wished the others to be satisfied with his genius — "I thought I was such a great mathematician," he was to say ruefully, looking back at this period — and, of course, to some extent they were satisfied. But when people inevitably wanted or needed more he found the strains unbearable.

22

A Special Friendship

Santa Monica, Summer 1952

Away from contact with a few special sorts of individuals I am lost, lost completely in the wilderness . . . so, so, so, it's been a hard life in many ways.
—JOHN FORBES NASH, JR., 1965

AFTER JOHN NASH LOST EVERYTHING — family, career, the ability to think about mathematics — he confided in a letter to his sister Martha that only three individuals in his life had ever brought him any real happiness: three "special sorts of individuals" with whom he had formed "special friendships."[1]

Had Martha seen the Beatles' film *A Hard Day's Night*? "They seem very colorful and amusing," he wrote. "Of course they are much younger like the sort of person I've mentioned. . . . I feel often as if I were similar to the girls that love the Beatles so wildly since they seem so attractive and amusing to me."[2]

Nash's first loves were one-sided and unrequited. "Nash was always forming intense friendships with men that had a romantic quality," Donald Newman observed in 1996. "He was very adolescent, always with the boys."[3] Some were inclined to see Nash's infatuations as "experiments," or simple expressions of his immaturity — a view that he may well have held himself. "He played around with it because he liked to play around. He was very experimental, very try-outish," said Newman in 1996. "Mostly he just kissed."[4]

Newman, who liked to joke about his past and future female conquests,[5] had firsthand knowledge because Nash was, for a time, infatuated with him — with predictable results. "He used to talk about how Donald looked all the time," Mrs. Newman said in 1996.[6] Newman recalled: "He tried fiddling around with me. I was driving my car when he came on to me." D.J. and Nash were cruising around in Newman's white Thunderbird when Nash kissed him on the mouth. D.J. just laughed it off.[7]

Nash's first experience of mutual attraction — "special friendships," as he called them — occurred in Santa Monica.[8] It was the very end of the summer of 1952, after Milnor had moved out and Martha had flown back home. The encounter must have been fleeting, coming in the last days of August, just before he was due to leave for Boston, and very furtive. But it was nonetheless decisive because for the first time he found not rejection but reciprocity. Thus it was the first real step out of his extreme emotional isolation and the world of relationships that were purely imaginary, a first taste of intimacy, not entirely happy, no doubt, but suggestive of hitherto unsuspected satisfactions.

The only traces of Nash's friendship with Ervin Thorson that remain are his description of him as a "special" friend in his 1965 letter and a series of elliptical references to "T" in letters in the late 1960s.[9] Few if any of Nash's acquaintances met him; Martha recalled a friend of Nash's who once spent the night on the couch of their Georgina Avenue apartment, but not his name.[10]

Thorson, who died in 1992, was thirty years old in 1952.[11] He was a native Californian of Scandinavian extraction. Nash described him to Martha as an aerospace engineer, but he may in fact have been an applied mathematician. He had been a meteorologist in the Army Air Corps during the war. Afterward, he earned a master's degree in mathematics at UCLA and went to Douglas Aircraft in 1951, just a few years after Douglas had spun off its R&D division to form the RAND Corporation.[12] At that time, Douglas was mapping the future of interplanetary travel for the Pentagon, and Thorson, who eventually led a research team, was very likely involved in these efforts.[13] His great passion, conceived twenty years before the United States launched *Viking,* was the dream of exploring Mars, his sister Nelda Troutman recalled in 1997.

Thorson was, his sister said, "very high strung, not a social person at all, very bright, knew a lot, very very academic."[14] Nash could easily have met him — given the close ties between Douglas and RAND, which was also heavily involved in studies of space exploration — at a talk or seminar, or perhaps even at one of the parties that John Williams, the head of RAND's mathematics department, gave.

If Thorson, who never married, was a homosexual, his surviving sister did not know it.[15] With his family, at any rate, he was unusually closemouthed, not just about his work, which was highly classified, but about all aspects of his personal life.[16] Given the mounting pressure to root out homosexuals in the defense industry during the McCarthy era, Thorson would have had to practice great discretion in any case; his career at Douglas was to last for another fifteen years.[17] When he abruptly resigned from Douglas in 1968, he apparently did so at the age of forty-seven because he feared dying. Several of his colleagues had recently died of heart attacks and Thorson, who had some sort of mild heart condition, decided he couldn't cope with the stress and overwork anymore. He moved back to his hometown of Pomona and became a virtual recluse except for an active involvement in the Lutheran church, living with his parents for the next twenty-five years until his death.

Whether Nash and Thorson saw each other again when Nash returned to Santa Monica for a third summer two years later or on one of his trips to Santa Monica during his illness in the early and mid-1960s is not known. But Nash continued to think of Thorson and to refer to him obliquely until at least 1968.

23
Eleanor

These mathematicians are very exclusive. They occupy a very high terrain, from which they look down on everyone else. That makes their relationships with women quite problematic. — ZIPPORAH LEVINSON, 1995

NASH WAS BACK in Boston in his old quarters by Labor Day. Number 407 Beacon Street was an imposing brick row house built before the turn of the century facing the Charles.[1] Its current owner, Mrs. Austin Grant, was the widow of a Back Bay physician. She liked to point out her home's opulent features to her lodgers, such as the carriage room where its original owners once waited for their horse-drawn carriages to be brought around. And she often bemoaned the neighborhood's decline. "Don't leave your bags on the street while you come in; they might not be there when you come out again," she said to Nash the day he moved in.

Nash occupied one of the front bedrooms, a large, comfortably furnished room with a fireplace. Lindsay Russell, a young engineer who had recently graduated from MIT, lived next door. Mrs. Grant regularly took Russell aside to remark on Nash's idiosyncrasies. Nash acquired a huge set of barbells and began lifting weights. When Nash made the dining-room chandelier, which hung directly below his bedroom, vibrate with his exertions, Mrs. Grant would say, "What does he think this is? A gymnasium?" Nash's mail also received comment, particularly the postcards from his mother expressing the hope, as Russell recalled, that "in addition to the pursuit of mathematics and other intellectual pursuits, he would make friends and engage in social activities."

With one single exception, however, Nash never had any visitors. Russell remembers once waking up in the middle of the night. There was a sound coming from Nash's room. It was a giggle. The giggle of a woman.

The pretty, dark-haired nurse who admitted Nash to the hospital on the second Thursday in September was named Eleanor.[2] He was due to have some varicose veins removed[3] and seemed awfully nervous — and young, more like a student than a professor.[4] Eleanor knew his doctor to be a notorious incompetent.[5] And a drunk. She was curious how an MIT professor had wound up with a quack like that. Nash told her that he'd chosen the doctor at random by closing his eyes and running his fingers down the list of physicians in the lobby. She felt, she recalled, rather protective of him.

Nash was on the ward for only a couple of days. Eleanor thought he was cute and sort of sweet, but when he left, she hardly expected to see him again. Somehow or other, they bumped into each other on the street not long afterward. It was a Saturday afternoon and Eleanor was on her way to meet a friend to buy herself a good winter coat. "I didn't chase him. He chased me. He kept pestering me," Eleanor recalled. "I wound up going shopping with him."[6]

They walked over to Jay's Department Store together. Nash followed her up to the coat department, which was on the second floor. He kept staring at her, not saying much, waiting for her to choose a coat. She started to enjoy herself. "John was very attractive," Eleanor recalled, laughing. "When I saw him, I thought he was something special." She began pointing to the ones she wanted to try on, and with elaborate courtesy he held out each coat for her to slip into. She thought she liked a purple one best. Nash started clowning around. He pretended he was her tailor, flung himself on his knees before her, loudly made believe he was measuring her coat for alterations — and generally made a fool of himself. Embarrassed, Eleanor blushed, protested, and tried to hush him up. "Get up quick!" she whispered. Secretly, however, she was quite thrilled.

At twenty-nine, Eleanor was an attractive, hardworking, tenderhearted woman. A friend of Nash's later described her as "dark and pretty, quite shy, a good person" of "ordinary intelligence," with "simple manners" and "a very peculiar way of speaking."[7] By that the friend meant that her accent was pure New England. Life hadn't been very kind to her. She'd grown up in Jamaica Plain, a dreary blue-collar section of Boston.[8] She'd had a hardscrabble childhood, a harsh mother, and the burden, far too heavy for a young girl, of caring for a younger half-brother. She missed a great deal of school as a result. She was, on the whole, grateful to be able to take up a profession, practical nursing, that she enjoyed and that provided her with steady work. Her mother died of tuberculosis when Eleanor was eighteen. Her early experiences endowed her with a soft heart. She had a deep appreciation, which stayed with her all her life, for what it was like to be poor and vulnerable. It brought out a tenderness in her, toward patients, neighbors, other people's children, and stray animals. She was the kind of woman who, later in life, would literally give coats to strangers and invite people who had nowhere else to stay into her home.[9]

Shy and lacking confidence, Eleanor also tended to be suspicious and guarded, especially around men. She said, in an interview, "I wasn't a bad girl. I didn't run around with a lot of men. In fact, I was really good. I was a little afraid of men. I didn't want to be involved with them sexually. I thought it was kind of disgusting."[10] But Nash disarmed her from the start. Yes, he was an MIT professor, yes, he came from an upper-class sort of background, yes, he did top-secret work for the government. But he was also very young, five years Eleanor's junior, and there was a sweetness about him, a lack of guile. She sensed, moreover, that he was, if anything, less experienced than she was.

After that Saturday afternoon, Nash took her out for cheap meals and drove her around in his beat-up car. He talked about himself, his work, the department, his friends — endlessly. He hardly asked her anything about herself, something that relieved rather than distressed her. She wasn't eager to share the rather dispiriting details of her modest background, particularly as Nash hinted that his own ancestry was rather distinguished. He pressed her to let him come up to her apartment. She wouldn't let him at first. She didn't want to seem easy. But she finally agreed to go to his place. She found him eager, ardent, but not frightening.

That Nash, who had preferred dancing with chairs to dancing with girls as an adolescent and who had given the pretty Ruth Hincks not so much as a real glance, progressed so swiftly and had so suddenly and at that particular moment found his way into a woman's arms suggests either love at first sight or some resolution "to take the plunge." The encounter with Thorson might have provided the impetus. Nash may have been looking to repeat a loving experience, or he may have been looking for confirmation of his own "masculinity." On a number of occasions he asked Eleanor to provide him with steroids. "There were always big bottles of stuff around the places I worked as a nurse," said Eleanor.[11] Although she later said that she never acceded to Nash's requests, she believed that "he delved into drugs" hoping that they "would make him more manly."[12] He wasn't proving his interest in women to the world, however; he kept his liaison with Eleanor a deep dark secret for years, even while he displayed his infatuation with various men more or less in public.

Caught up as he was with teaching, seminars, and work on his embedding problem that fall, Nash nonetheless managed to see Eleanor frequently. He confided in her. He enjoyed being alone with her. He liked going over to her place and having her cook him dinner. She cooked very well. She fussed over him. Most of all, she was womanly, full of warmth and artless affection. For Nash, who had never even known a woman other than his mother and sister, it was a novel experience.

As for the gulf between their educations and social statuses, what more time-honored formula for romance and eventual marriage than Eliza Doolittle meets Professor Higgins? For Eleanor, Nash was a chance for a life she could not possibly have achieved on her own; for Nash, she was the prospect of retaining, to put it bluntly, the upper hand. It was a compelling fantasy and a highly practical arrangement rolled into one. And the same thing went for the difference in temperaments. Matches between egocentric and childish men and self-abnegating and maternal women abound in the history of genius. Nash was looking for emotional partners who were more interested in giving than receiving, and Eleanor, as her entire life testified, was very much that sort.

Nash thought about introducing Eleanor to his mathematical friends and about taking her around to one of the department parties. But he decided against it. The fact that nobody at MIT knew that Eleanor existed made the affair even more delicious.

By election day in early November, Eleanor strongly suspected that she was pregnant. On Thanksgiving, when she invited Nash to come to her place, she was absolutely certain, having missed a second period by then.

Nash seemed, oddly enough, more pleased than panicked.[13] He seemed proud of fathering a child. In fact, he made it clear that he found the notion of progeny quite attractive. (Later, when such things became fashionable, he talked about joining a sperm bank for geniuses in California.)[14] He hoped that the baby would be a boy. He wanted the baby to be called John. He did not, however, say anything about marriage, Eleanor's future, or, for that matter, how she and the baby would manage.

Eleanor hardly knew what to make of his reaction. She had hoped, of course, that he would see the pregnancy as a crisis to be solved by an offer of marriage. When this was not forthcoming, she did her best to hide her disappointment from him. She comforted herself with the thought that he was, after all, a remarkable young man. She told herself that, of course, he loved her and would do the right thing "in the end." In any case, she found that the idea of having a baby made her feel quite sentimental. The subject of an abortion — illegal but available if one had the money — never came up.

Before long, however, the relationship between the lovers lost its playful and light-hearted quality. That winter, Eleanor was often tense and tired. She fretted a great deal about the symptoms of pregnancy and the long hours at the hospital. Nash's mind was, more often than not, elsewhere. Soon, he and Eleanor were engaged in a tug of war that occasionally turned quite ugly.

When Eleanor irritated him with her complaints, Nash would needle her. He called her stupid and ignorant. He made fun of her pronunciation. He reminded her that she was five years older. Mostly, however, he made fun of her desire to marry him. An MIT professor, he would say, needed a woman who was his intellectual equal. "He was always putting me down," she recalled. "He was always making me feel inferior."[15]

She, in turn, began to resent what she called his superior airs and lack of sensitivity. Their evenings together frequently degenerated into nasty spats. Eleanor, a friend of Nash's later reported, once complained that Nash had pushed her down a flight of stairs.[16]

But there were also tender moments — when, for example, Nash told her that he liked the way she looked with her big belly — and Eleanor's feelings about Nash were, on the whole, loving. She was convinced that he loved her and would do right by the baby, whom he seemed to be looking forward to with great eagerness. She still recalled that period of their relationship as "beautiful."[17] She excused his cruelty by telling herself that it was occasional, that "he didn't know how to live." She put it down to his having achieved extraordinary success at too young an age. "That can be overwhelming," she later said.[18]

In the late spring when she could no longer work, Eleanor moved into a home for unwed mothers. Around that time, Nash finally introduced her to one of his friends from MIT, a graduate student.[19] Eleanor took this as an encouraging sign.

John David Stier was born on June 19, 1953, six days after Nash's twenty-fifth birthday. Nash rushed to the hospital and was greatly excited when Eleanor presented him with their son.[20] He stayed as long as the nurses would let him and came back at every opportunity. But he did not offer to put his name on his son's birth certificate,[21] and he did not offer to pay for the baby's delivery.[22]

Mother and son came home to an apartment Nash had moved to on Park Drive. It wasn't a happy homecoming. Nash wouldn't buy any baby clothes, Eleanor recalled. "He didn't want us to stay," she said years later. Eleanor finally managed to find a live-in position with an employer who would let her keep her infant with her.[23] Despite the employer's insistence on "no male visitors," Nash came over frequently. "He wanted to be around him all the time," Eleanor recalled.[24] But he still did not offer to marry Eleanor or to support her, although his professor's salary and frugal habits surely would have made that possible.

His visits eventually resulted in Eleanor's being fired.[25] The simultaneous loss of her job and her living arrangements created an immediate crisis. With Nash still unwilling to care for her and the baby, Eleanor was finally forced to place John David in foster care.[26]

Like some hapless heroine of a Victorian melodrama, Eleanor left her baby with a series of families, one in Rhode Island, another in Stoneham, Massachusetts, and, finally, at an orphanage whose sentimental name, the New England Home for Little Wanderers, only underscored the Dickensian realities into which she and her son were plunged.[27] Founded during the Civil War, the home was on the southern outskirts of Boston, across the Charles River from the Veterans' Hospital, a good hour by bus from her apartment in Brookline. Eleanor visited her son on Saturdays and Sundays. John Stier remembers standing in the stairwell landing there, peering out of the window, feeling a terrible loneliness and homesickness.[28] Sometimes she brought him back to her apartment where she kept a large supply of toys and baby books.[29]

Being separated from the baby nearly drove Eleanor mad. More than anything that had gone on before, it made her feel real bitterness toward Nash, who, she believed, left all the anguish and the worry to her and gave no sign that he understood, even remotely, what such a separation might mean for a mother or her child. "I should have been home to take care of him," Eleanor said in 1995. "I worried. [Nash] never worried."[30]

• • •

Yet the affair continued. They visited the baby, wherever he was, on Sundays. Eleanor came over to Nash's apartment and cooked and, when he demanded it, cleaned for him. Nash also went around to her place for meals.[31] He continued to oscillate between sweetness and outbursts of cruelty. He continued to keep his affair with Eleanor under wraps, told no one at first except Jack Bricker, who was enjoined to keep the secret. "He never told anyone about us," said Eleanor, still unable to fathom his behavior.[32] Most of the MIT mathematics community, in fact, did not learn of the existence of his first family until years later.

When John David was a year old, Nash introduced Eleanor to another friend in the department, Arthur Mattuck, without, however, revealing the baby's existence.[33] He and Eleanor sometimes had Mattuck, who seemed to like Eleanor, over to dinner. They told Mattuck afterward that they always had a good laugh after he left because Mattuck never noticed all the baby things around the apartment. It was, to say the least, a strange state of affairs.

Or was it? Eleanor was in love with Nash. "People told me never to see him again," said she. "It's better if you have a normal man. Not one who's all puffed up by his own importance. One of my friends said that you didn't see a thing in his face. It was like a dead person. I didn't think so, though."[34] She mused many years later: "Did I love him? I wouldn't have gone with someone I didn't love. He was awkward. His awkwardness seemed standoffish. But . . . he could be very sweet. He was very attractive in a way. Love is foolish."[35]

As late as 1955 and 1956, after Nash introduced Mattuck to Eleanor, Eleanor's attitude toward Nash was "adoring." Mattuck recalled: "Eleanor realized Nash was a total egoist, but she was dazzled by his brilliance. He thought he was a genius. She was sleeping with one of the smartest men in America. Did he love her? She didn't know. She didn't ask. In those days, it wasn't 'Talk to me.' If you slept with a man, you assumed he loved you."[36]

Eleanor also continued to hope that Nash would marry her, if only for the sake of their son. Nash wasn't, she was sure, seeing another woman. Nash's failure to disappear from her life, despite his tantrums and complaints about her, must have seemed to Eleanor powerful evidence that he did, after all, love her, and would ultimately come around. How else to explain her passivity — her unhappy acceptance, but acceptance nonetheless, of his refusal to pay for her and the baby's support — until it was, as it were, too late, until a rival appeared on the scene? She might have threatened him with exposure, or with a lawsuit, but, because she believed he would marry her eventually, she feared alienating him and thus ruining her chances for good. It was only much later, in 1956, after Eleanor discovered that Nash was having an affair with an MIT physics student and concluded that he intended to marry the girl — possibly even before Nash himself reached that decision — that she took more aggressive action.

Nash's behavior is a bit more mysterious. Why did he keep coming around, even though he had reached the conclusion that Eleanor wasn't good enough for him or his social circle? Perhaps he simply hadn't made up his mind. In the late summer of 1954, for example, he was carrying a photograph of Eleanor and John

David in his wallet, and he told at least one person, "This is the woman I plan to marry and our son."[37] Perhaps he felt that the decision to have the child was strictly Eleanor's. Quite possibly, Eleanor's passivity in the face of his own bad behavior might have signaled to him that she was content to be his mistress and resigned to living apart from her child. Perhaps each, by his or her actions, misled the other.

Whether Nash ever intended to marry Eleanor is a matter of dispute. Arthur Mattuck believes he did, but that he was talked out of it by Bricker.[38] Bricker's recollection differs radically. He remembers having tried to persuade Nash but said that "Nash's mind was made up."[39] We aren't likely to learn which account is the more accurate. Perhaps both were, at different points in time. Nash didn't marry Eleanor, despite his stated intentions on at least one occasion.

One likely reason was Nash's snobbery, the roots of which went back to his Bluefield upbringing. Not for him a wife, however adoring, who pronounced words incorrectly, whose manners were simple, and whose sense of social inferiority would have made it difficult for her to mingle comfortably with the other wives in the Cambridge mathematical community. Unconventional as he was, Nash's obsession with class and surface propriety were as strong as his father's. This certainly was Eleanor's perception, and while that perception was no doubt colored by resentment, it seems accurate.

It wasn't only social snobbery, though. Nash didn't believe that Eleanor was educated enough to be a good mother to his children. His own mother was a schoolteacher who devoted a great deal of time to seeing that her children spoke grammatically, after all. Moreover, he may simply have found Eleanor boring, a thesis that Arthur Mattuck put forward and that gains some credence from the fact that Nash ultimately married a young woman who never cooked but possessed a degree in physics and career ambitions. Eleanor said as much: "He wanted to marry a real intellectual girl. He wanted to marry somebody in the same capacity as he was."[40]

Whatever went through Nash's mind regarding marriage in the four years that Eleanor was his mistress, he did at one point make a proposal that suggested that he had made up his mind he wouldn't marry her.

Nash suggested to Eleanor that she give John David up for adoption. He more or less told her openly that John David would be better off if she gave him up. "He wanted to have John adopted," Eleanor later said bitterly. " 'We'd always know where he was,' he'd say."[41]

It was a cold-blooded suggestion, and it all but killed any remaining love Eleanor felt for Nash. One only hopes that among Nash's considerations in putting it forward — apart from eliminating any financial responsibility he might face for his child, which prompted Eleanor to say that Nash "wanted everything for nothing" —

might have been a genuine belief that John David's chances in life would be greater with some middle-class couple than with his single, working mother.

"Everybody wanted him," Eleanor recalled. "Some people even offered me a lot of money to let them have him. It was frightening. There were these wealthy people who were taking care of John David. They were going to move to California. If they'd gone to California, I would never have seen him again."[42]

For the first six years of John David's life, during which time the little boy was shifted from home to home, father and son saw each other from time to time. One photograph, taken in what appears to be a city park, of the two-year-old with his long face framed by a woolen hat with funny flaps, standing tall like a little soldier, hand in hand with his sweet-faced, girlish-looking mother, bareheaded, wearing a trim woolen coat, smiling into the eyes of the camera held, no doubt, by her lover, evokes the flavor of these brief visits. "She shouldn't have had a baby, she shouldn't have been so gullible," John Stier later said,[43] but somehow, looking at the evidence of that scene, it is impossible for him, or anyone else, to deny the feeling that this little trio, out on a Sunday outing, was indeed a family in every sense but a legal one.

Nash displayed a rather curious inconsistency in his attitude and behavior toward his son. At the time of his birth, he had reacted in neither of the ways one might have expected of a young man confronted with the pregnancy of a woman with whom he has recently begun sleeping, eschewing both the high road that would have led to a shotgun wedding, as well as the more commonly elected low road of flat-out denying his paternity and simply vanishing from his girlfriend's life.

He doubtless behaved selfishly, even callously. His son and others later attributed his acknowledgment of paternity and desire to maintain a bond, even while failing to protect his child from poverty and periodic separation from his mother, to a pure narcissism. But even if this is partly true, it is natural to conclude that Nash, like the rest of us, needed to love and to be loved, and that a tiny, helpless infant, his son, drew him irresistibly.

In 1959, when Nash suddenly disappeared from John David's life altogether, a badly wrapped, broken-up package arrived one day containing a smashed but beautifully made wooden airplane, "a lovely thing," as John David later recalled. "There was no return address, or note or anything, but I knew it was from my father."[44]

24
Jack

Nash met Jack Bricker in the fall of 1952 in the MIT common room. Bricker, a first-year graduate student from New York, knew Newman and some of the others from City College's math table and quickly became one of the regulars in the common room.[1]

Just two years Nash's junior, Bricker was immediately dazzled by Nash. He was "mesmerized," "hypnotized," and "enamored," a few of the words contemporaries used to describe his reaction to Nash. Bricker "was overwhelmed by Nash's smartness," Mattuck said in 1997. "Nash was the smartest person he'd ever met. He worshiped Nash's intellect."[2] It wasn't only Nash's intellect, though. It was everything else too: the southern breeding, Princeton pedigree, good looks, and self-confidence.

Bricker, by contrast, was short, skinny, full of angst.[3] He had grown up poor in Brooklyn; he still dressed badly, was often broke, and fretted over his lack of experience with girls. Although he was undeniably bright—the logician Emil Post considered him the best mathematician in his class at City—his self-doubt bordered on the pathological. "There's no hope" and "It's useless" were his most-often-used expressions. Yet he was endearing in his own way. His sense of humor—dark, self-deprecating, very New York—was always on tap even when he was depressed, which was much of the time. People liked talking to him because he was interested, acute, and responsive. Awkward as he was, he had a way of putting others at their ease. He was, as Gus Solomon once described him, "the world's greatest audience."

Perhaps for this reason, Bricker caught Nash's eye. Nash, usually so disdainful of lesser minds, made a point of getting Bricker off by himself. Bricker liked to play Lasker—a board game named after a chess champion that became popular in the late 1940s—and Nash started playing with him. "We became Lasker partners," said Bricker in 1997. "That's how we got to know each other."[4] Soon they were taking long, aimless rides in Nash's Studebaker, with Nash behind the wheel, playing with the back of Bricker's neck as he drove.[5] They became friends—and then more than friends.

Donald Newman and the rest of the MIT crowd watched Nash and Bricker with amused tolerance and concluded that the two were having a romance.[6] "They were importantly interested in each other," Newman said; they made no secret of their affection, kissing in front of other people.[7] "Bricker hero-worshiped John," Eleanor recalled. "He was always hanging around. They were always patting each other."[8] Nash himself, in his 1965 letter, described his relationship with Bricker as one of three "special friendships" in his life.[9] The special friendship with Bricker lasted, on and off, for nearly five years until Nash married.

Once Nash had told Herta Newman, Donald's wife, that he realized "there was something that happened between people that he didn't experience."[10] What was missing from Nash's life, to a singular degree, was what the biographer of another genius called "the strong force that binds people together."[11] Now he knew what that was.

It was this sense of vital connection that Nash referred to in his letter to Martha when it dawned on him that away from special sorts of individuals, the Brickers in his life, young men who were "colorful," "amusing," and "attractive," he was "lost, lost, lost completely in the wilderness . . . condemned to a hard hard hard life in many ways."[12]

The experience of loving and being loved subtly altered Nash's perception of himself and the possibilities open to him. He was no longer an observer in the game of life, but an active participant. He was no longer a thinking machine whose sole joys were cerebral. Yet his was not a passionate nature. Love, though thrilling, did not suddenly banish detachment, irony, and the desire for autonomy, but merely served to modulate them. Nor did it banish other compelling imperatives such as his desire for fatherhood and family. Nash did not think of himself as a homosexual. Alfred Kinsey's report on the sexual behavior of white American men was published, amid great publicity, in 1948 when Nash was a graduate student at Princeton, and Nash was no doubt aware of its conclusion that a large fraction of heterosexual men had, at one time or another, same-sex relationships.[13] Besides, he was ambitious, and he wished to succeed on society's terms. He carried on as before. Even as his emotional involvement with Bricker grew, he continued to see Eleanor and continued to weigh the pros and cons of marrying her.

The relationship between Nash and Bricker was not an especially happy one. Nash revealed more of his private self to Bricker than he had to any human being. But each act of self-exposure stimulated a defensive, self-protective reaction. Nash wrapped himself, as he later wrote to Martha with considerable regret, in the mantle of his own superiority to Bricker, the mantle of "the great mathematician."[14] He took to belittling Bricker just as he belittled Eleanor. "He was beautifully sweet one moment and very bitter the next," Bricker recalled in 1997.[15]

For most of that first year, Bricker was completely unaware of Eleanor's existence, like everyone else at MIT. At the end of the spring term, Nash finally let Bricker in on his secret, telling him in somewhat melodramatic tones, "I have a mistress." Nash even engineered a meeting between the two, Bricker recalled, just weeks before Eleanor was due to give birth.

The revelation of a competitor for Nash's affections produced more strains. Among other things, Bricker grew increasingly disturbed by, and critical of, Nash's treatment of Eleanor, he later said. He, Eleanor, and Nash would have dinner together in Nash's apartment, and Bricker became a frequent witness to what he later called Nash's "mean streak" and temper tantrums. When Bricker tried to intervene, Nash would lash out at him. To make things even more difficult, Eleanor began turning to Bricker for sympathy and advice. She would call him to complain about Nash's treatment of her.

Nash could indulge in jealousy himself. Jerome Neuwirth had dinner with Nash and Bricker and some other mathematicians in Boston in early August 1956. Neuwirth, a graduate student, had arrived at MIT that day and was particularly pleased to see Bricker, whom he knew from City. He recalled the evening vividly: "They weren't embracing, but they were always looking at each other. Nash was very hostile. He kept throwing angry looks at me. He couldn't stand anyone talking to Bricker." [16]

The relationship with Nash "was a very disturbing thing" to Bricker, said Neuwirth. "Bricker didn't know what to do. He was having a terrible time." Mrs. Neuwirth advised him to see a psychiatrist.

And the very thing that had attracted him so powerfully in the first place, Nash's genius, only heightened Bricker's sense of inadequacy. That first year, Bricker managed to perform reasonably well in his courses. But later he was hardly able to work.[17] He dropped courses. He finally managed to pass his preliminary exams in November 1954, but his ability to concentrate on his courses had all but evaporated at that point. However, he waited until February 1957, by which time Nash was away on sabbatical, before dropping out of graduate school and relinquishing his dream of becoming an academic. Nash's game was just too painful to play any longer.

They saw each other for the last time in 1967 in Los Angeles, where Bricker was working in private industry. By that time Bricker was married, and Nash was terribly ill. "He was very wild," recalled Bricker in 1997. "He sent me a lot of letters. They were pretty disturbing." [18]

Only one postcard, unsigned and dated August 3, 1967, survived.[19] The only message is "No to No" and presumably came after Bricker had told Nash "No." After that, Nash's constant references to Bricker suggest both Bricker's importance —Bricker is always B to some power, 2 or 22—and Nash's resentment. "Dear Mattuckine, It has obviously been Mr. B who has caused me the largest personal injury," he wrote to Mattuck in 1968.[20] But even then, there are sad notes of regret. "All along since 1967 I've been afraid to write to Bricker except in an indirect fashion. As yet this trouble persists however the reasons why change. There is a feeling of impropriety, etc."

Traces of past affection, however, remained. In 1997, by which time Bricker himself was ill and in virtual isolation, his first questions were "How is Nash? Is he better?"[21] But he was unwilling to talk much about his past relationship with Nash. "I don't want to discuss it further," he said.[22]

25

The Arrest

RAND, Summer 1954

NINETEEN FIFTY-FOUR was to be Nash's last summer at RAND.[1] After an episode that captured some of the most vicious currents of an increasingly paranoid and intolerant era, RAND abruptly withdrew Nash's security clearance, canceled his consulting contract, and effectively banned him from the select community of Cold War intellectuals.

That August, *The Evening Outlook* was full of the Senate's censure of Joe McCarthy, the polio epidemic in the Malibu Bay area, and the news that LA's noxious smog resulted from the chemical action of sunshine on auto exhaust.[2] Meanwhile, a heat wave drew tens of thousands of Angelenos to the Santa Monica beaches.[3] Nash, too, was drawn to the beach.[4] He spent hours at a time walking on the sand or along the promenade in Palisades Park, watching the bodybuilders on Muscle Beach, the crowds on the pier, the surfers nearby. He rarely swam. He preferred to watch and ruminate. Quite often he would still be walking past midnight.

One morning at the very end of the month, the head of RAND's security detail got a call from the Santa Monica police station,[5] which, as it happened, wasn't far from RAND's new headquarters on the far side of Main. It seemed that two cops in vice, one decoy and one arresting officer named John Otto Mattson,[6] had picked up a young guy in a men's bathroom in Palisades Park in the very early morning. He had been arrested, charged with indecent exposure, a misdemeanor, and released.[7] The man, who looked to be in his mid-twenties, claimed that he was a mathematician employed by RAND. Was he?

The RAND lieutenant immediately confirmed that Nash was indeed a RAND employee. He took down the details of the arrest, thanked the cop for the back-channel heads-up, and, as soon as he'd hung up the phone, practically ran down the hall to the office of Richard Best, RAND's manager of security.

Best was a tall, good-looking Navy man who had survived the battle of Midway only to suffer a prolonged and nearly fatal bout of tuberculosis.[8] After his discharge, he wound up at RAND soon after RAND had moved to Fourth and Broadway and was assigned to the "front office" where RAND's handful of top executives was clustered. Discreet and capable, Best had an easy manner that made him popular both with his bosses and with RAND's rank and file. His first assignment was to set up RAND's library, but he quickly adopted the role of general factotum and troubleshooter. In 1953, after the new Eisenhower security guidelines were issued,[9] Best somewhat reluctantly agreed to accept the job of security manager. He disliked the McCarthy hysteria over spies and security leaks and thought all the poking around in individuals' private lives was nasty and not altogether necessary. But he felt he owed RAND, which had kept him on after he suffered a relapse of his illness, and he recognized that RAND couldn't afford any public-relations disasters.

Best listened carefully, but what was going to happen next was clear. Nash had a top-secret security clearance.[10] He'd been picked up in a "police trap."[11] He'd have to go. Best was a Truman liberal who didn't like the McCarthy witch hunts, and he couldn't understand what would make a young cop join a "dirty detail like vice." But he was responsible for enforcing the new security guidelines and the guidelines specifically forbade anyone suspected of homosexual activity to hold a security clearance. Criminal conduct and "sexual perversion" were both grounds for denying or canceling a clearance.[12] Vulnerability to blackmail—which was thought to apply to all homosexuals regardless of whether they were open or not—and, indeed, any behavior hinting at a "reckless nature indicating poor judgment"—were also grounds.[13]

In its early days, RAND had been rather nonchalant about security matters. It hired Nancy Nimitz, the admiral's daughter, even though she had gone to too many communist front meetings at Radcliffe and Harvard to have a prayer of working for the CIA as she had wished.[14] It had done its best to defend the mathematician Richard Bellman, a flamboyant character who not only had a wife who had been in the Communist Party but had somehow managed to befriend a cousin of the Rosenbergs on an airplane flight.[15] One of its top mathematicians in the late 1940s and the author of a book on game theory that is still cited was J. C. C. McKinsey, an open homosexual.[16] But McKinsey was one of the first victims of the increasingly suspicious and intolerant attitude. No matter that McKinsey was completely open about his homosexual lifestyle and that his research was highly theoretical, thus making him an unlikely target for blackmail. McKinsey was forced to leave RAND.[17] The de facto prohibition against homosexuals and suspected homosexuals was so strong, then and later, that the director of the national security program testified in 1972 that "it was conceivable that an ongoing [sic] homosexual might be granted a security clearance, but that he could not think of a single case where it had been granted" in the two decades since he had been in his job.[18]

Nash's arrest was a crisis that had to be dealt with on the spot. Best told Williams the bad news. Williams was genuinely regretful though not especially shocked. Best recalls Williams as being "very open, very relaxed, but appalled that such a valuable researcher as Nash would be lost to RAND." Williams told Best that Nash was "a nut, an eccentric," but an extraordinary mathematician, one of the most brilliant he had encountered. But he did not question for a minute that Nash would have to go.

Nash was not the first RAND employee to be caught in one of the Santa Monica police traps. Muscle Beach, between the Santa Monica pier and the little beach community of Venice, was a magnet for bodybuilders and the biggest homosexual pickup scene in the Malibu bay area.[19] In the early 1950s, the Santa Monica police were running regular undercover operations to entrap homosexuals with the aim of driving them out of town. "One cop follows a guy into the head and makes a remark. If he's accepted, a second cop comes in and arrests him," explained Best. The police rarely stopped at the arrest itself but, in an act of special vindictiveness, almost always notified the man's employer.[20] "We lost five or six people to police programs over a period of several years," said Best.

Normally the department head, in this case Williams, would fire the employee personally. However, Best and his boss, Steve Jeffries, went around to Nash's office and confronted him with the bad news themselves.[21] Nash, for a change, was at his desk. He did not ask what they were doing there but just stared at them. The two men closed the door and said they had something to discuss. Best's manner was unthreatening but direct and he proceeded calmly. RAND would be forced immediately to suspend Nash's Air Force clearance.[22] The Air Force would be notified.[23] And — this was the bottom line — Nash's consulting arrangement with RAND was over for good.

"You're too rich for our blood, John," he concluded.

Best was nonplussed by Nash's reaction. Nash did not appear shaken or embarrassed, as Best had anticipated. Indeed, he seemed to be having trouble believing that Best and Jeffries were serious. "Nash didn't take it all that hard," said Best. "He denied that he had been trying to pick up the cop and tended to scoff at the notion that he could be a homosexual. "I'm not a homosexual," Best quotes Nash as saying. "I like women." He then did something that puzzled Best and shocked him a little. "He pulled a picture out of his wallet and showed us a picture of a woman and a little boy. 'Here's the woman I'm going to marry and our son.'"

Best ignored the picture. He asked Nash what he'd been doing in Palisades Park at 2:00 A.M. Nash responded by saying that he had merely been engaging in an experiment. The phrase Nash kept repeating was something to the effect that he was "merely observing behavioral characteristics."[24] Best recalled retorting, "But John, the police picked you up. You were found doing such and so." Best repeated what he knew of the police report in detail. Recalling the incident in 1996, Best said: "Nash was charged with 'indecent exposure.' That's going into a public head and making a come-on to another man. That means taking out your penis and masturbating. That's the come-on." Best made it clear that it didn't really matter whether the cops were telling the truth or not. "The very act of charging you makes it impossible for you to continue here," he told Nash.

Jeffries and Best told Nash that he would have to leave his office right away.

They escorted him from the building. They would clear out his desk and send his personal papers and belongings, they said. It was all done very politely, with no hint of vindictiveness. Nash had the option of working in quarantine, the preclearance room located just beyond the main lobby. Or, if he preferred, he could finish up whatever he was working on at home.

What was Nash's reaction? Due to leave Santa Monica in another week or so anyway, he did not decamp immediately, though Best doesn't remember whether he returned to the RAND building. "He left in a week or two weeks. Not helter-skelter," Best recalled. What was going through Nash's mind in that interval? Was he angry? Depressed? Frightened? Was he thinking of approaching Williams or Mood with his version of events? Did he try to have RAND's decision reversed? Generally, of course, people did not. Fearful of scandal and aware of the contempt with which any hint of homosexuality was viewed, people in Nash's shoes were usually only too happy to slink away without a murmur of protest.

In the end, Nash did what he had learned to do in less extreme circumstances. He acted, weirdly, as if nothing had happened. He played the role of observer of his own drama, as if it were all a game or some intriguing experiment in human behavior, focusing neither on the emotions of people around him nor on his own, but on moves and countermoves. In his first postcard home that September, he described—with remarkable detachment—another kind of storm: "The hurricane was a fascinating experience."[25] At some point he told his parents he'd had trouble with his RAND security clearance, blaming it on the fact that his mentor at MIT, Norman Levinson, was a former communist who had been hauled before HUAC that year.

Meanwhile, the highly efficient RAND machinery ground on. Best said: "We withdrew his clearances and notified the Air Force of the charges that had been made." RAND negotiated with the Santa Monica police, who wound up dropping the charge in return for RAND's assurance that Nash had been fired and was leaving the state for good. According to Best, such deals were typical. In any case, the arrest did not make *The Evening Outlook* and any record of it has long since been expunged from police files and court records.

Alexander Mood didn't try to keep the arrest a secret—that was impossible given Nash's sudden eviction from his office—but he concocted a cover story to the effect that Nash had simply been strolling in Palisades Park trying to solve a mathematical problem when he was picked up. "He told the officers he was just thinking and . . . they finally learned that what he had told them was true," Mood said later.[26] Most RAND employees learned nothing different. It was after all close to Nash's normal departure date in any case. But his name was abruptly crossed off the list of consultants.[27] Nash never bothered to deny the arrest.[28] And Lloyd Shapley and others in the math division learned about it because Nash had called Shapley from the police station to bail him out.[29] Shapley later told another mathematician that Nash had been playing some kind of game.[30] In any case, with so many mathematicians shuttling back and forth between RAND, Princeton, and other universities, news of the arrest soon leaked back to Princeton and MIT,[31] adding to Nash's already considerable reputation for quirkiness, if not downright instability.

Nobody protested his treatment. He was not the easiest person to sympathize with, and few people, even in the mathematical community, questioned the government's attitude toward homosexuals. Homophobia was, after all, widespread in a society increasingly paranoid and fearful of nonconformity of any kind. Williams, true to form, used the incident in one of his homilies on managing mathematicians. In a memorandum to the mathematics division, written a year or two later, he asked the rhetorical question: "What can mathematicians do to hurt us?" One of his examples was alluded to only with a phrase — "He could get arrested for solicitation." Williams's punch line, however, was "the worst thing a mathematician could do to RAND is to leave." [32]

Although Nash appeared unscathed, the arrest was a turning point in his life. Aloof, ambitious, coolly indifferent to others as he often appeared, Nash was by no means a true loner. Living in a tolerant ivory tower, he had been lulled into believing that he could do as he liked. Now he learned, in a particularly brutal fashion, that the emotional connections he sought threatened to destroy all else that he valued — his freedom, his career, his reputation, success on society's terms. Contradictory imperatives can engender tremendous fear. And fear can be subtly destructive.

An individual's vulnerability to schizophrenia, researchers now believe, lies in his genes. But psychological stresses are thought to be catalysts. Psychologist Irving I. Gottesman at the University of Virginia, whose studies of twins helped discredit the old Freudian theories of schizophrenia, puts it this way: "Each case is different, with a different mix of genetic and psychological factors. Certain events are definite stressors, but it's not famine or war. It's idiosyncratic. It's things that get to the soul and self-identity and expectations of oneself." [33] Rather than a single trauma, a string of events from childhood through young adulthood produces strains that mount like straws on the proverbial camel's back. "It's things that build up, things that lead to a lot of brooding," says Nikki Erlenmeyer-Kimling, a professor of genetics and development at Columbia University.[34] Like the effects of the teasing he endured in childhood and adolescence, the damage from his arrest would only become apparent with time.

The arrest preceded the onset of Nash's illness by more than four years. Stories of other mathematicians who were caught up in the meanness and bigotry of those times illustrate how disequilibrating being harassed and humiliated can be. J. C. C. McKinsey committed suicide in 1953 within two years of being fired by RAND.[35] Alan Turing, the mathematical genius who cracked the Nazi submarine code, was arrested, tried, and convicted under Britain's anti-homosexual statutes in 1952; he committed suicide in the summer of 1954 by taking a bite of a cyanide-laced apple in his laboratory.[36] Others, less well known, less obviously brutalized, had breakdowns that led to their giving up mathematics and living on the margins of society.

The biggest shock to Nash may not have been the arrest itself, but the subsequent expulsion from RAND. His initial reaction after Best confronted him suggests that he simply assumed Williams would overlook the incident. He was after all, one of RAND's resident geniuses. But like McKinsey, Turing, and others, Nash learned that life was more precarious, and he was more vulnerable, than he had previously imagined — a dangerous lesson.

26
Alicia

She had this steely determination. I liked it. I found it very interesting. She always had some agenda, some goal. — EMMA DUCHANE, 1997

HAVING RETURNED TO Cambridge in an anxious, uneasy frame of mind that made the dull task of preparing his lectures even more impossible than usual, Nash escaped to the music library almost every afternoon.[1] The library, on the first floor of Charles Hayden Memorial, had an impressive collection of classical recordings and soundproofed, private cubicles where one could sit and play records, surrounded by deep-blue walls that made one feel as if one were floating in water.[2] Nash would go into one of these and listen to either Bach or Mozart for hours on end.

On his way into the library he would stop at the desk to exchange a few bantering remarks with the music librarians — a mode of interaction that kept people at a distance, much as in the games he liked to play. On one of the first afternoons, he was surprised to see a young woman who had been his student the previous year standing behind the librarian's desk. He had encountered her in the library from time to time before, but now it seemed she was actually working there. She too had seemed a bit startled when she saw him come in, but had given him a sweet smile and had greeted him by name. When he walked away from her he felt her eyes following him.

There was only a handful of coeds at MIT at the time, and the twenty-one-year-old Alicia Larde glowed like a hothouse orchid in this otherwise drab, barrackslike environment. Delicate and feminine, with pale skin and dark eyes, she exuded both innocence and glamour, a fetching shyness as well as a definite sense of self-possession, polish, and elegance.[3] Always perfectly groomed, she wore her short black hair like Elizabeth Taylor's in *Butterfield 8,* was almost always seen in very full skirts cinched tightly around her tiny waist and very, very high heels.[4] She carried herself like a little queen. The student newspaper, *The Tech,* once included a reference to her beautiful ankles in the annual feature on MIT coeds.[5] She was bright, vivacious, playful, and talkative — occasionally sarcastic and often very sharp — popular with the "little boys," as she called the male students, and mad about movies.[6] Her origins were exotic. One of her friends described her as "an El Salvadoran princess with a sense of noblesse oblige."[7]

The Lardes were, in fact, an aristocratic clan.[8] Their origins, like those of all the families which composed Central America's elite, were European, primarily French. Eloi Martin Larde, a wine grower in Champagne, escaped from France during the revolution and settled in Baton Rouge. His son Florentin Larde moved to Central America, first to Guatemala, and ultimately to San Salvador, where he, his wife, and son Jorge became hoteliers and, eventually, owners of a large cotton-growing hacienda.

The Larde men were handsome and the women exceptionally beautiful. A photograph of Alicia's father, Carlos Larde Arthes, and his nine siblings, taken a few days after their mother's death in 1911, might have been of the Romanovs. The family's history had romantic overtones. Alicia's uncle Enrique believed himself to be the bastard son of one of the Austrian Hapsburgs, Archduke Rudolf. Family legend also included a link with an aristocratic French family, the Bourdons.[9] The Lardes, mostly doctors, professors, lawyers, and writers, belonged to the intelligentsia rather than the landed oligarchy that dominated El Salvador's indigo and coffee economy. But they mingled with presidents and generals and, in Carlos Larde's generation, were prominent in public life. They were well educated, spoke French and English as well as Spanish, and traveled widely. Their interests ran to artistic and literary subjects as well as science and philosophy.

Carlos Larde got his medical training in El Salvador but spent several years studying abroad, in America and France, among other places.[10] His early career had been full of promise: He held a number of public posts, including that of head of El Salvador's Red Cross and, before World War II, was chairman of a League of Nations committee. Once he served as El Salvador's consul in San Francisco. His second wife, Alicia Lopez Harrison, came from a wealthy, socially prominent family; Alicia's maternal grandmother was the wife of an English diplomat. Mrs. Larde was not only beautiful but also warm, a wonderful cook, a charming hostess, and a popular aunt with her nieces and nephews.[11]

Alicia, or Lichi, as her family called her, was born on New Year's Day, 1933, in San Salvador. She was the second of Carlos and Alicia's children. Her brother Rolando, five years older, was eventually confined to an institution. A half-brother from her father's first marriage lived with them as well. Treated as an only child by her doting older parents, Lichi was by all accounts a lovely child, with blonde ringlets. She grew up, amidst aunts, uncles, cousins, and servants, in a lovely villa near the center of the capital.

The idyll ended abruptly a year before the end of World War II, when Alicia was eleven. In 1944, in the midst of a yearlong popular insurrection against dictator Hernandez Martinez,[12] Alicia's uncle Enrique had suddenly left for Atlanta with his wife and five young children one night, in the middle of bomb blasts, in a station wagon draped with a white sheet to signal their civilian status. Carlos Larde followed him not long afterward, leaving his wife, daughter, and two sons behind temporarily. He joined his brother in Atlanta, but then moved on to Biloxi, Mississippi, on the Gulf of Mexico, where he obtained a position as a staff doctor at a veterans' hospital. Some weeks later, Mrs. Larde and Alicia joined him, after making the long journey by train through Mexico and stopping in Atlanta to visit Enrique and his family.[13]

What motivated Carlos Larde to follow his brother to the United States at age forty-six isn't entirely clear. Possibly he feared the outbreak of a full-scale civil war. Possibly he saw a chance to revive his medical career, having apparently suffered a series of professional setbacks. But very likely a major reason for emigrating—and the one given Alicia by her parents—was his health. Carlos Larde was suffering from a number of increasingly debilitating physical ailments, among them a severe stomach ulcer, and working as a doctor in the United States would give him access to top-notch medical care. Whatever the reason, the move turned out to be permanent. Enrique went back to El Salvador after a few years, but Carlos Larde was to remain in this country until his death in 1962. Alicia Lopez-Harrison de Larde stayed for another decade after her husband's death.

Hot, dank, slightly seedy, Biloxi lay sprawled on that shallow, murky stretch of the gulf between Mobile and New Orleans, among its barrier islands and river mouths.[14] It was known for shrimp fishing, illegal gambling, and being a favorite wintering place for Chicago mobsters. Rationing made day-to-day life difficult. Carlos was often exhausted and ill and Alicia's mother was plainly distressed by their new surroundings and terribly homesick. Later, the mother of a friend of Alicia's would describe Mrs. Larde as a "very sad, very stoical person." Alicia learned English quickly and easily but suffered pangs of dislocation and isolation on top of the ordinary anxieties of early adolescence. It was not a happy time. For consolation, she turned to schoolwork and the movies.

The Lardes did not stay in Biloxi for long. Less than a year after the war ended, they followed Enrique's family to New York, where Enrique took a job as an interpreter at the United Nations. Once again, Alicia and her mother lived with Enrique's family until Carlos found a position at the Pollak Hospital for Chest Diseases in Jersey City and a house for them to live in. Alicia commuted to Prospect High School, a Catholic school in Brooklyn.

Alicia wasn't to stay trapped in the lower-middle-class environs of Prospect High for long. At the beginning of her sophomore year, the Lardes enrolled her at the Marymount School, an exclusive Catholic girls' school in New York.

Marymount, which was operated by one of the oldest European orders, the Sisters of the Sacred Heart, occupied three adjacent Beaux Arts mansions, on the southeast corner of Eighty-fourth Street and Fifth Avenue, directly across from the Metropolitan Museum of Art and Central Park. It was another world. The student body, mostly day pupils from the surrounding Upper East Side, were from New York's Catholic elite.[15] Many of the girls were daughters of celebrities like Joe DiMaggio,

Jackie Gleason, Paul Whiteman, and Pablo Casals. Alicia's best friends there included the daughter of an Italian count. Tuition was several times what most private universities charged at the time, easily equivalent, once inflation is taken into account, to $15,000 today. Admission was based strictly on families' social standing; the El Salvadoran ambassador wrote Alicia's letter of reference, attesting to the Larde family's social position.[16]

The school's atmosphere, appropriately to girls being groomed to become "wives of Catholic leaders," was cosmopolitan and cultured.[17] The girls' uniforms included stylish blazers and black high heels. Parents insisted that the school "keep up the social end of things." Alicia took riding and tennis lessons in Central Park, played basketball, helped out on plays and musicals, and went to parties. She went to her senior prom, and afterward to the Stork Club, with her friend Chicky Gallagher's brother.[18]

She looked, on graduation day, just like the other girls, only more beautiful, wrapped in the same white tulle and cradling the same three dozen long-stemmed roses, like a debutante before a coming-out ball. Much, however, separated Alicia from her wealthy schoolmates. Outwardly she was gay, charming, unruffled, and compliant, but her appearance veiled a keen intelligence, an outsider's ambition, and what a future friend called steely determination. Self-controlled and reluctant to confide her real feelings to anyone, a legacy of her Latin upbringing, she hid a great deal from view. As a woman who got to know Alicia several years later said, "You have to keep the times in mind. Women dissembled then. Alicia behaved like a fifties ditz, but that doesn't mean she was one. She was flirtatious but she was saying quite serious things. She always had some agenda, some goal."[19]

As a child, she'd dreamed of becoming a modern-day Marie Curie.[20] Alicia was twelve years old when she huddled with her father near the radio in their Biloxi apartment and listened with him to the broadcast about Hiroshima.[21] It was for her, as for so many scientifically inclined youngsters, a defining moment. Within weeks, the Japanese surrender and the War Department's revelation of the three hidden "atomic" cities in the southwestern desert turned anonymous men like Oppenheimer and Teller into public heroes. Instantly, the image of the "nuclear physicist" seized the popular imagination the same way that "rocket scientist" did after *Sputnik.* Alicia, already showing signs of her father's talent and interest in scientific subjects, knew what she wanted to be. "The world was physics. It was what kids with a talent for, and interest in, math and science aspired to," a fellow physics major at MIT said in 1997. "To Carlos Larde it was the top, and it was for Alicia too."[22]

Her aptitude for mathematics and science had long been evident and became more so at Marymount. By the late 1940s, the school was already something more than a fancy finishing school. It had always had an exceptionally well-trained faculty, lay and religious, but during Alicia's tenure the school was run by a forceful young Irish graduate of the London School of Economics — Sister Raymond — who was not only an ardent Keynesian, but a gifted educator determined to raise the educational standards of the place. Sister Raymond improved the caliber of students by introducing scholarships and gave more intellectual heft to the school's curriculum by adding serious science and mathematics courses. Alicia had a choice between a classical education emphasizing the arts and languages and one focusing on science and mathematics. She was one of the few girls who chose the latter and, as a consequence, took biology, chemistry, and physics as well as three years' worth of mathematics, often in tiny classes of two or three girls. Sister Raymond recalled her as a gifted and willing student: "Very intelligent. Not too pushy. Very very interested in her studies." [23]

By her senior year, Alicia was quite definite about wanting to pursue a career in science. "I wanted a career, so I wanted to study something definite," she said. [24] Carlos Larde, who was delighted by his daughter's ambitions, wrote an eloquent and touching letter to Sister Raymond urging her to make every effort to help Alicia realize her dream of becoming a nuclear scientist by helping her gain admission to a first-rate technical university. [25] Alicia was accepted at MIT, one of only seventeen women and two female physics majors in the class of 1955. [26]

The Lardes were no less thrilled than Alicia. Carlos Larde, who had studied at the University of Chicago and Johns Hopkins, particularly appreciated what an MIT degree would mean, but he drew the line at her going off to a virtually all-male engineering school on her own. Alicia's mother, it was decided, would accompany Alicia in order to watch over and take care of her. [27] Besides the natural protectiveness toward a precious daughter, the arrangement may have reflected a wish on the part of Alicia Lopez-Harrison de Larde to escape her ailing, difficult husband. Alicia's friends at MIT were struck, later, by the fact that mother and daughter never referred to Carlos Larde and that he never came to visit. [28] In any event, in the late summer of 1951, the two women rented a tiny furnished apartment in Boston [29] not far from Beacon Street where John Nash had just found a room, across the river from MIT and not far from the Harvard Bridge.

It was marvelous being an MIT coed in the early 1950s, an era famous for its celebration of mothers and dumb blondes, because the coeds were so special and had, as it were, the best of both worlds: it was serious, but there were lots of men. There were girls who wore cocktail dresses and high heels while dissecting rats in the lab. [30] A date wasn't going dancing and sipping Manhattans, it was going to a lecture and out to coffee afterward, or maybe having a boy take you to his parents' house and showing you, through a telescope, everything Galileo had seen.

Alicia was to tell her girlfriends that being there made her feel like a "Queen Bee." It was also a chance to meet, finally, other women who didn't think that having brains and ambitions was a major liability. "We were a self-selected group of fairly strong women," said Joyce Davis, a native New Yorker and the only other female physics major in the class of 1955. "We had our own culture. It wasn't normal American female culture, the 'you can't be as good as the boys' culture, which we were always trying to escape. And it wasn't the MIT boys' culture either."[31]

Alicia spent most of her time with the other coeds either at the dorm or on the campus. She studied with the other girls in the Cheney room, the coed lounge, ate breakfast and lunch with her friends at Pritchett lounge every day, and generally was up for whatever the girls felt like doing, whether it was playing basketball or organizing a charity fair.[32] She attended a great many concerts and plays, thanks to the coeds' wealthy patroness, a Mrs. McCormick, who showered them with tickets and even paid for them to take taxis across the Harvard Bridge in winter.

MIT's academic program was brutally demanding, especially for physics majors. Class schedules were heavy, spread over six days, and consisted mostly of required courses. All the girls lived in healthy fear of flunking out. Alicia, who had sailed through her science and math courses at Marymount on native ability, found that this was no longer enough. Much to her dismay, she had to struggle to maintain a C average (which was a respectable performance in those days before grade inflation turned a C into a subaverage mark). "You either had to buckle down or accept just getting by," said Joyce, Alicia's best friend. "Alicia never really buckled down."[33]

Alicia's ambition survived her freshman year intact, despite a fair amount of teasing, especially in her chemistry class, from boys and instructors who were sure that she would not make the cut. In a letter to Joyce, in the summer of 1952, Alicia wrote:

Dear Joyce,
By this time you must be wondering whether I'm dead, dying or have mearly [sic] been kidnaped judging from the amount of communication you have received from me; the sad truth of course is my laziness. Except for one week that I went to Canada with Betty Sabin and her parents I have spent the Summer working as a sales girl in a small store (I hate to say 5 + 10) behind the ribbon counter; I have done all but strangled the customers with "our" fine products. But life hasn't been all tears (I hate to think of my report card) we have fortunately moved to a new apartment half a block away from Kenmore Square. And so I will be able to walk home with you (the dorm is only about a block and $\frac{1}{2}$ away).
    By now you must be beginning to believe the malicious rumors that I bribe my English teachers; not to mention the grammar and the spelling is atrocious (get me!). My report card was the same as last term with the unhappy exception of a B in English; my cum. is still above 3 though; .02 above that is. I'm unhappy that we won't be in the same section this year but c'est la vie! I wanted to take French instead of German in order to make my life easier but I'm not sure I can because of my hope for a Ph.D. in physics . . . remember all

I was going to study this summer? Well, I've gotten to page 17 of the Physics book and that's all; I am however many movies wiser.

Give my regards to your mother and answer soon (do as I say not as I do).[34]

A profile, a look, a voice can capture a heart in no time at all. Alicia gave away hers in the space of a single calculus lecture. She was sitting, her best friend Joyce beside her, in the front row of M351, Advanced Calculus for Engineers, a course required of all physics majors. John Nash arrived late wearing a haughty and bored expression. Without so much as a glance or a word to the assembled, he closed all the windows, flipped open his copy of Hildebrand, and embarked on a lackluster exposition of the properties of ordinary differential equations.

It was mid-September, Indian summer weather, and as Nash droned on, the room got quite hot. First one, then several students interrupted Nash to complain and to ask that he let them open the windows. Nash, who had obviously shut the windows to prevent any outside noise distracting anyone, ignored them. "He was so wrapped up in himself that he wouldn't pay attention to what we wanted. His attitude plainly said, 'Shut up and take notes,' " Joyce recalled.[35] At that point, Alicia jumped up from her seat, ran over to the windows in her high heels, and opened them one after another, each time with a toss of her head. On her way back to her seat, she looked straight at Nash, as if daring him to reverse her action. He did not.

Joyce thought Nash an indifferent lecturer and insensitive besides. "He presented the material but that was it. He was sort of cold." Joyce transferred out of the section after the first class, but Alicia surprised her by staying. "She thought he looked like Rock Hudson," said Joyce.

To see Nash through Alicia's eyes during their first encounters as student and professor conveys much about the elementary force that was to bind her to him. In MIT's intellectual hierarchy — where "mathematics was the highest thing," as Joyce was to say — Nash was the closest thing to royalty.[36] It was his good looks, however, that made Alicia's heart beat faster. "A genius with a penis. Isn't that what we all want?" an actress once quipped, and the quip captures the combination of brains, status, and sex appeal that made Nash so irresistible. Herta Newman, Donald's wife, said the same thing in less bald terms: "He was going to be famous. He was also cute."[37] Emma Duchane, a physics major two years behind Alicia at MIT, said, "Alicia thought he was gorgeous. She thought he had beautiful legs."[38] Nash wasn't scruffy like many of the mathematicians. He was always neatly combed, pressed, and shined. His haughty manner and cool indifference only confirmed his desirability. His name, two monosyllables that advertised his Anglo-Saxon ancestry, added to his appeal. "He was very, very good-looking," Alicia later said. "Very intelligent. It was a little bit of a hero worship thing."[39]

Nash took no notice of her, but Alicia was quite prepared to woo him. All that year, she would seek him out. "Come with me to the music library, Joyce," or, "Come with me to Walker Memorial. I want to see Nash."[40] "She set her cap for him," Joyce recalled. "She had a campaign going."

Her grades suffered. She got two Ds and for the first time in her MIT career her grade point average slipped below a C. The following April, Joyce wrote to her parents: "Alicia is still not doing to [sic] well since she is in LOVE. She goes around with a faraway expression on her face."[41]

When the calculus course was over, Alicia got a job in Nash's favorite haunt, the music library. It is a measure of her lovesickness that she found it a far more interesting place to work than Lincoln Laboratories, where she also had a job. "Work here isn't very stimulating; what I do mostly is count 'tracks' seen thru a microscope," she wrote to Joyce during the summer. "I only work 15 hrs a week here but what tires me out is the overtime; I keep seeing the little monsters every time I close my eyes. *Music library proves more interesting, so far several strange boys have tried to pick me up."*[42]

Alicia was still playing the field, but with less enthusiasm than her letter to Joyce implied: "A few more weeks now and I expect to be seeing 'blondie' again. It seems peculiar but I feel so indifferent about him now."

She continued this letter a few weeks later:

I am writing in the music library now (obviously). Something funny {?} happened to me here the other day. A boy I know came to talk to me while one of the ones I am out "gunning" for was sitting out there; or so I thought. In order to seem attractive to the one out there I began pouring on the "charm" on my little friend; then in my loudest possible voice I announced my working hours in the ML; they must have heard me over the radio. Well, the persecuted one seemed to be getting the idea while I became bolder and bolder. Finally he came over. Then, boy, was I mortified. The moral of the story is "wear glasses." Needless to say he wasn't the "one."

Nash, of course, was at RAND most of that summer.

When Nash started coming around the library again that fall, Alicia engaged him in conversation and studied him as minutely as any fan studies his or her favorite star. She found out that he played chess. She found out that he was a science fiction fan. She made it her business to learn chess and, in addition to her job in the library, she took to sitting in the science library near the science fiction collection. "My activities besides the music library include the science library where I read science fiction (John likes it)," she wrote to Joyce.

Despite Alicia Larde's crush, which seemed to have erased the earnest student of science, she was playing a serious game. Her romantic dreams of becoming a famous scientist herself hadn't survived the harsh reality test provided by MIT. As she put it later, "I was no Einstein."[43] Pragmatically, she recognized that marriage to an illustrious man might also satisfy her ambitions. Nash seemed to fit the bill.

"John could give her a lot of things she didn't have," observed John Moore, a mathematician who fell in love with Alicia some years later.[44] Sadly, the romantic girl whose favorite song was "Lady of Spain" would most agonizingly disappear in just a few years.

27

The Courtship

NASH STARTED to make occasional references to "the music librarian" in his conversations with Mattuck.[1] He was at a crossroads. The dangers of his sexual experiments had become suddenly, devastatingly obvious. Marriage was a possible answer and he had, at his most frightened, almost convinced himself that he would marry Eleanor. Now that he was back in Boston and seeing her again, however, he could not bring himself to take any practical steps in that direction. Alicia came along at the right moment.

Moreover, Nash liked what he saw. The son of a beautiful mother *would* be drawn by the classical symmetry of Alicia's features and the slenderness of her frame. Alicia's aristocratic lineage and social ease appealed to his own sense of superiority. The effect of her intelligence on him should not be underestimated. Nash was easily bored. He found her interesting company, liked the fact that she set her own compass, and was amused by her flashes of sarcasm and irreverence.

It was part of Nash's genius to choose a woman who would prove so essential to his survival. He took her willingness to pursue him, to make every effort, not merely as flattery, to which he was no less immune than the next man, but as a sign that she was prepared to take him as he was. He saw her determination to have him as a real key to her character, suggesting that she knew what she was getting and expected nothing more.

They shared a good deal. Both were close to their mothers. Both had emotionally distant but intellectually stimulating fathers. Both had grown up in households where intellectual achievement and social status, rather than emotional intimacy, were the coin of the realm. Both, on account of their intellectual precocity, had somewhat delayed adolescences. Both felt that they were, in different ways, outsiders and compensated for this by seeking status for themselves. There was a coolness, a calculation, that guided their actions.

Nonetheless, the progress of the courtship was slow. Nash finally asked Alicia out during the spring. In July 1955 she wrote to Joyce that they were seeing each other "on and off."[2] She said that he had introduced her to his parents some three weeks earlier. But she made it clear that they were not sexually intimate. The significance of his having introduced her to his parents, given his mother's chronic concern over Nash's social life, wasn't clear. Alicia, who must have taken it as a hopeful sign, did not admit to taking it that way.

> I've been making slight progress with JFN but can't tell just yet if it's significant. I don't think he's really too interested but more or less can take me or leave me. About 3 weeks ago I met his parents who'd come up to visit him for a week. I've been seeing him on and off and last Saturday we went to the beach together — I had fun.[3]

Alicia hinted at one reason why Nash remained lukewarm: "He still thinks I'm too innocent but has now condescended to accept me as is and just let my 'sweet innocent little self' develop."

And in her own mind, Alicia was still playing the field, though it was clear that she was distracting herself and hoping in the process to pique Nash's interest.

> I've picked up a few admirers this summer including that Junior that Marolyn was talking about. I keep refusing dates with him but he doesn't seem to get the idea and just follows me around, so far he has written a couple of cute poems that I'm keeping as suveniers [sic]. I realize that I'm sounding quite egocentric with all this but not much else has been happening.

Whether because of preoccupation with Nash or simply because of a waning interest in physics, Alicia failed to graduate with her class. She had to stay on to make up a number of courses. But the shock of not graduating on time, and the unpleasant business of having to admit this to her father, did little to refocus her attention on her studies. She says in the letter to Joyce that she is making up M39 but that "so far I'm up to page 10 in Hildebrand."

Nash and Alicia saw more of each other in the fall. He took her to a math party. Then another. And out to the Newmans' house or to Marvin Minsky's. "Let's go Minskify," he would say to a group.[4] Sometimes they double-dated with one of Alicia's friends. On those occasions, he almost ignored her once they had arrived and the introductions were made, going off to join the circle of men talking about mathematics. Sometimes Alicia would stand at the edge of the circle listening to Nash say things like "Who are the great geniuses: Wiener, Levinson, and me. But I think maybe I'm the best." Other times she found herself among mathematicians' wives talking about their children. There was no flirtation, no going off in a corner to hold hands, but in fact the relationship was more intoxicating for those reasons. The other women treated her with the deference accorded to the genius consort, which made Alicia feel rather smug. As for Nash, he could not help but be aware that the other men, impressed and surprised, envied him this adoring, gorgeous creature.

Other times they would go out for lunch, usually with someone else. Bricker often joined them, and also Emma Duchane. Bricker recalled Alicia as "very bright" and "quite sarcastic."[5] Emma recalled, "She was not deferential at all. She never stopped talking."[6]

True, Nash was not especially nice to Alicia. Among other things, he called her unflattering nicknames, including "Leech," a nasty play on her childhood nickname, Lichi.[7] He never paid for her meals, dividing every restaurant check down to the penny. "He was not infatuated with her," Emma recalled in 1996. "He was infatuated with himself."[8]

To Nash, Alicia was part of the background, charming and decorative. He treated her the way other mathematicians treated their women. But Alicia wasn't looking for companionship either. Later Emma said: "We wanted intellectual thrills. When my boyfriend told me *e* to the *pi* times *i* equals negative 1, I was thrilled. I felt the absolute joy of the idea."[9] Nash was no less fun to be with than the other mathematicians.

A February 1956 letter from Alicia to a friend doesn't mention Nash at all. But at the end of that month Alicia's mother would move to Washington (Carlos Larde had gotten a position at Glendale Hospital in Maryland), a move that Alicia anticipated with some glee.

It was probably sometime that spring that Nash and Alicia began sleeping together, at the end of those evenings in company where they barely exchanged three words. Nash was still involved with both Bricker and Eleanor. Indeed, he may have continued, even at this late date, to think of Eleanor as his likely wife. Alicia and John were in bed one evening when his doorbell rang.[10] John answered the door. It was not Arthur Mattuck, who sometimes dropped by unannounced. It was Eleanor, indeed, an angry and shaken Eleanor. She said nothing but walked right past Nash into the apartment. She acted as if she'd come to talk things out with him.

When she realized Nash was not alone, she began shrieking and crying and threatening until finally she had cried herself out and Nash drove her home. Alicia, meanwhile, white-faced, left.

The next day, Nash went into Arthur Mattuck's office, told him the story, grabbed his head with both hands, and moaned, genuinely pained, over and over, "My perfect little world is ruined, my perfect little world is ruined."

Eleanor called Alicia and told her that she was stealing another woman's man. She told her about John David. She told her that Nash was planning to marry her and that she, Alicia, was wasting her time. Alicia invited Eleanor to her apartment for a meeting. Eleanor came; Alicia was waiting with a bottle of red wine. "She tried to get me drunk," Eleanor recalled. "She wanted to see what I was like. We talked about John."[11]

And, having met her, and realizing that Eleanor was an LPN, that she was practically thirty, that the affair had been going on for nearly three years, Alicia concluded that it wasn't going anywhere. She was not shocked. Men had mistresses, they even had children by them, but they married women of their own class. Of that she felt quite confident. Eleanor had called her up to complain. Alicia was pleased. She took it as a sign that, as her friend Emma said, "she was beginning to matter." [12]

Nash was due for a sabbatical the following year. He had won one of the new Sloan Fellowships, prestigious three-year research grants that would let the recipients spend at least one year away from teaching and, for that matter, away from Cambridge.[13] He could go where he liked. He was, perhaps unreasonably, still worried about the draft, as he had confided to Tucker in a letter a year earlier.[14] He decided to spend that year at the Institute for Advanced Study.[15] He was beginning to think seriously about various problems in quantum theory and thought that a year at the institute might stimulate his thinking.

Alicia meanwhile complained in a letter to Joyce that February that she was "just vegetating." She mentioned a vague desire (which she did not say was connected with Nash) "to get a job in New York instead of staying on at the Institute [MIT] to attend graduate school." [16]

At the end of the spring term, Nash took Alicia to the math department picnic in Boston. The picnics were always held during reading week and often on the commons. Wiener came, as did all the graduate students. It was an unusually warm day, and Nash was in high spirits. Nash did something curious that engraved itself on the memories of another instructor, Nesmith Ankeny and his wife, Barbara. It was, of course, Nash's notion of a joke. He wished to show everyone that he was the master of this gorgeous young woman, and that she was his slave. At one point, late in the afternoon, he threw Alicia to the ground and placed his foot on her neck.[17]

But despite this display of machismo and possessiveness, Nash left Cambridge in June without suggesting marriage or even that she move to New York.

Indeed, at the start of that summer, in June, another friend of Alicia's described Alicia as being in Cambridge and "in an unbelievable state of depression, due to a certain instructor at MIT." [18]

28

Seattle

Summer 1956

N ASH LEFT CAMBRIDGE for Seattle in mid-June with the light heart of a man making a temporary escape from a tangle of personal and professional dilemmas.[1] Travel always lifted his spirits and this trip was no exception. The month-long summer institute at the University of Washington was exactly what he wanted. A top-notch crowd of mathematicians working in differential geometry would be there: Ambrose, Bott, Singer, as well as Louis Nirenberg and Hassler Whitney. Nash expected that his embedding work would make him one of the centers of attention. And he was looking forward to hearing Busemann's seminar on the state of Soviet mathematics because everyone knew that the Russians were doing great things, but the authorities were no longer allowing even abstracts of their mathematics articles to be translated into English.

The signal event of the summer institute turned out to be the surprise announcement, within a day or two of the start of the meetings, of Milnor's proof of the existence of exotic spheres.[2] For the mathematicians gathered there, it had the same electrifying effect as the announcement of a solution of Fermat's Last Theorem by Andrew Wiles of Princeton University four decades later. It stole Nash's thunder.

Nash reacted to the news of Milnor's triumph with a display of adolescent petulance.[3] The mathematicians were all camping out in a student dormitory and eating their meals in the cafeteria. Nash protested by grabbing gigantic portions. Once he demolished a pile of bread. Another time, he threw a glass of milk at a cashier. And on one occasion, during a sailboat outing, he got into a shoving match with another mathematician.

Nash didn't immediately recognize Amasa Forrester, who looked like a shaggy bespectacled bear with the hint of a double chin, a haphazardly shaven face, and glasses, and who even walked like a bear with a slightly forward-leaning gait, when the latter buttonholed him after a talk.[4] Forrester had to remind Nash that they'd been at Princeton together, Forrester having been a first-year graduate student during Nash's final year. After they starting talking, however, Nash remembered Forrester as a Steenrod student who was always holding court in the Fine Hall common room, waving a water pistol around.

Despite his somewhat unprepossessing appearance, Forrester had interesting things to say. He was fast, aggressive, and seemed to know everything about everything that came up in their conversation. Forrester explained some of the details of Milnor's work to Nash. They also talked, then and later, about Nash's embedding papers, which Forrester appeared to know quite well.

Forrester invited Nash to come to see his living quarters, moored on Lake Union, between Lake Washington and Puget Sound in downtown Seattle.

To Nash, Forrester was "a different sort."[5] He would later refer to Forrester, who went by the name Amasa, in the same terms that he used when he compared Thorson and Bricker to the Beatles — "young," "colorful," "amusing," and "attractive" — someone who made him feel like "the girls who love the Beatles so wildly."

There was much to draw them together. Forrester, who had just turned thirty, was as brash and brilliant as Nash.[6] He'd had a stellar graduate-school career. Steenrod, who was on his dissertation committee, had given him spectacular references. He was disorganized and sloppy but he had a photographic memory and wide-ranging interests. He hadn't done much since arriving in Seattle in 1954 and, indeed, hadn't been able to publish his dissertation because it turned out to contain a substantive flaw, but he was still full of enthusiasm, or at least so it seemed to Nash. He shared Nash's predilection for insult and one-upmanship — at Princeton he'd been referred to as King of the Common Room for that reason — and was given to sweeping judgments of the kind Nash admired. Once, for example, when a listener tried to question him after a talk, he responded by claiming, "It's easier to predict what mathematicians will be talking about fifty years from now than what they'll be interested in next year."[7] His obvious eccentricity made him seem like a kindred spirit. This was a young man who had once managed to get himself permanently banned from the dining rooms of the Graduate College by Sir Hugh Taylor, the dean, for having deliberately broken dishes and crockery in the breakfast room. And his relationship with his mother was fodder for all kinds of stories. Former friends recall that a family record of worldly success and an overbearing mother both weighed heavily on him. Arthur Mattuck, who was at Princeton with Forrester, recalled: " 'Amasy, Amasy, Amasy!' his mother would say. 'Oh, mom, you know how much I love you,' Amasa would coo back in a falsetto."[8]

Forrester was also openly homosexual. It's unlikely that his graduate-school professors or Sir Hugh were aware of this, but "he was fairly open about his homosexuality at Princeton and everybody at the Graduate College knew," said John Isbell, a professor of mathematics at the State University of New York at Buffalo and a fellow graduate student at Princeton.[9] Initially, Forrester had been quite circumspect with his colleagues at the University of Washington, but by the time Nash ran into him—perhaps because things were beginning to loosen up even in Seattle—he had concluded that he no longer had to pretend to be what he was not. Robert Vaught, a retired logician at the University of California at Berkeley, shared a house with Forrester during their first year as instructors in Seattle. He recalled:

> It wasn't that he "discovered" his homosexuality then. It was very difficult for homosexuals then. In those days people thought the best thing to do was to get rid of it by some act of will. He sort of decided that he had to be a homosexual. Sometime during his third year in Seattle he bought himself a houseboat—there was a far-out group living on the waterfront—and gradually he began to let people know about his homosexuality.[10]

Nash always found the people who could give him what he needed. Forrester was the kind of smart, verbal, quick-witted man Nash was frequently attracted to. Forrester was also emotionally available. Under his eccentric, sometimes brash and loud exterior, Forrester was an exceptionally sweet man. "Kind and gentle, much loved by his students," was the description given by Albert Nijenhuis, another of Forrester's colleagues.[11] Forrester also had an unusual capacity for connecting with troubled individuals. When Vaught, who, as a student, had endured repeated hospitalizations for episodes of mania and depression, first came to Seattle, Forrester was amazingly kind. Vaught recalled: "He was a *very* fine man. I was a manic-depressive long before lithium came along. He was very helpful to me. Amasa encouraged me to find a psychiatrist in Seattle. I could talk to him."[12] In his first year at Seattle, Forrester "adopted" a mentally ill graduate student—a computer genius who had suffered some kind of psychotic breakdown—and tried to care for him, recalled John Walter, a mathematician at the University of Illinois who shared the house with Vaught and Forrester. "It was one of his projects."[13]

It would have been obvious to Forrester that Nash, arrogant and aloof as he might appear, would respond to his sympathetic interest. "Amasa was pretty sharp. He would have seen through the veil," said Walter.[14]

Nash and Forrester hardly had much time to spend together; Nash was in Seattle only a month. Although Nash referred to Forrester, either by name or simply by the letter *F*, in letters until the early 1970s, there is no evidence to suggest that

Nash and Forrester corresponded regularly or saw much of each other in subsequent years. Forrester stayed very much on Nash's mind, however. Eleven years later, on a pilgrimage that took him to Los Angeles and San Francisco, Nash spent nearly a month in Seattle.[15]

Forrester was still living in his houseboat with dozens of cats for company and was by then almost entirely cut off from his former mathematical friends.[16] He had never lived up to his early promise, had been denied tenure, and had left the University of Washington in 1961. He worked briefly at Boeing and later at the giant Atomic Energy Commission plant in Hanford, Washington, before dropping out of the mathematical community in the mid-1970s. Later, he made his living tutoring and, on one occasion, acting as a live-in tutor for some children on a ranch. Nijenhuis, who ran into him a final time at a mathematics congress in Vancouver, British Columbia, in 1974, recalled that Forrester had told him that he'd worked as a goatherd. For years he would drop by the mathematics and physics library, looking progressively more seedy and disheveled. He died in 1991. This once-promising mathematician did not even merit an obituary in the *Seattle Times*. If, for Nash, Forrester's was the road not taken, one would have to argue that Nash, on this occasion, was perceptive about human beings.

Nash knew immediately that something was wrong when someone fetched him from the dormitory. The Nashes communicated exclusively by letter and postcard. A long-distance telephone call indicated that something was amiss.[17]

John Sr. was on the line. He sounded unnaturally grave. Nash's first thought was that he was calling with some bad news about his mother or sister, but he heard anger rather than sorrow or anxiety in his father's voice.

Eleanor Stier had contacted them and revealed the existence of their grandson, John Sr. said. The shock was enormous.

"Don't come home," John Sr. told him sternly. "Go right to Boston and make this right. Marry the girl."

Nash was too stunned to argue. The secret he was so anxious to keep from his parents was out. There was nothing to be done now. He agreed not to come to Roanoke. In a postcard dated July 12, he wrote his parents that he was "thinking of going back to BeanTown."[18]

Nash did go back to Boston in mid-July and stayed for two weeks. He spent most of his time either with Bricker or working in his office late nights.[19] He turned to Bricker for advice on what to do about Eleanor. She had hired a lawyer. She wanted regular child support payments. The attorney, Nash found out, was threatening to go to the university. Nash, as Bricker recalled in 1997, was inclined to refuse to pay.

Bricker, as usual, found himself in the middle. Eleanor had been calling him regularly. She was devastated by Nash's abandonment and bitter over his refusal to support their son. Bricker remonstrated with Nash. "He didn't want to pay child support. I told him, This is terrible. This is your son. If nothing else, do it for your own future. If the university got wind of this it'll ruin your career. You owe it to her."[20] Nash, to Bricker's surprise, agreed to pay.

29
Death and Marriage

1956–57

ALTHOUGH NASH WAS TO SPEND the year at the Institute for Advanced Study, he decided to live in New York instead of Princeton.[1] Within a day or two of coming to the city in late August, he found an unfurnished apartment on Bleecker Street in Greenwich Village just south of Washington Square Park, a street lined with jazz clubs, Italian cafés, and secondhand book shops. The apartment was a typical railroad flat, small, dingy, and suffused with smells of his neighbors' cooking. Nash bought a few pieces of used furniture from a local junk dealer and sent his parents a postcard proclaiming a sentiment that they would be sure to approve, namely, that he'd rather save money than live luxuriously.[2]

But his reasons for choosing a five-story walk-up in downtown New York over a spartan flat on Einstein Drive in quasirural Princeton were more romantic than practical. The towering scale of the city, with its frenetic rhythms, ever-present crowds, and round-the-clock activity — "the wild electric beauty of New York"[3] — seemed wonderful to him, always had, from the first time Shapley and Shubik had invited him, when all three were living in the Graduate College at Princeton, to come up for a weekend. After he'd moved to Boston, he had seized every opportunity to return, sometimes staying with the Minskys,[4] just to reexperience that sensation of simultaneous connectedness and anonymity. The bohemian enclave around Washington Square had long been a magnet for those who were sexually and spiritually unconventional, and Nash too was attracted to its crooked streets, Old World charm, and implied promise of freedom.

If the decision to move to Bleecker Street meant that Nash was toying with adopting a different sort of life from the one he had hitherto imagined for himself, it was not to be. John Sr. and Virginia announced that they too were coming to New York.[5] John Sr. had some business to transact for the Appalachian. Nash feared that they would confront him again on the subject of Eleanor. But the Nashes were even more preoccupied with the precarious state of John Sr.'s health at that moment. When Nash met them at the McAlpin Hotel, a few blocks from Penn Station, he tried to demonstrate that he was a loyal son by urging his father, several times in the course of the evening, to consult a specialist in New York. He told his father he ought to consider an operation.[6] It was the last time Nash saw his father.

In early September, John Sr. suffered a massive heart attack.[7] Virginia had a difficult time reaching Nash, who had no telephone. By the time she got a message to him, his father was already dead. Thereafter, he would think of fall as a season of "misfortunes."[8]

John Sr., who was sixty-four at the time of his death, had been ill on and off all year. That Easter Sunday he had been feeling too unwell to go to Martha and Charlie's house for dinner (Martha had married in the spring of 1954). And in late summer when he and Virginia were in New York, he suffered from a spell of weakness and nausea in the hotel.[9] The news of his father's death shocked Nash. He couldn't fathom its suddenness, its finality. He was convinced that the death had not been inevitable, might have been prevented if only John Sr. had gotten better medical care, if only . . .[10]

Nash rushed to Bluefield to attend the funeral, which was held at Christ Episcopal Church on September 14, two days after John Sr. died.[11]

There was no outpouring of grief, no sign that Nash's unnatural calm was shaken.[12] But the death of his father produced another fissure in the foundation of Nash's "perfect little world." The loss of a parent before one has really stepped fully into one's own adult life in the same role is a one-two punch—losing the father and having to step into the father's shoes.

There was, for starters, a newfound sense of responsibility for Virginia's welfare. It may not have signified much in practical terms, given that Martha lived in Roanoke and, as the female offspring, would have been expected to look after Virginia, but emotionally Nash was now in the hot seat. Suddenly, his mother's wishes regarding him, in particular her intense desire that he adopt what she regarded as a "normal" life—that is, that he marry—weighed more heavily on him than at any time since he had left home for college.

For Nash this dilemma—and it was a dilemma, as his father's shoes were not exactly the ones that he felt prepared to step into—was compounded by the particular circumstances of the summer. Nash's misbehavior with regard to Eleanor and John David lay between him and Virginia. The thought that he had hastened his father's death must have occurred to him. Or, if it didn't—and this is certainly possible given Nash's inability to imagine how his actions affected other people—the thought surely occurred to Virginia, who may have communicated it, indirectly or directly, to Nash. Virginia was not just grief-stricken but deeply angry. She wrote Eleanor a letter accusing her of causing her husband's death. It is quite possible that she said something similar to her son, or implied as much.[13]

Such guilt would be a heavy burden to bear. More likely, it was not just the feeling of guilt, but also the more potent threat of losing his mother's love on the heels of the actual loss of his father, that would have placed tremendous pressure on Nash to act. Virginia felt that Nash was duty bound to legitimize his relationship to his son. John Sr. had an abhorrence of scandal and a strong belief in doing one's duty. Whether, by the time of her husband's death, Virginia still persisted in the demand that Nash marry Eleanor isn't clear. It may be that her contact with Eleanor — including the evidence of Eleanor's lower-class origins, her lack of education, or her threats to make trouble for Nash — convinced her that even a temporary marriage was out of the question. She may have feared that Eleanor would never agree to a divorce. Or simply, she may have realized that she had no way of forcing Nash to do something that he did not wish to do.

If Virginia reacted so to Nash's mistress and illegitimate son, how might she react to the far more disturbing facts of Nash's liaisons with other men? As a practical matter, the likelihood of her ever finding out about the arrest seemed negligible. Yet that too must have crossed Nash's mind. His confidence that he could keep his secret lives completely separate and keep his parents in the dark as well was jolted by Eleanor's betrayal. He must have felt on his neck the hot breath of other potential discoveries.

In addition to commuting to the Institute in Princeton, Nash was spending a good deal of time at New York University, whose campus began a block north of Bleecker Street, at the Courant Institute of Mathematical Sciences. One afternoon, very soon after his father's funeral, Nash stopped at the desk of the beautiful Natasha Artin, the wife of Emil Artin and one of Richard Courant's assistants. A famously gorgeous creature, Natasha had a doctorate from the University of Berlin, where she'd been a student of Artin's before they married. Everyone knew that she was the latest object of Courant's infatuation. Nash liked to chat with her on his way up to tea.

"I wonder how easy it is to get a divorce in New Jersey," he said out of the blue one day to her.[14] Natasha immediately took this for a declaration that he intended to get married. She found it quite typical of Nash to investigate the exit doors even as he was hovering near the entrance.

On another occasion, Nash gave a lecture at Chicago and had dinner afterward with Leo Goodman, a mathematician he knew from the graduate-school days in Princeton. He told Goodman that he thought Alicia would make a fine wife. Why? Because she watched so much television. That meant, he felt, that she wouldn't require much attention from him.[15] The exchange brings to mind Eleanor's oft-repeated remark about Nash: "he always wanted something for nothing."

Alicia has insisted that she cannot remember when Nash proposed or whether he did so in person or by letter.[16] They simply had an understanding, she said. But

Alicia's actions that fall belie her later account. After Nash had left Cambridge in June, Alicia stayed on, desperately unhappy. All this suggests the opposite of any "understanding."

Alicia's letter to Joyce Davis on October 23, 1956, does not mention Nash at all. Presumably, if they'd gotten formally engaged by that date, Alicia would have announced the fact to Joyce.

> As you might know I've been looking for a job in New York and had applied to several places. At first I was afraid things might prove difficult but so far I've already had offers from Brookhaven, as a junior physicist with the reactor group, and from the Nuclear Development Corporation of America also in the reactor field. I'm accepting the latter at $450 per month. I'm told I might get $500 some other place but I think N.D.C. offers good experience and I've always wanted to do nuclear physics specifically.[17]

It's possible that Alicia would have left school and gotten a job regardless of the state of her relationship with Nash. She was increasingly unenthusiastic about attending graduate school. "I'm tired of the studying and procrastinating routine. . . . All I know is I want to 'LIVE.' " Since she had gone to high school in New York, it would have been natural for her to think of returning there to work. But Alicia herself said later that she moved to New York on Nash's account. She may have gone there in the hopes of renewing her relationship with him. She may have gone at his express invitation.

Alicia moved into the Barbizon Hotel, the legendary hotel for young women that is the setting of Sylvia Plath's fifties novel *The Bell Jar.* References were required to obtain lodging there. And the rooms, tiny and white with metal beds, were only for sleeping, Alicia complained in a PS to Joyce.[18] "This hotel — the Amazon — was for women only," writes Plath, who spent the summer of 1952 in residence, "and they were mostly girls my age with wealthy parents who wanted to be sure their daughters would be living where men couldn't get at them and deceive them; and they were all going to posh secretarial schools like Katy Gibbs, where they had to wear hats and stockings and gloves to class, or . . . simply hanging around in New York waiting to get married to some career man or other."[19]

Whether or not Alicia came to New York as Nash's fiancée at the end of October, she visited Nash's family in Roanoke that Thanksgiving.[20] Nash did not give her a ring, however. He had some idea, typically odd and pennypinching, that he wanted to buy one in Antwerp, directly from a diamond wholesaler.[21]

Virginia found Alicia charming and dignified and was impressed by Alicia's obvious devotion to Nash, but at the same time she thought her quite different from the sort of girl she had imagined for her son's bride.[22] She thought the relationship between the two strange. Alicia was a physicist who talked about her job at a nuclear reactor company and displayed no interest in anything domestic, a young woman completely out of Virginia's ken. While Virginia and Martha busied themselves in the kitchen, Alicia and Nash spent most of Thanksgiving Day sitting on the floor of Virginia's living room poring over stock quotations. Martha's reaction was similar to her mother's. (At Virginia's insistence, and thinking it might turn Alicia's head in the right direction, Martha took Alicia shopping in Roanoke one afternoon to buy a hat.)

The wedding took place on an unexpectedly mild, gray February morning in Washington, D.C., at St. John's, the yellow-and-white Episcopal church across Pennsylvania Avenue from the White House.[23] Nash, by then an atheist, balked at a Catholic ceremony. He would have been happy to get married in city hall. Alicia wanted an elegant, formal affair. It was a small wedding. There were no mathematicians or old school friends present, only immediate family. Charlie, his brother-in-law, whom Nash hardly knew, was best man. Martha was matron of honor. Bride and groom were both late, having been held up at the portrait photographers. Nash and Alicia drove to Atlantic City for a weekend honeymoon on the way back to New York. It wasn't a success. Alicia hadn't been feeling well, Nash wrote in a postcard to his mother.[24]

In April, two months later, Alicia and Nash threw a party to celebrate their marriage. They were living in a sublet apartment on the Upper East Side, around the corner from Bloomingdale's. About twenty people came, mostly mathematicians from Courant and the Institute for Advanced Study and several of Alicia's cousins, including Odette and Enrique. "They seemed very happy," Enrique Larde later recalled. "It was a great apartment. They were just showing off their new marriage. He looked very handsome. It seemed very romantic."[25]

PART THREE

A
Slow
Fire
Burning

30

Olden Lane
and Washington Square

1956–57

*Mathematical ideas originate in empirics. . . . But, once they are so
conceived, the subject begins to live a peculiar life of its own and is better
compared to a creative one, governed almost entirely by aesthetical
motivations. . . . As a mathematical discipline travels, or after much "abstract"
inbreeding, [it] is in danger of degeneration. . . . whenever this stage is
reached, the only remedy seems to me to be the rejuvenating return to the
source: the reinjection of more or less directly empirical ideas. — JOHN VON
NEUMANN*

THE INSTITUTE FOR ADVANCED STUDY, nestled on Princeton's fringes on what
had been a farm, was a scholar's dream. It was bordered by woods and the Delaware-Raritan Canal, its lawns were immaculate, and one of its streets was Einstein
Drive. It was also blessedly free of students. The atmosphere in the Fuld Hall
common room resembled that of a venerable men's club, with its newspaper racks
and mingled scents of leather and pipe tobacco; its doors were never locked and its
lights burned far into the night.

In 1956, the Institute's permanent faculty were not many more than a dozen
mathematicians and theoretical physicists.[1] They were, however, outnumbered six-
fold by a host of distinguished temporary visitors from around the globe, prompting
Oppenheimer to call it "an intellectual hotel."[2] For young researchers, the Institute
was a golden opportunity to escape the onerous demands of teaching and administration, and, indeed, the tasks of everyday life. Everything was provided the visitor:
an apartment less than a few hundred yards from an office, an unending round of
seminars, lectures, and, for those so inclined, parties where the booze was plentiful
and where one could glimpse Lefschetz balancing a martini glass in an artificial
hand, or witness a very drunk French mathematician displaying his mountaineer-
ing skills by rope-climbing up and over the fireplace mantel.[3]

Some found the idyllic setting, carefully designed to remove all impediments
to creativity, vaguely disquieting. Paul Cohen, a mathematician at Stanford Univer-

sity, remarked, "It was such a great place that you had to stay at least two years. It took one year just to learn how to work under such ideal conditions."[4] By 1956, Einstein was dead, Gödel was no longer active, and von Neumann lay dying in Bethesda. Oppenheimer was still director, but much humbled by the McCarthyite inquisitions and increasingly isolated. As one mathematician said, "The Institute had become pure, very pure."[5] Cathleen Morawetz, later president of the American Mathematical Society, put it more bluntly: "The Institute was known to be about the dullest place you could find."[6]

By contrast, the Courant Institute of Mathematical Sciences at New York University was "the national capital of applied mathematical analysis," as *Fortune* magazine was soon to inform its readers.[7] Just a few years old and vibrant with energy, Courant occupied a nineteenth-century loft less than a block to the east of Washington Square in a neighborhood that, despite the university's growing presence, was still dominated by small manufacturing concerns. Indeed, Courant initially shared the premises — with its fire escapes and creaky old-fashioned freight elevator — with a number of hat factories.[8] Financing for the institute had come from the Atomic Energy Commission, which had been hunting for a home for its giant Univac 4 computer. At the time, this great mass of vacuum tubes, with its armed guard, occupied 25 Waverly Place.[9]

The institute was the creation of one of mathematics' great entrepreneurs, Richard Courant, a German Jewish professor of mathematics who had been driven out of Göttingen in the mid-1930s by the Nazis.[10] Short, rotund, autocratic, and irrepressible, Courant was famous for his fascination with the rich and powerful, his penchant for falling in love with his female "assistants," and his unerring eye for young mathematical talent. When Courant arrived in 1937, New York University had no mathematics worth speaking of. Undaunted, Courant immediately set about raising funds. His own stellar reputation, the anti-Semitism of the American educational establishment, and New York's "deep reservoir of talent," enabled him to attract brilliant students, most of them New York City Jews who were shut out of the Harvards and Princetons.[11] The advent of World War II brought more money and more students, and by the mid-1950s, when the institute was formally founded, it was already rivaling more established mathematical centers like Princeton and Cambridge.[12] Its young stars included Peter Lax and his wife, Anneli, Cathleen Synge Morawetz, Jürgen Moser, and Louis Nirenberg, and among its stellar visitors were Lars Hörmander, a future Fields medalist, and Shlomo Sternberg, who would soon move to Harvard.

The Courant Institute was practically on Nash's doorstep and, given its lively atmosphere, it was not surprising that Nash was soon spending at least as much time there as at the Institute for Advanced Study. At first Nash would stop by for an hour or two before driving down to Princeton, but he soon found himself staying the whole day.[13] He never came too early, for he liked to sleep late after working into the wee hours at the university library.[14] But he was almost always there for teatime in the lounge on the building's penultimate floor.[15]

As for the Courant crowd, a friendly, open group with little taste for the competitiveness of MIT or the snobbery of the Institute, it was happy to have him. Tilla Weinstein, a mathematician at Rutgers, who recalled that Nash liked to pace around on one of the building's fire escapes, said, "He was just a delight. There was a wit and humor about him that was thoroughly unstandard. There was a wonderful playful quality, a lightness."[16] Cathleen Morawetz, the daughter of John Synge, Nash's professor at Carnegie, assumed Nash was just another postdoctoral fellow and found him "very charming," "an attractive fellow," "a lively conversationalist."[17] Hörmander recalled his first impressions: "He wore a serious expression. Then he'd break out into a sudden smile. He was an enthusiast."[18] Peter Lax, who had spent the war at Los Alamos, was interested in Nash's research and "his own way of looking at things."[19]

At first, Nash seemed more interested in the political cataclysms of that fall — Nasser nationalized the Suez Canal, prompting an invasion by England, France, and Israel, the Russians crushed the Hungarian uprising, and Eisenhower and Stevenson were again battling for the presidency — than in pursuing mathematical conversations. "He'd be in the common room," one Courant visitor recalled, "talking and talking of his views of the political situation. From the afternoon teas, I remember him as voicing very strong opinions on the Suez crisis, which was going on at that time."[20] Another mathematician remembered a similar conversation in the institute dining room: "When the British and their allies were trying to grab Suez, and Eisenhower had not made his position unmistakably clear (if he ever did), one day at lunch Nash started in on Suez. Of course, Nasser wasn't black, but he was dark enough for Nash. 'What you have to do with these people is to take a firm hand, and then once they realize you mean it . . .' "[21]

The leading lights at Courant were very much at the forefront of rapid progress, stimulated by World War II, in certain kinds of differential equations that serve as mathematical models for an immense variety of physical phenomena involving some sort of change.[22] By the mid-fifties, as *Fortune* noted, mathematicians knew relatively simple routines for solving ordinary differential equations using computers. But there were no straightforward methods for solving most nonlinear partial differential equations that crop up when large or abrupt changes occur — such as equations that describe the aerodynamic shock waves produced when a jet accelerates past the speed of sound. In his 1958 obituary of von Neumann, who did important work in this field in the thirties, Stanislaw Ulam called such systems of equations "baffling analytically," saying that they "defy even qualitative insights by present methods."[23] As Nash was to write that same year, "The open problems in the area of non-linear partial differential equations are very relevant to applied mathematics and science as a whole, perhaps more so than the open problems in any other area of mathematics, and this field seems poised for rapid development. It seems clear however that fresh methods must be employed."[24]

Nash, partly because of his contact with Wiener and perhaps his earlier interaction with Weinstein at Carnegie, was already interested in the problem of turbulence.[25] Turbulence refers to the flow of gas or liquid over any uneven surface, like water rushing into a bay, heat or electrical charges traveling through metal, oil escaping from an underground pool, or clouds skimming over an air mass. It should be possible to model such motion mathematically. But it turns out to be extremely difficult. As Nash wrote:

> Little is known about the existence, uniqueness and smoothness of solutions of the general equations of flow for a viscous, compressible, and heat conduct- ing fluid. These are a non-linear parabolic system of equations. An interest in these questions led us to undertake this work. It became clear that nothing could be done about the continuum description of general fluid flow without the ability to handle non-linear parabolic equations and that this in turn required an *a priori* estimate of continuity.[26]

It was Louis Nirenberg, a short, myopic, and sweet-natured young protégé of Courant's, who handed Nash a major unsolved problem in the then fairly new field of nonlinear theory.[27] Nirenberg, also in his twenties, and already a formidable analyst, found Nash a bit strange. "He'd often seemed to have an internal smile, as if he was thinking of a private joke, as if he was laughing at a private joke that he never [told anyone about]."[28] But he was extremely impressed with the technique Nash had invented for solving his embedding theorem and sensed that Nash might be the man to crack an extremely difficult outstanding problem that had been open since the late 1930s.

He recalled:

> I worked in partial differential equations. I also worked in geometry. The problem had to do with certain kinds of inequalities associated with elliptic partial differential equations. The problem had been around in the field for some time and a number of people had worked on it. Someone had obtained such estimates much earlier, in the 1930s in two dimensions. But the problem was open for [almost] thirty years in higher dimensions.[29]

Nash began working on the problem almost as soon as Nirenberg suggested it, although he knocked on doors until he was satisfied that the problem was as important as Nirenberg claimed.[30] Lax, who was one of those he consulted, com- mented recently: "In physics everybody knows the most important problems. They are well defined. Not so in mathematics. People are more introspective. For Nash, though, it had to be important in the opinion of others."[31]

Nash started coming to Nirenberg's office to discuss his progress. But it was weeks before Nirenberg got any real sense that Nash was getting anywhere. "We would meet often. Nash would say, 'I seem to need such and such an inequality. I think it's true that . . .' " Very often, Nash's speculations were far off the mark. "He was sort of groping. He gave that impression. I wasn't very confident he was going to get through."[32]

Nirenberg sent Nash around to talk to Lars Hörmander, a tall, steely Swede who was already one of the top scholars in the field. Precise, careful, and immensely knowledgeable, Hörmander knew Nash by reputation but reacted even more skeptically than Nirenberg. "Nash had learned from Nirenberg the importance of extending the Holder estimates known for second-order elliptic equations with two variables and irregular coefficients to higher dimensions," Hörmander recalled in 1997.[33] "He came to see me several times, 'What did I think of such and such an inequality?' At first, his conjectures were obviously false. [They were] easy to disprove by known facts on constant coefficient operators. He was rather inexperienced in these matters. Nash did things from scratch without using standard techniques. He was always trying to extract problems . . . [from conversations with others]. He had not the patience to [study them]."

Nash continued to grope, but with more success. "After a couple more times," said Hörmander, "he'd come up with things that were not so obviously wrong."[34]

By the spring, Nash was able to obtain basic existence, uniqueness, and continuity theorems once again using novel methods of his own invention. He had a theory that difficult problems couldn't be attacked frontally. He approached the problem in an ingeniously roundabout manner, first transforming the nonlinear equations into linear equations and then attacking these by nonlinear means. "It was a stroke of genius," said Lax, who followed the progress of Nash's research closely. "I've never seen that done. I've always kept it in mind, thinking, maybe it will work in another circumstance."[35]

Nash's new result got far more immediate attention than his embedding theorem. It convinced Nirenberg, too, that Nash was a genius.[36] Hörmander's mentor at the University of Lund, Lars Gårding, a world-class specialist in partial differential equations, immediately declared, "You have to be a genius to do that."[37]

Courant made Nash a handsome job offer.[38] Nash's reaction was a curious one. Cathleen Synge Morawetz recalled a long conversation with Nash, who couldn't make up his mind whether to accept the offer or to go back to MIT. "He said he opted to go to MIT because of the tax advantage" of living in Massachusetts as opposed to New York.[39]

Despite these successes, Nash was to look back on the year as one of cruel disappointment. In late spring, Nash discovered that a then-obscure young Italian, Ennio De Giorgi, had proven his continuity theorem a few months earlier. Paul Garabedian, a Stanford mathematician, was a naval attaché in London. It was an Office of Naval Research sinecure.[40] In January 1957, Garabedian took a long car trip around Europe and looked up young mathematicians. "I saw some oldtimers in Rome," he recalled. "It was a scene. You'd talk mathematics for half an hour. Then you'd have lunch for three hours. Then a siesta. Then dinner. Nobody mentioned De Giorgi." But in Naples, someone did, and Garabedian looked De Giorgi up on his way back through Rome. "He was this bedraggled, skinny little starved-looking guy. But I found out he'd written this paper."

De Giorgi, who died in 1996, came from a very poor family in Lecce in southern Italy.[41] Later he would become an idol to the younger generation. He had no life outside mathematics, no family of his own or other close relationships, and, even later, literally lived in his office. Despite occupying the most prestigious mathematical chair in Italy, he lived a life of ascetic poverty, completely devoted to his research, teaching, and, as time went on, a growing preoccupation with mysticism that led him to attempt to prove the existence of God through mathematics.

De Giorgi's paper had been published in the most obscure journal imaginable, the proceedings of a regional academy of sciences. Garabedian proceeded to report De Giorgi's results in the Office of Naval Research's European newsletter.

Nash's own account, written after he had won the Nobel for his work in game theory, conveys the acute disappointment he felt:

> I ran into some bad luck since, without my being sufficiently informed on what other people were doing in the area, it happened that I was working in parallel with Ennio De Giorgi of Pisa, Italy. And De Giorgi was first actually to achieve the ascent of the summit (of the figuratively described problem) at least for the particularly interesting case of "elliptic equations."[42]

Nash's view was perhaps overly subjective. Mathematics is not an intramural sport, and as important as being first is, how one gets to one's destination is often as important as, if not more important than, the actual target. Nash's work was almost universally regarded as a major breakthrough. But this was not how Nash saw it. Gian-Carlo Rota, a graduate student at Yale who spent that year at Courant, recalled in 1994: "When Nash learned about De Giorgi he was quite shocked. Some people even thought he cracked up because of that."[43] When De Giorgi came to Courant that summer and he and Nash met, Lax said later, "It was like Stanley meeting Livingstone."[44]

Nash left the Institute for Advanced Study on a fractious note. In early July he apparently had a serious argument with Oppenheimer about quantum theory — serious enough, at any rate, to warrant a lengthy letter of apology from Nash to Oppenheimer written around July 10, 1957: "First, please let me apologize for my manner of speaking when we discussed quantum theory recently. This manner is unjustifiably aggressive."[45] After calling his own behavior unjustified, Nash nonetheless immediately justified it by calling "most physicists (also some mathematicians who have studied Quantum Theory) . . . quite too dogmatic in their attitudes," complaining of their tendency to treat "anyone with any sort of questioning attitude or a belief in 'hidden parameters' . . . as stupid or at best a quite ignorant person."

Nash's letter to Oppenheimer shows that before leaving New York, Nash had begun to think seriously of attempting to address Einstein's famous critique of Heisenberg's uncertainty principle:

> Now I am making a concentrated study of Heisenberg's original 1925 paper . . . This strikes me as a beautiful work and I am amazed at the great difference between expositions of "matrix mechanics," a difference, which from my viewpoint, seems definitely in favor of the original.[46]

"I embarked on [a project] to revise quantum theory," Nash said in his 1996 Madrid lecture. "It was not a priori absurd for a non-physicist. Einstein had criticized the indeterminacy of the quantum mechanics of Heisenberg."[47]

He apparently had devoted what little time he spent at the Institute for Advanced Study that year talking with physicists and mathematicians about quantum theory. Whose brains he was picking is not clear: Freeman Dyson, Hans Lewy, and Abraham Pais were in residence at least one of the terms.[48] Nash's letter of apology to Oppenheimer provides the only record of what he was thinking at the time. Nash made his own agenda quite clear. "To me one of the best things about the Heisenberg paper is its restriction to the observable quantities," he wrote, adding that "I want to find a different and more satisfying under-picture of a non-observable reality."[49]

It was this attempt that Nash would blame, decades later in a lecture to psychiatrists, for triggering his mental illness—calling his attempt to resolve the contradictions in quantum theory, on which he embarked in the summer of 1957, "possibly overreaching and psychologically destabilizing."[50]

31

The Bomb Factory

What's the matter with being a loner and innovative? Isn't that fine? But the
[lone genius] has the same wishes as other people. If he were back in high
school doing science projects, fine. But if he's too isolated and he's
disappointed in something big, it's frightening, and fright can precipitate
depression. — PAUL HOWARD, McLean Hospital

JÜRGEN MOSER had joined the MIT faculty in the fall of 1957 and was living
with his wife, Gertrude, and his stepson, Richy, in a tiny rented house to the west
of Boston in Needham near Wellesley College. Needham was then more exurb than
suburb, still predominantly rural, a lovely place for walking, boating, and stargaz-
ing, all of which Moser, a nature lover, was fond. That October and November,
Moser would go outside every evening at dusk with eleven-year-old Richy, climb a
great dirt mound behind their house, and wait for *Sputnik*—a tiny silvery dot
reflecting the sun's last rays—to pass slowly over Boston.[1] Having calculated the
satellite's precise orbit, Moser always knew when it would appear on the horizon.

Very often, he would still be thinking of the afternoon's conversation with
Nash. Nash drove out to Needham often. Despite their very different temperaments, Nash and Moser had great respect for each other. Moser, who thought
Nash's implicit function theorem might be generalized and applied to celestial
mechanics, was eager to learn more of Nash's thinking. Nash, in turn, was interested in Moser's ideas about nonlinear equations. Richard Emery recalled in 1996:
"I remember Nash being very much a part of our life. He used to come to the
house and talk with Jürgen. They would walk and talk together and spend time in
the study. The intensity of it was unimaginable. There could be no interruptions.
An interruption was an absolute sin, a violation most serious. It was met with real
wrath. When Jürgen and Nash met, it was very intense. I always had to be quiet."[2]

Returning to Cambridge in late summer, Nash and Alicia found an apartment with
some difficulty.[3] They each paid half the rent, for they had decided not to pool
their funds.[4] Alicia got a job as a physics researcher at Technical Operations, one
of the small high-tech companies that were springing up along Route 128.[5] She
also enrolled in a course on quantum theory taught by J. C. Slater.

They quickly settled into the pleasant private and social rituals of a newly married academic couple. Alicia almost never cooked. She would meet Nash on the campus after work, they would eat out with one or more of Nash's mathematics friends, and often spend the evening at a lecture, concert, or some social gathering.[6] Alicia made sure that they were always surrounded by amusing people, sometimes Nash's old graduate-student friends, including Mattuck and Bricker, sometimes Emma Duchane and whomever Emma happened to be dating, and, increasingly, other young couples like themselves, including the Mosers, the Minskys, Hartley Rogers and his wife, Adrienne, and Gian-Carlo Rota and his wife, Terry.

When they were with other people, Nash talked to the mathematicians, Alicia to the wives or Emma. Yet her attention was always focused on Nash: what he was saying, how he looked, how others reacted to him. He too, seemed always aware of her, even when he appeared to be ignoring her. That he wasn't especially nice to her, or generous, mattered less than that he was interesting and made things happen.

Their friends accepted Nash's new status as a married man with more or less good grace. Some found Alicia "ambitious, strong-willed," others quite the opposite. Rogers recalled in 1996 that "Alicia subordinated herself to John. She wasn't there to compete with him. She was totally dedicated to his support."[7] Some of their acquaintances found their relationship oddly cool, but others came away with the impression that marriage suited Nash well and that Alicia was having a good effect on him. "Somehow, he was relating a little better," Rogers recalled. Zipporah Levinson agreed: "John was awkward. Alicia made him behave."[8] Photographs of Alicia taken in those months show a radiant young woman. It was, as Alicia would say many years later, "a very nice time of my life."[9]

Nash continued to work on the problem he had solved the previous year at Courant. There were some small gaps in the proof, and the paper Nash had begun to write, laying out a full account of what he had done, was in very rough shape.[10] "It was," a colleague said in 1996, "as if he were a composer and could hear the music, but he didn't know how to write it down or exactly how to orchestrate it."[11] As it turned out, it would take most of the year, and a collective effort, before the final product — which some mathematicians regard as Nash's most important work — was finally ready to be submitted to a journal.

To complete it, Nash came as close as he ever had or would to an active collaboration with other mathematicians. "It was like building the atom bomb," recalled Lennart Carleson, a young professor from the University of Uppsala who was visiting MIT that term. "This was the beginning of nonlinear theory. It was very difficult."[12] Nash knocked on doors, asked questions, speculated out loud, fished for ideas, and at the end of the day, got a dozen or so mathematicians around Cambridge interested enough in his problem to drop their own research long enough to solve little pieces of his puzzle. "It was a kind of factory," Carleson, who contributed a neat little theorem on entropy to Nash's paper, said. "He wouldn't tell us what he was after, his grand design. It was amusing to watch how he got all these great egos to cooperate."[13]

Besides Moser and Carleson, Nash also turned to Eli Stein, now a professor of mathematics at Princeton University but then an MIT instructor. "He wasn't interested in what I was doing," recalled Stein. "He'd say, 'You're an analyst. You ought to be interested in this.' "[14]

Stein was intrigued by Nash's enthusiasm and his constant supply of ideas. He said, "We were like Yankees fans getting together and talking about great games and great players. It was very emotional. Nash knew exactly what he wanted to do. With his great intuition, he saw that certain things ought to be true. He'd come into my office and say, 'This inequality must be true.' His arguments were plausible but he didn't have proofs for the individual lemmas — building blocks for the main proof."[15] He challenged Stein to prove the lemmas.

"You don't accept arguments based on plausibility," said Stein in 1995. "If you build an edifice based on one plausible proposition after another, the whole thing is liable to collapse after a few steps. But somehow he knew it wouldn't. And it didn't."[16]

Nash's thirtieth year was thus looking very bright. He had scored a major success. He was adulated and lionized as never before.[17] *Fortune* magazine was about to feature him as one of the brightest young stars of mathematics in an upcoming series on the "New Math."[18] And he had returned to Cambridge as a married man with a beautiful and adoring young wife. Yet his good fortune seemed at times only to highlight the gap between his ambitions and what he had achieved. If anything, he felt more frustrated and dissatisfied than ever. He had hoped for an appointment at Harvard or Princeton.[19] As it was, he was not yet a full professor at MIT, nor did he have tenure. He had expected that his latest result, along with the offer from Courant, would convince the department to award him both that winter.[20] Getting these things after only five years would be unusual, but Nash felt that he deserved nothing less.[21] But Martin had already made it clear to Nash that he was unwilling to put him up for promotion so soon. Nash's candidacy was controversial, Martin had told him, just as his initial appointment had been.[22] A number of people in the department felt he was a poor teacher and an even worse colleague. Martin felt Nash's case would be stronger once the full version of the parabolic equations paper appeared in print. Nash, however, was furious.

Nash continued to brood over the De Giorgi fiasco. The real blow of discovering that De Giorgi had beaten him to the punch was to him not just having to share the credit for his monumental discovery, but his strong belief that the sudden appearance of a coinventor would rob him of the thing he most coveted: a Fields Medal.

Forty years later, after winning a Nobel, Nash referred in his autobiographical essay, in his typically elliptical fashion, to his dashed hopes:

It seems conceivable that if either De Giorgi or Nash had failed in the attack on this problem (or *a priori* estimates of Holder continuity) then that the lone climber reaching the peak would have been recognized with the mathematics' Fields medal (which has traditionally been restricted to persons less than 40 years old).[23]

The next Fields Medal would be awarded in August 1958, and as everyone knew, the deliberations had long been under way.

To understand how deep the disappointment was, one must know that the Fields Medal is the Nobel Prize of mathematics, the ultimate distinction that a mathematician can be granted by his peers, the trophy of trophies.[24] There is no Nobel in mathematics, and mathematical discoveries, no matter how vital to Nobel disciplines such as physics or economics, do not in themselves qualify for a Nobel. The Fields is, if anything, rarer than the Nobel. In the fifties and early sixties, it was awarded once every four years and usually to just two recipients at a time. Nobels, by contrast, are awarded annually, with as many as three winners sharing each prize. Tradition demands that recipients of the Fields be under forty years of age, a practice designed to honor the spirit of the prize charter, which stipulates that the purpose of the honor is "to encourage young mathematicians" and "future work."[25] The incentive, incidentally, is of an intangible variety, as the cash involved, in contrast to the Nobel, is negligible, a few hundred dollars. Yet since the Fields is an instant ticket in midcareer to endowed chairs at top universities, ample research funds, and star salaries, this seeming disadvantage is more apparent than real.

The prize is administered by the International Mathematical Union, the same organization that organizes the quadrennial world mathematical congresses, and the selection of Fields medalists is, as one recent president of the organization put it, "one of the most important tasks, one of the most taxing responsibilities."[26] Like the Nobel deliberations, the Fields selection process is shrouded in greatest secrecy.

The seven-member prize committee for the 1958 Fields awards was headed by Heinz Hopf, the dapper, genial, cigar-smoking geometer from Zurich who showed so much interest in Nash's embedding theorem, and included another prominent German mathematician, Kurt Friedrichs, formerly of Göttingen, and then at Courant.[27] The deliberations got under way in late 1955 and were concluded early in 1958. (The medalists were informed, in strictest secrecy, in May 1958 and actually awarded their medals at the Edinburgh congress the following August.)

All prize deliberations involve elements of accident, the biggest one being the composition of the committee. As one mathematician who took part in a subsequent committee said, "People aren't universalists. They're horse trading."[28] In 1958, there were a total of thirty-six nominees, as Hopf was to say in his award ceremony speech, but the hot contenders numbered no more than five or six.[29]

That year the deliberations were unusually contentious and the prizes, which ultimately went to René Thom, a topologist, and Klaus F. Roth, a number theorist, were awarded on a four–three vote.[30] "There were lots of politics in that prize," one person close to the deliberations said recently.[31] Roth was a shoo-in; he had solved a fundamental problem in number theory that the most senior committee member, Carl Ludwig Siegel, had worked on early in his career. "It was a question of Thom versus Nash," said Moser, who heard reports of the deliberations from several of the participants.[32] "Friedrichs fought very hard for Nash, but he didn't succeed," recalled Lax, who had been Friedrichs's student and who heard Friedrichs's account of the deliberations. "He was upset. As I look back, he should have insisted that a third prize be given."[33]

Chances are that Nash did not make the final round. His work on partial differential equations, of which Friedrichs would have been aware, was not yet published or properly vetted. He was an outsider, which one person close to the deliberations thought "might have hurt him." Moser said, "Nash was somebody who didn't learn the stuff. He didn't care. He wasn't afraid of moving in and working on his own. That doesn't get looked at so positively by other people."[34] Besides, there was no great urgency to recognize him at this juncture; he was just twenty-nine.

No one could know, of course, that 1958 would be Nash's last chance. "By 1962, a Fields for Nash would have been out of the question," Moser said recently. "It would never have happened. I'm sure nobody even thought about him anymore."[35]

A measure of how badly Nash wanted to win the distinction conferred by such a prize is the extraordinary lengths to which he went to ensure that his paper would be eligible for the Bôcher Prize, the only award remotely comparable in terms of prestige to the Fields. The Bôcher is given by the American Mathematical Society only once every five years.[36] It was due to be awarded in February 1959, which meant that the deliberations would take place in the latter part of 1958.

Nash submitted his manuscript to *Acta Mathematica,* the Swedish mathematics journal, in the spring of 1958.[37] It was a natural choice, since Carleson was the editor and was convinced of the paper's great importance. Nash let Carleson know he wanted the paper published as quickly as possible and urged Carleson to give it to a referee who could vet the paper in a minimum of time. Carleson gave the manuscript to Hörmander to referee. Hörmander spent two months studying it, verified all the theorems, and urged Carleson to get it into print as quickly as possible. But as soon as Carleson informed Nash of the formal acceptance, which was, in any case, largely a foregone conclusion, Nash withdrew his paper.

When the paper subsequently appeared in the fall issue of the *American Journal of Mathematics,* Hörmander concluded that Nash had always intended to publish the paper there, since the Bôcher restricted eligible papers to those published in American journals — or, worse, had submitted the paper to both journals, a clear-cut breach of professional ethics. "It turned out that Nash had just wanted to get a letter of acceptance from *Acta* to be able to get fast publication in the *American Journal of Mathematics.*"[38] Hörmander was angry at what he felt was "very improper and most unusual."[39]

It's possible, though, that Nash had simply submitted the paper to *Acta* before learning that doing so would exclude it from consideration for the Bôcher, but that upon discovering this fact, he was willing to antagonize Carleson and Hörmander in order to preserve his eligibility. He may therefore not have used *Acta* quite so unscrupulously. Withdrawing the paper after it had been promised to *Acta,* and after it had been refereed, would have been unprofessional, but not as clear a violation of ethics as Hörmander's scenario suggests. However, it still showed how very much winning a prize meant to Nash.

32

Secrets

Summer 1958

It struck me that I knew everything; everything was revealed to me, all the secrets of the world were mine during those spacious hours. — GERARD DE NERVAL

NASH TURNED THIRTY that June. For most people, thirty is simply the dividing line between youth and adulthood, but mathematicians consider their calling a young man's game, so thirty signals something far more gloomy. Looking back at this time in his life, Nash would refer to a sudden onset of anxiety, "a fear" that the best years of his creative life were over.[1]

What an irony that mathematicians, who live so much more in their minds than most of humanity, should feel so much more trapped by their bodies! An ambitious young mathematician watches the calendar with a sense of trepidation and foreboding equal to or greater than that of any model, actor, or athlete. *The Mathematician's Apology* by G. H. Hardy sets the standard for all laments of lost youth. Hardy wrote that he knew of no single piece of first-rate mathematics done by a mathematician over fifty.[2] But the age anxiety is most intense, mathematicians say, as thirty draws near. "People say that for better or worse you will probably do your best work by the time you are thirty," said one genius. "I tend to think that you are at your peak around thirty. I'm not saying you won't equal it. I would like to think that you could. But I don't think you will ever do better. That's my gut feeling."[3] Von Neumann used to say that "the primary mathematical powers decline at about twenty-six," after which the mathematician must rely on "a certain more prosaic shrewdness."[4]

Compounding the irony is that the act of creating new mathematics, which appears so solitary from the outside, feels from the inside like an intramural competition, a race. One never forgets the crowded field. And one's relative standing, vis-à-vis past and present competitors, is what counts. Again, Hardy best conveyed what motivates many mathematicians, including himself. He wrote that he could not recall ever wanting to be anything but a mathematician, but also that he could not remember feeling any passion for mathematics as a boy. "I wanted to beat other boys, and this seemed to be the way in which I could do so most decisively."[5] More ambitious than most, Nash was also more age-conscious than most — or perhaps simply more frank about it. "John was the most age-conscious person I've ever met," recalled Felix Browder in 1995. "He would tell me every week my age relative to his and everybody else's."[6] His determination to avoid the draft during the Korean War suggested not just a desire to avoid regimentation, but also an unwillingness to take time out of the race.

The most successful are the most vulnerable to the feeling that time is running out. Such fears may be exaggerated, but they are quite capable of producing real crises, as the history of mathematics amply attests. Artin, for example, switched frantically from field to field trying to catch hold of something that would equal his early accomplishments.[7] Steenrod slipped into a deep depression. When one of his students published a note on "Steenrod's Reduced Powers" — the reference was, of course, mathematical, not personal — other mathematicians smirked and said, "Oh, yes, Steenrod's reduced powers!"[8]

Nash's thirtieth birthday produced a kind of cognitive dissonance. One can almost imagine a sniggering commentator inside Nash's head: "What, thirty already, and still no prizes, no offer from Harvard, no tenure even? And you thought you were such a great mathematician? A genius? Ha, ha, ha!"

Nash's mood was odd. Periods of gnawing self-doubt and dissatisfaction alternated with periods of heady anticipation. Nash had a distinct feeling that he was on the brink of some revelation. And it was this sense of anticipation, as much as his fear, as he put it, of "descending to a professional level of comparative mediocrity and routine publication," that spurred him to begin working on two great problems.[9]

Sometime during the spring of 1958, Nash had confided to Eli Stein that he had "an idea of an idea" about how to solve the Riemann Hypothesis.[10] That summer, he wrote letters to Albert E. Ingham, Atle Selberg, and other experts in number theory sketching his idea and asking their opinion.[11] He worked in his office in Building Two for hours, night after night.

Even when a genius makes such an announcement, the rational response is skepticism. The Riemann Hypothesis is the holy grail of pure mathematics. "Whoever proves or disproves it will cover himself with glory," wrote E. T. Bell in 1939. "A decision one way or the other disposing of Riemann's conjecture would probably be of greater interest to mathematicians than a proof or disproof of Fermat's Last Theorem."[12]

Enrico Bombieri, at the Institute for Advanced Study, said: "The Riemann Hypothesis is not just a problem. It is *the* problem. It is the most important problem in pure mathematics. It's an indication of something extremely deep and fundamental that we cannot grasp."[13]

Whole numbers that are evenly divisible only by themselves and one — so-called prime numbers — have exerted a fascination for mathematicians for two thousand years or more. The Greek mathematician Euclid proved that there were infinitely many primes. The great European mathematicians of the eighteenth century—Euler, Legendre, and Gauss—began a quest, still under way, to estimate how many primes there are, given a whole number n, less than n.[14] And since 1859 a string of mathematical giants—G. H. Hardy, Norman Levinson, Atle Selberg, Paul Cohen, and Bombieri, among others—have attempted, unsuccessfully, to prove the Riemann Hypothesis.[15] George Polya once gave a young mathematician who had confided in him that he was working on the Riemann Hypothesis a reprint of a faulty proof of the conjecture by a Göttingen mathematician who thought he'd solved the problem. "I think about it every day when I wake up in the morning," the young mathematician had said, and Polya delivered the reprint the following morning with a note: "If you want to climb the Matterhorn you might first wish to go to Zermatt where those who have tried are buried."[16]

Before World War I, a German banker endowed a prize, lodged in Göttingen, for whoever proved or disproved the hypothesis. The prize was never awarded and, indeed, vanished in the inflation of the 1920s.[17]

Nash's first encounter with Georg Friedrich Bernhard Riemann and his famous conjecture took place when Nash was fourteen, probably lying on the den floor in front of the radio, reading Bell's *Men of Mathematics*.[18]

Riemann, the sickly son of an impoverished Lutheran minister, was also fourteen and preparing to follow in his father's footsteps when a sympathetic headmaster, who sensed that the boy was more suited to mathematics than the ministry, gave him a copy of Legendre's *Théorie des Nombres* to read.[19] As Bell tells it, the young Riemann returned the 859-page work six days later, saying, "That is certainly a wonderful book. I have mastered it." This episode, which took place in 1840, was likely the origin of Riemann's lifelong interest in the riddle of prime numbers and, as Bell theorizes, Riemann's Hypothesis may have originated in his later attempt to improve upon Legendre.

In 1859, at the age of thirty-three, Riemann wrote an eight-page paper, *"Ueber die Anzahl der Primzahlen unter einer gegebenen Groesse"* ("On the number of prime numbers under a given magnitude"), in which he laid out his famous conjecture—"one of the outstanding challenges, if not the outstanding challenge to pure mathematics."

Here is how Bell explains the conjecture:

The problem concerned is to give a formula which will state how many primes there are less than any given number n. In attempting to solve this Riemann was driven to an investigation of the infinite series $1 + \frac{1}{2^s} + \frac{1}{3^s} + \frac{1}{4^s} + \ldots$ in which s is a complex number, say $s = u + iv$ $(i = \sqrt{-1})$, where u and v are real numbers, so chosen that the series converges. With this proviso the infinite series is a definite function of s, say zeta(s) (the Greek zeta is always used to denote this function, which is called "Riemann's zeta function"); and as s varies zeta(s) continuously takes on different values. For what values of s will zeta(s) be zero? Riemann conjectured that *all* such values of s for which *u* lies between 0 and 1 are of the form $\frac{1}{2} + iv$, namely, all have their real part equal to $\frac{1}{2}$.[20]

When Riemann died of tuberculosis at thirty-nine, he left behind a vast legacy, including the abstract, four-dimensional geometry that Einstein would employ in formulating his general theory of relativity. Just as geographers had to go from two-dimensional plane geometry to three-dimensional solid geometry to create an undistorted map of the earth, Einstein, to map the cosmos, went from three-dimensional to four-dimensional geometry. But it was for his tantalizing conjecture that Riemann is best remembered. Proving or disproving it would settle many extremely difficult questions in the theory of numbers and in some fields of analysis. As Bell put it, "Expert opinion favors the truth of the hypothesis."[21]

It is impossible to say how long Nash had been contemplating his own attempt, but it seems likely that his interest crystallized sometime toward the end of his year in New York. Jack Schwartz recalled conversations with Nash on the subject in the Courant common room.[22] Jerome Neuwirth, a second-year graduate student at MIT in 1957–58, remembered that Nash had developed a very proprietary feeling about the problem around that time.[23] Neuwirth recalled that Newman, perhaps to tease Nash, told Nash that Neuwirth, too, was working on the Riemann Hypothesis. Nash came roaring into Neuwirth's office. "How dare you?" he said. "What's a guy like you doing?" It quickly became a running joke. Every time Nash saw Neuwirth he'd say, "Well, did you get anywhere yet?" And Neuwirth would answer, "Almost got it. I'd tell you about it, but I've got to run."

As Stein recalled it, Nash's idea was "to try to prove the hypothesis by logic, by internal consistency of the system. Some proofs are based on analogies, on rules of logic whereby something is proved [indirectly]. If one could show that the structure of two problems was in some sense identical, one could show that the logic of one proof had to apply to the other. It's a proof by logic and it doesn't relate to the real context. It's not proving that one object is related to another object."[24]

Stein was dubious. "He told me this very sketchy thing. It was an idea of an idea about how he was going to prove this thing. He was going to find another number system in which it was true. I thought, 'It's wild, it doesn't hang together.' This struck me as simply unbelievable. This was as opposed to my earlier conversations with him about parabolic equations, which struck me as daring but probably right."[25]

Richard Palais, a professor of mathematics at Brandeis University, recalls some particulars: "Nash was considering so-called pseudoprime sequences, i.e., increasing sequences $p_1, p_2, p_3, \ldots$ of integers that have many of the same distribution properties as the sequence 2, 3, 5, 7, . . . of prime numbers. For each of these one can associate in a natural way a 'zeta function,' which for the case of the true primes reduces to the Riemann zeta function. As I recall, Nash claimed to be able to show that for 'almost all' of these pseudoprime sequences the corresponding zeta function satisfied the Riemann Hypothesis." [26]

Bell warned that "Riemann's Hypothesis is not the sort of problem that can be attacked by elementary methods. It had already given rise to an extensive and thorny literature." [27] By the time Nash turned to it seriously, that literature had grown several-fold. Both Ingham and Selberg, possibly others as well, warned Nash that his ideas had been tried before and hadn't led anywhere. [28] Eugenio Calabi, who was in touch with Nash in this period, said: "For a person who is not a library hound, it's a very dangerous area to go into. If you have a flash of an idea with a scenario and think you may get a result, in the first flash of illumination you think you have a revelation. But that's very dangerous." [29]

There was, as Nash suggested, nothing absurd in his attempting to solve *the* outstanding problems in pure mathematics and theoretical physics. The skepticism with which his early formulations were greeted was, after all, merely a replay of the skepticism voiced by experts toward his earlier efforts, and has no doubt been exaggerated in hindsight. When those problems *are* solved it will be by a young mathematician who attacks them with the hubris, originality, raw mental power, and sheer tenacity that Nash brought to bear on his greatest work.

Yet the timing of Nash's decision to pursue these problems, just as he turned thirty and while he was licking various wounds to what he would later call his "merciless superego," [30] suggests that a fear of failure lay behind his willingness to take unusual risks. Stein's impression of Nash during their conversations about the Riemann problem is interesting: "He was a little . . . on the wild side. There was something exaggerated about his actions. There was a flamboyance in the way he talked. Mathematicians are usually more careful about what they will assert to be true." [31] But, of course, hubris is not exactly uncommon. As Hörmander, who went on to win a Fields Medal in 1962, put it: "It's part of life that not all things one works on work out. You overestimate your own abilities. After solving a big problem, nothing smaller is good enough. It's very dangerous." [32] Later, quite possibly because of the effects of shock treatments, Nash had absolutely no recall of his attempt to solve Riemann's conjecture. [33] But, as it was, Nash's compulsion to scale this most difficult, most dangerous peak proved central to his undoing.

There were other signs that Nash felt, at that particular juncture, a growing pressure to prove himself—as well as a newfound taste for taking risks. Nash had always been obsessed with money, even trivial amounts. Nash had made friends with Samuelson, Solow, and a number of other young economists at MIT. Samuelson recalled in 1996 that Nash told him about a bank with no checking charges at all.

"Do they give you stamped, self-addressed envelopes too?" Samuelson shot back. Nash, who didn't get the joke, immediately replied: "No. Do you know a bank that gives you stamped self-addressed envelopes?"[34] Privately, Samuelson thought it was all a bit pathological. Norman Levinson, who complained to Samuelson about Nash's parsimony, apparently once told him "to cut out his cheese-paring ways." Levinson said: "One extra theorem will earn you more than all that stuff." (Not everyone thought it was weird. Nash was able to convince Martin and a few others in the math department to switch their accounts to the Peoples National Bank of Rocky Mount, Virginia, which charged no fees on checking accounts!)[35]

That summer Nash's somewhat compulsive attitude toward money blossomed into an obsession with the stock and bond markets. Solow recalled: "It seemed he had a notion that there might be a secret to the market, not a conspiracy, but a theorem — something that if you could only figure it out, would let you beat the market. He would look at the financial pages and ask, 'Why is this happening? Why is that happening?' as if there had to be a reason for a stock to go up or down."[36] Martin, the chairman of the mathematics department, also recalled that "Nash liked to chat about the stock market. He had the idea you could get rich."[37] Nash had some notion of arbitraging July 1999 bonds against September 1999 as well as various ideas about over-the-counter stocks.[38] Solow was aghast to learn that Nash was investing his mother's savings. "I was horrified," he recalled. "That's something else," said Samuelson. "It's vanity. It's like claiming you can control the tides. It's a feeling that you can outwit nature. It's not uncommon among mathematicians. It's not just about money. It's me against the world. A lot of traders start that way. It's about proving yourself."

In late July, against this backdrop of grand designs, the Nashes, who had not yet gone on a proper honeymoon, left Cambridge for Europe. They sailed from New York on the *Île de France*.[39] Their ultimate destination was Edinburgh, where the World Congress of Mathematics was to take place in the second week of August. Nash was giving a lecture on nonlinear theory. Many colleagues from MIT and Princeton would be there, and Nash was able to pay for his trip partly out of Sloan funds.

But first they went to Paris. There, having calculated that importing a used car from Europe was a bargain, Nash purchased an olive-green Mercedes 180 diesel. He and Alicia then drove south over the Pyrenees to Spain, back to Italy, and up to Belgium. The trip was a success. "We were young," Alicia recalls. "It was fun."[40] Another of his plans was to buy Alicia the diamond that he had promised her. Antwerp was the center of the world diamond market, and Nash had the idea that it would be advantageous to buy a stone directly from a wholesaler.[41] Eli Stein's father had been a diamond merchant there before the war and that is what may have given Nash the idea in the first place. If Nash had hoped for a bargain, he was disappointed; the yellow stone that he purchased was no cheaper than it would have been in the States, he recalled in 1996. From Belgium, they drove to the North Sea, crossed over into Sweden, and visited Lund and Stockholm before crossing back to England.

They rendezvoused with Felix and Eva Browder in London and drove to Scotland with them. The men ignored the women, who sat together in the backseat gossiping (at that time, Eva recalled, "Nash wouldn't talk to women").[42] On the second, rainy day of the drive, Felix managed to dent the Mercedes, prompting Nash to repeat incessantly for the rest of the trip that "this car has been Browderized."[43]

There were, as Alicia later said, "lots of famous people around."[44] Nash seemed very much his usual self. He pouted a bit when Milnor gave his invited half-hour lecture, a great honor. He got into a loud argument with Olga Ladyshenskaya from the University of St. Petersburg, an expert on *a priori* estimates of parabolic equations and the leading female mathematician of her generation. Nash was picking her brains and she, somewhat paranoid, reacted rather violently.[45] The Nashes held a party in their hotel room. Nash raised eyebrows by complaining at great length that Alicia took too long to get dressed and that she was always late.[46] But he showed no emotion when, as he and Alicia sat in the balcony with the Browders, Moore, Milnor, and others, the Fields prizes were awarded.

33
Schemes

Fall 1958

The growing consciousness is a danger and a disease.
— FRIEDRICH NIETZSCHE

THE NASHES WERE BACK in Cambridge and Nash was already teaching when Alicia discovered, half with joy, half with dismay, that she was pregnant. Alicia, who liked her job and her paycheck, would have preferred to wait a few years. It had been Nash's wish that they start a family right away.[1] He stopped short of saying that his desire for another child had been his motive for marrying, but he reminded Alicia often that the whole purpose of marriage, in his view, was to produce children.[2] Now that his wish was to be realized, Nash was on the whole rather pleased, passing the great news on to Albert Tucker in a postscript to a letter in early October by referring to "a 'new addition' that we are expecting."[3]

He demanded that Alicia stop smoking. When she lit up at a math party he told her to put out her cigarette and made a scene after she refused.[4] But otherwise, all seemed to be well. Nash was teaching a graduate course. The course number — M711, a sly reference to craps — was Nash's idea and helped draw enough students to fill a small amphitheater.[5] Nash's first assignment also reflected his high spirits. He asked his students to invent a way to grade each other's papers so that he, Nash, wouldn't have to be bothered.

Nash was at that moment preoccupied with his own future and feeling increasingly restless. Martin had assured him that he was coming up for tenure that winter.[6] The promise of a decision mollified him somewhat: Nash wrote to Tucker that the situation at MIT had "reached a modus vivendi condition *which is an improvement over early 1958.*"[7]

But the sense that others were deciding his future oppressed him. And he was more convinced that he didn't belong at MIT. "I do not feel this is a good long-term position for me," he wrote to Tucker, saying that he was afraid of becoming isolated within the department like Wiener. "I would rather be one of a smaller number of more nearly equal colleagues."[8] His sister Martha recalled that "he had no intention of staying at MIT. He wanted to go to Harvard because of the prestige."[9]

Meanwhile, the University of Chicago was putting out feelers about Nash's possible interest in moving there.[10] Chicago had gone a long time without making any senior hires, even after Andre Weil had left for the Institute for Advanced Study. Now the math department had a new chairman, Adrian Albert, and some cash.[11] Albert was looking at a young Harvard professor, John Thompson, who had done brilliant work in group theory,[12] and also at Nash, who had a number of strong supporters in the department, including Shiing-shen Chern.

Nash felt the pressure from these decisions acutely and decided, in any case, that he wanted to get away the following year for a separate sabbatical. He wanted to spend the fall term of 1959 in Princeton at the Institute for Advanced Study and the spring term in Paris at its French equivalent, the Institut des Hautes Études Scientifiques, which, like the Institute, was dominated by mathematicians and theoretical physicists. Around the end of October, he began the process of applying for various grants, including those from the National Science Foundation, the Guggenheim Foundation, and the Fulbright program. He also applied to the Institute for a membership. He wrote: "This is part of the plan. The other part is to learn French."[13]

Albert Tucker was supportive. He wrote to the Fulbright program on October 8 that "Nash is eager to talk mathematics with others he thinks are up to snuff. . . . He is often rather rough on those less able . . . but this is standard practice in France . . . Nash should do well with energetic give and take . . . benefit from relationship with Leray."[14] His letter of recommendation to the National Science Foundation called Nash "one of the most talented and original mathematicians in the US . . . in his final year of a Sloan fellowship. One of two or three best men who ever got a Sloan."[15] His November 26 letter to the Guggenheim Foundation was couched in similarly laudatory terms.[16]

What Nash planned to work on isn't clear. He was at the time thinking about several different problems, including quantum theory and the Riemann Hypothesis. His desire to go to Paris may or may not have been motivated by Leray's presence at the Collège de France. Gian-Carlo Rota recalled: "He was bragging that he had enough fellowships to survive three or four years."[17]

One particularly unpleasant episode occurred in the early fall. His investments had proved disastrous,[18] to say the least, and he had to confess his failure to Virginia. He also had to promise to repay her. "I'll forward my debt," Nash was forced to write Virginia that fall. The amount wasn't huge, but the whole thing was quite upsetting.[19]

Everything, in short, seemed suddenly to be in flux—which may be why Nash found himself drawn to another young man. That summer a brilliant mathematician, six years Nash's junior, turned up at MIT. By the mid-1960s, Paul Cohen would be famous for solving a logical puzzle posed by Gödel—a result so stunning that *The New York Times* reported it[20]—and would win both a Fields and a

Bôcher.[21] But in the fall of 1958, Cohen was a fiercely ambitious, enormously frustrated upstart.

Cohen, who had grown up poor in New York, had been on the math team at Stuyvesant High School, and had just earned his Ph.D. at the University of Chicago.[22] But his thesis had not been well received and as a consequence he had been unhappily marooned at the University of Rochester. Desperate to get away, he had begged his old friend from Stuyvesant, Eli Stein, to help him get an instructorship at MIT.[23] This Stein had managed to do, and Cohen had come to Cambridge as soon as classes ended at Rochester.

Big, slightly feline in his movements, his eyes burning with fiery intensity under a high dome of a forehead, Cohen was self-obsessed, suspicious, aggressive, and charming by turns. He spoke several languages. He played the piano. His ambitions were seemingly unlimited and he spoke, from one moment to the next, of becoming a physicist, a composer, even a novelist. Stein, who became a close friend of Cohen's, said: "What drives Cohen is that he's going to be better than any other guy. He's going to solve the big problems. He looks down on mathematicians who do mathematics for the sake of making incremental improvements in the field."[24]

He was as fast as Newman, ambitious as Nash, arrogant as the two put together, and he very quickly fell in with the other two. Cohen was competitive — "wildly competitive," as one fellow instructor put it. "He was good at tearing people down," Adriano Garsia recalled in 1995.[25] They challenged each other with problems. "Well, Nash what kind of garbage are you working on now?" Cohen would say. "What wrong theorems did you prove today? Okay . . . you want a real problem? I'll give you a problem!" They ragged the chess players mercilessly. As Garsia recalls, "They were always eager to show how much better they were at whatever game it was that other people were playing. They engaged in horseplay . . . playing tunes on beer bottles." D.J. and Paul typically got the better of Nash, but not always. Cohen was the more articulate. But occasionally Nash could shut them up. "He could say an enormous amount in three words," said Garsia.

They delighted in ganging up on a graduate student struggling with a dissertation, dissecting a problem that some poor guy had been working on for two years and springing their own solution on him. They liked to argue that theirs was more powerful, but in fact they abjured elegance for brute force. "They wanted to solve it any way at all," said Garsia.

Nash "cultivated" Cohen, according to the latter. It was "unusual," Cohen recalled. "Maybe I liked him because he liked me. He'd ask me to lunch. He was not a friend of mine, though. I don't know that he had any friends."[26] Still, Cohen was intrigued. He used to go to dinner with the Nashes, speaking Spanish to Alicia, wondering how Nash had won this beautiful girl, and aware that Alicia was somehow "concerned" about Nash's paying so much attention to Cohen.

Nash never made any advances or ever said anything personal to Cohen. But he dropped hints. He'd say things like "So and So was a homosexual," Cohen recalled. Or he'd say a word and ask Cohen if he knew what it meant. If Cohen said no, Nash would come back with "Oh, you don't know what so and so means." People around the department were soon gossiping that Nash was in love with Cohen.[27]

Cohen was flattered, even fascinated, by Nash's interest, but he took special delight in rubbing Nash's face in the disparity between the grandiose claims and reality. He was critical, to the point of viciousness, of Nash's hubris. Later, Cohen would say, "Mathematically I didn't interact with him. I didn't feel I could talk to him about mathematics."

But they did talk a good deal about Nash's ideas on the Riemann Hypothesis. "Nash thought he could work on any problem he wanted," said Cohen in a tone of mild outrage. "He wrote a letter to Ingham, and he passed it around. I shot it down. What he was trying to do, you couldn't do. I would have been very unsympathetic to Nash's notion. The Riemann Hypothesis can't be solved as stated. He came by with this letter. But any expert would have said these ideas are naive. What I admired is the enormous self-confidence to even conjecture. If he's right, this guy's intuition is in the stratosphere. But it turned out to be just another wrong idea."

A year later, after he had been hospitalized, some blamed disappointed love and the intense rivalry with a younger man for Nash's breakdown.[28] Ironically, Cohen's career wound up mirroring Nash's. After his great success, he turned to the Riemann Hypothesis and physics. He did publish, but rarely and never anything that rivaled the work he did before age thirty. "Nothing was worthy of his notice," said a mathematician who knew him at MIT. "He sat in glorious isolation."[29]

34

The Emperor of Antarctica

There is a kindling. A slow fire burning. — JOSEPH BRENNER, *psychiatrist,
Cambridge, Massachusetts, 1997*

SOMEONE WAS CALLING, "It's time to play charades. It's time to play charades."[1] A crowd of costumed guests filled the entire ground floor of the Mosers'
small frame house in Needham. Outside, snow had been falling for hours. Inside,
the atmosphere was thick with smoke, liquor, jazz. Everyone was talking, laughing
a little louder than usual, heads close together, waving cigarettes, posing for the
camera, still a bit self-conscious but already loosening up in the carnival-like atmosphere. The Mosers were dressed as pirate and Indian squaw. Karin Tate, Artin's
musician daughter, was dressed as a black cat. Her husband, John, the algebraist,
came as the Vector Space Man, wearing a metal cap with bobbing antennae and
arrows all over his chest. Gian-Carlo Rota looked as elegant as ever in his monk's
tunic, his dark-haired wife, Teresa, dashing in her Spanish bolero and slim black
pants.

Richy Emery, the Mosers' son, watched through the dining-room window as
a big dark car pulled into the driveway and a virtually naked man got out. There
was a pounding on the kitchen door and Richy ran to open it. As Nash came
striding into the room, followed by Alicia, heads turned, eyebrows shot up, and
conversation suddenly quieted. Alicia was laughing excitedly and Nash wore a
smirky smile as they surveyed the astonished guests. He was barefoot and entirely
naked except for a diaper and a sash, which was draped across his powerful chest,
that had the numerals *1959* written on it. Having stolen the show, Nash grinned
and bowed, waved a baby bottle full of milk at the assembled company, which was
laughing loudly at this point — and then sauntered into the living room to join in
the game of charades.

Jürgen and Gertrude were just dividing the guests into two teams. Nash was
on one team, Richy on the other. When it was Richy's turn, Nash walked over to
him and whispered in his ear the name of the character that he was supposed to
act out. Richy was delighted. He adored Nash, who was much younger and more
animated than most of Jürgen's math friends. Richy's pantomime initially mystified
everyone. Finally a woman, the best player in the room, read his eleven-year-old

mind: *The Critique of Pure Reason*! Richy looked over at Nash, who shrugged his shoulders and gave him a big grin.

Between that New Year's Eve, December 31, 1958, and the last day in February, as his fellow mathematicians and friends looked on in puzzlement, Nash would undergo a strange and horrible metamorphosis. But on New Year's Eve, he was, by all accounts, simply his flamboyant, eccentric, and slightly off-key self, playful and mischievous. Alicia was in high spirits as well. The idea for Nash's costume had been hers.[2] She was the one who sewed it, draped his sash, and choreographed the entrance a moment past midnight. There is no hint of unease or premonition in the photograph of Nash sprawling somewhat drunkenly, with a laughing, gleeful Alicia on his lap, her arm on his shoulder. Most of the evening, though, it was Nash who was curled up in Alicia's lap. Some of the other partygoers found it extremely bizarre, "really gruesome," "disturbing."

Nash had already crossed some invisible threshold. The feverish activity and the fierce competition with Cohen and Newman in the common room, so noticeable in the early fall, had already slowed. He seemed a trifle more withdrawn, a little spacier. A graduate student who had just come into Nash's orbit recalled his not being able to keep up with Cohen and Newman. Paul Cohen recalled in 1996 that that fall Nash would make little jokes, little offhand remarks about world affairs, interesting license numbers, and the like. They were funny—Nash was always very bright and very witty—but they showed that something was not all right. "I'd think, 'That's going a little too far,' " Cohen said.[3]

Nash started singling out individuals. One was a senior named Al Vasquez, who had never taken a course from Nash and was something of a protégé of Paul Cohen's. "I'd see him in the common room. He'd say something. It wasn't a conversation. More like a monologue. He gave me preprints of his articles and asked me strange questions about them."[4]

But none of this was especially alarming or suggested outright illness, just another stage in the evolution of Nash's eccentricity. His conversation, as Raoul Bott put it, had "always mixed mathematics and myth."[5] His conversational style had always been a bit odd. He never seemed to know when to speak up or shut up or take part in ordinary give and take. Emma Duchane recalled in 1997 that Nash always, from their earliest acquaintance, which dated back to Nash and Alicia's courtship, told interminable stories with mysterious, off-key punch lines.[6]

In his game-theory course, Nash behaved like his usual self, according to students who were in the class.[7] On the first day, he said to the class, "The question occurs to me: Why are you here?," a remark that caused one student to drop the course. Later, he gave a midterm without announcing it in advance. He also paced a great deal and he sometimes fell into reveries in the middle of lecturing or answering a student's question. Just before Thanksgiving, Nash had invited his TA from the game theory course, Ramesh Gangolli, and Alberto Galmarino, a student from the course whom he was helping to choose a dissertation topic, to accompany him on a walk.[8] As they walked over the Harvard Bridge on the Charles River late one afternoon, Nash embarked on a lengthy monologue that was difficult to follow for the two, who had just come to the United States. It concerned threats to world peace and calls for world government. Nash seemed to be confiding in the two young men, hinting that he had been asked to play some extraordinary role. Gangolli recalled that he and Galmarino were quite disturbed and that they wondered briefly if they should inform Martin that something was not quite right. Awed as they were by Nash, and new as they were to America — and so reluctant to form any judgments — they decided to say nothing.

Also around that time, Atle Selberg, one of the masters of analytic number theory, gave a talk in Cambridge. Nash, who was in the audience, seemed to think that Selberg knew some secret that he was holding back. Selberg recalled, "He asked some questions I thought were in a sense, to my way of thinking, somewhat inappropriate to the subject. He seemed to see something quite different than what I had intended. . . . [His] questions were formulated as if I had some hidden, not fully disclosed, agenda that he wanted to discover. The lecture was about the rigidity of several locally symmetric spaces. He asked some questions that seemed to imply I had a hidden, secret motive. He suspected it had something to do with the Riemann Hypothesis, which of course it did not. I was rather taken aback. This was something that had nothing to do whatsoever [with the Riemann Hypothesis]."[9]

After the New Year's party, people around the department started talking about Nash. Classes resumed January 4. A week or ten days later, Nash asked Galmarino to teach a couple of his classes. He was going away, he said. Galmarino, who was flattered by Nash's confidence in him, readily agreed. Nash showed up at Rota's apartment on Sacramento Street on his way out of town. Then he disappeared.[10]

Cohen disappeared at around the same time. After a few days, the scuttlebutt among the graduate students was that Nash and Cohen had run away together.[11] As it happens, Cohen had gone to visit his sister. He was terribly upset when he returned to hear what the others had been saying about him and Nash. Nash, meanwhile, had driven south, ultimately to Roanoke, but perhaps also to Washington, D.C.

A couple of weeks later Nash slouched into the common room. Nobody bothered to stop talking. Nash was holding a copy of *The New York Times.* Without addressing anyone in particular, he walked up to Hartley Rogers and some others and pointed to the story on the upper left-hand corner of the *Times* front page, the off-lede, as *Times* staffers call it.[12] Nash said that abstract powers from outer space, or perhaps it was foreign governments, were communicating with him through *The New York Times.* The messages, which were meant only for him, were encrypted and required close analysis. Others couldn't decode the messages. He was being allowed to share the secrets of the world. Rogers and the others looked at each other. Was he joking?

Emma Duchane recalled driving with Nash and Alicia. She recalled that "he kept shifting from station to station. We thought he was just being pesky. But he thought that they were broadcasting messages to him. The things he did were mad, but we didn't really know it." [13]

Nash gave one of his graduate students an expired license, writing the student's nickname — St. Louis — over his own. He called it an "intergalactic driver's license." He mentioned that he was a member of a committee and that he was putting the student in charge of Asia. The student recalled, "He seemed to be joking around." [14] His manner took on a certain furtiveness. Another student, an undergraduate, recalled, "I have this impression of him darting about. I'd walk into a stairwell and he'd disappear as if he'd been lurking there." [15]

Nash showed up at the apartment of John and Karin Tate one evening. Everybody was horsing around and finally they settled down to play a game of bridge. Nash's partner was Karin Tate. His bidding was bizarre. At one point he bid six hearts when, as it turned out, he held no hearts at all. Karin asked him, "Are you crazy?" Nash responded quite calmly, explaining that he somehow had expected her to read his bids. "He expected me to understand. He genuinely thought I could understand. I thought he was pulling my leg, but it became obvious that he wasn't. I thought he was doing some sort of experiment." [16] Some people continued to think Nash was engaged in some elaborate private joke. There was a lot of discussion about it.

Nash's recollections of those weeks focus on a feeling of mental exhaustion and depletion, recurring and increasingly pervasive images, and a growing sense of revelation regarding a secret world that others around him were not privy to. He began, he recalled in 1996, to notice men in red neckties around the MIT campus. The men seemed to be signaling to him. "I got the impression that other people at MIT were wearing red neckties so I would notice them. As I became more and more delusional, not only persons at MIT but people in Boston wearing red neckties [would seem significant to me]." [17] At some point, Nash concluded that the men in red ties were part of a definite pattern. "Also [there was some relation to] a crypto-communist party," he said in 1996.

Things started happening fast. Alicia Nash later compared Nash's disintegration to that of a man who is conversing quite normally at a dinner party, suddenly starts arguing loudly, and finally has an all-out temper tantrum. [18]

He told Cohen: "People are talking about me. You've heard them. Tell me what they're saying." Cohen recalled: "It had a nasty edge. I told him I didn't know what he was talking about, that I hadn't heard anything." [19]

Nash was still working on the Riemann problem. Once Nash accused Cohen of rifling through his trash can. Was he trying to steal Nash's ideas about Riemann? Again, it sounded like a bit of an over-the-top joke, but it upset Cohen sufficiently so that he repeated the incident to a student.[20]

In mid-February, Harold Kuhn, who was on a Fulbright in London with Estelle and his children, spent a few days in Paris where he visited a French mathematician, Claude Berge. Berge showed Kuhn a letter from Nash, written in four colors of ink, complaining that his career was being ruined by aliens from outer space.[21]

Possibly, the event that triggered Nash's strange letter to Berge was the announcement of the winner of the 1959 Bôcher Prize, Louis Nirenberg, the Courant professor who had suggested the partial differential equation problem to Nash. Paul Cohen later recalled that Nash's reaction was furious. He told Cohen that he deserved the prize and that the fact that an older mathematician had won it was merely a sign that these things were "political." [22]

Nash also approached Neuwirth about his work. "He said he was giving this lecture on the Riemann Hypothesis," Neuwirth recalled. "But when he started talking it was gibberish. Probability is everything!!! I knew that was crazy. I mentioned it to Newman, who brushed it off." [23]

On yet another occasion, Nash wandered into Moser's office, unannounced as always. Moser, always affable, suppressed a feeling of irritation and waved him in. Nash stood at the blackboard. He drew a set that resembled a large, wavy baked potato. He drew a couple of other smaller shapes to the right. Then he fixed a long gaze on Moser. "This," he said, pointing to the potato, "is the universe." Moser nodded. Moser was at that time engaged in trying to apply Nash's implicit function theorem to certain problems in celestial mechanics. "This is the government," Nash said, in the same tone that used to say, "This is an elliptic equation." "This is heaven. And this is hell." [24]

Ted and Lucy Martin had been in Mexico on a winter vacation. When Martin returned, Levinson took him aside and told him that Nash was having a nervous breakdown. "Tell me about it," said Martin, who said later that he "almost didn't believe in these things." Martin recalled, "Levinson said, 'He's very paranoid. If you go down to his office, he won't want you between him and the door.' Sure enough, when I went down to his office that Sunday night, Nash edged himself over between me and the door." [25]

Strange letters began turning up in the department mail. Ruth Goodwin, the department secretary, would put them aside and show them to Martin.[26] They were addressed to ambassadors of various countries. And they were from John Nash. Martin panicked. He tried to retrieve the letters, not all of which were addressed and most of which weren't stamped, from mailboxes around the campus.

What was in the letters? None have survived, but various people recalled hearing from Martin that Nash was forming a world government. There was a committee that consisted of Nash and various students and colleagues in the department. The letters were addressed to all the embassies in Washington, D.C. The letter said he was forming a world government. He wanted to talk to the ambassadors. Later he would talk to the heads of state.[27]

Martin was in a most awkward position. The faculty, after some internal dissension, had just voted on Nash's promotion, and it was now before the president of the university. He dithered and delayed.

Meanwhile, Adrian Albert, the chairman of the mathematics department at the University of Chicago, called Norman Levinson. What was Nash's state of mind? he asked Levinson. Chicago had made an offer of a prestigious chair to Nash, Nash was scheduled to give a talk, and now he had received a very odd letter from Nash.[28] It was a refusal of the Chicago offer. Nash had thanked Albert for his kind offer but said he would have to decline because he was scheduled to become Emperor of Antarctica. The letter, Browder recalled in 1996, also contained references to Ted Martin's stealing Nash's ideas. The affair came to the attention of MIT president Julius Stratton, who, upon seeing a copy of Nash's letter, is supposed to have said, "This is a very sick man."

The spring term began February 9. Shortly after Washington's birthday, Eugenio Calabi, who was a member that year at the Institute for Advanced Study in Princeton, gave a seminar at MIT. Undergraduates, even very bright ones, didn't normally attend departmental seminars, but Al Vasquez, a senior, decided he would go. He put on a sport coat and tie for the occasion. Feeling rather self-conscious, he sat a few rows from the rear and hoped that he looked less conspicuous than he felt.

He had noticed, as he sat down, that Nash was sitting in the row behind him. In the middle of Calabi's lecture, Nash started speaking rather loudly, although he did not appear to be addressing Calabi. After a few moments, Vasquez realized that Nash was talking to him. "Vasquez, did you know that I'm on the cover of *Life* magazine?" Nash kept repeating until Vasquez turned around.[29]

Nash told Vasquez that his photograph had been disguised to make it look as if it were Pope John the Twenty-third. Vasquez, he said, also had his picture on a *Life* cover and it too was disguised. How did he know that the photograph, apparently of the pope, was really of himself? Two ways, he explained. First because John wasn't the pope's given name but a name that he had chosen. Second, because twenty-three was Nash's "favorite prime number."

Almost the strangest thing, Vasquez later recalled, was that Calabi kept on lecturing as if nothing untoward were happening, and the rest of the audience too ignored the interchange, although it must have been audible to everyone in the room.

• • •

Nash and Calabi knew each other from their graduate-school days at Princeton. Before Calabi had come up to Cambridge, Nash had telephoned him at his apartment on Einstein Drive and asked whether the Calabis could put him and Alicia up for a few days.[30] He wanted to spend a few days at the institute consulting with Atle Selberg, the number theorist, and preparing a talk that he was scheduled to give at the upcoming regional math society meeting.

Calabi and the Nashes went out to dinner after Calabi's talk. Both Nashes seemed unusually nervous, Calabi recalled. "At one point, Nash made a wrong turn and Alicia began yelling hysterically. He was somewhat anxious."

The next day, the Nashes left for Princeton while Calabi stayed on in Cambridge. A day or two later, Calabi got a call from his wife, Giuliana, who said that Nash was behaving very strangely and would he come home?

On one occasion, Nash had walked into another apartment, used the bathroom, and walked out again. All the apartments on Einstein Drive looked virtually identical from the outside and mistakes were commonplace, but even afterward Nash didn't seem to be aware that he had been in the wrong apartment.

On the afternoon of February 28, Nash was even more agitated. Calabi had just returned. "He was acting much more nervous than usual. Very agitated. At the moment of leaving, he was misplacing notes, running back and forth between the car and the house. Alicia was trying to calm him down." Calabi watched, full of misgivings. Speaking of Nash's mathematical investigation, he said, "I knew in that area that problem was not going to yield to a flash of inspiration."[31]

Nash's consultations with Selberg apparently came to naught. Selberg had merely been irritated by Nash's persistence, as he later recalled, and told Nash, in even harsher terms, that the probabilistic approach he was pursuing had been tried before and had already been demonstrated to be fruitless.[32]

One can only imagine the fear and confusion that Nash felt that afternoon as he stood before the 250 or so mathematicians who came to his lecture, sponsored by the American Mathematical Society, in a Columbia University auditorium.[33]

Harold N. Shapiro, a professor at the Courant Institute and a number theorist who had known Nash since the summer they spent together at RAND in 1952, introduced Nash.

There was in fact an air of tremendous expectation in the hall. Regional AMS meetings were essentially job meetings. The audience consisted both of job seekers and established mathematicians, among them many who knew Nash and his work intimately. "Here was a great young mathematician with a proven ability for tackling the most difficult problems about to announce what he felt was a likely solution to the deepest problem in all of mathematics," recalled Shapiro. "I remember hearing that he was interested in prime numbers. Everybody's reaction was that if Nash turns to number theory, number theorists better watch out. There was a buzz."[34]

Peter Lax, a professor at the Courant Institute, described it as "a very strange adventure."

Lipman Bers reminded me, as we were listening to Nash's talk, that Heifetz gave his first concert at Carnegie (accompanied by the pianist Godowski). An older violinist, turning to the musician seated next to him, said, "It's very hot in here." "Not for the pianist," came the answer. It must have been hot in there, but only for the number theorists in the audience. It was work in progress. I couldn't judge it. Mathematicians don't usually present unfinished work.[35]

At first, it seemed like just another one of Nash's cryptic, disorganized performances, more free association than exposition. But halfway through, something happened. Donald Newman recalled in 1996:

One word didn't fit in with the other. I was at Yeshiva. Rademacher, who had worked on the Riemann Hypothesis, was present. In fact, he wrote a brilliant paper on How Not to Solve the Riemann Hypothesis. It was Nash's first downfall. Everybody knew something was wrong. He didn't get stuck. It was his chatter. The math was just lunacy. What does this have to do with the Riemann Hypothesis? Some people didn't catch it. People go to these meetings and sit through lectures. Then they go out in the hall, buttonhole other people, and try to figure out what they just heard. Nash's talk wasn't good or bad. It was horrible.[36]

Cathleen Morawetz, who had enjoyed joking around with Nash at Courant two years earlier, ran into Nash in the stairwell after the talk: "He was laughed out of the auditorium," she recalled. "I felt terrible. I said something nice to him, but I was disturbed. He seemed very depressed." (Later Cathleen used the phrase "heaping scorn on him" to describe the audience reaction.)[37]

Nash had been invited to give a talk at Yale as well on his way back to Cambridge. It was his second talk at Yale that year, but he couldn't find his way there. He kept calling Felix Browder, then teaching at Yale, and telling him that he couldn't understand how to get off the Merritt Parkway.

Nash talked about the Riemann Hypothesis just as he had at Columbia. Again, it was a disastrous performance, as recalled by Browder, who contrasted his performance with the earlier one. "The preceding year there was no hint of trouble. That is when he finished the parabolic equations proof. [In fact] he completed the proof during a talk. I [had] asked him if he wanted to come and give another talk at Yale. It wasn't coherent. I thought something was wrong."[38]

35

In the Eye of the Storm

Spring 1959

It was like a tornado. You want to hang on to what you have. You don't want to see everything go. — ALICIA NASH

DESPITE ALICIA'S apparent elation on New Year's Eve, her state of mind in the preceding months had been anything but carefree. Since returning from their European holiday, her starry-eyed view of her new life had given way to a darker, more somber perspective. She and Nash had moved out to West Medford, a small industrial city north of Cambridge, and Alicia felt cut off and isolated. Her goal of establishing a career seemed more distant than ever. Her feelings about her pregnancy were ambivalent, and her initial hopes that it would draw her and Nash closer were disappointed. Her husband had become, if anything, more cold and distant. As the weather turned colder and the days shorter, she felt more and more dispirited, anxious, and alone — so much so that she was thinking of consulting a psychiatrist.[1]

That had been before Thanksgiving. Since then, Nash's behavior, rather than her own low mood, had become her chief source of distress. Several times, Nash had cornered her with odd questions when they were alone, either at home or driving in the car. "Why don't you tell me about it?" he asked in an angry, agitated tone, apropos of nothing. "Tell me what you know," he demanded.[2] He behaved as if she knew some secret but wouldn't share it with him. The first time he said it, Alicia thought Nash suspected her of having an affair. When he repeated it, she wondered whether he might not be having an affair himself. That would account for his growing secretiveness and air of abstraction. Might he not be trying to deflect attention from himself by accusing her?

By New Year's Day, the day she turned twenty-six, Alicia was sure that "something was wrong."[3] Nash's behavior had become more and more peculiar. He was irritable and hypersensitive one minute, eerily withdrawn the next. He complained that he "knew something was going on" and that he was being "bugged." And he was staying up nights writing strange letters to the United Nations. One night, after he had painted black spots all over their bedroom wall, Alicia made him sleep on the living-room couch.[4]

Alarmed, Alicia searched for explanations rooted in their day-to-day life. Her first thought was that Nash was unduly worried about the impending tenure decision. She suspected that the prospect of a baby, with all the new responsibilities that implied, was another source of pressure. And she wondered whether marriage to someone "different" like her wasn't proving too much of a strain for a southern WASP.[5]

Alicia vainly tried to reassure Nash. She told him, over and over, that his worries about tenure were unfounded, that he was the department's fair-haired boy, that Martin, after all, was confident that the decision would be favorable. She reasoned with him, pointing out that the letter writing "could undermine his professional credibility" and might even jeopardize his tenure. When that failed, she remonstrated with him. "You can't act silly," she would say. Then Nash did a number of things that frightened her—and made inescapable the conclusion that he was suffering some sort of mental breakdown.

He started to threaten to take all of his savings out of the bank and move to Europe.[6] He had some idea, it seemed, of founding an international organization. And he began to stay up, night after night, long after she had gone to bed, writing. In the morning, his desk would be covered with sheets of paper covered in blue, green, red, and black ink. They were addressed not just to the U.N. but to various foreign ambassadors, the pope, even the FBI.

It was in mid-January, while classes were still in session, that Nash took off for Roanoke in the middle of the night after a wild scene. Seeing no alternative, Alicia broke her silence and telephoned Virginia to warn her. She told her mother-in-law very little, though, as Martha recalled, other than that Nash was suffering from stress and was behaving somewhat irrationally. When he arrived in Roanoke, Virginia and Martha were frightened by his agitated state. At one point, he struck Virginia on the arm.[7]

When Nash returned, he continued to badger Alicia in private. Once he threatened to hit her "if you don't tell me."[8]

Alicia was initially more worried about Nash and their future together than about any physical threats to herself. Her immediate, overwhelming instinct was to prevent the university from finding out about Nash's difficulties. "I didn't want the bad things to get out."[9]

She quit her job at Technical Operations and took one at the Computer Center on campus. She began to watch Nash all the time, to stick very close to him, to keep him more to herself. She would stop by the mathematics department every afternoon after work and pick him up. She no longer invited others to join them when they ate out. She particularly tried to avoid Paul Cohen, although Nash's insistence sometimes made this impossible. "Alicia wanted to save his career and preserve his intellect," a friend of Alicia's later recalled. "It was in her interest to keep Nash intact. She was extremely tough." [10]

Until the Roanoke episode, Alicia had confided in no one. Now she consulted a psychiatrist from the MIT medical department, a Dr. Haskell Schell.[11] She also asked Emma to have lunch with her alone a few times and, although reluctantly and holding much back, told her friend some of what had been happening.

At the beginning, it seemed to Alicia that her psychiatrist was more intent on asking her questions — about her upbringing, her marriage, her sex life — than on offering practical advice on how to cope. "At first Alicia trusted them because it was MIT," Emma recalled. "But it was a very Freudian time. The psychiatry department was ultra-Freudian. They wanted to treat Alicia. She wanted practical help." Emma continued:

> They asked Alicia a lot of questions. She got very impatient. Nash was threatening to go off to Europe, to withdraw all their money, to start an international organization. She was looking into the laws. She found out that you could have somebody committed for a limited time with the signature of two psychiatrists. To keep them longer, you had to have a court hearing.[12]

Emma was working with Jerome Lettvin, a former psychiatrist who was now pursuing research in neurophysiology at MIT. She asked Lettvin what Alicia should do. The result was that Alicia got very conflicting advice. On the one hand, Lettvin was urging her, through Emma, to consider shock treatments. "Lettvin's idea was that when somebody was delusional the sooner he was shocked out of it the better," Emma recalled. On the other hand, Schell was recommending that Nash go to McLean Hospital, an ultra-Freudian institution that eschewed shock treatments in favor of psychoanalysis and new antipsychotic drugs like Thorazine. Alicia rejected the notion of shock treatment. "She was very concerned with preserving his genius," Emma stated in 1997. "She wasn't going to force anything on him. She also wanted there to be nothing that would interfere with his brain. No drugs. No shock treatments."[13]

In January, the department voted to give Nash tenure. A few weeks later, Martin, now aware that Nash was suffering some sort of "nervous breakdown," decided to relieve Nash of his teaching duties for the coming semester.[14] Although distressed that the university had found out about Nash's problems, Alicia was greatly relieved. She hoped that this move would lift some of the pressures on Nash and that he would improve spontaneously.

Deciding what, if anything, to do was so difficult because Nash often seemed quite normal. The on-again, off-again nature of his symptoms also convinced some of his colleagues and graduate students in the department that nothing was seriously wrong. Gian-Carlo Rota recalled that Nash's personality "didn't seem very different," although "his mathematics no longer made sense."[15] Some days everything looked just as it always had, and Alicia found herself wondering, until the next outburst of bizarre behavior, whether she had been exaggerating, unnecessarily alarmed, premature in her judgments.

In mid-March, two weeks after the disastrous New York trip when Nash had given his lecture on the Riemann Hypothesis, Nash was writing reassuring letters home. "My talk in New York went reasonably well," he wrote Virginia on March 12, urging her to come up to Boston to visit him and Alicia.[16] On the same day, he even wrote a long letter to Martha in which he complained of boredom. Nash wrote, "Since she has become pregnant Alicia does not like to go out. She enjoys TV and movie magazines. These things tend to bore me. The level is too low."[17]

But these periods of lucidity and calm soon gave way to an eruption that Alicia later compared to a "tornado."[18] The episode that convinced Alicia that she had no choice but to seek treatment for Nash occurred around Easter. Nash took off for Washington, D.C., in his Mercedes. He was, it appeared, trying to deliver letters to foreign governments by dropping them into the mail slots of embassies.[19] This time Alicia went with him. Before they left, she telephoned her friend Emma and asked her to contact the university psychiatrist if they did not return within a week or so. Emma recalled in 1997 that Alicia was afraid Nash might harm her. Curiously, her concern, at least in Emma's recollection, was less for herself than for Nash: "She wanted the world to know that Nash was mad. She was worried about Nash. She worried that if she came to harm that he'd be treated like a common criminal, so she wanted to be sure that everyone knew that he was insane."[20]

When Emma did call Schell he refused to come to the telephone and had a nurse tell her that "Dr. Schell doesn't discuss his patients." She added, "I was interviewed at Lincoln Labs about Alicia. I was asked whether she was afraid of her husband. But she wasn't. He was just very sick."[21]

Emma's impressions to the contrary, Alicia was afraid, though she managed to hide her fear from almost everyone. Paul Cohen, however, recalled that "she was afraid of him."[22] A few weeks later she would tell Gertrude Moser, who questioned her decision to have Nash hospitalized, that, in Gertrude's words, "Something had happened in the middle of the night and she had to save herself and the child."[23] It was fear for her own safety, as well as her psychiatrist's warning that Nash would continue to deteriorate unless he got treatment, that prompted her to seek commitment, at least for observation. She wished, however, to conceal what he would inevitably regard as an act of treachery. So she turned to her mother-in-law and asked her to come to Boston.

George Whitehead, one of Nash's colleagues, had temporarily moved to Princeton with his wife, Kay. In mid-April, the Whiteheads drove up to Boston to have their car, which was still registered in Massachusetts, inspected. It was an annual ritual. That evening they went to a party at the home of Oscar Goldman in Concord. Most of the MIT mathematics department was there. Kay recalled in 1995: "The word was 'Tomorrow, Alicia is having John committed.' Obviously, there was a lot of talk about it." [24]

36
Day Breaks
in Bowditch Hall

McLean Hospital, April–May 1959

This is the way day breaks in Bowditch Hall at McLean's. — "Waking in the Blue," Life Studies, ROBERT LOWELL

W HEN A STRANGER in a suit knocked on Paul Cohen's office door to inquire whether he had seen Dr. Nash that afternoon, the man's slightly unctuous, self-important manner made Cohen wonder whether this was the psychiatrist who was going to have Nash "locked up." [1] For days the younger people in the department had been speculating—based on hints dropped by Ambrose and some of the other senior faculty—that Nash's wife was about to have him committed. Furious controversies had broken out over whether Nash was truly insane or merely eccentric, and over whether, insane or not, anyone had the right to rob a genius like Nash of his freedom. [2] Cohen, who felt that he had been somehow unfairly implicated in the whole affair, had pretty much steered clear of these debates, but he nonetheless felt a certain morbid fascination. To the stranger, however, he merely said no, he hadn't seen Dr. Nash all day.

So when Nash showed up at Cohen's door not very long afterward, seemingly oblivious to whatever machinations were under way, Cohen was more than a little surprised. Nash wanted to know if Cohen would like to go for a walk with him. Cohen agreed, and the two wandered around the MIT campus for an hour or more. As they walked, Nash spoke in a fitful monologue while Cohen listened, perplexed and uncomfortable. Occasionally Nash would stop, point at something, and whisper conspiratorially: "Look at that dog over there. He's following us." [3] He frightened Cohen a bit by talking about Alicia in a way that made the younger man feel that she might be in danger. After they parted, Cohen learned later, Nash was picked up and taken to McLean Hospital.

• • •

It was not difficult to get someone into McLean even if they did not want to go. Nash's involuntary commitment to a mental hospital for observation was likely arranged by MIT's psychiatric service, probably in consultation with the president of the university as well as Martin and Levinson.[4] Given Nash's acute paranoia, his bizarre letter writing, his inability to teach, and the potential that he might carry out his threats to harm Alicia, the pressure to intervene would have been great. One imagines that before taking the drastic measure of involuntary commitment, one of the psychiatrists in MIT's employ attempted to convince Nash to obtain treatment voluntarily first. Merton J. Kahne, a professor of psychiatry at MIT who ran McLean's admissions ward during the 1950s, said in 1996:

> They would have tried to figure out how to get him into therapy without coercion. A lot of heads would have been put together to try to find a solution. In those days, there was an attempt to maintain some respect for the human being, whether they were crazy or not. They weren't interested in peremptorily putting someone in the hospital against their will. The stigma was enormous.

The decision was an especially tricky one because of Nash's prominent position at the university, and because, as is often the case, it was inherently controversial. As Kahne put it, "The more powerful or exceptional the individual, the more controversial the decision."

The mechanics, however, were fairly straightforward. Any psychiatrist could apply to a mental hospital to have a patient taken for a ten-day observation period. A university psychiatrist would have signed a temporary care order—a so-called pink paper—asking McLean to take Nash on the grounds that he was a danger to himself or others (although a simple inability to care for oneself was sufficient grounds). The pink paper gave MIT the right to pick Nash up and transport him to McLean. Technically, it was the hospital that made the decision to hold a patient, initially for a ten-day period.

That April evening, some hours after Nash and Cohen parted company, two Cambridge policemen arrived at the Nash's West Medford house. As Nash recalls, "they as if arrested me. . . ."[5] The use of police officers was, by all accounts, an extreme measure; it suggested that the university psychiatrists were expecting trouble. Most cases of involuntary commitment involving university personnel were handled far more discreetly, in a manner designed to avoid scandal and humiliation, by out-of-uniform campus police driving a gray Chevrolet station wagon, marked only with maroon lettering, whose interior was equipped as an ambulance.[6] As it happened, Nash refused to go and a scuffle ensued. "I actually struggled with them in resistance at first," he recalled. Resistance was useless, however. Big and strong as he was, Nash was quickly overpowered and bundled into the back of the police cruiser. The drive from West Medford to Belmont took less than half an hour.

• • •

One Hundred Fifteen Mill Street, Belmont, Massachusetts, was, and still is, a verdant 240-acre expanse of rolling lawns and winding lanes and a scattering of buildings of old brick and ironwork nestled among majestic trees or perched airily on rises—a precise copy, that is to say, of a well-manicured New England college campus of late-nineteenth-century vintage.[7] Many of its smaller buildings were designed to resemble the homes of wealthy Boston Brahmins—long the bulk of McLean's clientele. A psychiatrist who reviewed the hospital for the American Psychiatric Association in the late 1940s recalled, "There were all these little two-story homes with suites—kitchen, living room, bedroom. They had suites for the cook, the maid, the chauffeur."[8] Upham House, a former medical resident recalled, had four corner suites per floor and on one of its floors all four patients turned out to be members of the Harvard Club!

McLean was, as it still is, connected to Harvard Medical School. So many of the wealthy, intellectual, and famous came there—Sylvia Plath, Ray Charles, and Robert Lowell among them[9]—that many people around Cambridge had come to think of it less as a mental hospital and more as a kind of sanatorium where high-strung poets, professors, and graduate students wound up for a special kind of R&R.

The resident on duty that evening urged Nash to sign a "voluntary paper." Nash refused. There was a great movement for world peace, he said, and he was its leader. He called himself "the prince of peace."[10] He was informed of his legal rights, including his right to file a petition for release. A tentative diagnosis was made, but this was not discussed with him. And a document applying to a judge for a ten-day commitment was filled out. He was then escorted to the admissions ward in Belnap One, a low brick building on the north side of McLean's campus, just beyond the administration building.

Nash used the pay telephone in the lounge. He did not call a lawyer, but rang Fagi Levinson instead. "John wanted to know how he could get out of there," she said. "He said he wanted a shower. 'I stink,' he said."[11]

Virginia Nash traveled up from Roanoke to see her son. She was devastated. She wept and wept, Emma Duchane recalled, saying over and over that she could not "bear to see Johnny in this situation."[12] She seemed close to a breakdown herself. She did not offer Alicia any help, financial or otherwise. Alicia, who was very short of funds, about to give birth, and mad with worry, was bitterly disappointed. She had counted on Virginia for support, but it was obvious that Virginia needed even more help than she did.

• • •

Nash was soon transferred to Bowditch Hall, a low white frame building at the edge of the McLean campus. Bowditch was a locked facility for men. Within a couple of weeks, Robert Lowell, the poet, joined him there.[13] Lowell was already famous, a dozen years older than Nash, and a manic depressive who was now enduring his fifth hospitalization in less than ten years. For Lowell, it was "a mad month" spent "rewriting everything in my three books," translating Heine and Baudelaire, reworking Milton's "Lycidas," which he believed he had himself written, feeling "I had hit the skies, that all cohered."[14]

"Thrown together like a bundle of kindling, [unable] to escape," as Lowell's widow, Elizabeth Hardwick, later put it,[15] Lowell and Nash spent a good deal of time together. When Arthur Mattuck came to visit Nash, he found fifteen or twenty people crowded in Nash's narrow shoebox of a bedroom.[16] In what turned out to be an oft-repeated scene, Lowell was sitting on Nash's bed, surrounded by patients and staff sitting at his feet on the floor or standing against the walls, delivering what amounted to a long monologue in his unmistakable voice — "weary, nasal, hesitant, whining, mumbling." Nash was hunched over beside him. Mattuck recalled in 1997: "I don't remember anything of the conversation except that it was general. In other words, only one person spoke at a time and that was most of the time Lowell. Basically he was holding forth on one topic after another, and the rest of us were appreciating this brilliant man. Nash said very little, like the rest of us."

Once a women's residence where no man had "apparently entered since perhaps 1860," Bowditch was, in Lowell's words, now designated for "ex-paranoid boys"[17] — the ones who thought there was nothing wrong with them and couldn't be trusted not to bolt. As such, it was oddly genteel. At Bowditch, Nash and his fellow inmates were treated "to a maze of tender fussy attentions suitable to old ladies."[18] The crew-cut Roman Catholic nurses, many of them Boston University students, brought him chocolate milk at bedtime, inquired about his interests, hobbies, and friends, and called him Professor.[19] "Hearty New England breakfast[s]" were followed by ample lunches and homey dinners; everybody got fat. Nash had a private room "with a door that shut," a "hooded night light," and a view. There were no screams, no violent episodes, no straitjackets. His fellow patients, "thoroughbred mental cases," were polite, full of concern, eager to make his acquaintance, lend him their books, and clue him in to "the routine." They were young Harvard "Cock[s] of the walk" slowed down by massive injections of Thorazine, yet "so much more intelligent and interesting than the doctors," as Nash confided to Emma Duchane when she came to visit.[20] There were also old Harvard types "dripping crumbs in front of the TV screen, idly pushing the buttons." (Nearly half of McLean's patients were geriatric, like Lowell's "Bobbie/Porcellian '29," who strutted around Bowditch late at night "in his birthday suit.")[21]

Yet, there Nash was, stripped to his underwear, his belt and shoes taken away, standing before a shaving mirror that was not glass, but metal. As for his view the next morning, in Lowell's words, "Azure day/makes my agonized blue window bleaker." The days must have seemed very long: "[H]ours and hours go by." Above all, there was the terrible awareness when visitors came that they were free to go back through the locked doors through which they had come while he could not. It was in no way horrible; he was merely, as another inmate of a mental hospital once put it, "considered beyond reasoning with . . . and treated like a child; not brutally, but efficiently, firmly, patronizingly." [22] He had merely relinquished his rights as an adult human being. Like Lowell, he must have asked himself, "What good is my sense of humor?"

Alicia urged everyone they knew to visit Nash. [23] Fagi Levinson organized a visitor's schedule. [24] The feeling was that with the support of friends, Nash would soon be on his feet again. "Everyone at MIT felt responsible for trying to make Nash better," recalled Fagi in 1996. "At McLean, all felt the more companionship and support he had, the quicker he would recover."

One afternoon, Al Vasquez ran into Paul Cohen, who was extremely upset. He had been out to McLean to visit Nash. And he'd been turned away. What had happened, he told Vasquez, was that McLean had some sort of list of verboten visitors. "He was on the list," Vasquez recalled. "And I was on it too. I was really shocked." [25] Vasquez — along with most of the students in the department — hadn't even known that Nash was in the hospital.

> It was a list of some sort of committee. I remember Cohen being very upset. That was the first time I was aware that Nash had been hospitalized. I have a memory of about twenty people [on the list], almost all of whom were in the math department. Cohen must have told me some of the names. It was the hospital that wouldn't let people on the list see Nash. I called it "The Committee to Rule the World."

At first, Nash, who found it strange shuffling around without his shoes, was furious. "My wife, my own wife . . . ," he said to Adriano Garsia, one of the first to visit. He threatened to sue Alicia for divorce, to "take away her power." [26] Jürgen and Gertrude Moser recall a similar conversation. "He was very resentful," Moser remembered, "[but] otherwise not very different. Gertrude was initially very sympathetic and somewhat outraged at the way Nash was being treated. 'He doesn't seem crazy,' she said." [27] Emma Duchane, who also visited Nash in Bowditch, recalled that Nash was nicer to her than he had ever been. "He was saying such reasonable things," she said. [28] When Gian-Carlo Rota and George Mackey, a Harvard professor, came, Nash joked about the oddness of locked doors, remarked how strange it was to be held there, and told them, in the most rational tone, that he was aware that he had been having delusions. [29] When Donald Newman came to visit him, Nash asked him half-jokingly, "What if they don't let me out until I'm NORMAL?" [30] To Felix Browder, Nash complained that staying in the hospital was too expensive (the daily rate that spring was thirty-eight dollars). [31]

Some of his visitors wondered what he was doing there. Donald Newman was the most vehement that Nash was sane. "There's no discontinuity!" he kept repeating.[32] Garsia recalled in 1995: "I was totally appalled by the fact that his wife had done this. I couldn't believe my idol was under the thumb of some stupid nurse who had total power over him."[33]

The medication — initially, an injection of Thorazine immediately upon admission — calmed Nash down, made him drowsy and slow of speech — but did nothing to dispel "the deep underlying unreality."[34]

Nash told John McCarthy, who also came out, despite his horror of hospitals and illness, "These ideas keep coming into my head and I can't prevent it."[35] He told Arthur Mattuck that he believed that there was a conspiracy among military leaders to take over the world, that he was in charge of the takeover. Mattuck recalled, "He was very hostile. When I arrived, he said, 'Have you come to spring me?' He told me with a guilty smile on his face that he secretly felt that he was the left foot of God and that God was walking on the earth. He was obsessed with secret numbers. 'Do you know the secret number?' he asked. He wanted to know if I was one of the initiated."[36]

For the first two or three weeks — during which time McLean had applied to a judge for an extension of the observation period for another forty days — Nash was watched, studied, and analyzed.[37] A biography was written. A young psychiatrist was assigned to construct Nash's life story, a complete catalog of his personality covering no fewer than 205 separate topics. All that led up to this disaster was included: family, childhood, education, work, past illnesses, and so forth. When it was done, the history was presented to a case conference attended by McLean's senior psychiatrists, and a more definitive diagnosis was arrived at.

From the start, there was a consensus among the psychiatrists that Nash was obviously psychotic when he came to McLean.[38] The diagnosis of paranoid schizophrenia was arrived at very quickly. "If he was talking about cabals," said Kahne, "it would have been almost inevitable."[39] Reports of Nash's earlier eccentricity would have made such a conclusion even more likely. There was some discussion, of course, about the aptness of the diagnosis. Nash's age, his accomplishments, his genius would have made the doctors question whether he might not be suffering from Lowell's disease, manic depression. "One always fudged it. One couldn't be sure," said Joseph Brenner, who became junior administrator on the admissions ward shortly after Nash's hospitalization.[40] But the bizarre and elaborate character of Nash's beliefs, which were simultaneously grandiose and persecutory, his tense, suspicious, guarded behavior, the relative coherence of his speech, the blankness of his facial expressions, and the extreme detachment of his voice, the reserve which bordered at times on muteness — all pointed toward schizophrenia.

Everyone was talking about which events the psychiatrists believed had produced Nash's breakdown. Fagi recalled that Alicia's pregnancy was thought to be the culprit: "It was the height of the Freudian period — all these things were explained by fetus envy."[41] Cohen said: "His psychoanalysts theorized that his illness was brought on by latent homosexuality."[42] These rumored opinions may well have been held by Nash's doctors. Freud's now-discredited theory linking schizophrenia to repressed homosexuality had such currency at McLean that for many years any male with a diagnosis of schizophrenia who arrived at the hospital in an agitated state was said to be suffering from "homosexual panic."[43]

Nash wasn't privy to any of this. His psychiatrist wouldn't have told him, even if Nash had pressed. But it would have been easy enough for Nash to figure — by going to McLean's library or talking with his fellow inmates — what his doctors were thinking.

Everyone was very upbeat. The optimism was part of that "heavily psychoanalytic" era at McLean. Lowell's doctors were telling his wife, Elizabeth Hardwick, that the most serious illnesses, psychotic illnesses, the kind that produced the chronic cases like Lowell's Bobbie, were now susceptible to "permanent cures."[44]

Alfred H. Stanton had been charged by McLean's trustees in 1954 to modernize McLean.[45] Before Stanton arrived in the early 1950s, as Kahne recalled, "The nurses were spending all their time classifying fur coats and writing thank you letters." Moreover, patients spent most of the day lying in bed as if they were suffering from some physical ailment. Stanton hired a large number of nurses and psychiatrists, expanded the medical residency program, instituted an intensive psychotherapy program, and organized social, educational, and work activities.

McLean's treatment philosophy boiled down to the notion that "it was impossible to be social and crazy at the same time."[46] The staff was dedicated to encouraging all new patients, no matter what the diagnosis, to relate. Along with this "milieu" therapy, as it was called, intensive, five-day-a-week psychoanalysis was the main mode of treatment.[47] Nobody thought of Thorazine as anything but an initial aid in preparing the way for psychotherapy. "Stanton's attitudes harked back to early days of 'moral treatment' of patients," said Kahne, "which included having expectations of them and having staff become close to patients. The idea was to involve patients in decision-making and to abolish some of the hierarchy of medical institutions."

Stanton was a student of Harry Stack Sullivan, a leading American disciple of Freud, and had helped run Chestnut Lodge, a private hospital outside Washington, D.C., where psychoanalysis was being used to treat psychotic disorders. He also put an end to the use of lobotomies and shock therapies at McLean. "Freudianism was pretty strong at McLean," said Brenner. "It was the dawn of psychopharmacology. We were desperately creating cures in all good faith."[48]

· · ·

"Our knowledge of schizophrenia was negligible," Fagi remembered sadly. "I was a dope. All he needed was a good shrink and support and everything would be over soon. Everyone at MIT pretended that Nash was going to recover in a flash. At McLean they would cure him with advanced therapy. Norbert was the only one who sensed the tragedy. He expressed his heartfelt sympathy. 'It's very difficult,' he said to Virginia. She was tearful, shaken, trying to keep herself in check. She wanted to know as much as possible. Wiener's eyes filled up with tears."[49]

Isadore Singer and Alicia came to visit Nash one evening. There was no one else in the large, rectangular common room. Singer recalled the scene:

> We were the only visitors. Robert Lowell, the poet, walked in, manic as hell. He sees this very pregnant woman. He looks at her and starts quoting the begat sequences in the Bible. Then he started spinning quotes with the word anointed. He decided to lecture us on the meaning of anointed in all the ways it was used in the King James version of the Bible. In the end I decided that every word in the English language was a personal friend of his. Nash was very quiet and almost not moving. He wasn't even listening. He was totally withdrawn. Mrs. Nash was sitting there, pregnant as hell. I focused mostly on the wife and the coming child. I've had that picture in my mind for years. "It's all over for him," I thought.[50]

Perhaps it was the Thorazine, perhaps the confinement, perhaps the overwhelming desire to regain his liberty, but Nash's acute psychosis disappeared within a matter of weeks.[51] On the ward, he behaved like a model patient — quietly, politely, tolerantly — and was soon granted all sorts of privileges, including the freedom to walk around McLean's grounds without supervision.[52] In his therapy sessions, he stopped talking about going to Europe to form a world government and no longer referred to himself as the leader of the peace movement. He made no threats of any kind, except divorce. He readily agreed, if asked, that he had written a great many crazy letters, had made a nuisance of himself to the university authorities, had otherwise behaved in bizarre ways. He denied emphatically that he was experiencing any hallucinations. The two young residents who were assigned to him — Egbert Mueller, a highly regarded German psychoanalyst, and Jacqueline Gauthier, a more junior French-Canadian — noted that his symptoms had all but "disappeared," although privately they agreed that he was likely merely concealing them.[53]

This was so. In his heart, Nash felt that he was a political prisoner and he was determined to escape his jailers as quickly as possible. With the help of other patients, he quickly figured out the rules of the game. If a patient wished to leave, the law placed the burden of proof on the hospital. Nash's psychiatrists would have had to show convincingly that he was likely to harm himself or someone else. In practice, a patient who was hallucinating or was obviously delusional wouldn't stand much chance of getting out. (Later, he would take the position, with respect to his younger son, that it was quite possible for a so-called schizophrenic to control both his delusions and his behavior.)[54]

He hired a lawyer, Bernard E. Bradley, to petition for his release.[55] Bradley worked in the public defender's office at the time, but Nash, who was hardly destitute, was likely his private client. At Nash's suggestion, Bradley hired A. Warren Stearns, a prominent Boston psychiatrist, to examine him and to support his petition for release. Stearns was a prominent researcher as well as a major figure in state mental health and prison policy.[56] He had, at various points in his long career, been dean of Tufts medical school, director of prisons for the state of Massachusetts, and associate mental health commissioner. At the time Nash had Bradley contact him, he was founder and head of Tufts's sociology department. His views on crime anticipated those of James Q. Wilson: He held that most crimes were committed by a small slice of the population, namely, young men between the ages of eighteen and twenty-three. His book on the subject, *The Personality of Criminals,* was considered a classic. Stearns had been involved in all sorts of famous criminal cases, including that of Sacco and Vanzetti.

Stearns went to visit Nash twice, once on May 14 when he was able to see Nash for only a few minutes and a second time, a few days later, when the two men talked for some time. Nash neither spoke of any delusions nor admitted to hallucinations. "I couldn't say he's psychotic," Stearns wrote to Bradley. "He was straightforward and frank and of course is anxious to get out."[57] Around May 20, ten days before the second, forty-day, phase of Nash's commitment was due to expire, Stearns went back a third time to study the commitment papers and the record of Nash's hospital stay.[58] He talked with Mueller and Gauthier, who — in spite of their conviction that Nash was merely concealing his delusions — admitted that they "doubted Nash was committable" any longer.[59] "I still do not know what is the matter with him," Stearns, who was being paid one hundred dollars for rendering his opinion, wrote to Bradley on May 20.[60] He added, however, "I certainly recommend his discharge."[61]

Mueller and Gauthier nonetheless recommended that Nash remain in the hospital. At that point, Alicia told them she was unwilling to sign another petition for commitment although she agreed to make arrangements for her husband to be treated by a psychiatrist after his release from McLean.[62] Accordingly, on May 28, after fifty days of incarceration, just over one week after the birth of his son, Nash was once again a free man.

37
Mad Hatter's Tea

May–June 1959

AFTER NASH WAS COMMITTED, Alicia couldn't face staying at the West Medford house by herself, and in any case, the lease was due to expire May 1. Alicia telephoned Emma and asked whether they might live together.[1] "One day Alicia just called me up and said she wanted to share an apartment with me," Emma recalled. At first Emma was reluctant because she was afraid Alicia would insist on their finding an expensive place, but then it occurred to her that they might rent a house owned by their mutual friend Margaret Hughes. So, on May 1, Alicia and Emma moved into a tiny saltbox at 18½ Tremont Street, in Cambridge, halfway between MIT and Harvard.

Alicia indulged in no tears, hysteria, or unnecessary confidences. She accepted what help she could get. She had very little faith that anyone would come to her aid. She was well aware that everyone, including close friends like Arthur Mattuck, considered Nash her responsibility. She defended herself against criticism of her decision to commit Nash, but only when pressed, as, for example, by Gertrude Moser, who, after visiting Nash at McLean, began to doubt that he was insane and demanded that Alicia justify her decision to have Nash locked up. For a young woman whose husband was in a lunatic asylum, threatening to hurt her, to divorce her, and to take their money and run off to Europe, she maintained a remarkable calm. The apparently flighty young woman who had, in the throes of lovesickness, sat in the science fiction section of the library, hoping her idol would come in, had reserves of strength that she would need to draw on the rest of her life.

Another young woman might have thrown up her hands and gone home to her parents. But Alicia told herself that John's mind and career could be saved. She focused on the crisis at hand as best she could and put herself in the capable hands of Emma and Fagi Levinson. Her ability to focus on her own agenda, her iron self-control, sense of entitlement, deep conviction that her own future depended on this man — and perhaps also the combined energy, optimism, and ignorance of youth — all came to her aid in this very dark hour. All her attention was focused on a single task — not the task of giving birth, but that of saving John Nash.

"She never talked about the baby, only about Nash," Emma recalled. "She regarded the pregnancy as a problem. Just a danger to Nash. She was worried that it would interfere with her ability to take care of [him]."

There was no waiting nursery, no layette, no dog-eared copy of Dr. Spock's new best-selling baby manual sitting on the night table. Alicia had no time or attention for such things. She wished for the pregnancy to end, but she had not looked beyond it. She had vaguely assumed that her mother would come and help her, but hadn't bothered to make the arrangements. Nor had she asked Virginia to come again. She barely paid any attention at all, in fact. Even after the baby kept her awake nights with its vigorous kicks, she never talked about it.

Emma recalled, "The observation period [with Nash at McLean] was coming to an end. The psychiatrists were telling Alicia that the crisis was precipitated by her pregnancy. She asked her doctor to induce her labor. He wouldn't."

On May 20, when Alicia's labor began, Nash was still in McLean and she was still living with Emma at 18½ Tremont Street. The pains began in her lower back. Eventually she crawled into bed. Emma was there. The two of them couldn't decide whether the labor had started. Later when her sister was about to give birth, Emma would buy an obstetrics textbook and discover that back labor was in fact quite common. But at that moment, the two MIT women were in the dark about such things. Finally, when the pains became more insistent and closer together, either she or Alicia telephoned Fagi, who confirmed that, yes, indeed, it sounded like labor and said she would jump into her car right away and drive over. She did and, after taking one look at Alicia, who was by now looking quite scared, told her to get into the car and they'd drive to the hospital immediately.

Alicia gave birth to a baby boy that night. He weighed nearly nine pounds and was 21.5 inches long. She did not give the baby a name. She felt that the naming would have to wait until his father was well enough to help choose one. As it happened, the baby remained nameless for nearly a year.

Alicia had still to bear Nash's anger. The day after the birth, Nash came to the Boston Lying-In Hospital to visit his wife and new son, having gotten permission to leave McLean for the evening. Although Fagi Levinson does not remember doing so, one imagines that it was she who arranged this. Another friend came to see Alicia halfway through Nash's visit. Alicia was lying in bed, looking tiny and wan. Nash was sitting beside her. Her dinner tray was on the table next to the bed. At some point, Nash carefully took the napkin, stood up, and went over to a sign on the wall with the name of the hospital on it and covered up the "In" in the hospital's name so that it read "Boston Lying Hospital." The visitor recalled, "The implication was that it was Alicia who was lying. She observed what he was doing. I made no comment. I certainly didn't want the situation to escalate into speech."[2]

Nash's sense of humor had in no way deserted him. On the afternoon of his release one week later, Nash went directly to the mathematics common room. He strolled in, greeted everyone, and said he'd come straight from McLean. "It was a wonderful place," he told the graduate students and professors who were sipping tea. "They had everything but one: freedom!"[3]

A day or two later, Nash was back in the department. He carefully posted hand-printed notices in the hallways announcing a "coming out party." The notices read: "All the people who are important in my life are invited! YOU KNOW WHO YOU ARE!" Over the following week, he went around to everyone's office and asked each member of the department if he were coming. If the person said "Yes," he asked them "Why?"[4]

He referred to the party as a "Mad Hatter's Tea," and he asked people to dress up in costumes.[5] Whether the event was his idea or Alicia's isn't clear. Fagi Levinson, Norman's wife, thought that Alicia — who was home with a week-old baby — had organized it for the purpose of thanking all of those who had visited Nash in McLean.[6] One graduate student, who said he went to New York that weekend to avoid it, remembered that it was held at Mattuck's apartment. Mattuck doesn't remember it at all. Very likely, it took place at 18½ Tremont Street. Fagi remembered it as a "big party."

The Nashes held at least one dinner party too. The mystified guest was Al Vasquez, who was about to graduate on June 12, and he remembers it as a sad and depressing event. In 1997, he recalled:

> It was one of the most bizarre evenings I've ever spent. I went there and there was Alicia, the baby, and Alicia's mother. John was behaving very oddly. Whenever John got up, Alicia's mother would get up and place herself between him and the baby. It was a pretty strange dance. It lasted a couple of hours. Alicia had no idea who I was. Everybody tried to act like everything was normal. The weirdness of this was overwhelming. Nash couldn't sit still. He'd bolt up and as soon as he did, Alicia's mother would jump up and fuss over this and that. But she wouldn't let him get anywhere near the baby.[7]

Nash was determined to leave for Europe as soon as possible. He wrote to Hörmander on June 1 asking whether Hörmander would be in Stockholm during the summer. He was thinking of traveling to Sweden that summer, he wrote, and was looking for "(nominal) mathematical associations" to justify the trip.[8] And he wrote to Armand and Gaby Borel, who were in Switzerland at the time, to ask that they help him obtain Swiss citizenship.[9]

Nash was also determined to resign his MIT professorship. Furious that MIT had connived in his involuntary hospitalization, Nash "dramatically"—as he later put it—submitted a letter of resignation[10] and simultaneously demanded that MIT release a small pension fund that had accumulated from the time he joined the full-time faculty.[11] Levinson was aghast. With Martin and others, he tried to persuade Nash that what he wished to do was mad. He told Nash that MIT would not accept his resignation. Levinson acted in the most altruistic fashion. He was well aware of the heavy expenses of medical treatment, and he was anxious for Nash to retain the insurance coverage that MIT provided its faculty members. "Norman tried to convince him not to do it," Fagi said. "He felt responsible for him."[12]

Martin recalled, "It was a very difficult period. By the time he resigned, he couldn't meet his classes and people felt that he had no hope of any recovery. We were on the spot. I couldn't even talk to him. There was no having a coherent conversation with him. Levinson always backed Nash to the hilt. There was no pressure on me either [from the administration to accept Nash's resignation]."[13]

But Nash was intransigent. At Levinson's urging, the university administration tried to prevent Nash from withdrawing his pension money, but here too Nash prevailed. On June 23, James Faulkner, a physician affiliated with MIT, telephoned Warren Stearns on behalf of MIT's president, James Killian, to say that the university was extremely concerned about Nash's future.[14] According to Paul Samuelson, Stearns once again took the position that Nash was not insane and was fully competent, in a legal sense, to make such decisions.[15] The amount was negligible, but once the check was issued, Nash's last formal tie to MIT was cut.

Shortly after his resignation, he ran into one of his former students from the game-theory course, Henry Wan, telling him that he was now engaged in a study of linguistics. When Wan expressed surprise, Nash said that mathematicians had a unique ability to "abstract the essence of a field. That is why we can move from one area to another."[16]

Nash said that he was sailing on the *Queen Mary* in early July. Alicia tried to dissuade him, but when it became clear to her that he would go, she made up her mind to accompany him and to leave their son behind in her mother's care.

Nash had an invitation to spend the year in Paris at the College de France, the leading French center of mathematics. Alicia hoped that a few months abroad, away from the pressures of Cambridge and among new faces, would let Nash forget his dreams of world peace, world government, and world citizenship; he might settle down to work again. To Nash, however, the journey seemed to promise a more permanent escape from his old life. He talked as if they were never to return.

They drove down to New York and said their good-byes to Alicia's cousins. The occasion was uneventful except that Nash had refused to eat facing the huge mirror opposite the dining table.[17] They left their Mercedes, its trunk full of old issues of *The New York Times,* in the Institute parking lot in Princeton. Nash wished to bequeath both car and newspapers to Hassler Whitney, the mathematician whom he most admired.[18] They left their baby — not yet named and therefore referred to as Baby Epsilon, a little mathematical joke — behind as well. Alicia's mother had already taken the infant home with her to Washington.[19] Mrs. Larde, they had agreed, would join them in Paris with the baby as soon as they were settled.

PART FOUR

The
Lost
Years

38

Citoyen du Monde

Paris and Geneva, 1959–60

I have a difficult task ahead of me and I have dedicated my whole life to it.
— *K, in* The Castle, *by FRANZ KAFKA*

I seem as in a trance sublime and strange
To muse on my own separate fantasy.
— *PERCY BYSSHE SHELLEY, "Mont Blanc"*

S HORTLY AFTER Independence Day, Nash and Alicia left from New York
harbor on the *Queen Mary,* standing by the rail with the rest of the throng. They
watched the pier, then the skyline, then the Statue of Liberty move away from
them as they sailed slowly toward the open sea. They looked very much as they
had a year earlier when they'd embarked on their honeymoon voyage—he tall,
well dressed, and handsome, she slender, small, and delicate—but less animated,
more subdued. They were both lost in their own thoughts.

The Nashes reached London on July 18 after a "restful" crossing.[1] Two days
later they were in Paris.[2] The beauty of Paris overwhelmed them just as it had a
year earlier, "verdure everywhere . . . with the giant blue Paris pigeons bolting
above it, two by two."[3] For a few hours after they emerged from the Gare St-Lazare
and made their way to a modest Left Bank hotel incongruously named the Grand
Hôtel de Mont Blanc, the leaden weight of the miserable months in Cambridge
seemed to lift from their shoulders and they felt, briefly, as light as air again. They
set out, that afternoon, for the American Express Office to buy francs and to inquire
if they had any mail. As always during the summer, the Place de L'Opéra was
crowded with American tourists. To their delight, they immediately spotted the
familiar face of John Moore, a mathematician Nash knew from MIT, who would
soon become co-chairman of the mathematics department at Princeton. Moore
was sitting outside the Café de la Paix, reading, when he looked up and saw the
Nashes. "I was surprised, but not surprised," Moore recalled in 1995. "A lot of
mathematicians come to Paris. We talked about Edinburgh. I noticed nothing
unusual."[4]

What their real plans were at the time, Alicia was later unable to say. She had followed Nash to Europe, not because she hoped that Paris would provide a cure for his troubles, but because she had no way of stopping him, and, that being the case, she had not been able to bear seeing him go off to a strange land, alone, without someone to watch over him. But, in those first few days in Paris, the Nashes behaved as if this would be their new home for some time. Alicia enrolled in a French-language course at the Sorbonne and looked around for more permanent lodgings.[5] Her twenty-year-old cousin Odette, who was planning to spend the year at the University of Grenoble, happened to be in Paris, too. The two young women went house hunting together until they found a pretty, clean, and spacious flat for the Nashes at 49 Avenue de la République, in a nondescript but perfectly respectable blue-collar neighborhood on the Right Bank.[6]

Paris, indeed all of Europe, was sizzling hot that July. The newspapers were full of heat-wave stories, including one about a parked car that had burst into flames, a seemingly genuine case of spontaneous combustion. The rear windshield had apparently acted like a magnifying glass and some papers left on the rear dashboard had ignited.[7] The mood of Paris, always a magnet for alienated and disaffected Americans and full of self-declared exiles of the Silent Generation, was hot as well. The war in Algeria raged on, with its right-wing terrorist bombings, its civilian massacres, its tortures. The city reverberated with mass demonstrations, strikes, and explosions. And the latest word on the nuclear arms race—the American announcement that it now could match Russia's ICBMs, missile for missile—left open the question of whether the world wasn't in for another, more deadly case of spontaneous combustion.

If the heat and high political theater influenced Nash's mood, they induced not torpor, but a heightened sense of purpose. Acting on "special" knowledge, Nash was animated by a desire to cut himself off from all vestiges of his former social self. In the rightness of this he believed with absolute certainty, resisting any and all attempts by Alicia to persuade him to give up his "silly" notions. Having resigned his professorship, having left not only Cambridge but the United States, and having given up mathematics for politics, he wished, quite simply, to shed the layers of his old identity like so many outworn articles of clothing.

Ideas of world government, and the related concept of world citizenship, were at their heyday during Nash's Princeton graduate-school days and permeated the 1950s science fiction that Nash devoured as a student and afterward. Founded after the collapse of the League of Nations in the 1930s, the one-world movement exploded into the national consciousness within a few years of the end of World War II. Princeton was a center of that movement, largely because of the presence of physicists and mathematicians—notably Albert Einstein and John von Neumann—who acted as midwives to the nuclear age.[8] One of Nash's contemporaries in graduate school, John Kemeny—a brilliant young logician, the assistant to

Einstein, and later the president of Dartmouth College — was a leader of the World Federalists.

However, the one-worlder who fired Nash's imagination was a loner like himself, the Abbie Hoffman of the one-world movement. In 1948, Garry Davis, a leather-jacketed World War II bomber pilot, Broadway actor, and son of society band leader Meyer Davis, had walked into the American embassy in Paris, turned in his U.S. passport, and renounced his American citizenship.[9] He then tried to get the United Nations to declare him "the first citizen of the world."[10] Davis, "sick and tired of war and rumors of war," wished to start a world government.[11] "Every paper headlined the story," the columnist Art Buchwald recalled in his Paris memoir.[12] Albert Einstein, eighteen members of the British Parliament, and a slew of French intellectuals, including Jean-Paul Sartre and Albert Camus, had come out in support of Davis.[13]

Nash intended to follow in Davis's footsteps. In the overwrought, hyperpatriotic atmosphere of the America he was leaving behind, Nash was choosing the "path of most resistance," and one that captured his radical sense of alienation. Such "extreme contrariness" aimed at cultural norms has long been a hallmark of a developing schizophrenic consciousness.[14] In ancestor-worshiping Japan the target may be the family, in Catholic Spain the Church. Motivated as much by antagonism to his former existence as by an urge for self-expression, Nash particularly desired to supersede the old laws that had governed his existence, and, quite literally, to substitute his own laws, and to escape, once and for all, from the jurisdiction under which he had once lived.

While the motivation may have been highly abstract, the plan itself was strangely concrete. To effect his makeover, he wished to trade his American passport for some more universal identity card, one that declared him to be a citizen of the world.

On July 29, a little over a week after his arrival in Paris, Nash went by train to Luxembourg.[15] He chose Luxembourg as the site for the renunciation of his American citizenship for prudent reasons, possibly at the advice of the Paris-based World Citizen Registry, an organization founded by Davis. The smaller and more obscure the country, the less likely that turning in his American passport would result in immediate arrest and deportation. France was a notoriously bad site for protests of this sort. When Nash arrived at the Central Station in the city of Luxembourg, he walked to the American embassy at 22 Boulevard Emmanuel Servais, demanded to see the ambassador, and announced that he no longer wished to be an American citizen.

Section 1481 of the 1941 Immigration Act contains a clause that permits American citizens to give up their citizenship.[16] It was intended, of course, to allow citizens to resolve cases of dual citizenship. By 1959, some dozens of Americans, also inspired by Garry Davis, were making use of the provision for protest purposes.[17] The law is quite clear. It delineates an oath, which must be taken in a foreign country, right hand raised, in the presence of an American diplomat: "I desire to make a formal renunciation of my American nationality . . . and pursuant thereto I hereby absolutely and entirely renounce my nationality in the United States and all rights and privileges pertaining, and abjure all allegiance and fidelity to the United States of America."[18]

Nash's announcement was greeted as one might expect. An embassy official — not the ambassador! — made a number of strongly worded arguments to convince Nash that what he wished to do was unwise. Somewhat surprisingly, given the strength of Nash's conviction at that moment, the diplomat convinced Nash to take back his passport. It was a sign, perhaps, of a vacillation and indecisiveness that would become more pronounced with time.

The official's argument made sense to him. As Nash said in his 1996 Madrid lecture: "I wouldn't have been able to leave Luxembourg and return to Paris because I no longer had a passport. They allowed me to retract my action as irrational and insane."[19]

When the news of his first attempt to give up his American citizenship reached Virginia and Martha in Roanoke and his former colleagues at MIT, it proved to them that the confinement at McLean had done little to halt the galloping illness. Virginia, who had been deeply depressed on her return from Boston, had been drinking heavily and was headed for a breakdown herself. (She would be hospitalized in September.)[20] When Armand Borel got back to Princeton from Switzerland at the end of the summer and inquired about Nash, one of his colleagues told him simply: "There is trouble."[21]

The plan's having been aborted did little to suppress Nash's high spirits when he returned to Paris two days later. The mere fact of having attempted to act sufficed to make him feel that he was, as he wrote on a postcard to Virginia, mailed July 31, well "on the way to becoming a world citizen."[22] His mind was full of other aspects of his intended transformation. He was visiting the "Bibliotek," that is, the Bibliothèque Nationale, which is the French equivalent of the Library of Congress, he wrote to Virginia, and was working on learning French ("part of the plan," as he had written to Tucker nearly a year earlier).[23] He also confided in his mother that he wished "to take up painting."

Before long, however, Nash was afire with a new plan. His objectives, somewhat obscure even to himself until now, were suddenly much clearer. As Paris emptied for the August vacation, Nash decided that he preferred to be in Switzerland, a country he associated with neutrality, world citizenship, and Einstein.[24] Einstein, who liked to refer to himself as a world citizen, had adopted Swiss citizenship. Possibly the fact that several European nations had been conducting the longest summit on record that summer in Geneva influenced his thinking.[25] But it appears that the Nashes did not leave Paris as soon as Nash intended. The actual departure was delayed by protests on Alicia's part over the sudden move after having just rented an apartment.

Nash's desire to go to Geneva was based, he later said, on his having heard that Geneva was "the city of refugees." [26] This was absolutely true, in both a historical and a contemporary sense. Hugging the southern shore of the crescent-shaped Lac Leman, set against a panorama of glaciers, the snowy ridges of Mont Blanc visible on all but the foggiest days, Geneva had once been the beacon of the Protestant Reformation and the refuge of French Protestants as well as freethinking intellectuals, including Voltaire and Rousseau. [27] Mary Wollstonecraft Shelley had spent the summer of 1816 in the suburb of Cologny writing *Frankenstein or The Modern Prometheus.* [28] In the twentieth century, Geneva had become the site of the ill-fated League of Nations and was a major international banking center. The European headquarters of the United Nations and other international enterprises such as the Red Cross were located there.

In 1959, Geneva was an overnight train trip from Paris. When the Nashes arrived, they took a room at the Hôtel Athenée in the Rue Malganou. [29] Alicia, however, did not stay long. She left almost immediately for Italy where she met Odette and remained for several weeks.

Alone for the first time in his life, Nash was "without parents, home, wife, child, commitment or appetite . . . and the pride that might be taken in these," [30] and thus completely free to dedicate himself single-mindedly to his quest. His objectives, as suggested by his choice of venue, were shifting. He now wished not only to shed his American citizenship, but to obtain official refugee status —to be declared a refugee from "all NATO, Warsaw, Middle East and SEATO pact countries." [31] Presumably, these alliances were now fused in his mind with threats to world peace, but the desire for refugee status also reflected an expanding feeling of alienation, a sense of persecution, and fear of incarceration. He saw himself as a conscientious objector in danger of being drafted and as an opponent of the kinds of military research American mathematicians were expected to do. [32]

He spent most of his evenings in that loneliest of places, a small blank hotel room in a distant and nondescript part of the city, writing letters that would never be answered, filling out endless forms, applications, and petitions that would be filed away. His days were spent haunting various anterooms and offices.

For five solitary months, Nash's ambiguous and self-annulling efforts resembled nothing so much as the anti-quest of the land surveyor in Kafka's novel *The Castle,* probably the most compelling rendering of the schizophrenic consciousness in all of literature. Known only as K, Kafka's hero's sole object in life is to penetrate "the shadowy heart of the Castle" which looms high over a mazelike village K reaches but cannot get beyond. [33] In Kafka's novel, K, a man whose job it is to measure and estimate, seeks to enter a clouded locus of authority, not because he desires "to lead an honored and comfortable life," but in order to "gain acceptance by the higher perhaps celestial powers and thereby to discover the reason of things."[34]

Nash's lifelong quest for meaning, control, and recognition in the context of a continuing struggle, not just in society, but in the warring impulses of his paradoxical self, was now reduced to a caricature. Just as the overconcreteness of a dream is related to the intangible themes of waking life, Nash's search for a piece of paper, a carte d'identité, mirrored his former pursuit of mathematical insights. Yet the gulf between the two recognizably related Nashes was as great as that between Kafka, the controlling creative genius, struggling between the demands of his self-chosen vocation and ordinary life, and K, a caricature of Kafka, the helpless seeker of a piece of paper that will validate his existence, rights, and duties. Delusion is not just fantasy but compulsion. Survival, both of the self and the world, appears to be at stake. Where once he had ordered his thoughts and modulated them, he was now subject to their peremptory and insistent commands.

Like K, Nash found himself trapped in a "farce of endless paper shuffling . . . a vast soulless mechanism for the circulation of papers . . . a world cluttered with paper, the white blood of bureaucracy . . . doomed by forces beyond his control ('they're playing with me'), yet also distracted through an inner confusion of desires."[35]

Nash appealed to many authorities. Yet he seemed unable to make much progress. The American consulate, he discovered, was not prepared to accept his passport or to allow him to take the oath of renunciation.[36] Smiling, kindly, but seemingly obtuse diplomats dissuaded and deflected him, offering him excuses and rationales. Confused and weakened by their lengthy explanations, Nash would go away again, only to return the next day.

The U.N. High Commission for Refugees, on which he pinned his hopes, sent him away. It appeared that the commission, its promising name notwithstanding, had rules that precluded cases like his. One could claim refugee status only in connection with "events occurring in Europe before 1 January 1951" and "owing to a well-founded fear of being persecuted for reasons of race, religion, nationality, membership of a particular social group or political opinion, [and only if one] is outside the country of his nationality and is unable or, owing to such fear, is unwilling to avail himself of the protection of the country."[37] The officials of the commission suggested he contact the Swiss police.

At that time, the Swiss federal police handled all requests for asylum, of which there were perhaps a dozen a year that fell into the category of "unusual" in the sense that they involved individuals from countries that typically produced no refugees. Since Nash claimed to be a conscientious objector who was fleeing the draft, the police referred him to the military authorities. These authorities cautiously turned to Berne for advice, and Berne, in turn, consulted Washington.[38] In September, the Geneva military authority sent a letter to Berne saying of Nash that "en renonçant à son passeporte américain, et cela pour la seule raison qu'il ne désire pas être appelé à faire service dans les forces armées des U.S.A., ni même prêter aux organisations officielles de son pays sons concours en qualité de mathematicien, craignant que sa collaboration puisse aider les autorités de son pays à maintenir la guerre froide ou préparer la guerre" (he is renouncing his American passport, for the sole reason that he doesn't want to be drafted into the United States Army, nor lend to official organizations his services as a mathematician, fearing that his collaboration might aid the authorities of his nation in maintaining the Cold War or in preparing for war).[39]

In November, the Geneva authorities were informed that Nash was, for all practical purposes, far beyond the American draft age and that he was in no way obligated to do defense-related research. Moreover, Nash had committed none of the acts that would provoke the American government to strip him of his citizenship: "Au surplus, la simple declaration de renonciation au passeport américain n'a en soi pas d'effet juridique."[40] In other words, since he had not signed the oath of renunciation, he was still technically an American citizen. At that point, the police began threatening Nash with deportation.

His sense of himself was now full of the starkest contradictions. On the one hand, Nash's most intimate thoughts and actions seemed to be those of another, controlling psyche — "I am the left foot of God on earth." On the other, he felt himself to be at the epicenter of the universe, with outer reality simply a projection of his mind. At times his posture was that of an abject petitioner, at other times that of a "religious figure of great, but secret, importance."[41] He spent a great deal of time opening various bank accounts — usually under false names, including one that he later said was "mystical," and wiring money to various countries. "I shifted money from one bank to another," Nash recalled in his Madrid lecture in 1996. "I opened an account at a Swiss bank. It was Credit Andorra. The account was in Swiss francs. But I didn't have very much money."[42] Many years later, in a limousine going to the center of Stockholm where he would attend the Nobel ceremonies, Nash pointed out a bank in passing to Harold and Estelle Kuhn, saying that he had wired money there as part of an effort to organize a defense against "an invasion of aliens."[43]

Such self-contradiction is also characteristic of schizophrenia, every symptom being matched by a "countersymptom." John Haslam — in what is widely regarded as the first psychiatric description of schizophrenic thinking — focused, early in the nineteenth century, on this peculiar combination of omnipotence and impotence: The person is "sometimes an automaton moved by the agency of persons . . . at others, the Emperor of the whole world," the tendency toward megalomania mixed with feelings of persecution, powerlessness, inferiority.[44]

He maintained both positions at the same moment, often, it seems, apparently untroubled by the apparent inconsistency — a flouting of what Aristotle considered the fundamental rule of reason: "The identity principle or law of contradiction that states the impossibility of affirming both p and not p."[45] It was a cruel, cosmic joke. The man who produced a compelling theory of rational behavior no longer thought in terms of either/or.

It is not true, however, that Nash had lost all contact with reality. The clearest evidence that reality in fact pressed heavily and unpleasantly on him is that the frustrations of his situation were beginning to oppress him. His expectant mood turned slowly and inexorably into one of deep disappointment and depression. Nash spent long hours walking around the city, mostly in the parks and along the lake, waiting, endlessly waiting. At the end of September, he wrote to Virginia and Martha: "My life is not exciting at present. . . . Waiting for favorable developments. I'm somewhat disillusioned with a great many of my former associates, colleagues, friends, etc."[46]

His dark mood may have reflected more than his difficult current situation. Martha had written that Virginia had had "a nervous breakdown and spent two weeks in the hospital."[47] Nash found the news virtually unbelievable. He simply could not imagine his forceful mother ill in this fashion, but he must have sensed, from the tone of Martha's letter, that his mother's distress was linked, in some way, to his own.

Finally, in September or October, in a fit of desperation, Nash destroyed or threw away his passport. Alicia later recalled that he had merely "lost" it and while that is certainly possible, later events suggest otherwise.[48] When the consulate became aware of this action, an effort was made to persuade Nash to apply for a new one.[49] This he refused to do.

In his own mind, Nash was now stateless, a man without a country; in the eyes of the authorities, he was a man without proper documents, which placed him in a vulnerable situation. Nash had, as he later wrote to Lars Hörmander, "requested refugee status. This produced difficulties."[50] On October 11 he wrote to Virginia and Martha that he was no longer able to travel "because of certain legal formalities," a reference, presumably, to his lack of a passport.[51] In the same letter, he enclosed a long free-style poem about feeding the gulls on the shores of Lac Leman. He did, however, manage to visit nearby Liechtenstein, where he considered requesting citizenship, on account of the fact that Liechtenstein didn't levy income taxes on foreign residents.[52]

During her Roman holiday, for a few short weeks, Alicia recaptured — for the last time, it turned out — a bit of her old lighthearted, girlish self. Odette recalled in 1995 that Alicia, once again, seemed "fun-loving."[53] These two exceptionally pretty, stylish young women had quite a holiday. They visited the Vatican, where they had an audience with Pope John XXIII. Odette fainted and had to be carried out of the chamber by two young Italian medics who afterward showed the two women around the city. They went to nightclubs, shopped, and were admired and pursued, by Americans as well as Italians, wherever they went. After Rome, they visited Florence and Venice. In Venice, the two young women had a photograph taken of themselves, Odette looking like a young Audrey Hepburn, Alicia like a young Elizabeth Taylor, standing in their high heels and bouffant hairdos in the Piazza San Marco surrounded by pigeons.

At the end of August, Alicia returned to Paris and began making arrangements for her mother and baby to come to France. She may have gone to Geneva first, but if so, she stayed there only briefly. She wrote to Nash urging him to come to Paris and contacted the American embassy for help in getting Nash back from Switzerland. "Alicia is in Paris expecting 'e,' " Nash wrote in early November — "e," of course, was John Charles, whom Nash called Baby Epsilon.[54] ("Baby Epsilon" was a tongue-in-cheek reference to a well-known mathematical anecdote about a famous mathematician who believes that all infants are born knowing the proof of the Riemann Hypothesis and retain that knowledge until they are six months of age.)[55]

It was Nash's first mention of the baby in his letters to Roanoke, yet he gave no indication that he intended to join them. While she waited for her mother and son to arrive, Alicia visited Odette in Grenoble. "We'd go to my room and eat pastries, baba au rhum," Odette recalled. "We'd gossip about the other students. We went skiing."[56]

Back in Washington, Baby Epsilon was finally christened with his grandparents and Martha in attendance.[57] The baby, dressed in a little sweater on a bright fall day when leaves littered the ground, was named John Charles Martin Nash. The christening took place at St. John's in Lafayette Square, the same church where Nash and Alicia had exchanged marriage vows. (It is not clear who settled on the name John. Nash's first son, of course, was already called John. It was as if the Nashes and Lardes wished to obliterate, through replacement, the first child.)

In early December, when the frigid north wind called *le bise* swept across Lac Leman and made walking along its shores a misery, Nash's mood was bleaker than ever. One can almost feel his "sense of helplessness in an ice-cold universe."[58] His efforts to renounce his citizenship and to obtain refugee status had been, for reasons baffling to him, frustrated. He spent most of his time indoors writing letters. His feeling of having chosen to escape from Cambridge was replaced by one of having been exiled. He wrote to Norbert Wiener:

> I feel that writing to you there I am writing to the source of a ray of light from within a pit of semi-darkness. . . . It is a strange place where you live, where administration is heaped upon administration, and all tremble with fear or abhorrence (in spite of pious phrases) at symptoms of actual non-local thinking. Up the river [a reference to Harvard], slightly better, but still very strange in a certain area with which we are both familiar. And yet, to see this strangeness, the viewer must be strange.[59]

The letter was decorated with silver foil, a newspaper photograph of a Lenin-like character, a story about Nehru's seventieth birthday containing a reference to Khrushchev, and ticket stubs from a trolley.

Even while he described himself as someone capable of inspiring fear in others on account of his "non-local thinking," Nash's reference to "administration . . . heaped upon administration" suggests a growing sense of vulnerability, a free-floating anxiety, and a belief that the authorities were toying with him. Shortly afterward, for reasons unknown, Nash changed hotels, moving now to a cheaper and more remote one — the Hotel Alba in the Rue de Mont Blanc.[60]

In this claustrophobic hotel room during what would turn out to be Nash's final week in Geneva, the true dimensions of his tragedy would become clear. He was in Switzerland, free of Alicia, free of external restraint, but as thoroughly immobilized as the hero of another Kafka story, "The Metamorphosis," who wakes up one morning to discover that he has become a cockroach lying helplessly on its back.[61] Kafka never wrote the final chapter of *The Castle,* but confided to his friend and biographer, Max Brod, that he had envisioned a scene in which K is lying on his bed in the inn exhausted to the point of death. "K was not to relax his struggle, but was to die worn out by it."[62] Nash did not relax his struggle either, but he was defeated all the same.

James Glass, a political scientist at the University of Maryland who has studied the delusions of schizophrenia, writes, "Delusion provides a certain, often unbreakable identity, and its absolute character can maneuver the self into an unyielding position. In this respect, it is the internal mirror of political authoritarianism, the tyrant inside the self . . . an internal domination as deadly as any external tyranny."[63]

On December 11, Nash had been held for several hours by the police — apparently in an effort to convince him that "deportation was unavoidable" — and released "under surveillance," requiring him to report to a police station two or three times every day.[64] According to a telegram, dated December 16, from the American consul in Geneva, Henry S. Villard, to Secretary of State Christian A. Herter, the Swiss authorities had issued a deportation order naming Nash as an "undesirable alien" on December 11.[65] Throughout, the Swiss authorities evidently were acting with the "full knowledge of Dr. Edward Cox, assistant science advisor" and presumably with tacit approval at higher levels of the State Department.

The final curtain came down on December 15. Nash was arrested, for the second time.[66] He adamantly refused, as he had at the time of his first arrest, to return to the United States, and continued to demand to sign the oath of renunciation. On the morning of the fifteenth, Cox, a kindly, avuncular retired chemistry professor from Swarthmore College,[67] now serving as assistant science attaché in Paris, arrived in Geneva by overnight train. He was accompanying an exhausted and apprehensive Alicia Nash.[68] Together they hoped to persuade Nash to return directly to the United States. Neither knew what to expect, and both, in their separate ways, feared the worst.

Secretary Herter was being apprised of the situation in daily cables, as was the State Department's science adviser, Wallace Brode. On the fifteenth, a cable to Washington from Ambassador Amory Houghton in Paris informed them: "RECEIVED WORD FROM GENEVA TO EFFECT NASH DESPITE ALL EFFORTS TO DISSUADE HIM DETERMINED TO SIGN OATH OF CITIZENSHIP RENUNCIATION."[69]

Even in jail, Nash refused to return to the United States, refused furthermore to cooperate in the issue of a new passport, and continued to demand that he be permitted to take the oath of renunciation.

At this point, Alicia agreed to take Nash back to Paris with her where they had, after all, an apartment. The consul general agreed to issue Alicia a new passport that included Nash. Nash protested it all. He did not wish to go even to Paris. It was useless. The police escorted Nash to the train station. He was hustled onto the train and, at 11:15 P.M., it pulled out of the covered station into the open air. The police inspectors reported that "at train time Nash [was] still reluctant [to] leave Geneva but no force [was] required."[70]

Nash and Alicia celebrated Christmas at 49 Avenue de la République. It was, as Nash was to write to Virginia, "interesting."[71] Alicia's mother was there and so was the eight-month-old John Charles. There was a Christmas tree, perhaps the first one that the Nashes had ever had, decorated in the German manner with tiny lady apples and red wax candles. When they lit them, it scared Alicia's mother terribly. "We kept a bucket of water nearby," Odette, who had come to Paris for the holidays, recalled.[72] Alicia, who had occupied herself that fall with learning to cook, served French hors d'oeuvres. There were presents for the baby, Nash jealously noted, adding in a letter to Virginia and Martha that "he seems a little attention spoiled now."

On St. Etienne's Day, the day after Christmas, Alicia gave a party attended by several mathematicians, American as well as French. Shiing-shen Chern, a mathematician who had met Nash at the University of Chicago and was in Paris for the semester, came. He recalled "an interesting idea" that Nash had then, namely that four cities in Europe constituted the vertices of a square.[73] The most striking visitor at 49 Avenue de la République, however, was Alexandre Grothendieck, a brilliant, charismatic, highly eccentric young algebraic geometer who wore his head shaved, affected traditional Russian peasant dress, and held strong pacifist views.[74] Grothendieck had just taken a chair at the new Parisian mathematics center, the Institut des Hautes Études Scientifiques (modeled after Princeton's Institute for Advanced Study), and would win a Fields Medal in 1966. In the early 1970s, he founded a survivalist organization, dropped out of academia altogether, and became a virtual recluse in an undisclosed location in the Pyrenees.[75] In 1960, however, he was dynamic, voluble, and immensely attractive. Whether he was mainly interested in the beautiful Alicia or felt an affinity for Nash's anti-American sentiments is not clear; in any case, Grothendieck was a frequent visitor at the

Nashes' apartment and on a number of occasions attempted to help Nash obtain a visiting position at the IHES.

That January, Odette and Alicia would sit around the apartment smoking and gossiping about Odette's boyfriends, including thirty-four-year-old John Danskin, a mathematician at the Institute for Advanced Study who had met the entrancing Odette at the Nashes' wedding party in New York. He wooed Odette by letter, ultimately proposing to her by telegram in Russian. Nash would sit in the corner of the living room poring over a Paris telephone directory, saying little except to occasionally object to the smoke, which he abhorred, or to ask a question. Odette recalled:

> We were having a wonderful time. We just laughed and gossiped, tried French cooking and met the people who Alicia invited into her apartment. We'd be chattering. We'd talk about boys. John Nash wouldn't even notice. Alicia used to smoke. He used to complain about it. He couldn't bear it. Occasionally he would interrupt with a question: "Do you know what Kennedy and Khrushchev have in common? No. Both their names start with a K."[76]

Odette soon returned to Grenoble and Alicia's mother left Paris as well, leaving her daughter and grandson behind. Alicia struggled to care for the baby and to cope with her husband, finding both overwhelming.[77] She desperately wanted to return to the United States and continued, as best she could, to obtain the help of the American authorities.

A concerted effort was, in fact, under way, led by the State Department's Brode, who dispatched his deputy, Larkin Farinholt, to Paris.[78] Farinholt, a chemist who would subsequently become the director of the Sloan Foundation's fellowship program, vainly tried to convince Nash to return to America voluntarily. The effort was inspired not just by the government's desire to avoid embarrassment, but by a genuine wish that Nash not be lost to the scientific community nor suffer the consequences of his own seemingly irrational behavior.

Nash's legal situation was increasingly tenuous. After his deportation from Switzerland, he had been issued a three-month temporary residency permit by the French. His status in France, as he explained to Hörmander in a letter in late January, was "of Swiss resident or domicilee."[79] As Nash explained in his Madrid lecture, he had wanted to be declared a refugee from all NATO countries, but since he found himself in France he had — "so as not to be inconsistent" — to settle for declaring himself "only a refugee from the USA."[80] Once again, he applied for asylum. When it became clear that the French were not going to grant it, Nash attempted to obtain a Swedish visa. This, too, was refused. He then turned to Hörmander, who in turn consulted the Swedish foreign ministry and was told that without an American passport Nash had no hope of obtaining a visa. Hörmander, now impatient, wrote back: "Personally I would strongly advise you to reconsider your views concerning NATO and other countries."[81]

Nash then managed a rather extraordinary feat. In early March, he traveled, alone and without passport, to East Germany.[82] Hard as it is to believe that an American without documents could get into the DDR in 1960, Nash confirmed in 1995 that he had indeed traveled there, explaining that in his "time of irrational thinking" he had gone "places where you didn't need an American passport."[83] What actually must have happened, given the tremendously tight security at the border at that time, was that Nash applied to the DDR for asylum and was then permitted by the authorities to enter the country until the request was decided. In any case, he went to Leipzig and stayed with a family named Thurmer for several days. According to a card he sent Martha and Virginia, he was able—presumably as a guest of the government—to attend a famous propaganda event that happened to be taking place at the time, the Leipzig industrial world fair, which was the Iron Curtain's answer to the Brussels world fair. Later, mathematicians in America would hear from Farinholt that "Nash tried to defect to the Russians" but that the Russians had refused to have anything to do with him.[84] That story, repeated by Felix Browder, is very probably based on Nash's Leipzig adventure. At least no evidence has turned up that Nash ever approached the Soviets. By that point, everyone involved—the Americans, the French, and presumably the DDR—was aware that Nash's actions were those of a very sick man. Apparently, however, the incident would prompt the FBI to raise questions about Alicia's security clearance in the early 1960s when she was working at RCA.[85] In any case, Nash was eventually asked to leave East Germany—or quite possibly Farinholt got him out—and returned to Paris where he wrote to Martha and Virginia that he was "thinking of returning to Roanoke" but was worried about coming back to the United States when he had no guarantee that he would be able to leave again.[86]

As in Geneva, Nash spent much of his time sitting in the apartment writing letters. Michael Artin, the son of Princeton's Emil Artin, found a letter from Nash, after the death of his father, in his father's files. "It started out plausibly about mathematics," Artin recalled. "But it was stamped all over, with [Metro] tickets and tax stamps pasted on it. By the end of the letter it was obvious that it was completely fantastic. It was about Köchel's numbers for Mozart symphonies. Köchel had catalogued all of Mozart's works, more than five hundred. It was very graphic. It must have affected my father very much because he had kept it for all those years."[87] Al Vasquez, the MIT undergraduate Nash had gotten to know in his final year in Cambridge, recalled: "His letters were filled with numerology. I didn't keep them. They weren't just letters. They were collages, pastiches. Full of newspaper clippings. Very clever. I was always showing them to people. They contained some insights. Little patterns, puns."[88] Cathleen Morawetz recalled that her father, John Synge, who had taught Nash tensor calculus at Carnegie, received postcards from Nash at this time and was frightened by them. They reminded him, he told her, of his brilliant brother Hutchie, who suffered from schizophrenia and had quit Trinity College in order to settle in the bohemian enclaves of Paris before the First

World War. Morawetz said, "The letters were about things like Milnor's differential structure of spheres. Nash would quote a theorem. Then he'd derive a political meaning for it."[89]

Money was a growing worry. The Nashes' lodgings were cheap by American standards, but living, particularly food, was not. Nash was greatly preoccupied with trying to sell his Mercedes, still in the Institute for Advanced Study's parking lot. The mathematician with whom he had left his car, Hassler Whitney, had called John Danskin and asked him to deal with it.[90] John Abbat, a Frenchman who had invented a kind of bowling pin and was married to Odette's older sister Muyu, got involved as well. The book value, Danskin recalled, was $2,300, but Nash was determined to get $2,400 or $2,500. "He was absolutely unreasonable," Danskin recalled. "I didn't sell it. It was still there when he got back." From time to time, Nash asked Martha to send Eleanor money.[91] He also asked Warren Ambrose to visit John David, or perhaps Ambrose offered. Eleanor recalled that John David, now nearly seven, was frightened of Ambrose.[92]

Nash's hair had by now grown long, and he had a full beard. In early April, he sent Martha a photograph of himself, taken in a Chinese restaurant, which he asked her to return to him, labeling it "Picture of Dorian Gray."[93] He referred to an "autorisation de séjour" for April 21 and said that he was planning to leave soon for Sweden.[94] On April 21, Virginia received a telegram from the State Department requesting funds to bring Nash back to the United States.[95] She wired the money. Nash was taken from the apartment on Avenue Rue de la République by the French police, who escorted him, under guard, all the way to Orly.[96] Nash would later tell Vasquez that he had been brought back from Europe, "on a ship and in chains, like a slave,"[97] but Alicia recalled quite definitely that they came back on a plane.[98] While the departure repeated the trauma of Geneva, it was also a mirror image of their journey to France the previous summer. This time it was Nash who was the unwilling one. Ironically, in this, too, he was walking in Davis's path, for Davis, too, was once forcibly placed on the *Queen Mary* and sent back to America confined in first-class quarters.[99]

39

Absolute Zero

Princeton, 1960

T HE OLIVE-GREEN MERCEDES 180 was still in the institute parking lot in Princeton. Nash had come straight there while Alicia and the baby went to Washington to stay with the Lardes.[1] He hung around Princeton. In June, having heard that his sister had had a baby, Nash drove down to Roanoke to visit Martha in the hospital. She remembered being frightened by his appearance and concealing from him her son's due date, June 13. "I was worried that he would put some meaning in it," she recalled in 1995.[2] Her recollection is that Nash stayed in Roanoke with Virginia for several weeks.

Alicia, meanwhile, was looking for work and had enlisted, among others, John Danskin — now married to Odette — to help her.[3] Danskin was now teaching at Rutgers, and the newlyweds lived on the outskirts of Princeton. Alicia was apparently considering staying in Washington, presumably so that her parents could help with the baby. She was also thinking of moving back to New York. During the summer, Alicia stayed with her old MIT friend, Joyce Davis, by now living in Greenwich Village and working in the city, and interviewed for various computer programming jobs. As she told Joyce in a note she left at her apartment on the day that she returned to Washington, she got offers from IBM and also from Univac but was undecided over whether to accept them, saying, "Now I've got a real problem, work in NY or Wash?"[4]

Odette urged Alicia to move to Princeton.[5] Nash was also in favor. Alicia thought that her husband would benefit from being around other mathematicians again and hoped that he would be able to find work in Princeton. The upshot was that Alicia turned down the offers to work in New York City and instead took a position with the Astro-Electronics Division of the Radio Corporation of America, which had a big research facility on Hightstown Road between Princeton and Hightstown.[6] Alicia left John Charles in her mother's care once more and rented a small apartment at 58 Spruce Street, on the corner of Walnut, about a mile from Palmer Square. Nash joined her there at the end of the summer.

• • •

Initially, at least, Princeton seemed to offer a respite after the anxious final months in Paris. Alicia and Nash were very much part of a crowd that had gathered around John Danskin and Odette in the charming enclave near the Delaware-Raritan Canal. Griggstown consisted at that time of Tornquist's, a general store, and a few picturesque houses, including the former cider mill where the Danskins lived. It was especially beautiful in the summer, the air heavy with the scent of honeysuckle. Napthali Afriat, a game theorist who worked with Morgenstern at the time, lived there, as did Jean-Pierre Cauvin, a graduate student in French at Princeton, and a couple that worked at Rutgers, Agnes and Michael Sherman.[7] The Danskins held frequent parties at which the Milnors, Ed Nelson and his wife, and Georg Kreisel, a logician, were also frequent visitors.[8] The parties lasted long into the night, with Beethoven sonatas, a great deal of wine, barbecued steaks and shish kebab, nighttime swims in the canal, and bright conversation led by the convivial, cultivated, mercurial Danskin. Cauvin remembered John Nash very vividly.

> He had a kind of childlike air and disposition, a gentleness, this very vulnerable quality, a kind of helplessness. It blew my mind that someone who gave this appearance of being so simple could be a genius. He was subdued and rather passive. He always spoke very softly and in a monotone. I don't recall him ever initiating a conversation. He would respond to a question or remark after a little momentary hesitation. Alicia was very attentive to him.[9]

Alicia was learning to drive. Danskin and Milnor were both giving her lessons, with haphazard success.[10] They invited her along to a Thursday-night folk dance group at Miss Fines's School on Route 206 that Danskin and Milnor belonged to.[11] "She was very pretty, very quiet. I remember her pulling out a photograph of a cute little boy," said Elvira Leader.[12] Her husband, Sol, danced with Alicia: "She was weightless," he recalled.[13]

Danskin would bring the dancers home afterward. He remembered talking with Nash about mathematics. They'd been drinking by then. Danskin was trying to prove a theorem:

> He immediately hit you with the hardest point. He was still very sharp. He understood what I was doing. I wanted to avoid the hard way and he caught me. Who in the hell would ask that? You would if you were proving it yourself, but he was just listening. And understanding.[14]

Danskin spearheaded an effort to find Nash a job. Danskin was doing some consulting work for Oskar Morgenstern and Morgenstern, it seemed, was willing to hire Nash as a consultant. That fall, Nash was given a one-year consulting contract, with a ceiling of two thousand dollars. Morgenstern indicated to the university that he was making the offer under "a small charitable pressure" but that he felt "Nash could contribute strongly to his program if he was able to pull out of his present mental depression and utilizes his faculties to their greatest extent."[15]

The university balked, "fearing that the appointment might be based on human kindness, rather than on realistic, technical needs."[16] It was decided to review Nash's performance after two months. The contract was dated October 21, 1960.[17]

Nash, however, was talking about returning to France. He contacted Jean Leray, who was visiting at the Institute for Advanced Study, asking Leray to invite him once more to the College de France.[18] This time Alicia, much alarmed, intervened. She asked Donald Spencer — the mathematician at Princeton who had helped Nash work out the final version of his paper on algebraic varieties in 1950 and 1951 — to write to Leray to ask that Leray discourage Nash from going to France again so soon. "Her advice is not to invite John to France at the present time since she feels it will only stir him up again. . . . If this job [with Oskar Morgenstern] materializes it will have a quieting effect on her husband. She feels that remaining in Princeton for a time might possibly bring him back to mathematical work."[19]

By now, Nash had been in the grip of unremitting psychotic illness for nearly two years. It had transformed him. The change in Nash's appearance and manner made it surprising that his old friends from the mathematics department recognized him at all. The man who walked up and down the main street of Princeton in the stifling summer of 1960 was clearly disturbed. He would go into restaurants with bare feet. With dark hair to his shoulders and a bushy black beard, he had a fixed expression, a dead gaze. Women, especially, found him frightening. He looked no one in the eye.

Nash spent most of his time hanging around the university, including Fine Hall. Most days he wore a smocklike Russian peasant garment.[20] He seemed, as one graduate student at the time remembered, to "talk to the squirrels." He carried around a notebook, a scrapbook entitled ABSOLUTE ZERO in which he pasted all sorts of things, presumably a reference to the rock-bottom temperature at which all activity ceases.[21] He was fascinated by bright colors.

He was often in the common room where he "liked to spectate, to watch people playing Kriegspiel, and to make cryptic little remarks."[22] On one occasion, when William Feller was standing nearby, for example, Nash said, to no one in particular: "What would we do with an overweight Hungarian?"[23] On another, "What do Spain and the Sinai have in common?" (This was after Israel's takeover of the Sinai.) He answered his own question, "They both start with *S*."[24]

Everyone around Fine knew who he was, of course. The senior faculty tended to avoid him, and the Fine Hall secretaries were slightly afraid of him, as his size and strange manner gave him a somewhat threatening air. On one occasion, Nash disquieted the formidable Agnes Henry, the departmental secretary, by asking her for the sharpest pair of scissors she possessed.[25] Henry was taken aback and consulted Al Tucker about what to do. Tucker, who was walking with a cane at that time and would hardly have been Nash's match, said, "Well, give it to him and if there's trouble I'll handle it." Nash grabbed the scissors, walked over to a phone book that was lying out, and cut out the cover, a map of the Princeton area in primary colors. He pasted it in his notebook.

He found a few graduate students to talk to. Burton Randol, then a first-year mathematics graduate student, recalled: "I wasn't bothered by his strangeness and I wasn't afraid of him physically. I was willing to have conversations with him. In some sense we enjoyed each other."[26] He and Nash would take long, rambling walks around Princeton, and Randol particularly recalled Nash's wry sense of humor, which he remembered as "intentional, self-referential, and self-deprecating. He knew he was crazy and he made little jokes about it."

He referred to himself, obliquely and usually in the third person, as one Johann von Nassau, a mysterious figure whose name was curiously similar to John von Neumann's and suggested a connection with Nassau Street, the main street of Princeton, as well as Nassau Hall, the main building on the university campus. He talked, in rather lofty terms, of world peace and world government, making it clear he was in touch with these ideas on some very grand scale — though he rarely, if ever, alluded to his actual experiences in Paris and Geneva.

The job with Morgenstern fell through. As Danskin recalled, Nash refused to fill out the necessary W-2 forms, claiming that he was a citizen of Liechtenstein and not subject to taxes.

> I got him a job in the economic research group by calling Oskar Morgenstern. Oskar said fine. I got an application. It called for his social security number and asked whether he was a citizen of the U.S. He wouldn't cooperate, so he didn't get the job.[27]

Whether this was why the contract was canceled in early December, or whether by then it was obvious that Nash was far too sick to work, is unclear.

Nash was also writing all sorts of letters to people. When he heard that Martin Shubik was applying game theory to the theory of money, he sent Shubik a Richie Rich comic book.[28] He sent Paul Zweifel, his friend from Carnegie, postcards in care of the French chargé d'affaires at the French embassy in Washington.[29]

Nash was also making a great many telephone calls, usually, as Martha recalled, using fictitious names. Ed Nelson recalled, "I did my part by talking to John on the telephone during those years.[30] He used to call me a lot." And Armand Borel recalled: "I got unending phone calls from Nash. Harish-Chandra also often got calls. It was unending. It was all nonsense. Numerology. Dates. World affairs. This was really painful. It was very often."[31]

Nash's bizarre behavior was attracting the attention of university officials. Danskin recalled:

He was irritating the president of the university. He was talking about something that was going on in the Gaza Strip. He was playing hopscotch on campus. Goheen's secretary called me. He wasn't threatening anyone, but he was behaving crazily. He would go into the offices. The young women would be frightened. At my house, he'd play with my stereo and screw it up. He frightened people. But he was the gentlest person imaginable.[32]

Alicia was beside herself. She had become quite depressed. Members of the folk-dancing group remember her sad expression, her showing them pictures of her baby, and her sadness at being separated from her son. She began seeing a psychiatrist at the Princeton Hospital, Phillip Ehrlich, who urged her to have her husband hospitalized, against his will if necessary. He recommended a nearby state hospital.[33] Odette recalled, in 1995: "It was awful that such a strong and handsome man should be locked up. Alicia had some guilt trips. We talked it over, back and forth. The doctors advised her. She didn't understand. It was very painful."[34] Alicia had initially asked John Danskin to commit Nash. Danskin refused. She then turned to Virginia and Martha.

A day or two before the police picked Nash up, Nash showed up on campus covered with scratches. "Johann von Nassau has been a bad boy," he said, visibly terrified. "They're going to come and get me now."[35]

40

Tower of Silence

Trenton State Hospital, 1961

Reposing in the midst of the most beautiful scenery in the valley of the Delaware, combining all the influences which human art and skill can command to bless, soothe, and restore the wandering intellects that are gathered in its bosom. — First annual report of the New Jersey State Lunatic Asylum, 1848

I'm as if left to rot in a "Tower of Silence," with anti-Promethean vultures gnawing away at my vitals. — JOHN NASH, 1967

AT THE END OF JANUARY, ten months after Nash's return from Paris, a much-aged Virginia Nash and her daughter Martha boarded a train in Roanoke and traveled north all day, arriving in Princeton in the late afternoon.[1] The last time they had made this trip together was a decade earlier, to attend Johnny's graduation, and the contrast between that trip and the present one was much on their minds. As they disembarked, tearful and weary, John Milnor, now a full professor in the Princeton mathematics department, was waiting for them. It was nearly dark and already snowing lightly. After a few awkward exchanges, Milnor showed them his car, turned over the keys, and gave them directions to West Trenton.

Martha took the wheel and the two women drove in silence down Route 1, the car slipping and sliding on the thin layer of slick ice that now covered the road. They were almost thankful for the distraction. They dreaded what lay ahead. Johnny was already at the Trenton State Hospital. He had been picked up earlier in the day by the police, taken first to Princeton Hospital, a small general hospital, and then transported by ambulance to Trenton State. Now they were going down to talk to the doctors, sign the necessary forms, and, if possible, see Johnny. They would see Alicia, at whose apartment they were staying, afterward.

Full of doubt and self-reproach, they felt they had little choice but to accede to another commitment. Whatever hope they had that Johnny's settling in Princeton, in familiar surroundings and among old mathematical acquaintances, would bring about some improvement in his condition had been shattered weeks before. Alicia's telephone calls had become increasingly frantic. The psychiatrist whom Alicia had been in touch with had tried, without success, to convince Johnny to go into the hospital on his own. Johnny had been dead set against the idea. Finally, the three women had agreed among themselves that there was no other way. He would have to go.

And this time it wouldn't be to a private hospital. As Martha recalled in 1995: "At first, we had thought that thirty days at McLean would straighten him out. By then we knew there were no short-term answers. We were concerned that John's illness would eat into Mother's capital and that she couldn't afford a private hospital."[2]

In the moonlight and freshly fallen snow, the gray stone building, with its white marble dome and tall columns, set atop a gentle wooded slope, looked reassuringly solid and respectable. Institutions like the Trenton State Hospital owed their existence to the same mid-nineteenth-century reform movements that opposed slavery and advanced women's suffrage.[3] Many, in fact, owed their existence to the efforts of Dorothea Dix, a fiery, single-minded Unitarian who made the appalling plight of the insane — condemned to almshouses, prisons, and the streets — her life's crusade.[4] When she was old, ill, and penniless, Dix lived on the ground floor of Trenton's administration building in an apartment set aside for her by the trustees of Trenton State until her death in 1887.

Like all such institutions, Trenton hardly evolved as its founder anticipated. In particular, it was soon overwhelmed by the sheer numbers of people who sought — or whose families sought on their behalf — shelter there. During World War II, Trenton State, long since expanded from a single large building into a large complex, had an average of four thousand patients.[5] The census dropped sharply after the war, but was rising rapidly in the late 1950s. By 1961, there were nearly twenty-five hundred patients, ten times as many as at a private hospital like McLean. Staffing was minimal, and consisted mostly of young foreign residents. The six hundred patients in the so-called West hospital, for example, were cared for by six psychiatrists; the five hundred chronic patients in the annex — predominantly senile or epileptic — were cared for by just one doctor. The presence of a large number of chronic patients obscured the fact that most patients who came to Trenton stayed a relatively short time, perhaps three months.

"You really were not close to patients," said Dr. Peter Baumecker, who worked at both the hospital's insulin unit and the rehabilitation ward during Nash's stay. The poorest and sickest patients wound up at Trenton. "I remember very few patients specifically," Baumecker said. "There was one patient who gouged out the eye of another. There was another patient who'd lost his eye when the police beat him up after he'd killed his father. But that was very exceptional."[6]

"There were good wards and bad wards. Trenton was not as plush as other places. As a matter of fact, Trenton was pretty crummy," recalled Baumecker in 1995. "But I remember a lot of warmth, a lot of caring. We helped an awful lot of people."[7]

Later Nash would recall, with great bitterness, the fact that he was assigned a serial number at Trenton, as if he were an inmate of a prison.[8] To occupy a room shared by thirty or forty others, to be forced to wear clothes that are not your own, to have no place, not even a locker, for your things, even your own soap or shaving cream, is an experience that few people can imagine. Yet this is how Nash—a man who craved, because of his nature and the nature of his illness, solitude and mobility— lived for the next six months, surrounded by strangers. If he had dreaded military duty, what must this have been like for him?

Nash would have been brought to Payton One, the men's admitting ward, on the ground floor of Payton, off to the right of the main administration building. Baumecker was in charge of admissions then and conducted the initial interview. "Nash was my patient," said Baumecker. "He didn't like me because my name started with a 'B.' He had something against the letter B."[9]

The admission interview took place in a small admitting room that had a cot, a couple of chairs, a desk, and a small window. Baumecker asked Nash the usual questions, such as "Do you hear voices?" He tried to find out whether Nash had delusions and whether they were elaborate. He watched his expressions to see whether the emotions he showed were appropriate to what he was saying. The hijacking of a Portuguese ocean liner, the Santa Maria, off Caracas that week— and the subsequent efforts of the hijackers, who turned out to be anti-Salazar rebels, to obtain asylum in Brazil—was, it seemed, very much on Nash's mind; he had his own private theory about it.[10]

The following morning, Nash's "case" was presented to the staff, and he was interviewed in the dormitory before a group of residents. That was when the preliminary diagnosis was reached, treatment was decided upon, and he was as- signed a psychiatrist.

One wound up in Trenton if one had no money or insurance, or was too sick for a private institution to handle. The decision to commit Nash to an overcrowded, underfunded, and understaffed state institution seems puzzling in retrospect. Alicia had at least some insurance coverage through her position at RCA, and Virginia, although by now worried that her son's treatment would eat into her capital, was surely able to pay for some private care. Martha and Virginia certainly had their misgivings: "We went down to talk to them, to beg them to put a red flag on the case and pay special attention to John. It was the only state hospital that John ever stayed in."[11]

John Danskin recalled:

I had heard he was in Trenton. I called his family and said, for God's sake, do something. I drove down to Trenton State. I wanted to find out what the hell happened. I was shocked. It wasn't brutal but he was being treated rather roughly. The attendant kept calling him Johnny.

I told the people there: "This is the legendary John Nash." He was all right too. He gave me no sign at all of being out of his mind. I kept thinking, my God, these shrinks! Who's going to figure out what's wrong with a genius? I resented them.[12]

News that Nash had been committed to a state hospital spread quickly around Princeton. One person deeply disturbed by the notion that a genius like Nash was incarcerated at a state hospital, notorious for its overcrowding and aggressive medical treatments — including drugs, electroshock, and insulin coma therapy — was Robert Winters.[13] Winters, a Harvard-trained economist who happened to be the business manager of the physics department at the time, was friendly with both Al Tucker and Don Spencer. Winters contacted Joseph Tobin, the Institute for Advanced Study's psychiatric consultant and director of the Neuro-Psychiatric Institute in Hopewell, which is a few miles from Princeton, calling him in late January to say, "It is in the national interest that everything possible be done to bring Professor Nash back to his original productive self." [14] Tobin suggested that Winters contact Harold Magee, Trenton's medical director at the time. Winters did so and won an assurance from Magee, as he later wrote to Tobin, that "there would be a thorough study of Dr. Nash's condition before any treatment was started at the state hospital." [15]

In truth, this was too much to expect. As Seymour Krim, a beat writer in New York, wrote in 1959 in his essay "The Insanity Bit" about his own experiences in mental hospitals, that work "in a flip factory is determined by mathematics; you must find the common denominator of categorization and treatment in order to handle the battalions of miscellaneous humanity that are marched past your desk with high trumpets blowing in their minds." [16]

Very soon after that assurance was given, or perhaps even before, Nash was transferred from Payton to Dix One, the insulin unit.[17] Ehrlich, the psychiatrist at Princeton Hospital who had recommended Trenton, was convinced that Nash would benefit from the treatments available at Trenton.[18] Whether Alicia, Virginia, or Martha gave explicit consent for insulin coma therapy is not clear. "I don't remember whether the family had to give further permissions beyond the commitment," Baumecker recalled. "In those days you could do just about anything without asking anybody." [19] Martha recalled that she was consulted: "That was a drastic decision. We were extra wary of anything that might affect his mental abilities. We discussed this with doctors." [20]

The insulin unit was the most elite unit within Trenton State Hospital.[21] The unit had two separate wards—one with twenty-two male beds, the other with twenty-two female beds.[22] Danskin later described it as looking like "the inside of the Lincoln Tunnel."[23] Its chief had the eye and ear of the hospital's directors. It had the most doctors, the best nurses, the nicest furnishings. Only patients who were young and in good health were sent there. Patients on the insulin unit had special diets, special treatment, special recreation. "All the best of what the hospital had to offer was showered on them," said Robert Garber, who was a staff psychiatrist at Trenton in the early 1940s and later president of the American Psychiatric Association. He said, "The insulin patients got a hell of a lot of TLC. In the family's eyes, insulin had great appeal. Patients' relatives were overwhelmed."[24]

For the next six weeks, five days a week, Nash endured the insulin treatments.[25] Very early in the morning, a nurse would wake him and give him an insulin injection. By the time Baumecker got to the ward at eight-thirty, Nash's blood sugar would already have dropped precipitously. He would have been drowsy, hardly aware of his surroundings, perhaps half-delirious and talking to himself. One woman used to yell, "Jump in the lake. Jump in the lake," all the time. By nine-thirty or ten, Nash would be comatose, sinking deeper and deeper into unconsciousness until, at one stage, his body would become as rigid as if it were frozen solid and his fingers would be curled. At that point, a nurse would put a rubber hose through his nose and esophagus and a glucose solution would be administered. Sometimes, if necessary, this would be done intravenously. Then he would wake up, slowly and agonizingly, with nurses hovering over him. By eleven in the morning, Nash would be conscious again. And by the late afternoon, when the whole group would walk over to occupational therapy, he would be among them, the nurses bringing along orange juice in case anyone felt faint.

Very often, during the comatose stage, patients whose blood-sugar levels dropped too far would have spontaneous seizures—thrashing around, biting their tongues. Broken bones were not uncommon. Sometimes patients remained in the coma. "We lost one young man," recalled Baumecker. "We'd all become very alarmed. We'd call in experts and do all kinds of things. Sometimes patients would get very hot and we'd pack them in ice."[26]

Good, firsthand accounts of the experience are difficult to find, in part because the treatment destroys large blocs of recent memory. Nash would later describe insulin therapy as "torture," and he resented it for many years afterward, sometimes giving as a return address on a letter "Insulin Institute."[27] A hint of how unpleasant it was can be gleaned from the account of another patient:

> Breaking through the first sodden layers of consciousness . . . the smell of fresh wool . . . they make me come back every day, day after day, back from the nothingness. The sickness, the taste of blood in my mouth, my tongue is raw. The gag must have slipped today. The foggy pain in my head . . . this was my unbroken routine for three months . . . very little of it is clear in retrospect save the agony of emerging from shock every day.[28]

It's true, as Garber said, that insulin patients were coddled compared to others at Trenton. Insulin patients got richer and more varied food. They got special desserts. They had ice cream every night at bedtime. Most had ground privileges and permission to go out on weekend visits. All the patients gained weight. That was considered a good sign. The doctors on the ward were proud that their patients were in good physical health. "People would put on a lot of weight because of the insulin," recalled Baumecker. "The low blood sugar would make it necessary to give them a lot of sugar and the sugar had a lot of calories. For some of these spindly, skinny schizophrenics it wasn't such a bad thing."[29] But patients often hated it. Nash's subsequent obsession with his diet and weight may well have stemmed from this experience of being "force-fed."

Treating schizophrenic patients with insulin coma was the idea of Manfred Sackel, a Viennese physician who thought of it during the 1920s and used it on psychotic patients, especially ones with schizophrenia, in the mid-1930s.[30] His notion was that if the brain were deprived of sugar, which is what keeps it going, the cells that were functioning marginally would die. It would be like radiation treatments for cancer. Some practitioners who used it in the 1950s, when the first effective antipsychotic drugs became available, took the view that insulin shock was more effective than antipsychotics, especially with regard to delusional thinking.[31] No one understood the mechanism, but two large-scale studies in the late 1930s found that insulin-treated patients had better and more lasting outcomes than untreated individuals, but evidence for insulin's efficacy was hardly overwhelming.[32]

It was in any case riskier and far more involved than electroshock, and by 1960, insulin shock therapy had been phased out by most hospitals as too dangerous and expensive when compared with electroshock. The conclusion was that insulin wasn't worth the investment of time and money or the risks.

The treatments produced at least temporary improvement in many patients, according to Garber:

They'd see everybody hovering over them, very concerned about them, a feeling of loving camaraderie. I always thought that was very therapeutic. For the first time, somebody cared. Patients became more outgoing, more active. They got to go out on weekend visits. They got ground privileges. I think it helped. Patients were brighter, more alert, more conversational.[33]

While Nash later blamed the treatments for large gaps in his memory,[34] he also told his cousin Richard Nash, whom he visited in San Francisco in 1967, that "I didn't get better until the money ran out and I went to a public hospital."[35]

• • •

As dangerous and agonizing as it was, insulin was one of the few treatments available for serious illnesses like schizophrenia which, until the middle of the century, often meant lifelong incarceration. And, like other state hospitals, Trenton was a laboratory for every "cure" that came along. Before the war, Garber recounted:

> [We] treated all patients with the tools that were available. Colonic irrigation was still used. So was fever therapy. We had a strain of malaria that we would inoculate patients with. Later on we used a typhoid strain. We'd inject a typhoid vaccine and within hours patients would experience nausea, vomiting, diarrhea and fevers of 104 to 105. We'd do that for eight or ten weeks, two or three days a week. We did it to take the starch out of disturbed patients.
>
> At Trenton the first order of the day, when I arrived at the hospital supervisor's office at 8 A.M. was to see who could be moved out of seclusion to make room for another eight to fifteen patients who needed to be secluded. [The rooms] were ten by twelve, lined with glazed tiles, with terrazzo floors. There was a toilet and a sink and a drain in the middle of the floor so that if a patient, say, smeared feces around the room, we could hose it down.
>
> You would do anything to give yourself a handle to bring the patient under control.[36]

After six weeks, Nash, whose insulin treatments were judged to be effective, was transferred to Ward Six, the so-called rehab or parole ward.[37] There was group therapy every day, some recreation, and occupational therapy. "This was the cream of the patient crop," Baumecker recalled. "There were only about fifteen beds. Other wards had thirty patients per room. Patients got individual attention, went on trips, and were allowed to go home on visits."[38]

Nash actually began to work on a paper on fluid dynamics while he was on Ward Six. Baumecker recalled, "The patients made fun of him because he was always so up in the clouds. 'Professor,' one of them said on one occasion, 'let me show you how one uses a broom.'"[39] Alicia visited Nash every week. Once he was allowed out on passes, she took him to her folk-dancing group and out to Swift's Colonial Diner.[40] It was the highlight of Nash's week.

He seemed to be in remission, clearly no longer a threat to himself or others. Baumecker recommended him for discharge, pointing out that, contrary to the popular belief, "We had to discharge people as fast as we could to get the census down."[41] He was discharged on July 15, a month after his thirty-third birthday.[42] A few months after Nash got out, Baumecker called the Institute for Advanced Study and asked to speak to Oppenheimer about whether Nash was now sane. Oppenheimer replied, "That's something no one on earth can tell you, doctor."[43]

41

An Interlude of Enforced Rationality

July 1961–April 1963

When I had been long enough hospitalized . . . I would finally renounce my delusional hypotheses and revert to thinking of myself as a human of more conventional circumstances. — JOHN NASH, Nobel autobiography, 1995

A MAN EXPERIENCING a remission of a physical illness may feel a renewed sense of vitality and delight in resuming his old activities. But someone who has spent months and years feeling privy to cosmic, even divine, insights, and now feels such insights are no longer his to enjoy, is bound to have a very different reaction. For Nash, the recovery of his everyday rational thought processes produced a sense of diminution and loss. The growing relevance and clarity of his thinking, which his doctor, wife, and colleagues hailed as an improvement, struck him as a deterioration. In his autobiographical essay, written after he won the Nobel, Nash writes that "rational thought imposes a limit on a person's concept of his relation to the cosmos."[1] He refers to remissions not as joyful returns to a healthy state but as "interludes, as it were, of enforced rationality." His regretful tone brings to mind the words of Lawrence, a young man with schizophrenia, who invented a theory of "psychomathematics" and told Rutgers psychologist Louis Sass: "People kept thinking I was regaining my brilliance, but what I was really doing was retreating to simpler and simpler levels of thought."[2]

It is possible, naturally, that Nash's feeling reflected an actual dulling of his cognitive capacities relative not just to his exalted states, but to his abilities before the onset of his psychosis.[3] The consciousness of how much his circumstances in life, not to mention his prospects, were altered compounded his distress. At thirty-three, he was out of work, branded as a former mental patient, and dependent on the kindness of former colleagues. Excerpts from a letter to Donald Spencer written around the time of Nash's release from Trenton on July 15 suggest how modest Nash's view of reality had become:

In my situation and anticipated situation a fellowship . . . with the idea being that I am expected to be doing research work and studies, etc. seems a better prospect . . . than a standard academic teaching position. For one thing, much of the conceivable worry over . . . the implications of my having been in a state mental hospital would be thereby by-passed.[4]

With the help of Spencer, who was on the Princeton faculty, and several members of the permanent mathematics faculty at the Institute for Advanced Study — Armand Borel, Atle Selberg, Marston Morse, and Deane Montgomery — a one-year research appointment at the institute was arranged.[5] Oppenheimer found six thousand dollars of National Science Foundation money to support Nash.[6] Nash's application, dated July 19, 1961, stated that he wished to "continue the study of partial differential equations" and mentioned "other research interests, some related to my earlier work," as well.[7]

In late July, Alicia's mother brought John Charles, a big, handsome two-year-old, to Princeton. Nash called the reunion "a big occasion for me since I haven't seen our little boy all during 1961!"[8] Then, at the beginning of August, Nash attended a mathematics conference in Colorado where he ran into a number of old acquaintances and went on a day-long excursion with Spencer, an enthusiastic mountaineer, to climb Pike's Peak.[9]

Nash and Alicia were living together once more, but not especially happily. The turbulence of the two previous years had produced an accumulation of hurts and resentments, and the resulting coldness lingered and was exacerbated by new conflicts over money, childrearing, and other issues of daily living. None of this was made easier by the fact that Nash's in-laws now lived with them. Carlos Larde's health had deteriorated markedly, and he and his wife Alicia moved to Princeton that fall. The two couples shared a house at 137 Spruce Street.[10] It was a great help that Mrs. Larde cared for Johnny while Alicia went to work, but living together created another layer of strain, especially for Alicia.

They tried to make the best of it. Nash attempted to care for his son, picking him up at nursery school and the like. They socialized with the Nelsons, the Milnors, and a few others. Once or twice, they drove up to Massachusetts to visit John and Odette Danskin, who had moved there the previous fall, and to see John Stier.[11] The visits were rather fraught and Eleanor used to call John Danskin afterward to complain about Nash. On one visit, apparently, Nash had come with a bag of doughnuts. "Eleanor kept saying, 'How cheap!' " Odette recalled.[12]

In early October, Nash attended a most historic conference in Princeton.[13] The conference, organized by Oskar Morgenstern, and attended by virtually the entire game-theory community, amounted to a celebration of cooperative theory. There was little mention of noncooperative games or bargaining. But John Harsanyi, a Hungarian, Reinhard Selten, a German, and John Nash, dressed in odd mismatched clothing, mostly silent, were all there.[14] This was the first time these three men had met, and they would not meet again until they traveled to Stockholm a quarter of a century later to accept Nobel Prizes. Harsanyi remembers asking one of the Princeton people why Nash said so little during the sessions. The answer, Harsanyi recalled, in a conversation in Jerusalem in 1995, was "He was afraid he would say something strange and humiliate himself."[15]

Nash was able to work again, something he had not been able to do for nearly three years. He turned once more to the mathematical analysis of the motion of fluids and certain types of nonlinear partial differential equations that can be used as models for such flows. He finished his paper on fluid dynamics, begun while he was in Trenton State hospital.[16] It was titled "Le Problème de Cauchy Pour Les Equations Differentielles d'une Fluide Générale" and published in 1962 in a French mathematical journal.[17] The paper, which Nash and others have described as "quite a respectable piece of work"[18] and which the *Encyclopedic Dictionary of Mathematics* called "basic and noteworthy," eventually inspired a good deal of subsequent work on the so-called "Cauchy problem for the general Navier-Stokes equations." In the paper, Nash was able to prove the existence of unique regular solutions in local time.[19]

"After Nash's hospitalization he came out and seemed OK," Atle Selberg recalled. "It was good for him to be at the IAS. Not everybody on the Princeton faculty was very friendly. It's true that he didn't speak. He wrote everything on blackboards. He was perfectly articulate in writing. He gave a lecture on Navier-Stokes equations — which concern hydrodynamics and partial differential equations — something I don't know much about. He seemed fairly normal for a while."[20]

He was most at ease in one-on-one encounters where his sense of humor came to his aid. Gillian Richardson, who was on the staff of the institute's computer center from 1959 to 1962, recalled eating lunch with Nash in the institute dining hall and Nash's saying all sorts of dry, wry things about psychiatrists. One time he asked, "Do you know a good psychiatrist in Princeton?" — adding that his own psychiatrist " 'sat on a throne way above' him, and he wondered if I knew one who didn't share that peculiarity."[21]

Nash showed up in French 105, the third-semester French course at the university, one day and asked Karl Uitti if he could audit it. He struck the French professor as "the typically dreamy and out-to-lunch mathematician."[22] Nash attended quite regularly and kept up with the work. He seemed less interested in picking up conversational "tourist French" than in acquiring "a sense of French structure," Uitti recalled, adding, "He was quite pro-French. He liked the language and the people."

Uitti and Nash became rather friendly and met outside class, and on a number of occasions with Alicia. At some point, Uitti asked Nash why he was learning French. Nash answered that he was writing a mathematical paper. "There was only one person in the world who would be able to understand it and that person was French. He wanted, therefore, to write the paper in French," Uitti said. Uitti could not recall Nash's intended audience; chances are it was either Leray, who was at the institute that year, or Grothendieck. After the paper was published, Nash gave it to another member of the Institute to read. The next time he saw the man, Nash asked him, "Did you detect the sexual overtones?"[23] Uitti commented in 1997:

That was the time that de Gaulle was in power and strong pressure was being exerted on French scientists to deliver their papers in French. Nash always struck me as very well-bred, very courteous. I'm certain that there was in his mind a sense of respect for whomever he was writing the paper for. It was sweet of him and I liked him for it.[24]

Nash asked Jean-Pierre Cauvin to edit a draft of the paper.[25] Cauvin, who was doing quite a bit of translation work at the time, recalled Nash's telling him that "Paris was the center for this kind of mathematics." Nash also turned to a French undergraduate, Hubert Goldschmidt, for help.[26]

Nash had not given up the idea of returning to France. He submitted the Cauchy paper to the *Bulletin de la Société Mathématique de France* on January 19. He was, Cauvin thought, more withdrawn and subdued than ever, and in retrospect it is clear that he was thinking a great deal about leaving Princeton. Very likely, he got in touch with Grothendieck at the Institut des Hautes Études Scientifiques. In April Oppenheimer wrote to Leon Motchane, director of the IHES, to ask Motchane to formally invite Nash to spend the first half of the academic year 1963–64 there.[27] Oppenheimer also asked Leray, who was at the institute that year, to see if he could provide a grant from the Centre de la Recherches Nationale Scientifiques for the second half of the year.[28] At the same time, he noted that Nash would have been welcome to continue at the Institute for a second year: "If [Nash] asked to stay here for the autumn, I think that my colleagues would probably accede; but that is not his choice."

Nash did not suggest that Alicia go with him to France, and this time Alicia did not try to dissuade him. Nor did she offer to go. It was clear that, by some mutual and unspoken agreement, the marriage was over and they were going to go their separate ways.

That winter, Nash spent more and more time in the Fine Hall common room, usually showing up at teatime and staying until evening. "He wore baggy, rumpled clothes," Stefan Burr, then a graduate student, recalled. "He didn't seem at all aggressive. In some ways his manner was not that different from a lot of mathematicians'."[29] For a while, Burr and Nash were playing endless games of Hex. The board in Fine had been drawn years before on heavy cardboard and was so worn that the lines had constantly to be redrawn with a ballpoint pen.

He was beginning to seem less well again. Borel recalled, "He was not quite right. He seemed to me very diminished. His mathematics was not at the same level. I found him odd, unpredictable, nonsensical. It was very painful. The secretaries were afraid of him. He was someone to avoid. You never knew what he would do or say."[30]

One time the Borels had Alicia and Nash over for tea. "We served tea and cookies," said Borel. "Nash went into the kitchen. I followed him. 'What do you want?' I asked. 'Well, I'd like some salt and pepper.' "[31] Gaby Borel added: "After he put salt and pepper in his tea, he complained that the tea tasted awful."[32]

During the spring, his state of mind had become more angry and restless, and he was beginning again to harp on his old obsessions. He decided, rather suddenly, to travel to the West Coast, where he saw, among others, Al Vasquez, who had graduated from MIT and was now a graduate student at Berkeley, Lloyd Shapley, and Al Tucker's former wife, Alice Beckenback, and her new husband. Vasquez recalled:

> I just walked into the common room [at Berkeley] and he was there. He was as surprised to see me as I was to see him. He didn't announce his visits in advance. I had no idea where he was staying. But he was around for more than just a day or two. He hadn't been looking for me. I had the impression that he'd been in Europe, the East Coast, and that he was traveling around. He talked a lot. He quite explicitly talked about [insulin] shock therapy. He described shock therapy as extremely painful. He also said he was taken back from Europe on a ship and in chains. Slavery was a word he used a lot. He was very bitter about his experiences.
>
> He was pretty disoriented. He wasn't able to talk about anything else but his obsessions. I was put off. It was odd. I never did understand why he talked to me. He knew me. He wasn't really trying to communicate. He wanted to talk elusively. [Yet] it wasn't gibberish. It was even clever at times, full of puns and allusions.[33]

Shapley, to whom Nash had written a great many letters, also found Nash's appearance in Santa Monica distressing. "He thought of me as a close friend. One had to put up with it. He would send me postcards in colored inks. It was very sad. They were scribbled with math and numerology, as if he were not expecting a reply. I was much on his mind. He had decayed in a very spectacular way," Shapley recalled in 1994. "He was groping."[34] Shapley remembered Nash telling him, "I have this problem. I think I can straighten it out if I can figure out which members of the Math Society did this to me." He didn't stay long, Shapley said, adding:

> It was a bit frightening. We had two young children. What was clear was that there was no way to talk to him or even follow what he was saying. He'd switch from topic to topic. It's very hard to be a good mathematician if you can't hold a thought in your mind.[35]

In June, Nash left for Europe. He was due to attend a conference in Paris in the last week in June and the World Mathematical Congress in Stockholm in early August. He went to London first, where he stayed at the Hotel Russell in Bloomsbury, which he described as "very grand."[36]

He got himself a private postal box and was once again writing letters, some on toilet paper, in green ink, in French. He was also sending drawings, including one of a prostrate figure pierced with arrows. One, postmarked June 14, contained a scrap of paper with the following written on it in green ink: $2 + 5 + 20 + 8 + 12 + 15 + 18 + 15 + 13 = 78$.

The conference at the College de France in Paris was a small and intimate affair, very much dominated by Leray, who was very excited at that time about nonlinear hyperbolic equations. Ed Nelson, who had become quite friendly with Nash over the academic year, recalled Leray's saying that it was a scandal that there were no global existence theorems. "The feeling he conveyed," Nelson said, "was that we had better get to work, or the world might come to an end at any moment."[37] Most of the speakers gave their talks in English. Lars Hörmander, who was also there, recalled that "1962 was very different from earlier visits."[38] But Nash insisted on giving his lecture in what he called his "pidgin French."[39] He did not speak extemporaneously but read from his notes in his very soft voice and with his very strong American accent. Hörmander recalled: "Nash's paper was respectable mathematically. It was a surprise to all of us [that he could have produced it at all]. For us it was like seeing somebody rise from the grave."[40]

His behavior, however, was decidedly odd, Hörmander later said:

> Malgrange, the official conference organizer, had a dinner for the participants. At the table, Nash exchanged his plate with the person next to him. Then he traded yet again until he was satisfied that his food wasn't poisoned. Everybody was very aware of his bizarre behavior but nobody said a word.
> Malgrange had bought a nice big jar of caviar which was being passed around. When the jar came to Nash, he tipped the entire thing upside down onto his plate. Everybody was very well-behaved and said nothing.[41]

• • •

While Nash was still in Paris, on July 2, his father-in-law died suddenly.[42] Alicia attempted, through Milnor and Danskin, to contact Nash but was not successful. Carlos Larde was buried in the churchyard of St. Paul's on Nassau Street.

Nash, meanwhile, went back to London. What drew him to London is not clear, since his original plan had been, presumably, to spend the summer, except for the congress in Stockholm, as well as the following academic year, in Paris. In any event, Nash was still in London on July 24 when he wrote to Martha from the Hotel Stefan on Talbot Square.[43] He apparently still intended to travel on to Stockholm. Addressing her as E-me-line, Martha's middle name, he wrote that he was merely passing the time, with little to do, until the mathematical congress in Stockholm and was considering seeing a psychologist or visiting some sort of clinic.

Danskin recalled that someone went looking for Nash and finally found him hanging around the Chinese embassy in London.[44] The head of the MIT economics department took a group of business management people to London that summer. He suddenly saw John Nash and asked him, "Where are you now?" Puzzled, Nash replied, "Where are you?"[45]

The International Mathematical Congress took place in the third week of August in Stockholm.[46] Among the plenary speakers were Armand Borel, John Milnor, and Louis Nirenberg. The Fields Medals were awarded to Milnor and Lars Hörmander, both of whom had been notified in May and instructed to tell no one, leaving each to sit on his secret while others around them speculated on the year's likely winners.

Nash, who felt that he should have been one of those honored, did not, however, go to Stockholm. He went to Geneva instead, returning to the Hotel Alba where he had spent his final week in December 1959 and writing in French to Martha "chez Charles L. Legg."[47] The letter made it clear that he was again thinking about the question of his identity! He drew an identity card with Chinese characters labeled "Des Secrets." He wrote "Could you sign this carte d'identité . . . a man all alone in a strange world," he wrote underneath. He sent Virginia another postcard with a picture of Geneva but mailed it from Paris.

When Nash returned to Princeton at the end of summer 1962, he was extremely ill. A postcard addressed to Mao Tse-tung c/o Fine Hall, Princeton, New Jersey, arrived in the mathematics department. Nash had written only a cryptic remark in French about triple tangent planes.[48]

Alicia let him move back in. He spent much of the fall at home with John Charles watching science-fiction programs on television, like Rod Serling's *Twilight Zone*.[49] He was writing a great many letters and making many phone calls to mathematicians in Princeton and elsewhere.

He was still obsessed with the idea of asylum. A letter to Martha and Charlie, postmarked November 19, reads: "Maybe you will say that I'm mad . . . request to St. Paul's in Princeton for sanctuary."[50] Nash apparently walked past St. Paul's every day. The letter referred to the Ecumenical Council and previous letters he had written to the pastor of St. Paul's earlier in the month. The letter ended with a reference to "past misfortunes, especially in the fall season." In contrast to his letter to Martha from London, Nash no longer interpreted his difficulties as a sign of illness but rather as the results of machinations by the Ecumenical Council. By January, his letters to Martha and Charlie had become nearly incomprehensible, the thoughts skipping from Albanians to Stalin to "secrets can't reveal" and "wood and nails of the true cross."[51]

Exhausted and dispirited by three years of turmoil and convinced that Nash's condition was more or less hopeless, Alicia consulted an attorney and instituted divorce proceedings. She had married someone who she thought could look after her but couldn't, who resented her bitterly, and who accused her of having malevolent intentions. To Martha and Virginia she wrote that being married was helping to create Nash's problems and that she felt that being freed from the marriage would be better for him as well.[52]

Alicia's attorney, Frank L. Scott, a genial Princeton divorce lawyer with an office on Nassau Street, filed for a divorce the day after Christmas 1962.[53] Alicia had given the formal go-ahead in a deposition a week earlier. According to the petition, Nash was still living with her at 137 Spruce Street. Alicia, meanwhile, temporarily rented a separate apartment on Vandeventer Street.[54]

Alicia's formal complaint read:

> On or about March 1959 it was necessary for the Plaintiff herein to cause the defendant to be committed to a mental institution from which the defendant was released on or about June 1959. Despite the fact that said committal was in the best interest of the defendant, the defendant became very resentful of the Plaintiff for causing his commitment, and declared he would no longer live with the Plaintiff as man and wife. Consistent with the defendant's vow not to again live with the plaintiff as her husband, the defendant did in fact move into a separate room and refused to have marital relations with the plaintiff. In January 1961 defendant was caused to be committed to Trenton State Hospital by his mother from which he was released in June 1961. The defendant's resentment of his wife and insistence that they no longer have marital relations continued after his release from the aforementioned commitment, as it had prior to said commitment, and has continued against the wishes of the plaintiff to the present date. The time during which defendant has thus deserted plaintiff and during which defendant was not confined to any institution but fully able to voluntarily resume marital relations, which he has not done, exceeds two years past and such desertion has been wilful, continuous and obstinate. Moreover defendant has failed to properly support plaintiff.[55]

Nash was served with a summons. Scott visited Nash the following day. On April 17, Scott once again talked to Nash, who, he said, had "no plans for changing either his residence or his occupational status." The judgment was rendered without a trial, granting a divorce and awarding Alicia custody of John Charles on May 1, 1963.[56] Final judgment was rendered August 2, 1963.[57]

There is no evidence that Nash was opposed to the divorce. While the petition was a lawyer's document and not necessarily true in its particulars — the Danskins, for example, maintained that Nash and Alicia never stopped sleeping together — Nash's animosity toward Alicia was no doubt very real. He blamed Alicia for engineering his hospitalizations, he had threatened to divorce her while at McLean, and probably afterward as well, and he had made plans to live in France without her.

Nash's increasingly disturbed state, and rumors of his impending divorce, prompted a number of mathematicians to rally around him that spring. That Nash desperately needed treatment was not a subject of controversy this time. Once again, Donald Spencer and Albert Tucker approached Robert Winters.[58] James Miller, a friend of Winters from Harvard, was in the psychiatry department at the University of Michigan and was connected with a university-sponsored clinic run by Ray Waggoner.[59] Through Miller, Winters succeeded in making a unique arrangement whereby Nash would be treated at the clinic and also have an opportunity to work as a statistician in the clinic's research program.

Tucker at Princeton and Martin at MIT decided to set up a fund to make the Michigan plan feasible.[60] Anatole Rappaport and Merrill Flood at the University of Michigan, Jürgen Moser at NYU, Alexander Ostrowski of Westinghouse, and others committed themselves to raise funds among mathematicians on Nash's behalf.[61]

The Ann Arbor group felt that a stay of two years was necessary. The cost for out-of-state patients was $9,000 a year or $18,000 for the entire stay. Virginia Nash offered to guarantee $10,000 and the group of mathematicians arranged, through the American Mathematical Society, to set up a fund-raising drive for the remaining $8,000. "If we are successful probably most of it will have to come from mathematicians who have known Nash," Martin wrote. "If anything can be done which will enable Nash to return to mathematics, even on a very limited scale, it would of course be very fine not only for him but also for mathematics."[62]

Albert E. Meder, Jr., the society's treasurer, was enthusiastic about the proposal, saying that "it would seem to me that it would be altogether appropriate for the AMS to receive contributions for the purposes set forth in [Martin's] letter of March 25. . . . I would be inclined to go ahead."[63]

Nash's increasingly bizarre behavior was triggering complaints, including some at the Institute for Advanced Study. Mostly these had to do with Nash's writing mysterious messages on the institute blackboards and making annoying telephone calls to various members. But one day the switchboard operators, who sat in an office immediately as one entered Fuld Hall, were all abuzz because each person who was coming through the door was being doused with water. The institute's dining hall was then on the fourth floor of Fuld, and it turned out, upon investigation, that Nash had been pouring water from the window above the main door.[64]

It was Donald Spencer, a man who could not stand to see anyone in trouble without intervening, who was elected to try to convince Nash to accept the Michigan offer and enter the clinic voluntarily.[65] Spencer chose, as he usually did, a bar as his venue. He invited Nash for some beers in Nassau Tavern, where Nash had once celebrated passing his generals. They sat in the booth for hours, Spencer downing warm martinis, Nash nursing a single beer. Spencer talked and talked; Nash appeared to be listening but said very little except to remark, at various intervals, that he wasn't interested in doing statistical work. It was no use. Nash didn't believe that he was ill, and he wasn't prepared to enter another hospital.

Years later, Winters wept when he recounted the story:

> I thought I had worked out a perfect solution to a most unusual problem. I thought I could save a very worthwhile person. I'm very emotionally tied to this. I thought I was doing something really wonderful. Jim Miller told me *never* let Nash get shock treatments. It takes the edge of genius off. Somebody sent him to Carrier, where they gave him shock treatments [sic], and I think it turned him into a zombie for many years. I consider that one of the worst failures of my life. When I look at the human race all over the world I think there's zero reason for humanity to survive. We're destructive, uncaring, thoughtless, greedy, power hungry. But when I look at a few individuals, there seems every reason for humanity to survive. He was worth doing the very best for.[66]

Meanwhile, Alicia, Virginia, and Martha had agreed among themselves that Nash would have to be committed involuntarily. This time they chose a private clinic near Princeton. Martha wrote to Spencer:

> The only reason it has not been done before now is that my mother and I are waiting to hear from Alicia when she has arrangements made. . . . We really had thought we would do this in March.
>
> We were very hopeful that we could persuade John to go to the University of Michigan and take advantage of the opportunities for research and treatment there. Unfortunately John will not agree that he needs treatment. Since we feel that something must be done for him, we have placed him in Carrier. . . .
>
> He was simply not going to enter ANY hospital voluntarily. Once we were convinced of this we had no choice but to commit him to a hospital in New Jersey.[67]

42

The "Blowing Up" Problem

Princeton and Carrier Clinic, 1963–65

T HE CARRIER CLINIC, formerly a sanatorium for the senile and retarded, was one of only two private mental hospitals in New Jersey. Located in the picturesque hamlet of Belle Meade, amidst rolling hills and lush farmland, Carrier was just five miles north of Princeton. Despite its easy proximity, however, it was generally avoided by Princetonians. As Robert Garber, a former president of the American Psychiatric Association who was Carrier's medical director at the time, recalled: "They didn't want to be in a psychiatric facility close to home. It was a disgrace, a terrible stigma, nothing like today. The idea was to get as far away as possible." [1]

Princetonians regarded Carrier, which had the look of a slightly seedy boarding school, with some distaste for another reason as well. Carrier had none of the prestige of top-of-the-line institutions like McLean, Austin Riggs, or Chestnut Lodge, whose academic affiliations, psychoanalytical orientation, and long-term approaches based on the "talking cure" were regarded, especially by academics, as more humane and appropriate, especially for the well-educated. Popular views of psychiatry were being shaped by *One Flew Over the Cuckoo's Nest, I Never Promised You a Rose Garden,* and the libertarian views of Thomas Szasz, who held that insanity was a social construct rather than a symptom of disease.[2] At the time when these views were gaining popularity, especially on campuses, Carrier had a reputation for the aggressive use of "chemical straitjackets" and electroshock, and short-term cookie-cutter approaches tailored to the time limits set by insurance policies.

The Carrier staff, well aware of such attitudes, defended itself by arguing that its approach was more practical and worked better. "McLean, Austin Riggs, Chestnut Lodge, Shepherd Pratt, and Institute for Living, these were all much fancier," said William Otis, a psychiatrist on Carrier's staff. "We were very clinical. None of us had any fancy training. None of us were stars. But the ironic thing is that if you were sick you were much better off at Carrier."[3] Garber said: "At Carrier we were proud of the fact that we set ourselves up as a short-term treatment center. That's why we were so successful. We were able to treat the patients and get them out, in contrast to McLean and Chestnut Lodge, which were notorious for having schizophrenic patients there for four, five, and seven years."[4]

It was Alicia who, despite the impending divorce, felt responsible for Nash, and therefore had to face the decision.[5] It took a great deal of courage, as anyone who has had to make such a decision knows. As one psychiatrist at Carrier said, "Commitments always created terrible conflicts in the family. It was very hard to find somebody who wanted to take the responsibility."[6] Alicia, like everyone else around Nash, abhorred the idea of involuntary commitment and feared that treatment, besides being uncertain of success, carried the risks of irreparable harm. But she also knew that Nash was on a disastrous course and was convinced that failure to act would almost certainly lead to further deterioration. The psychoanalysts at McLean had failed, the effects of the shock treatments at Trenton had proved short-lived. She was prepared to try something new. She recognized that the most prestigious hospitals were unaffordable. At Carrier, patients' families paid a flat fee of eighty dollars a day plus hourly fees for group and individual therapy; Virginia was able to pay that. Besides, it was important to Alicia that Nash be close by, so that she and his old acquaintances at Princeton could visit him.

So in the third week of April, after it had become all too clear that Nash was unprepared to enter treatment at Michigan, she went ahead with arrangements to have Nash taken to Carrier. Once again, she asked Martha and Virginia to come up to Princeton and sign the commitment papers.

From the outset, however, Alicia drew the line at electroshock.[7] "We debated electroshock therapy," Martha recalled. "But we didn't want to mess with his memory."[8]

At Carrier, electroshock was frequently used for schizophrenic patients, who generally got three times as many treatments — twenty-five versus eight — as patients suffering from depression.[9] Garber said, "What we were trying to do was to gain control of that patient — to break through his excitement, panic, depression — in the shortest possible time."[10] Generally, psychotic patients were initially treated with Thorazine, and those whose disturbed behavior didn't improve quickly were also treated with electroshock. Some of the psychiatrists at Carrier felt that the shock treatments were effective and produced fewer side effects than neuroleptic drugs. In any case, despite the nearly universal belief around Princeton that Nash received electroshock treatments at Carrier, he apparently did not.

Nash spent most of the next five months of 1963 in Kindred One, the only locked ward at Carrier. He said later that he made efforts to overturn his commitment; if so, they were not successful. Frank L. Scott recalled that Nash went AWOL from Carrier at least once — presumably after he got ground privileges — and that he had to track him down and return him to the hospital.[11]

Compared to Trenton, however, Carrier was, if no country club, at least more like a reform school than a prison. There were just eighty patients, the majority of whom came from comfortable middle-class homes, many from New York and

Philadelphia, and most of whom suffered from alcoholism, drug addiction, and depression rather than from psychotic illnesses.[12] Carrier had a dozen psychiatrists on its staff, a more adequate nursing staff than at Trenton, and a reasonable complement of medical doctors, psychologists, and social workers.

Kindred One had single and double rooms. Nash, it seems, had a room to himself. He had access to a telephone. He was allowed to wear his own clothing. Patients were addressed by their titles and last names, so he was Dr. Nash, not Johnny as he was at Trenton. Nash's wishes regarding his vegetarianism—which "doesn't exclude animal products, for example, milk, but only the animal products which become available only at the death (execution of the animal)"—were apparently respected.[13] Alicia visited regularly, as did a number of others from Princeton, among them Spencer, Tucker, and the Borels.[14]

Probably the best thing that happened to Nash at Carrier was that he met a psychiatrist, Howard S. Mele, who was to play an important and positive role in his life for the next two years.[15] The psychiatrist, who happened to be on duty the night that Nash was brought to Carrier, was assigned to care for him. A short, soft-spoken, dapper man of Italian descent who got his medical degree at Long Island College of Medicine and did his residency at Mt. Sinai Hospital in New York City, Mele was quiet and careful.[16] Described by his former colleagues as "conventional," "cautious," "not an exciting man," Mele was, as later events showed, competent and caring.[17] He was respected by the nursing staff. Belle Parmet, the institute's social worker at the time, said of Mele and the other staff psychiatrists: "They weren't just pill pushers or prescription writers. They were all humanistic."[18]

Nash responded quite quickly to his initial treatment with Thorazine. If someone responds at all to what are now called "typical" neuroleptics, dramatic changes are usually evident within a week, and the full effect becomes apparent within six weeks. Two weeks after his commitment, Nash wrote a relatively lucid letter to Norbert Wiener, saying, among other things, "My problems seem to be essentially problems of communications. I don't know how they can be resolved. Perhaps I shall be able to approach their solution as a result of begging for aid. (However, this isn't a begging letter!)"[19]

At this point, Nash was seeing Mele for therapy sessions and also participating in group therapy, which Mele particularly favored.[20] There was, however, no thought of releasing him quickly. As Garber said, "Paranoid schizophrenics are not that responsive. Once you do get them under control, you have to satisfy yourself that they've stabilized. You don't want a relapse, especially if there's been a commitment because then you and the family would have to start all over."

By August, Nash was beginning to look forward to getting out of Carrier. He wrote to Virginia that he was anticipating Alicia's visit on the weekend and was "thinking of getting out."[21] He added that "Mele thinks it depends on having a job." Nash admitted that he was ill and in need of treatment but said that "Michigan might have been a better deal." He asked Milnor for help in getting a job. On September 24, Nash wrote again saying that Sunday was "a sad day" because Alicia had to work overtime and couldn't come to take him out. He said that the Institute for Advanced Study had decided to offer him a position.[22] A week later, upbeat again, he wrote that he was thinking of buying a car and that there were "good propects for a reconciliation" with Alicia.[23]

It is a discouraging but well-documented fact that people who suffer from schizophrenia face an extremely high risk of suicide, comparable to those who suffer from severe depressions and one hundred times that of the general population.[24] This risk is greatest not when the person is sickest, but shortly after a course of treatment has been declared a success. Though no one else can truly know the state of mind that leads someone to take his life, one can imagine that this is a time when the absence of delusions allows other feelings, including very painful ones, to emerge and that hopes that one has been nurturing for months collide with harsh reality.

Louisa Cauvin, who married Jean-Pierre Cauvin in the summer of 1963, has a haunting memory, which likely dates from that summer, the only time she ever talked with Nash.[25] They met at a party. (Presumably he was home from Carrier on a pass.) Nash told Louisa that he didn't feel life was worth living and saw no reason why he should not do away with himself. There is no evidence to show that Nash ever came close to acting on this thought. But he was certainly depressed. His hope for a reconciliation with Alicia, for example, proved overly optimistic. Alicia insisted that Nash live apart from her and Johnny (as John Charles was now called), so, instead of moving back to Spruce Street, Nash found himself in a rented room at 142 Mercer Street, a few doors down from the house occupied by Einstein during his Princeton years.

Once again, Borel and Selberg had arranged a one-year membership at the Institute for Advanced Study, although this time they did so with less hope.[26] The 1963–64 membership was probably a rescue mission. Borel later said, "All members are voted by the whole school of people. I did the legwork. It was only to present the case to my colleagues."[27] Oppenheimer decided this time to use the Institute's own funds, saying in a note to Selberg, "This enterprise seems to me not too suitable for contract funds," implying that, in contrast to the previous 1961–62 appointment, this one was more clearly a charitable exercise.[28]

Meanwhile, Nash's old friends outside Princeton had not lost interest in his progress. A letter from David Gale to Deane Montgomery at the Institute, with copies to Milnor and Morgenstern, gives a flavor of the level of interest in and concern about Nash's situation:

We got onto the subject of John Nash and wondered what his present situation was, in particular with regard to his state of his mind. It turned out that none of us knew what was going on medically nor did we know of any one else who knew. We had all heard rumors varying from "the doctors say there is no hope" to "he's doing mathematics again."

The thing that disturbed us was not our own lack of knowledge about Nash's condition but the thought that perhaps everyone in the mathematical community was in the same position we were and that consequently Nash might not be getting the best possible medical attention. It is certainly true that the mathematical community has provided fellowships and jobs of various sorts for Nash whenever he has needed them. This is as much as we should be expected to do, provided some other competent, informed and adequately endowed person or persons are looking after the medical situation. Since Nash is now at the Institute, I thought you might be in a position to know whether such a person exists and to reassure us that everything that can be done is being taken care of. If it should turn out that for lack of money, for instance, Nash was not getting the care he ought to have, I'm confident that we could get together a friends of Nash group to see what could be done about it.[29]

To come out, to go through the motions of starting over, to see one's old friends and colleagues again was not easy. Nash stayed out of sight at the Institute. Few of that year's visitors recalled seeing him there. He complained in the fall of "feeling lonely."[30] He and Alicia still attended parties together, but she resisted any idea of their resuming their marriage. She was having difficulties at her job and found her son hard to handle. But when her mother took John Charles to El Salvador for several months that winter, she missed him terribly. Nash tried to be sympathetic, writing in March that "Alicia is seeing a psychiatrist. She is very depressed. She was crying."[31]

Yet he also said that he was "learning new things" and then, in December, that Selberg was trying to arrange visiting positions for him either at MIT or Berkeley.[32] He continued to hope for a reconciliation; he and Alicia continued to socialize as a couple. Nash seemed, as the fall unfolded, to be in far better shape than he had been during his previous interlude at the Institute. As he said in his Madrid lecture, he "had an idea which is referred to as Nash Blowing UP which I discussed with an eminent mathematician named Hironaka."[33] (Hironaka eventually wrote the conjecture up.)[34] William Browder, who was also visiting at the Institute that year, recalled: "Nash was working on real algebraic varieties. Nobody else had been thinking about these problems."[35]

During the winter, Milnor, by now chairman of the department, and his colleagues became greatly impressed by "some extremely interesting ideas [of Nash's] in algebraic geometry."[36] The new work sparked a wave of optimism and renewed a desire to help Nash. There was a growing feeling, both at the institute and at the university, that Nash might well be able to resume his interrupted career.

Milnor decided to offer Nash a one-year post as research mathematician and lecturer. In April 1964, Milnor tentatively proposed that Nash teach one course the following fall and perhaps two in the spring.[37]

Milnor consulted Nash's psychiatrist, Howard Mele, who confirmed on March 30 that Nash was seeing him regularly for psychotherapy, noting that this was the first time that Nash had agreed to seek outpatient treatment since the onset of his illness.[38] Garber recalled: "[Mele] tried to keep him on medication. He also helped Nash initiate relationships with other people. In my experience, positive relationships plus medication does wonders. 'Someone likes me': that's an experience that's almost impossible for a schizophrenic to have."[39]

Mele felt that Nash's recovery was permanent and that he could handle one or two courses without difficulty during the next academic year. He went on to say: "I cannot guarantee his future mental health (any more than I could my own or that of anyone else), but I do feel strongly that a recurrence is unlikely in his case."[40]

Dean of Faculty Douglas Brown wrote to President Goheen, saying, "This is a special situation," adding that Nash "is now recovered. . . . He needs a chance to get back into teaching gradually and to re-establish his status."[41] Brown said that the mathematics department unanimously supported the proposal. "I am strongly inclined to go along. It is a part of our job, I feel, in putting one of our most brilliant Ph.D.s back into top productivity." The appointment was made officially on May 1.[42]

Sadly, just when things looked brightest, and despite all of Nash's hard work, Mele's support, and the outpouring of goodwill on the part of colleagues and the university, another storm was gathering. As early as February, Nash began complaining of sleeplessness and of his "mind [being] filled with the thought of performing imaginary computations of a meaningless sort."[43] A comment, made in early March, that he had "avoided falling back into delusions" suggests that Nash was already being besieged by such thoughts.[44] And by the end of that month, Nash, who said he still hoped for a reconciliation with Alicia, mentioned that he felt he might have to leave Princeton.[45]

By the time the Princeton job was offered, Nash was already convinced that he ought to return to France, clear evidence that he was nowhere near as well as his behavior suggested.[46] His letters home were sufficiently strange to alarm Martha, who contacted Mele.[47] Mele was at first reassuring; he wrote back that Nash was no longer taking medication, but that Nash was still in therapy and that the therapy seemed to be working well.[48] Nash also wrote reassuringly, apparently in reply to questions from an anxious Virginia, that he was still seeing Mele.[49]

But around that time, Nash paid an unexpected call on his former French professor Karl Uitti. He appeared "rather anxious," Uitti recalled. "He said, 'I'm interested in getting the addresses of Jean Cocteau and André Gide. I have to write them letters.' I gently informed him that both Gide and Cocteau were dead and that writing letters to them would be impossible. Nash was very, very disappointed."[50]

By May, Nash was complaining that he was having trouble working: "I have some ideas but many of them don't seem to work out."[51]

Nash had apparently been in touch with Grothendieck once more. Grothendieck evidently responded with an invitation to the IHES for the following year. At the beginning of the summer, Nash wrote to a colleague in Europe, saying that he wished to spend the following year in France rather than stay in Princeton and accept the university's offer.[52]

Nash complained of finding himself in a "troubled situation," saying that he had difficulty when he tried to work on mathematics, and also that his relations with various faculty and students at the university were troubled as well. It is not clear to whom or what he was referring — the job offer from the mathematics department had been supported unanimously by Milnor and the rest of the faculty and Nash's contacts with students were presumably limited to the Fine Hall common room. He wrote that he expected something to change by June 1, but that he wasn't certain of that, adding: "Si ma situation reste essentiellément la même comme c'est de maintenant" (If my situation remains essentially the same as it is now), drawing a circle in the middle of the page accompanied by the parenthetical remark, "(Ici-compris ma situation de famille, etc., etc.)" (Including my family situation). He went on, "Et si je peux travailler effectivement aux mathématiques par le temps de l'automne, je pense que je devrais accepter l'offre de Grothendieck plutôt que l'offre de l'Université, s'il pourra encore me donner cet offre d'emploi" (And if I can work effectively at mathematics by the fall, I think I should accept Grothendieck's offer over the offer from the university, if he will still extend me this offer of employment).

As far as the institute knew, Nash was planning to spend the entire summer at Fuld Hall, with the exception of about three weeks, before going to France in the fall. On May 24, in response to a note from Oppenheimer granting him funds for the summer "with the understanding that you will remain at the Institute during the summer," Nash wrote that he planned to be away from June 22 through July 19 at a conference in Woods Hole on Cape Cod, organized by John Tate, on the theory of singularities, classifications of surfaces and modules, Grothendieck cohomology, zeta-functions, and arithmetic of Abelian varieties.[53] According to Tate and other participants, Nash never went to the conference.[54] Instead, he went to Europe.

He sailed on the *Queen Mary,* stopped briefly in London, and went to Paris.[55] There he tried to get in touch with Grothendieck, who evidently wasn't in town.[56] After hanging around a few more days, Nash flew to Rome. He was, as he later said, thinking of himself as a "great but secret religious figure."[57] This may have accounted for his desire to be in Rome, where, as he later said, he visited "the Forum and the catacombs but avoided the Vatican."[58] The Pope was, in any case, not in Rome at the time.

He was standing in front of the Forum when he began to hear voices "like telepathic phone calls from private individuals."[59] They seemed to him, at the time, he said in Madrid in 1996, to be the voices of "mathematicians opposed to my ideas." He wrote in a letter later in the 1960s: "I observed the local Romans show a considerable interest in getting into telephone booths and talking on the telephone and one of their favorite words was pronto. So it's like ping-pong, pinging back again the bell pinged to me."[60] Something odd was happening, he concluded. Harold Kuhn later said, "The stream of words was obviously being fed into a central machine where they were translated into English. The machine inserted the words, now in English, into his brain."[61]

Nash, however, did send a postcard from Rome, dated September 1, saying that he was returning to Paris and that he had attempted to contact Grothendieck and other mathematicians.[62] He said he would be staying at the Grand Hôtel de Mont Blanc, where he and Alicia had stayed five years earlier. Two days later, he was back in Paris, but had not yet managed to see Grothendieck, who was apparently away.[63] The staff at the IHES "suggested contacting Jean-Pierre Serre," but Serre does not remember Nash's ever getting in touch with him.[64] Nash's next postcard home was a collage: a card devoid of any writing, with a Parisian scene and a French coin and a long number for a return address.[65]

Meanwhile, Nash had not informed the mathematics department at Princeton that he was not intending to take their offer. Finally, on September 15, Tucker sent a terse note to Dean Brown, canceling the appointment and saying that Nash had gone to the University of Paris.[66]

Nash hung around Paris a few more weeks until he finally gave up. In mid-September, he wrote to Virginia from Paris that he would be returning on the *Queen Mary* on the twenty-fourth, adding a postscript: "Situation looks dismal."[67]

Back in Princeton, Nash took to calling people again and turning up at the Institute to write strange messages on the blackboards of various seminar rooms. Atle Selberg recalled one such message involving several Social Security numbers. "He tried to find mysterious patterns," Selberg recalled. "He claimed that he was born in a county named Mercer that had a town named Princeton. He seemed to find this a mysterious sign."[68]

By mid-December, Nash was back in Carrier. Once again, it was Alicia who had to make the painful decision. A letter written to John Milnor shows how fast Nash's thoughts were racing and how one association prompted another — even as Nash was conscious that Milnor would find the letter mad. Labeled "crazy letter for your entertainment," it was a fantastic monologue, skipping from slave calendars and lunar eclipses to advertising jingles and equations from Milnor's papers.[69]

Mele once again took over Nash's care and Nash once again responded quickly and dramatically to antipsychotic drugs. He was well enough in early April 1965 to leave Carrier for the day to attend a banquet with John Danskin at another game-theory conference in Princeton.[70] As Danskin recalled, "Nash's name was being mentioned a lot at the meeting. I thought it would be nice to produce him."[71] Once Nash learned that he would be going, he telephoned Harold Kuhn and asked him to bring a couple of game-theory books to Carrier, which Kuhn did, recalling that "it was a barracks-like place, not much privacy."[72] Nash stayed on at Carrier until midsummer, his departure delayed until Mele was confident that both a job and a psychiatrist were waiting for his patient.

In April Richard Palais, a mathematician at Brandeis, drove down to the institute to turn in a manuscript. "That day Borel said why not have lunch with Jack Milnor and me. We had lunch," he recalled.[73] Halfway through they started talking about Nash. Milnor and Borel thought Nash was much better now. They thought it would be a good thing for him to gradually get back to academic life. They believed Boston would be a good place. MIT and Harvard would be too difficult after he had insisted on resigning from MIT and threatened to sue the university. The Harvard department was too small. There was no way they were going to hire him. The Institute in those days didn't have five-year memberships, and it was almost unheard of to have someone more than two years.[74]

Norman Levinson, who had been in contact with Mele, Milnor, and Borel, offered to support Nash with his ONR and NSF grants. He felt that it was too soon for Nash to have an office at MIT. Palais recalled:

> I had a feeling they were on the level in helping him get back to the mainstream and that it would be better for him to be in Cambridge, away from Princeton. It was very late. I'm surprised we were able to do anything. But the [Brandeis] administration really liked the math department and Joe [Kohn, then chairman] would go and get what we wanted.
>
> There was a lot of that feeling [about Nash]. People were expecting an awful lot from this guy. In any four- or five-year span, there are one or two young bright people who are recognized as special. Everybody tries to get them. He was coming into that category. He was very special.[75]

When Nash got out of Carrier this time, in mid-July, he spent a couple of nights at John Milnor's house and then took a train to Boston.[76] He was, once again, hopeful and, in contrast to a year earlier, accepted the likelihood that he might have to start a new life without Alicia.

43

Solitude

Boston, 1965–67

IT WAS STRANGE to be back in Boston alone and after an absence of half a dozen years. The city had changed almost as much as Nash himself. Sundays were the bleakest. Nash's "traditional Sunday[s]" as he called them, were spent alone, sitting in one of the libraries trying to work, or, more often, walking for hours at a time, and then stopping to watch the ice skaters and hockey players in the Public Garden.[1] The evenings were given over, more often than not, to writing letters, one to Alicia, one to Virginia, and one to Martha, with whom Nash had lately developed a warmer, more confidential relationship.[2] Mailing the letters provided an excuse for a final nighttime stroll.

Weekdays, when he commuted to Waltham in a ratty old Nash Rambler convertible purchased on his arrival in Boston, were better. He was almost enjoying being at Brandeis. The place was undeniably lively, full of former students and acquaintances from the old days in Cambridge, former MIT undergraduates like Joseph Kohn, now chairman of the math department, and Al Vasquez, now an assistant professor. He liked having an office again, going to seminars, eating lunch with other mathematicians, tossing around ideas and mathematical gossip.

But he was terribly lonely. He missed Alicia and John Charles. He felt his new, humbler status in the mathematical hierarchy most acutely. But he also could see, perhaps for the first time since the onset of his illness, that there was, after all, a future for him, and he entertained hopes of reestablishing himself as an academic and even of finding someone new to share life with.

He had left Princeton almost immediately after being released from Carrier on July 29, traveling to Boston by train and staying in a Cambridge hotel while he found an apartment and a car.[3] He had seen Norman Levinson, who, in his gruff, taciturn, immensely tactful way, had let Nash know that he would be paying Nash's salary with National Science Foundation and Navy grants, and that he hoped Nash would be able to pursue his own research ideas, as before. He would have no teaching responsibilities, at least in the fall, which was a relief.[4]

He started to see a thirty-three-year-old psychiatrist, Pattison Esmiol. An affable

Coloradan with a medical degree from Harvard, Esmiol had just left the Navy to open a private practice in Brookline. Esmiol prescribed an antipsychotic drug, Stelazine, similar to Thorazine. Nash didn't like the drug and its side effects, worrying that they would prevent him from thinking clearly enough to resume mathematical work. But Esmiol, sympathetic to his client's concerns, kept the doses as low as possible, and Nash was grateful for the dependable human contact of his weekly appointments.

Nash was seeing Eleanor and John David, now a tall, handsome boy of twelve, every week or so.[5] Nash was glad for the dinners Eleanor cooked him and glad to have the company. The three of them spent Halloween together, he wrote to Virginia.[6] However, the old tensions in his relationship with Eleanor quickly surfaced again, and there were new and unanticipated tensions between himself and John David. Nash described Halloween as a "sad" occasion, for example, although it was not clear whether the sadness stemmed from friction that arose during the evening, or simply from a realization that his long separation from his son had produced a gulf that he could see no obvious way of bridging. John David was a particularly beautiful boy, musical and obviously bright. But Nash found it difficult to hide his dismay over his son's faulty grammar and indifferent performance in school — all John David had to do was to let a "you was" slip out and Nash would be all over him;[7] this, of course led to flare-ups with Eleanor and a rekindling of all the old resentments. John Stier recalls his father's visits as "frustrating." "He was always humming," Stier said. "He'd eat. He'd chill out. He'd leave. He never helped me with my homework or asked how I was doing. He was just very aloof."[8]

Before he became a teenager and he and Eleanor began living in Hyde Park, John Stier lived in two dozen different places, with and without his mother.[9] They included, between infancy and six, a series of foster homes in Massachusetts and Rhode Island, an orphanage on the outskirts of Boston, and when finally reunited with Eleanor, the Charden Home for Women and Children, a home for the destitute (no boys over age nine allowed!). In some school years, he attended three new schools and was deemed a "behavior problem." On one occasion, he was held back. The moves were prompted by the calamities that are regular events in the lives of poor families: lost jobs, ill health, lack of childcare, fear of crime. On one occasion, Eleanor recalled, "I had a woman taking care of him. She said John had been bad to her little boy. So she hit him and gave him a black eye. I didn't work for a while. I was always on edge."[10]

It was, as he said, "a miserable childhood, a shitty childhood."[11] His mother loved him, of course, but was herself desperately unhappy. Eleanor was often ill, suffering at times from severe anemia, frequently lost jobs, and when she was working often held two jobs. John David's illegitimacy was a dirty secret; Eleanor concocted a tale to explain away his fatherlessness and the child was forced to tell it at the different schools and neighborhoods, while living in constant dread of discovery. "There was a real stigma," John Stier said. "I had to lie."

In John David's eyes, however, his father's sudden reappearance in his life was a fine thing. Being corrected for the way he spoke and being admonished to work harder in school conveyed not just criticism, but fatherly interest. Nash also promised to pay for John David's college education, explaining that "his educational background will shape the whole future course of his life." Nash sometimes took pains to please his son. On Saturdays, he would take John Stier and a friend bowling. Afterward, they'd go to a Chinese restaurant for dinner. On John Stier's thirteenth birthday, Nash surprised him by taking him to a neighborhood bicycle shop and buying him a ten-speed racer. The next year, perhaps partly inspired by his father's interest in him, John Stier worked extremely hard in school, took a citywide examination, and got a place in one of Boston's elite "exam" schools.

In January, Nash wrote that "I have less time for Eleanor," hinting perhaps that he felt his early dependence on her company easing and feeling some relief on this account.[12] This would have given Eleanor new grounds for grievance; she may well have felt that he was once again using her without much intention of giving her very much in return. But at the end of February, Eleanor and John David were "among my few social contacts."[13] There were repeated flare-ups. "Eleanor was not nice to me," he wrote after they went to a restaurant together.[14] In April when Eleanor moved to a new apartment, several days went by before she was willing to give him her new telephone number.[15] In May there is another reference to Eleanor's not being nice, which again made Nash feel rather "sad."[16] If Nash's reappearance in Boston raised again the possibility of his marrying Eleanor—either in her mind or his—there is no hint of this in Nash's letters to Martha. Nash still had not completely given up hope of a reconciliation with Alicia.

On that sad Halloween, he had been thinking a great deal of Alicia. "I was very fond of her," he wrote to Virginia.[17] His sadness on that night probably had a good deal to do with the fact that she was discouraging him from visiting her in Princeton, as he had hoped to do, on Thanksgiving. She apparently put him off with excuses, citing among other things "propriety."[18] Nash persisted and Alicia continued to discourage him, so that a week before the holiday Nash said that he still had no invitation. Alicia was now talking of his coming down at Christmas, but it is not clear that the visit took place. In and amongst it all, perhaps because he was now aware of John David's discomfort around him, he expressed fear that his younger son, John Charles, was "forgetting his father."[19] It was not all that easy to renew his old acquaintanceships, though he saw a bit of Arthur Mattuck and his wife, Joan, as well as Marvin and Gloria Minsky.[20] People were kind but busy. He was anxious for anything to fill his evenings and went to a great many movies, plays, and concerts by himself.[21] Alicia, who continued gently to discourage any possibility of reconciliation, was encouraging him to find some female companionship. He wrote to Martha: "Alicia doesn't leave much hope." [22] In January, Nash was making awkward inquiries about dating. [23] He thought of inviting the Mattucks to his house for a meal and "making it a foursome." Jean Mattuck reintroduced him, apparently, to Emma Duchane, who later could recall none of this. [24] He pursued Emma for several weeks, saying to Martha, "She's a good conversationalist, but she isn't pretty really," before discovering that Emma had a fiancé.

After seeing *A Hard Day's Night* one Sunday afternoon in early November, he was seized by a terrible sense of regret that he poured into a poignant and introspective letter to Martha, full of references to the struggle between his "merciless superego" and "old simple me." This is the letter in which Nash referred to the "special friendships" in his life and his realization, in 1959, of "how things had been." He admits that "away from contact with a few special sorts of individuals I am lost, lost completely in the wilderness. . . ."

Brandeis was lively. A post-*Sputnik* infusion of money and a commitment on the university's part to building a serious graduate program in mathematics had attracted eight or nine young comers, all in their thirties. "We had lots of research money. We had plenty of money to pay for research associates and part-time instructors. We did everything together," recalled Richard Palais. [25] The atmosphere was friendly and informal, and Nash felt welcome there. "Everybody was well aware that he was a first-class mathematician," said Palais, adding:

> I ate with him most lunches. It was nice to see him more or less back. He was pretty sane. He was being treated with antipsychotic drugs. He was a much nicer person after he got sick than before. I kind of knew him when I was an instructor at Harvard, but not personally. I'd ask him a question. He'd be all snotty, proud of himself. You'd be afraid to ask him anything. He'd put you down without a thought. Typically, I'd say, "I have this problem," and Nash would shoot back, "Oh my God, how can you ask me this question? How stupid are you? How come you don't know this?" Afterward, he was nice, gentle, lots of fun to talk to. This old ego stuff was gone.

Vasquez has similar memories: "When Nash first showed up at Brandeis he was pretty zombielike. At the beginning, he said nothing. That changed over the course of the year. He got more and more normal. He started interacting with people. We mostly talked about mathematics. He never talked about his personal life." [26]

Nash's renewed appetite for life was most evident in the energy with which he was able to work that year. During that fall at Brandeis he wrote a long paper, "Analyticity of Solutions of Implicit Function Problems with Analytic Data,"[27] that pursued to their natural conclusion his ideas about partial differential equations. He circulated his draft for comments and submitted the paper to the *Annals of Mathematics* in early January.[28] Armand Borel, one of the editors, sent it to Jürgen Moser to referee. After a few telephone consultations between Borel and Nash, Nash quickly revised the paper and got a final acceptance from the *Annals* on February 15. Nash was thrilled, writing to Martha on Washington's birthday that the *Annals* was "the most prestigious American mathematical journal."[29]

His renewed productivity produced a rush of self-confidence. He went to see Oscar Zariski at Harvard to discuss some new ideas — and possibly to inquire about a visiting position. He made friends with a young German mathematician, Egbert Brieskorn, who was visiting at MIT that year. He showed Brieskorn his just-completed paper and talked over ideas for future work. Brieskorn was doing some interesting work in singularities. "Nash had interesting ideas," Brieskorn recalled. "He was always making propositions about what one could do. But I always got the feeling that he either couldn't or wouldn't do them himself."[30] A touch of Nash's old arrogance returned. There was some talk, apparently, of his teaching at Northeastern in the spring. "I'd rather be at a more famous place," he confided to Martha. He thought he would apply for a position at MIT instead. He wrote Martha that he felt MIT ought to reinstate him, adding, "Of course, MIT isn't the most distinguished . . . Harvard ranks much higher."[31] Throughout the spring he would fret about being forced to take a position at a second-rate institution: "I hope to avoid stepping down in social status because it may be difficult to come up again."

As early as the beginning of February, Nash had an idea for a second paper, but two weeks later he wrote to Martha that he was "sad because part of my new math idea fell apart."[32] He was able, however, to take the disappointment in stride, and by early April he was already working on another paper on the "canonical resolution of singularities." Many years later he would call this effort "more interesting" than his 1966 *Annals* paper. In May he gave a seminar on the subject at Brandeis, and by the end of the month he had completed a draft that he showed to Brieskorn for comments.[33] Nash quite likely submitted this paper to the *Annals* as well, but it was never published.[34] A copy finally wound up in Fine Hall Library at Princeton in September 1968. It was regularly cited in the succeeding years and was ultimately published in the *Duke Journal of Mathematics* in 1995 in a special issue in honor of Nash.

The quality of these two papers — the first of which geometer Mikhail Gromov calls "amazing"[35] — constitutes the single strongest reason for questioning Nash's diagnosis of paranoid schizophrenia.[36] Producing papers that broke new ground was a remarkable feat for someone who had, by 1965, been psychotic for most of six years and suffered substantial memory impairment.[37] Unlike manic depression, paranoid schizophrenia rarely allows sufferers to return, even for a limited period, to their pre-morbid level of achievement, or so it is believed.[38] However, at least one other mathematician with chronic schizophrenia was able, during a brief remission, to produce excellent work,[39] and Nash's papers, though superb, were not as ambitious as those that he had planned to write before he became ill.

At the end of June, Nash moved into Joe Kohn's apartment at 38 Parker Street in a two-family house not far from Harvard Square.[40] Kohn was off for a year's sabbatical in Ecuador. The sublet was arranged by Fagi Levinson, who recalled: "Everybody wanted to help Nash. His was a mind too good to waste."[41]

Nash enrolled in Operation Match, a Cambridge computer dating service. He was going on blind dates, acutely aware that "I'll need to learn how to behave properly and be polite etc." He wrote that he was "hopeful and optimistic": "I think I'll develop some good friends and I'll get remarried if not to Alicia and then I'll have a happy family life."[42] He had an appointment at MIT lined up for the fall: Ted Martin had offered to let him teach a senior seminar in game theory. In May Nash wrote to Kuhn saying that he wanted to "collect appropriate materials and learn about the more recent developments" in game theory and asking Kuhn for suggestions.[43]

Something, however, was no longer quite right. Some of his colleagues at Brandeis recalled an abrupt change sometime in the late spring. Palais recalled: "He sort of lost his balance completely. He went completely haywire."[44] Vasquez remembers a more gradual unraveling: "He went right past normal and became hyper. At some point, he wouldn't stop talking and he didn't make any sense. By the summer, he wasn't able to interact any more."[45] It's hard to say what triggered his relapse. Possibly, Nash had become overconfident and had stopped taking his medication.

He evidently spent the summer in Cambridge. By September, his letters to Martha were distinctly delusional. In one he referred to "the Indian wheel of life.... If a person is always correct and right . . . there is good reason to hope."[46] Alarmed, Martha wrote to Esmiol saying that her brother sounded "optimistic but not well."[47] She quoted him saying that "I have put my delusions aside" but she was sure that the delusions were now back in full force.[48] Esmiol wrote back in early October saying that he had seen Nash and that "he was about the same as last time." He urged her to express her concern directly to her brother.[49] A day later, Nash wrote to Martha reassuring her that his optimism was well-founded but admitting there "are always dangers to worry about." But in the next breath, he went on to say that he'd had an "interesting" letter from Alicia about "a large gift of money."[50] Martha later recalled that Nash, in his delusional periods, was always hinting that "something great was about to happen."[51]

By November, the tone of his letters had become paranoid, as in one to Virginia: "I'm very disillusioned in the past . . . hoping also that my future relations with all the relatives and especially you and Martha will be much better."[52] At

Thanksgiving he wrote: "I didn't have much to be thankful for this Thanksgiving." He planned to go to Roanoke for Christmas and to spend New Year's—Alicia's birthday—in Princeton.[53]

Vasquez, who had an apartment near Nash's, was running into Nash wandering around Harvard Square the way he later wandered around Princeton:

> He was concerned with the politics of Mao Tse-tung, that sort of thing. In Harvard Square, he was talking about a committee that was communicating with foreign governments who manipulated the news in *The New York Times* in order to send messages to him. He had this idea that with this information he could find out how negotiations between various powers were going.[54]

Nash was still attending the Harvard math colloquium on Thursdays. "He was very peculiar," Vasquez recalled. "He believed that there were magic numbers, dangerous numbers. He was saving the world."

Soon Kohn was getting letters from his neighbors, the landlords of the house, complaining that Nash wasn't taking out the garbage and that his apartment was full of piles of newspapers.[55] Fagi recalled feeling horribly embarrassed and responsible. "Joe wanted to give up the apartment. He tried to reach Norman. He couldn't, so he called me. So I called Nash every hour on the hour. I was worried. I got this crazy idea to call up this minister he had been seeing. The minister told me Nash was out of town."[56]

Just after the New Year, Nash left Boston for the West Coast. He traveled first to San Francisco where he spent several days visiting his cousin Richard Nash. He called his cousin first, who, in turn, called Martha. "He blamed Martha for hospitalizing him," recalled Richard Nash. "It was very hard for her to take."

> He came to my office. He was good-looking, very muscular. He was soft-spoken but his voice was much stronger than now. He was a lot of fun to talk to. He liked to talk a lot late into the night. Sometimes he spoke rationally, almost poetically. He was very concerned about not being able to contribute. "I started out so well," he said. "I think of myself as a valuable person. But I'm not contributing." Other times he made no sense. He had these things he was concerned about. He went to see a Catholic priest in San Francisco. I said, "I thought you were an atheist."[57]

Richard Nash, a broker, would drive to work in San Franciso and take Nash with him. Once there, "He'd get on the bus and go all around." Dick Nash expressed astonishment that Nash mastered complex schedules, went all over, but always managed to meet Dick at the appointed place for the return trip at exactly the right time.

After that, Dick Nash recalled, "John called me at odd hours. He had no awareness of time. I told him to stop calling me after bedtime. Then I'd get calls with just breathing. I was rude. I wish I'd been nicer."

After leaving San Francisco, Nash went next to Seattle, arriving there on February 3.[58] He almost certainly went there to visit Amasa Forrester, the only person he knew in Seattle. He seems to have spent nearly a month with Forrester, because he did not arrive in Santa Monica, his next destination, until Easter, which fell in mid-March that year.[59] There, apparently, Shapley and other acquaintances from RAND refused to see him. Nash visited Jacob Bricker in Los Angeles as well. Bricker recalled that Nash "was acting really wild."[60]

Nash apparently called Esmiol from time to time, although he disregarded Esmiol's pleas that he return to Boston and resume his treatment. Martha also called Esmiol a number of times that month. Esmiol's idea was to use the promise of a job at MIT as a lever to get Nash back into treatment.[61]

Martin was talking about letting Nash teach a section of linear algebra the following fall.[62] Levinson, still hopeful, was planning on Nash's being at MIT. He solicited a letter of recommendation from Armand Borel at the Institute. Borel's letter, dated May 17, was a strong endorsement:

> In the last eight years or so, he has been very much hampered by his health problems. Even then, he has managed to produce some interesting work. . . . Nash is clearly one of the most individualistic among the presently active mathematicians. He does not work systematically at long range programs, whose progress along more or less foreseen lines can be rather confidently expected but is more the pioneer type who proceeds along new paths. He is thus rather unpredictable; but in a way it makes it appear more likely that he might score new successes in spite of his ups and downs in health. Any contribution in mathematics on the level of his past work would be extremely valuable, and so I feel strongly that he should be supported.[63]

It's not clear exactly when Nash returned to Cambridge. But when he did, he was extremely ill. After a terrible scene, John David locked him out on the porch on a freezing night.[64] Nash told Palais at some point that he'd stopped taking medication. "Why, when they were making you well, did you stop taking drugs?" He answered, "If I take drugs I stop hearing the voices."[65]

A letter from Nash to Moser captures something of Nash's state of mind when he returned to Cambridge in late May. Nash gives his return address as Heilwig-klang University, Harbin, Manchuria.

> The Oblast in Russia, on the Manchurian border . . . there's the city of Birbidzhan. . . . If all the atomic powers of the security council of the United Nations did an action, and they were numbered 0, 1,2,3,4 then one would be able to say nobody did it, everybody did it, all did it . . .

The letter was signed "Chiang Hsin (New River)." [66]

Fagi ran into John on the subway. His manner was slippery, shady, shy, almost ashamed, a peculiar smile pulling at the corners of his mouth. She asked where he was going. He answered: "Home to Roanoke to stay with my mother for a while." [67] Nash left Cambridge on June 26, leaving his apartment in a shambles. He drove to Princeton, stayed in a hotel "for propriety" rather than with Alicia and John Charles, and proceeded to Roanoke a few days later.[68]

Fagi called Joe Kohn and said she'd get a moving van and send Nash his furniture. "I felt so guilty that I said to myself, I'll get his stuff moved out. I did, too, everything except the bathroom scale. I never even went into the bathroom." [69] Anna Rosa, Kohn's wife, went into the Parker Street apartment: "There were folded bags, one upon another, and cereal boxes. Not awful, but signs of compulsion." [70] A few days later, Norman Levinson wrote to Martha:

> For the past two years John has been employed as a research associate on my contract. John doesn't want to live here and I couldn't convince him to stay. A few days ago John left 38 Parker Street. There were piles of rubbish. Hints of bank accounts. Also other accounts here and abroad. John was very disturbed this past year. But in 1965–1966 he functioned very well and did fine work.[71]

44

A Man All Alone
in a Strange World

Roanoke, 1967–70

> *And then a Plank in Reason, broke,*
> *And I dropped down, and down —*
> *And hit a World, at every plunge. . . .*
> — *EMILY DICKINSON, Number 280*

THE SUMMER NASH TURNED forty, in 1968, he looked into the mirror in the bathroom of his mother's apartment and saw what he later called "a cadaver, almost."[1] Hollow-cheeked, sunken-eyed, gray-haired, with his shoulders hunched forward, he looked more like an old man than one just entering middle age. He wrote to a friend: "You should pity me . . . aging and drying processes have taken their toll."[2] Images of death-in-life crowded his mind: in a letter to another friend he invoked the images of the Parsee "Towers of Silence" in Bombay, where followers of Zoroaster leave their dead to be devoured by vultures.[3]

He had been living in Roanoke for nearly a year. He still had his Rambler and some savings, but eight years of illness had exhausted his former wife and friends and ruined much of his credit with the world. He had nowhere else to go. For him, Roanoke — a pretty little city at the foot of the Appalachians and the headquarters of the Norfolk & Western Railroad — was the end of the line.

He lived with Virginia in a small garden apartment on Grandin Road.[4] Martha and Charlie lived a few streets away. No one knew him there. The existence of someone with schizophrenia has been compared to that of the person living in a glass prison pounding on the walls, unable to be heard, yet very visible.[5] Martha recalled in 1994: "Roanoke was not a good place to be. There were no intellectuals there. He'd be too much alone. He would wander around town whistling."[6]

On many days, he simply paced round and round the apartment, his long fingers curled around one of Virginia's delicate Japanese teacups (a souvenir of her long-ago summer in Berkeley), sipping Formosa oolong, whistling Bach.[7] The sleepwalker's gait and fixed, faraway expression gave few hints of the vast and unending dramas unfolding in his mind. "Apparently I am simply passing time visiting my mother," he wrote, "but actually I've been under persecutions which I'm hoping will ease."[8]

His daily rounds extended no farther than the library or the shops at the end of Grandin Road, but in his own mind, he traveled to the remotest reaches of the globe: Cairo, Zebak, Kabul, Bangui, Thebes, Guyana, Mongolia. In these faraway places, he lived in refugee camps, foreign embassies, prisons, bomb shelters. At other times, he felt that he was inhabiting an Inferno, a purgatory, or a polluted heaven ("a decayed rotting house infested by rats and termites and other vermin"). His identities, like the return addresses on his letters, were like the skins of an onion. Underneath each one lurked another: He was C.O.R.P.S.E. (a Palestinian Arab refugee), a great Japanese shogun, C1423, Esau, L'homme d'Or, Chin Hsiang, Job, Jorap Castro, Janos Norses, even, at times, a mouse. His companions were samurai, devils, prophets, Nazis, priests, and judges. Baleful deities — Napoleon, Iblis, Mora, Satan, Platinum Man, Titan, Nahipotleeron, Napoleon Shickelgruber — threatened him. He lived in constant fear of annihilation, both of the world (genocide, Armageddon, the Apocalypse, Final Day of Judgment, Day of Resolution of Singularities) and of himself (death and bankruptcy). Certain dates struck him as ominous, among them May 29.

Persistent, complex, and compelling delusions are among the defining symptoms of schizophrenia.[9] Delusions are false beliefs, beliefs that constitute a dramatic rejection of consensual reality. Often, they involve misinterpretations of perceptions or experiences. They are thought, nowadays, to arise primarily because of the gross distortions in sensory data and the way thought and emotion are processed deep in the brain. Thus, their convoluted and mysterious logic is sometimes seen as the product of the mind's solitary struggle to make sense of the strange and uncanny. E. Fuller Torrey, a researcher at St. Elizabeth's in Washington, D.C., and author of *Surviving Schizophrenia,* calls them "logical outgrowths of what the brain is experiencing" as well as "heroic efforts to maintain some sort of mental equilibrium."[10]

The syndrome we now call schizophrenia was once called "dementia praecox," but, in fact, the delusional states typical of schizophrenia often have little in common with the dementia associated with, for example, Alzheimer's disease.[11] Rather than cloudiness, confusion, and meaninglessness, there is hyperawareness, over-acuity, and an uncanny wakefulness. Urgent preoccupations, elaborate rationales, and ingenious theories dominate. However literal, tangential, or self-contradictory, thought is not random but adheres to obscure and hard-to-understand rules. And the ability accurately to apprehend certain aspects of everyday reality remains curiously intact. Had anyone asked Nash what year it was or who was in the White House or where he was living, he could no doubt have answered perfectly accurately, had he wished to.

Indeed, even as he entertained the most surreal notions, Nash displayed an ironic awareness that his insights were essentially private, unique to himself, and bound to seem strange or unbelievable to others. "This concept that I want to describe . . . will perhaps sound absurd," is the sort of preface of which he was quite capable.[12] His sentences were filled with phrases like "consider," "as if," "may be thought of as," as if he were conducting a thought experiment or realizing that someone reading what he wrote would have to translate it into another language.

Like all other manifestations of the syndrome, delusions are not unique to schizophrenia; they can be present in a variety of mental disorders, including mania, depression, and a variety of somatic illnesses. But the types of delusions that Nash suffered from are particularly characteristic of schizophrenia, specifically of paranoid schizophrenia, the variant of the syndrome from which Nash apparently suffered.[13] Their content was, as it often is, both grandiose and persecutory, often shifting from one to the other in the space of moments or even including both at the same time. At different times, as we know, Nash thought of himself as uniquely powerful, as a prince or an emperor; at other times he thought of himself as extraordinarily weak and vulnerable, as a refugee or a defendant in a trial. As is quite typical, his beliefs were what is called referential, in that he believed that a host of environmental clues — from newspaper passages to particular numbers — were specifically directed at him and that he alone was capable of appreciating their true meaning. And his delusions were multiple, a particularly common feature of paranoid schizophrenia, although all were organized, in subtle ways, around coherent themes.

Bizarreness is thought to be especially characteristic of schizophrenic delusions. Nash's delusions were clearly implausible, difficult to penetrate, and not obviously derived from life experiences. Yet they were less bizarre, on the whole, than many delusions reported by other people with schizophrenia, and their connections to Nash's life history and his immediate circumstances, though indirect, were often discernible (or would have been had anyone who knew him well been willing to study in the same spirit as the loyal wife of Balzac's Louis Lambert). Many people with schizophrenia believe that their thoughts have been captured by outside forces, or that outside forces have inserted thoughts into their minds, but such beliefs did not seem to play a predominant role in Nash's thinking. Occasionally, as in Rome, he might think that thoughts were being inserted directly into his mind via machines, or, as in Cambridge in early 1959, that his actions were being directed by God. But, by and large, Nash maintained a sense of himself, or selves, as the primary actor. And many of his beliefs — such as that he was a conscientious objector in danger of being drafted; that he was stateless; that mathematicians belonging to the American Mathematical Society were ruining his career; that various persons, posing as sympathizers, were conspiring, with malevolent intent, to have him incarcerated in a mental institution — were no more implausible than, say, a belief that one is being spied on by the police or the CIA. Thus, in a sense, the breakdown of reality and boundaries between self and outside world had limits for him, even in Roanoke.

In particular, although Nash later referred to his delusional states as "the time of my irrationality," he kept the role of the thinker, the theorist, the scholar trying to make sense of complicated phenomena. He was "perfecting the ideology of liberation from slavery," finding "a simple method," creating "a model" or "a theory." The actions he referred to are mostly feats of mind, or involve language. At most, he was "negotiating" or "petitioning" or trying to persuade.

His letters were Joycean monologues, written in a private language of his own invention, full of dreamlike logic and subtle non sequiturs. His theories were astronomical, game theoretical, geopolitical, and religious. And while, years later, Nash often referred to pleasant aspects of the delusional state, it seems clear that these waking dreams were extremely unpleasant, full of anxiety and dread.

Before the 1967 Arab-Israeli war, he explained, he was a left-wing Palestinian Arab refugee, a member of the PLO, and a refugee making a "g-indent" in Israel's border, petitioning Arab nations to protect him from "falling under the power of the Israeli state." [14]

Soon afterward, he imagined that he was a go board whose four sides were labeled Los Angeles, Boston, Seattle, and Bluefield. He was covered with white stones representing Confucians and black stones representing Muhammadans. The "first-order" game was being played by his sons, John David and John Charles. The "second-order," derivative game was "an ideological conflict between me, personally and the Jews collectively." [15]

A few weeks later he was thinking of another go board whose four sides were labeled with cars that he had owned: Studebaker, Olds, Mercedes, Plymouth Belvedere. He thought it might be possible to construct "an elaborate oscilloscope display . . . a repentingness function." [16]

It seemed to him also that certain truths were "visible in the stars." He realized that Saturn is associated with Esau and Adam, with whom he identified, and that Titan, Saturn's second moon, was Jacob as well as an enemy of Buddha, Iblis. "I've discovered a B theory of Saturn. . . . The B theory is simply that Jack Bricker is Satan. 'Iblisianism' is a frightening problem connected to the Final day of Judgement." [17]

At this point, the grandiose delusions in which Nash was a powerful figure, the Prince of Peace, the Left Foot of God, and the Emperor of Antarctica were no longer in evidence; instead, the theme became predominantly persecutory. He discerned that "the root of all evil, as far as my personal life is concerned (life history) are Jews, in particular Jack Bricker who is Hitler, a trinity of evil comprised of Mora, Iblis and Napoleon." These were, he said, simply "Jack Bricker in relation to me." [18] At another point, he said, referring to Bricker, "Imagine if there would be a person who pats a guy on the back . . . with compliments and praises, while at the same time stabbing him in the abdomen with a deadly rabbit punch." [19] Seeing the picture so clearly, he concluded that he must petition the Jews and also mathematicians and Arabs "so that they have the opportunity for redress of wrongs," which must, however, "not be too openly revealed." He also had the idea that he must turn to churches, foreign governments, and civil-rights organizations for help.

In the story of Jacob and Esau, told in Genesis, Nash saw a parable full of meaning for his own life.[20] Jacob and Esau are brothers, the sons of Isaac and Rebekah, who love each other. Esau is the elder, and his father, Isaac, loves him, but Rebekah, their mother, loves Jacob more. As the story unfolds, Esau is twice supplanted by Jacob. First, Jacob tricks Esau into making a bad bargain and selling his birthright. Then, Jacob steals the blessing of the now blind Isaac, who had intended it for Esau. He does so by impersonating his brother. When Esau discovers Jacob's deception, Isaac rejects his claim: "See, away from the fatness of the earth shall your home be/and away from the dew of heaven on high./By your sword you shall live,/and you shall serve your brother;/but when you break loose,/you shall break his yoke from your neck." Esau, full of hatred for his brother, tells himself, "The days of mourning for my father are approaching; then I will kill my brother Jacob."

Nash believed that he had been cast out ("I've been in a situation of loss of favor") and ostracized. He was constantly threatened with bankruptcy and expropriation: "If accounts are held for a trustee, in effect, who is as good as defunct, through lack of 'rational consistency.' . . . It's as if accounts are held for persons suffering in an Inferno. They can never benefit from them because it's as if they were supposed to come from the Inferno — to the bank offices — and collect, but they need, as it were, a revolutionary ending of the Inferno before having any sort of possibility of benefiting from their accounts."[21]

There is a presumption of guilt. Punishment, penitence, contrition, atonement, confession, and repentance are constant themes — along with fears of exposure and the need for indirection and secrecy — and seem directly connected, but not limited, to his feelings about homosexuality. He refers to "the really dubious things that I have done in all the history of my personal life," including "draft dodging, truancy."[22]

Arrests, trials, and imprisonment were also recurring themes. Like Joseph K in Kafka's novel *The Trial,* Nash imagined that he was on trial "sufficiently complete in absentia." He recognizes that "it is as if the accused is his own chief accuser . . . the road of self-accusation is a road that leads to death not redemption."[23] He thinks of a "court of inquiry" investigating "the life histories and . . . interactions" of Jacob and Esau, whom he identifies as Bricker and himself.[24]

These are guilty, fearful dreams. Nash's state of imprisonment did not, it seems, refer to his illness, for he did not regard himself as ill except physically. It was existential. To Eleanor he wrote, "U see, U must sympathize more with the true needs of liberation, liberation from slavery, liberation from 'castration,' liberation from prison, liberation from isolation . . . I'm a refugee, in fact, from false symbols and dangerous symbols."[25] At times, he felt that he was in danger of crucifixion.

His own needs, he said, were "to be free, and to be safe and for friends."[26] He was always, he said, "in fear of 'death' (Indian style) through an Armageddon with Iblis . . . at the Day of Judgement." Even in these very dark hours he clung to a vision of liberation — which later became, more concretely, a wish for sexual liberation. "I'm hoping fervently to be saved (delivered) before reaching 40 in age," he had written a few weeks before his birthday. "One cannot substitute free life and love of the 40s for the lost possibilities of the 20s and 30s and also teens."[27]

Nash was acutely aware of the passage of time. "It does seem to me that I've been as if the victim of an excessively long wait for liberation. . . . It's as if there wasn't a ransom forthcoming, as if from Kuwait, which would have really substantially shortened the time of waiting for me."[28]

He was waiting for deliverance: "I see, it seems surprisingly clearly, how there's as it were, a time of grace before that time, a precious time of grace which is forever lost if not seized carpe diem and fully effective in its significance."[29] Nash was also hearing voices, voices that frightened him: "My head is as if a bloated windbag, with Voices which dispute within."[30]

Hallucinations can involve any of the senses — hearing, smell, taste, touch, sight — but voices, one or several, familiar or strange but distinct from one's own thoughts, are the most characteristic of schizophrenia.[31] These are quite distinct from the hallucinations that are part of religious experience, or the humming inside one's head, hearing one's name called occasionally, or hallucinations that occur while falling asleep or waking up. The content of schizophrenic hallucinations can be benign, but they usually involve ridicule, criticism, and threats, typically related to the content of the delusional theme. The integration of voices with thought can produce an acute sense of reality.

The so-called negative symptoms of schizophrenia are, most clinicians agree, even more crippling than the delusions and hallucinations. The terms used to describe them are derived from the Greek: affective flattening, alogia, and avolition. There was no trace of the sharp looks, the enthusiastic gesturing, the brash body language that announced, "I'm Nash with a capital N." His face was blank, his eyes empty, as if the fires of delusion had consumed everything that was once alive and left an empty husk.

One would feel comforted if one could believe that Nash, at this terrible time in his life, was at least spared the sight of his own condition. One of the consequences of chronic schizophrenia, noted long ago and verified since by numerous studies, is a curious insensitivity to physical pain. This insensitivity is often so great that there are high rates of premature deaths from physical illnesses among schizophrenics, at least in the era when such people spent most of their lives in institutions. Might there not be a similar dulling that would anesthetize one to psychic pain? Possibly. But for Nash there were moments of lucid self-knowledge, unbearable in their sadness: "So long a time has passed. I feel there are many sad tragedies. Today I feel very sad and depressed."[32]

It is often difficult to distinguish the effects of disease from those of its treatment. But Nash's condition during the two and a half years he spent in Roanoke was probably almost purely the consequence of his disease. Six years had passed since Nash had received insulin treatments and well over a year since he had been taking neuroleptics regularly. While some of his memory loss was, no doubt, a result of the insulin treatments of the first half of 1961 and some of his extreme quietness in the early months following his return to Cambridge no doubt reflected the side effects of Stelazine, his condition in Roanoke is a strong testament that lassitude, indifference, and the peculiarities of his thought were primarily the consequences of his illness and not of the early attempts to treat it. The popular view that antipsychotics were chemical straitjackets that suppressed clear thinking and voluntary activity seems not to be borne out in Nash's case. If anything, the only periods when he was relatively free of hallucinations, delusions, and the erosion of will were the periods following either insulin treatment or the use of antipsychotics. In other words, rather than reducing Nash to a zombie, medication seemed to have reduced zombielike behavior.

Nash was clearly among the majority of those with schizophrenia who benefited from traditional antipsychotics. These drugs were the only ones available between 1952 and 1988, when the more effective Clozapine arrived on the scene.[33]

Peter Newman, an economist at Johns Hopkins, was editing a volume of important contributions to mathematical economics. He wanted to include Nash's NAS note on Nash equilibrium.

> The first problem was finding him. I found him teaching or something at a small women's college near Roanoke. I wrote to him there to ask his permission to reprint the article. What I got back was an envelope on which my address was written in different-colored crayons. There was also a list of "yous" in different languages: Du, Vous, You, etc., and a plea for universal brotherhood. There was nothing inside the envelope at all. I then asked the in-house editor at the Johns Hopkins Press to call Nash. He did and he said it was the strangest telephone conversation he'd ever had in his life. Then we tried Solomon Lefschetz, since he was the one who sponsored the note. Calling Lefschetz wasn't easy either. Lefschetz only said, "Ah yes. He is not what he was." So I had to give it up. Later, when the book was reviewed, reviewers chided me for not including the Nash equilibrium.[34]

• • •

Nash was constantly fearful that Martha and Virginia would hospitalize him again. As he said in one letter, "It is the mechanism of how all the persons involved would collaborate in hospitalizing me which endangers me and which I fear."[35]

Most letters from this period end with a paragraph like the following:

> Let me beg (humbly) of U that U will favor the view that I ought to be guarded against the danger of hospitalization in the mental hospital (involuntarily or "falsely"), . . . simply for personal intellectual survival as a "conscious" and "reasonably conscientious" human being . . . and "good memory retention."[36]

For Virginia, Nash's illness was something that Martha later called, in her tactful and understated way, "a private sorrow."[37] Virginia never talked about it with the few acquaintances she had in Roanoke, mostly people she had met playing bridge, and only rarely with Martha. Her friends couldn't possibly have understood what it was like for her. It was also a practical nightmare. Nash was making so many long-distance telephone calls that Virginia had to put a lock on her phone.

Martha, whose second child was born in 1969, was at least angry. "It was so frustrating day by day. You wondered, is this ever going to get any better?" She realized, at least, that Roanoke was not a kind environment. "Only one time did I ask for help," recalled Martha. "The minister stopped me after church and told me I should be helping my mother more. He didn't ask whether I needed help. Later on I called and asked would he come to call. He didn't come. The retired minister came but he wasn't the one I wanted."

Virginia and Nash were nearly evicted from their apartment at one point. Martha's voice is still full of outrage thirty years later. There had been a fire that started in the incinerator. Nash was home at the time. He called the fire department. "The landlord accused John of setting it," Martha recalled. He had talked to the neighbors, who were up in arms. They found this large, strange man who walked around the grounds of the apartment complex alarming. It was only by begging that Martha was able to convince the landlord to let Virginia and Nash move back in.

Virginia died shortly before Thanksgiving in 1969. Afterward Nash was sure there was something sinister about her death. He also felt that perhaps he had done wrong by going to the corner store to buy her whiskey. Martha recalled, "When Mother died, it was not a good time. We weren't close. He felt threatened. He felt that I would put him in a hospital."

At this point, Eleanor got a court order to force Nash to continue child-support payments. When his money had run out, Virginia had taken over the payments. She also left small legacies for both her grandsons.

Nash then lived briefly with Martha and Charlie, but Martha found it impossible to cope with her brother. "Once Mother was gone, I couldn't clean with him in my home. I was here with the children and he's wandering around drinking tea and whistling. He'd take ideas and twist them into something strange."

Martha arranged to have Nash committed right after Christmas:

> After Mother died, I was afraid he'd leave town. I was hoping to get the hospital to appoint a committee so he could get Social Security and also get it for his son.
>
> We went to a judge. We got a court order. The court sent the police to pick him up. We had my mother's lawyer, Leonard Muse. You could get someone committed for observation. You didn't have to establish anything very drastic. In the hospital they decided whether to keep somebody. De Jarnette decided that John had paranoid ideas but that he was capable of maintaining himself.

Nash was released from DeJarnette State Sanitorium in Staunton, Virginia, in February. He wrote a final letter to Martha, breaking off all relations with her because of her role in his hospitalization. Then he boarded a bus for Princeton.

45

Phantom of Fine Hall

Princeton, 1970s

> *Much Madness is divinest Sense —*
> *To a discerning Eye. . . .*
> *— EMILY DICKINSON, Number 435*

AN IMPERSONAL NEW GRANITE-CLAD TOWER, built with defense dollars at the height of the Vietnam War, had replaced the old Fine Hall and neighboring Jadwin Hall.[1] Math and physics majors spent most of their waking hours below ground where the architects had situated the library — which had formerly occupied the highest floor of Old Fine — as well as the new computer center. Within a few days or weeks, the embryo scientist or mathematician would discover "a very peculiar, thin, silent man walking the halls, night and day," "with sunken eyes and a sad, immobile face."[2] On rare occasions, they might catch a glimpse of the wraith — usually clad in khaki pants, plaid shirt, and bright red high-top Keds — printing painstakingly on one of the numerous blackboards that lined the subterranean corridors linking Jadwin and New Fine.[3] More often, students would emerge from an 8:00 A.M. lecture to find an enigmatic epistle written the night before: "Mao Tse-Tung's Bar Mitzvah was 13 years, 13 months and 13 days after Brezhnev's circumcision," for example.[4] Or "I agree with Harvard: There is a brain flat."[5] Or a letter from Nikita Khrushchev to Moses with arcane mathematical statements involving the factoring of very long, ten- to fifteen-digit numbers into two large primes.[6] "Nobody knew where they came from," recalled Mark Reboul, who graduated in 1977. "Nobody knew what they meant."[7]

Eventually, some sophomore or junior would clue in the newcomer that the author of the messages, aka the Phantom, was a mathematical genius who had "flipped" while giving a lecture; while trying to solve an impossibly difficult problem; after discovering that someone else had scooped him on a major result; or upon learning that his wife had fallen in love with a mathematical rival.[8] He had friends in high places at the university, the older student would add. Students were not to bother him.[9]

Among the students, the Phantom was often held up as a cautionary figure:

Anybody who was too much of a grind or who lacked social graces was warned that he or she was "going to wind up like the Phantom."[10] Yet if a new student complained that having him around made him feel uncomfortable, he was immediately warned: "He was a better mathematician than you'll ever be!"[11]

Few students ever exchanged a word with the Phantom, although some of the brasher ones occasionally bummed a cigarette or asked for a light, for the Phantom was now a heavy smoker. One new physics student once erased two or three of the messages only to encounter the Phantom in front of the blackboard writing a few days later, "sweating, trembling, and practically crying." The student never erased another.[12]

Students and young faculty members studied the Phantom's messages and sometimes copied them down verbatim. The messages created an aura around the Phantom and confirmed the legends of his genius. Frank Wilczek, a physicist at the Institute for Advanced Study who lives in Einstein's old house on Mercer Street, was an assistant professor at the university at the time. He remembered feeling "intrigued and impressed" and "in the presence of a great mind."[13] Mark Schneider, a physics professor at Grinnell who was a graduate student in 1979, recalled: "We all found the remarkable connections, level of detail, and breadth of knowledge . . . exceptional, which is why I . . . collected a few dozen of the best of these."[14]

Shortly after Hironaka won a Fields prize for his brilliant proof of the resolution of singularities, one of Nash's messages read:

$$N^5 + I^5 + X^5 + O^5 + N^5 = 0$$

Can Hironaka resolve this singularity?[15]

Some of the messages seemed purely mathematical, at least until one looked at them more closely, as in this 1979 message:

> Open Letter to Prof. Heisuke Hironaka
>
> $$0 = E_1^5 + V^{22} + E_2^5 + R^{18} + E_3^5 + T_1^{19} + T_2^{20}$$
>
> The above algebraic variety of dimension 6, represented in affine 7-space is singular, having a point singularity at the origin $(0,0,0,0,0,0,0)$ of the coordinates.
>
> The question is: How singular comparatively, is the above 6-variety, that is, what is the comparative degree of its singularity, compared with other singularities of such a sort as to provide standards of comparison?[16]

Others contained indirect references to past events:

Indian Limbo

$$B = (RX)^7 + (MO)^6 + (OP)^5 + (QU)^4 + (ME)^3 + (OT)^2 + AAP$$

OT suggests "Occupational Therapy"[7] as in Dr. O.T. Beetle, M.D.

$$AAP = PR\,(2) - 1, \text{ as a number.}[17]$$

And still others were slyly humorous:

True or False Question

Statement: President Jimmy Carter is suffering from the disease of xanthochromatosis, the same disease which previously affected the careers of Nixon and Agnew, so that the disease has presumably jumped the gap of the apparently immune northern republicans Ford and Rockefeller and reinfected Air Force One via the person of Jimmy Carter.
The above statement is true.
The above statement is false.[18]

During one period, all the messages featured a commentator named Ya Ya Fontana who made mysterious pronouncements about current events, principally in the Middle East.[19] In another period, Alexandre Grothendieck's name appeared frequently.[20] In still another, Diophantine equations—equations like $x^n + y^n = z^n$—dominated.[21]

Margaret Wertheim, author of *Pythagoras' Trousers,* a history of mathematics, has pointed out that "people look to the order of numbers when the world falls apart."[22] Nash's romance with numerology blossomed when his world was falling apart, suggesting once again that delusions—like "mystical, cultic religious efflorescence" —aren't merely the ravings of madmen but conscious, painstaking, and often desperate attempts to make sense out of chaos.

Nash was making up numbers out of names and was often extremely worried about what he found. "He was quite agitated when he thought that the numbers were portents of something serious," recalled Peter Cziffra, the head librarian at Fine Hall. Hale Trotter, a mathematician on the Princeton faculty, recalled, "I'd say hello and he'd initiate a conversation. I remember one in which he was very concerned about the similarity of the telephone number of the United States Senate and the telephone number of the Kremlin. He was doing the arithmetic correctly but the reasoning for it was crazy."[23]

Nash did a lot of telephoning in those years. Early on, Peter Cziffra remembers, Nash tried to call public figures as well as people at the university: "It was a little odd. . . . He wanted to talk about something that had been in the paper. A crisis in Russia that he wanted to talk about with somebody."[24]

William Browder, who was now chairman of the mathematics department, recalled:

Nash was the greatest numerologist the world has ever seen. He would do these incredible manipulations with numbers. One day he called me and started with the date of Khrushchev's birth and worked right through to the Dow Jones average. He kept manipulating and putting in new numbers. What he came out with at the end was my Social Security number. He didn't say it was my Social Security number and I wouldn't admit that it was. I tried not to give him satisfaction. Nash was never trying to convince anyone of anything. He was doing things from a scholarly point of view. Everything he talked about always had a very scientific flavor. He was trying to gain an understanding of something. It was pure numerology, not applied.[25]

One has a distinct sense that Nash's condition had stabilized. To go to the blackboard took courage. To share ideas that Nash felt were important, and yet that might seem crazy to others, implied a willingness to make connections with the community at large. To stay in one place and not to run away, to labor at articulating his delusions in a way that attracted an audience that valued them must be seen as evidence of some progression back to consensual forms of reality and behavior. And, at the same time, to have his delusions seen not just as bizarre and unintelligible, but as having an intrinsic value, was surely one aspect of these "lost years" that paved the way for an eventual remission.

As James Glass, the author of *Private Terror/Public Places* and *Delusion,* put it upon hearing about Nash's years in Princeton: "It seemed to serve as a containing place for his madness."[26] It is obvious that, for Nash, Princeton functioned as a therapeutic community. It was quiet and safe; its lecture halls, libraries, and dining halls were open to him; its members were for the most part respectful; human contact was available, but not intrusive. Here he found what he so desperately wanted in Roanoke: safety, freedom, friends. As Glass put it, "Being freer to express himself, without fearing that someone would shut him up or fill him up with medication, must have helped pull him out of his disastrous retreat into hermetic linguistic isolation."[27]

Roger Lewin, a psychiatrist at Shepherd Pratt in Baltimore, said, "It seems that Nash's schizophrenia diminished in the way it appeared to others and that his madness became confined to intellectual and delusional projections rather than to wrapping him completely in behavioral expressions."[28] These are descriptions similar to those Nash himself has given of these years in Princeton: "I thought I was a Messianic godlike figure with secret ideas. I became a person of delusionally influenced thinking but of relatively moderate behavior and thus tended to avoid hospitalization and the direct attention of psychiatrists."

• • •

The immense effort—the reading, computations, and writing—of producing the messages may have played a role in preventing Nash's mental capacities from deteriorating. The messages had their own history and evolved over time. At some point, probably starting in the mid-1970s, Nash began writing epigrams and epistles based on calculations in base 26.[29] Base 26, of course, uses twenty-six symbols, the number of letters in the English alphabet, just as the base 10 of everyday arithmetic employs the integers zero through nine. Thus, if a calculation came out "right," it produced actual words.

Here was Nash, who as a boy had delighted in inventing secret codes, with his great mathematical ability and mystical preoccupations, and with plenty of time on his hands, taking names, converting them into numbers based on the letter-number correspondence, factoring the resulting numbers, and then comparing the primes in the hope of discovering "secret" messages. Daniel Feenberg, a graduate student of economics who ran into Nash at the computer center around 1975, recalled: "Nash had an obsessive concern with Nelson Rockefeller. He would take the letters, assign numbers to each letter, get a very large number, and then analyze that number for hidden meaning. It had the same relationship to mathematics as astrology to astronomy."[30] This, of course, is not only time-consuming but remarkably difficult, and the odds of finding meaningful words or combination of words minute.

Nash worked on one of those old-fashioned Friden-Marchant calculators with a tiny, glowing, green CRT.[31] He must have written an algorithm for doing base 26 arithmetic. Performing these calculations would have been tremendously tedious and would have required writing down intermediate results as he went along, since these calculators had very little storage capacity and weren't programmable. Generating the equations that constituted the core of his blackboard messages was not just fancy arithmetic, however. As one of the former physics students remarked, "It would have taken deep abstraction of the sort that real mathematicians perform."[32]

On one occasion, Feenberg wrote a computer program for Nash:

> He asked me if computer programming was something he should do. He'd seen me working with computers. He wanted to factor a twelve-digit number, which he felt was a composite number. He had already tested it against the first seventy thousand primes on a desk calculator. He had done it twice. He'd found no mistake, but he hadn't found a factor. I said we could do it. It took only about five minutes to write the program and test it. The answer came back: His number was a composite number that was the product of two primes.[33]

Nash was beginning to develop an interest in learning how to use the computer. (If one spent time in the computing center one had to sit at those ancient desk calculators by the hour, shuffling decks of computer cards.) Hale Trotter, who was working half-time in the computer center in those days, described it: "It was the old days. We fed cards into the computer. There was a large 'ready room' with a big counter, a card reader, table, and chairs and another room with a calculator. There was always lots of paper around."[34]

At the time, Trotter recalled, he kept track of people's computer time but nobody was billed. At some point the administration decided that he had to charge individual research accounts. Students and faculty alike had to open accounts and get passwords. Trotter initially told Nash that Nash could use his account number. At weekly meetings, the subject of regularizing the situation with Nash came up. Some students were wondering what was going on with Trotter's name on Nash's output. Someone suggested, said Trotter, "Why not give him his own account?" Everybody agreed to give him a free account. "He never, never made any trouble. If anything, he was embarrassingly diffident. Sometimes if one was having a conversation with Nash, it was hard to break away."

For most of the 1970s, Nash conducted his elaborate researches in the reference room of Firestone Library, where he was known to successive generations of students as "the library crazy man" and later as "the mad genius of Firestone."[35] In the late 1970s, he was often the last to leave the library at midnight. He spent evenings in the reference room, his floppy golf hat on the broad wooden table with a neat pile of books. He could spend two or three hours standing at the card catalog.

Charles Gillespie, a historian of science and editor of the *Dictionary of Scientific Biography,* had an office on the third floor of Firestone Library. Every day Nash would arrive at Firestone, marching down the walk, eyes straight ahead and briefcase in hand. He almost always headed for the third floor stacks, in a section of the library devoted to religion and philosophy. Gillespie always said good morning. Nash was always silent.[36]

Nash did, however, occasionally strike up acquaintanceships, as when he got to know two Iranian students during the summer of 1975. Amir Assadi, a big, smiling bear of a man, now on the mathematics faculty at the University of Wisconsin, recalled:

My brother spent the summer with me while I was studying for my generals. He used to wait for me in the common room. I'd seen Nash around and heard about him, but one day when I walked in he and my brother were talking intensely and I joined him. After that, I always said hello and we talked occasionally. He was extremely gentle and very shy. He seemed just so lonely. We were among the few people who talked to him. But he spoke freely to my brother. I suppose he saw a lonely foreigner.

Usually the conversations were quite short, but sometimes he would go on and on. It seemed scholarly to us. He didn't act bizarre. He used to read the *Encyclopaedia Britannica.* He had enormous knowledge. Nash was interested in Zoroastrian religion. Zarathustra was an ancient Iranian prophet. He wasn't mad. He wasn't someone who "had a yellow camel [i.e., crazy]." The religion he founded was based on three principles: good deeds, good thoughts, good expressions. Fire was holy. Light and darkness were always locked in struggle. Fires always burn in Zoroastrian temples. They are mono-theists. Nash would ask us to verify this and that. Occasionally we went and really read something.

In Iran the sense of sympathy and deep regret for a person being lonely is very great. We felt sorry.[37]

Nash's daily rounds in those years followed a predictable pattern. He would get up, not too early, and ride the Dinky into town, buy a copy of *The New York Times,* walk over to Olden Lane, eat breakfast or lunch at the Institute, and wander back to the university, where he could be found either in Fine or in Firestone. For some time, he became a regular at Fine Hall teas. The year Joseph Kohn became chairman of the math department, 1972, Kohn spent "many sleepless nights" over Nash. Some of the math department secretaries had come to him at various times saying that Nash's behavior worried them.[38] Kohn couldn't remember exactly what the behavior was but guessed that it involved staring. In any case, he brushed the women's complaints aside, saying that there was nothing to worry about, but privately he wasn't so sure.

With a few exceptions, such as Trotter, the faculty tended to avoid him. Claudia Goldin, who was on the economics faculty at the time, recalled:

He was an intriguing mystery. He just seemed to be around. Here was this giant and all of us were standing on his shoulders. But what kind of shoulders were they? For academics, there's always this fear. All you have is your brain. The idea that anything could go wrong with it is so threatening. It's threatening for everybody, of course, but for academics that's all of it.[39]

Mostly it was students who knew a bit of his legend, who generally found him nonthreatening, who sought him out. Feenberg, for example, had lunch with Nash. "Everyone knew he was a great man and just having lunch was an interesting experience. It was sad also. Here was this presence, this very famous person in our midst that people outside of Princeton often thought was dead."[40]

In 1978, largely thanks to the kindness of his old classmate from graduate school and RAND, Lloyd Shapley, Nash was finally awarded a mathematical prize. He was awarded the John von Neumann Theory Prize by the Operations Research Society and the Institute for Management Science jointly with Carl Lemke, a mathematician, of Rensselaer Polytechnic Institute.[41] Nash won for his invention of noncooperative equilibrium; Lemke for his work in computing Nash equilibria.[42]

Lloyd Shapley was on the prize committee. It was his idea. "I felt sentiment and nostalgia," he recalled.[43] Shapley, having received the honor himself the year before, thought: "Here's a chance to do something for Nash." He was motivated, he later said, by the hope that honoring Nash would somehow help Alicia and Johnny. "My sentiment, such as it was, was based on picturing him growing up. Here's this kid growing up and his dad isn't there. This might do something to increase his self-esteem. His father isn't there, but he's great, his work is being recognized."[44]

Nash was not, however, invited to the prize ceremony in Washington.[45] Instead, Alan Hoffman, a mathematician at IBM and the second member of the prize committee, went down to Princeton to present Nash with the award.[46] He said: "We gathered in Al Tucker's office. Al and Harold Kuhn were there, so we chatted a while. Nash was sitting in the corner. Let me tell you, seeing this man who was a genius and now functioning at subadolescent level really was tragic. There's a difference between knowing and seeing."[47]

46
A Quiet Life

Princeton, 1970–90

I have been sheltered here and thus avoided homelessness. —JOHN NASH,
1992

WHEN ALICIA OFFERED to let Nash live with her in 1970, she was moved by pity, loyalty, and the realization that no one else on earth would take him in. His mother was dead, his sister unable to accept the burden. Alicia was, divorced or no, his wife. Whatever her reservations about living with her mentally ill ex-husband, they played no role in her thinking: She was simply not prepared to turn her back on him.

Alicia also was moved by the conviction that she had something more to offer Nash than physical shelter. She believed, perhaps somewhat wishfully, that living in an academic community among his own kind, without the threat of further hospitalization, would help him get well. She took Nash's own assessment of his needs — for safety, freedom, and friendship — literally. In a letter to Martha written at Nash's request in late 1968, when he was convinced that his mother and sister planned to hospitalize him again, Alicia had argued that hospitalization was unnecessary and harmful: "Much of his past hospitalization I now feel was a mistake and had no beneficial permanent effects, rather the opposite. If he is to make a lasting adjustment, I think this has to be done under normal conditions."[1]

In 1968, Alicia had attributed her change of heart not just to the fact that Nash had relapsed despite aggressive treatment but, more important, to her own experiences since her divorce, which gave her new insights into Nash's plight. She wrote to Martha, "I feel that I now understand his difficulties much better than I ever did in the past, having experienced some of his type of problems personally."[2] Like many of those who tried to help Nash, Alicia was moved by a very personal and direct identification with his suffering.

Alicia's beauty and vulnerability, a mix made even more potent because of her history of personal tragedy, made it likely that someone would fall in love with her.

Forty-something, a professor of mathematics, John Coleman Moore might have inhabited the pages of an F. Scott Fitzgerald novel rather than an office at Fine Hall. His dark good looks, formal manners, and custom-made suits distinguished him from the rather scruffy ranks of fellow mathematicians. And his command of French and intimate knowledge of his native New York and assorted European capitals lent him a sophisticated aura. A bachelor, Moore was also a ladies' man.

When they returned from their separate years in Paris, Moore, Nash, and Alicia sometimes had dinners à trois. But it wasn't until after the Nashes' divorce, in mid-1963, and after Moore, described by a former girlfriend as "rigid and prim,"[3] suffered a devastating mental collapse of his own that the relationship turned romantic. Plagued by alcoholism and severe depression, Moore was hospitalized at a swank, psychoanalytically oriented hospital outside Philadelphia.[4] During two and one-half lonely years in which Moore remained in the hospital, other than Donald Spencer and George Whitehead, his thesis adviser from MIT, Alicia was his only regular visitor. Whitehead, who ran into Alicia a few times there, recalled: "There were lots of people in P-town who didn't come and see him. He was remarkably thankful for visitors."[5]

The friendship, born out of shared experiences and mutual sympathy, blossomed into romance.[6] Moore returned to Princeton and his teaching duties in the summer of 1965, about the same time that Nash moved to Boston. He became Alicia's regular escort at Princeton dinner parties, concerts, and the like. Whether it was a great love match, as her marriage to Nash had been, isn't clear. Moore, for all his charm and kindness, had little of the sort of charisma that had attracted Alicia so wildly to Nash. She yearned for someone who could take care of her, though. And for some time it appeared that they would marry.

At the time that Nash left Princeton, Alicia was still working at RCA. Her mother, who moved in with her after the death of her husband, kept house for Alicia as she had done in Cambridge years earlier. Mrs. Larde also helped take care of Johnny, who had grown into an extremely bright and altogether adorable boy, tall, sweet-faced, and still very blond.

Things started to unravel when Alicia suddenly lost her job at RCA. The company's space division had been periodically buffeted by contract cancellations and layoffs. Alicia, who was frequently absent, often late, or simply too depressed when she was at work to be effective, was particularly vulnerable.[7] She found another job fairly quickly, but it didn't last. She could not seem to get on her feet again. For a grim period that lasted several years, she drifted from job to job and was frequently unemployed, a fact to which she alluded obliquely in her letter to Martha. Alicia was determined to get a job that matched her educational credentials, but few aerospace companies were hiring female engineers in that era, and Alicia was turned down for more than thirty such positions. "There were times when I was going to interviews every day all day," she later recalled. "But I never got any offers. It was very depressing."[8]

Things got so bad after her unemployment benefits ran out that she was forced to go on welfare and to use food stamps.[9] Her hope of marrying Moore came to nothing. He backed away, finding the prospect of taking on a stepson as well as a wife "too much."[10] Her mother "held everything together," as Alicia later said, but it was very hard.[11]

Alicia and her mother were forced to give up the nice house they were sharing on Franklin Street in the heart of Princeton proper.[12] Alicia found a tiny nineteenth-century frame house in Princeton Junction, long ago swathed in Insulbrick, to rent. It was in poor repair, but cheap and convenient for commuting, since it was literally across the road from the railroad station. Johnny, who was twelve by this time, was extremely unhappy over having to leave his school and friends. But Alicia had little choice.

Nash moved to the Junction with her, contributing some of his small income from the trust left by Virginia to pay the rent and household expenses. Alicia referred to him as a "boarder,"[13] but in fact they ate meals together and Nash spent a fair amount of time with Johnny, sometimes helping him with his homework or playing chess with him.[14] Alicia had taught her son, who would later become a chess master, how to play.

Nash was very withdrawn, very quiet. "He was not a troublemaker," Odette recalled.[15] Haphazardly dressed, his gray hair long, his expression blank, he would wander up and down Nassau Street. Teenagers would taunt him, planting themselves in his path, waving their arms, shouting rude things directly into his startled face.[16] Alicia was a proud woman, always sensitive to appearances; her loyalty and compassion outweighed her concern for what others might think.

She was patient. She bit her tongue. She made very few demands on Nash. Looking back, her gentle manner probably played a substantial role in his recovery.[17] Had she threatened or pressured Nash, he very well might have wound up on the street. This point was made by Richard Keefe, a psychiatrist at Duke University. Contrary to conventional wisdom, which held that families of the mentally ill should "let it all out," more recent research suggests that people with schizophrenia are no more able to tolerate the expression of strong emotion than patients recovering from a heart attack or cancer surgery.[18]

Alicia is a scrupulously honest person. She says of the role she has played in protecting Nash simply, "Sometimes you don't plan things. They just turn out that way."[19] She does see that it helped him, though, saying, "Did the way he was treated help him get better? Oh, I think so. He had his room and board, his basic needs taken care of, and not too much pressure. That's what you need: being taken care of and not too much pressure."

In 1973, Alicia's circumstances started to improve. She had filed a sex discrimination suit against Boeing, one of the companies that had turned her down for a job in the late 1960s.[20] It was a feisty thing to do, and the suit, which eventually netted her a modest out-of-court settlement, helped boost her morale. She got a programming job at Con Edison in New York City, where her old college friend Joyce Davis was working.[21] It wasn't easy. She got up every morning at four-thirty to make the two-hour commute from Princeton Junction to Con Edison's Gramercy Park headquarters in downtown Manhattan and came home well past eight every evening. She often felt frustrated by the work itself, her boss, Anna Bailey, another acquaintance from MIT, recalled. She felt that her brains and education weren't being sufficiently recognized.[22]

But now that she was making a good salary again, she was able to enroll Johnny in the Peddie School, a private preparatory school in Hightstown, about ten miles west of Princeton.[23] Johnny, who had become moody and difficult at home, was nonetheless an excellent student. By the end of his sophomore year, when he won a Rensselaer Medal in a national competition, he had a 4.0 average.[24] And he was showing a marked interest in and a talent for mathematics. "John talked to Johnny a lot about mathematics when he was growing up," Alicia later recalled, adding, "If his father hadn't been a mathematician, Johnny would have been a doctor or a lawyer."[25]

Johnny started hanging around the Fine Hall common room to play chess and go and talk mathematics with various graduate students. Amir Assadi remembered him as "gentle, a nice kid, a tiny bit awkward, like other mathematicians . . . until they find their context."[26] Johnny was obviously gifted. Assadi recalled that he was studying "very high-powered math books." Sometimes father and son would come to Fine Hall together. Johnny didn't seem embarrassed, but neither did he ever refer to his father when talking to the students. Assadi recalled, "He disappeared one day. When he came back he'd shaved his head and had become a born-again Christian."

In 1976, Solomon Leader was visiting his friend Harry Gonshor — the same Gonshor who had been part of Nash's crowd at MIT, now a professor on the Princeton faculty — at the Carrier Clinic.[27] As the orderly ushered Leader through the locked door of the ward, a tall, wild-eyed young man suddenly loomed before him. "Do you know who I am?" he shouted right into Leader's face. "Do you want to be saved?" Leader noticed he was clutching a Bible. Afterward, Gonshor told him that the man was the son of John Nash.

By the time Johnny was hospitalized at Carrier at his mother's initiative, he had been truant for nearly a year.[28] He had dropped all of his old friends. For many months, he had refused to leave his room. When his mother or grandmother tried to intervene, he lashed out at them. He had begun reading the Bible obsessively and talking about redemption and damnation.[29] Soon he began hanging out with members of a small fundamentalist sect, the Way Ministry, and handing out leaflets and buttonholing strangers on street corners in Princeton.[30]

It was not immediately obvious to Alicia or her mother that Johnny's troubling behavior was anything more than an outburst of adolescent rebellion. In time it became clear that Johnny was hearing voices and that he believed that he was a great religious figure. When Alicia tried to get him into treatment, he ran away. He stayed away for weeks and Alicia had to go to the police for help in tracking him down and bringing him back. And then, when her son was in Carrier, Alicia learned that the thing she most dreaded, had dreaded all along, was true. Her brilliant son was suffering from the same illness as his father.[31]

Johnny seemed to improve quickly after the first hospitalization. But he did not return to school for three years.[32] Alicia never talked about him at work except when she was forced to ask for time off.[33] She never told anyone at Con Edison that John Nash was living with her again. Like Virginia Nash a decade earlier, she treated her woes as her private sorrow. She tried to cope with Johnny's refusal to take medication, his constant running away, his periodic need for hospitalization, and the terrible drain on her slender resources without giving in to her own depression. "You sacrifice so much, you put so much into it, and then it all goes," she said later.[34]

As the trouble with Johnny overwhelmed her, Alicia turned to her friend Gaby Borel for support. Gaby accompanied Alicia on visits to Carrier, and later to Trenton Psychiatric, talked with her on the telephone, and invited the Nashes to dinner.[35] Moore confirms this: "Gaby is the closest female friend Alicia has around here. Gaby is very good. Nobody else was around consistently."[36]

Gaby's tribute to Alicia's stoicism holds true to this day: "At first, you cannot tell anything about her. You do not realize who she is. She has put a sort of shield around herself. But she is a very brave and faithful woman."[37]

In 1977, John David Stier made a cameo appearance in Nash's life.[38] Father and son had been in touch by letter at least since 1971, John David's senior year in high school. Nash had become quite concerned about his son's college plans, and Alicia had written Arthur Mattuck to ask him to advise John David.[39] John David enrolled at Bunker Hill Community College and supported himself by working as an orderly.[40] Four years later, he applied to a number of four-year schools, was offered several scholarships, and in 1976 transferred to Amherst, one of the most elite liberal arts colleges in the country.

That fall Norton Starr, a professor of mathematics at Amherst, hired a student to do some yard work for him.[41] Afterward, Starr invited him into the house for a cold drink. As they chatted, the young man learned that Starr had done his Ph.D. at MIT. Had he known a mathematician there named John Nash? Only by sight and reputation, Starr replied. "He's my father," the young man said. Starr looked at him searchingly. He looked at the young man again. "My God, you do look just like him," he said. Shortly afterward, John David drove down to Princeton to visit his father. Alicia was friendly. He met his brother, Johnny, for the first time.

•   •   •

The following Christmas, Johnny came up to Boston to stay with Eleanor and John David. Eleanor welcomed him warmly, cooked him nice meals, fussed over him. He came without a winter coat, so Eleanor bought him a down jacket. Johnny was well-behaved around his older brother, but could turn nasty when he was alone with her. At the end of the holiday, Eleanor recalled, "he didn't want to let John go. So John took him back to school with him."[42]

The reunion between Nash and John Stier did not lead to a lasting reconciliation. "It just sort of petered out," John Stier recalled. His father was more interested in talking about his own problems than his son's. "When I asked him for advice, he'd answer with something about Nixon," he said.[43] Nash's confidences were unsettling. Nash had some idea that his son, having attained his majority, would play "an essential and significant personal role in my personal long-awaited 'gay liberation.' "[44] He had waited a long time, as he said at the time, to "tell him about my life and problems and life history." Eleanor Stier recalled that he did so.[45] John David eventually stopped returning his father's calls. The two would not meet again for seventeen years. "I haven't always wanted to have contact with him," John David said. "Having a mentally ill father was rather disturbing."

More often than commonly realized, schizophrenia can be an episodic illness, especially in the years following its initial onset. Periods of acute psychosis may be interspersed with periods of relative calm in which symptoms diminish dramatically either as a result of treatment or spontaneously.[46] This was the pattern for Johnny.

In 1979, on the first day of the fall semester at Rider College in Lawrenceville, New Jersey, Kenneth Fields, the chairman of the mathematics department, was asked to talk with a freshman who had made a pest of himself at the math orientation session, questioning everything and protesting that the presentation was not rigorous enough.[47] "I don't need to take calculus," the young man said when he arrived in Fields's office. "I'm going to major in math." Since Rider rarely attracted students with an interest or background in mathematics, Fields was intrigued. Quizzing the student as they walked around the campus, he quickly concluded that no mathematics course at Rider was advanced enough for this young man and offered to tutor him personally. "By the way, what's your name?" he finally asked. "John Nash," the student replied. Seeing Fields's look of astonishment, he added, "You may have heard of my father. He solved the embedding theorem." For Fields, who had been an undergraduate at MIT in the 1960s and was familiar with the Nash legend, it was an amazing moment.

Fields proceeded to meet with Johnny weekly. Johnny took a while to buckle down, but he was soon plowing through difficult texts in linear algebra, advanced calculus, and differential geometry. "It was obvious that he was a real mathematician," said Fields. He was also bright and friendly, a fundamentalist Christian who made friends with other religious, intellectually precocious students. He talked to Fields, who has several relatives who suffer from schizophrenia, about his mental illness. Occasionally he would do a riff on extraterrestrials, and on one occasion he threatened a history professor. By and large, said Fields, Johnny's symptoms seemed to be under control. He got straight A's and won an academic prize in his sophomore year.

Fields soon concluded that Johnny was wasting his time at Rider and belonged in a Ph.D. program. In 1981, despite his lack of a high school or college diploma, Johnny was accepted at Rutgers University with a full scholarship. Once there, he breezed through his qualifying examinations. From time to time he would threaten to drop out of school and Fields would get frantic calls from Alicia begging him to talk to Johnny. When Fields did, Johnny would answer, "Why do I have to do anything? My father doesn't have to do anything. My mother supports him. Why can't she support me?" But he didn't drop out. He succeeded brilliantly.

Melvyn Nathanson, then a professor of mathematics at Rutgers, liked to assign what he called simple versions of unsolved classical problems in his graduate course on number theory.[48] "I gave one the first week," he recalled. "Johnny came back with the solution the following week. I gave another one that week and a week later he had that solution too. It was extraordinary." Johnny wrote a joint paper with Nathanson that became the first chapter of his dissertation.[49] He then wrote a second paper on his own, which Nathanson called "beautiful" and which also became part of the thesis.[50] His third paper was an important generalization of a theorem proved by Paul Erdős in the 1930s for a special case of so-called B sequences.[51] Neither Erdős nor anyone else had succeeded in proving that the theorem held for other sequences, and Johnny's successful attack on the problem would generate a flurry of papers by other number theorists.

When Johnny got his Ph.D. from Rutgers in 1985, said Nathanson, he seemed poised for a long and productive career as a first-rate research mathematician. An offer of a one-year instructorship at Marshall University in West Virginia seemed like the first of the usual steps that eventually carry new mathematics Ph.D.'s to tenured positions somewhere in academia. While Johnny was in graduate school, Alicia Larde returned to El Salvador for good and Alicia Nash moved to a job as a computer programmer at New Jersey Transit in Newark.[52] Things seemed rather hopeful.

PART FIVE

The
Most
Worthy

47
Remission

As you know, he has had his illness, but right now he's fine. It's not attributable to one or several things. It's just a question of living a quiet life.
—ALICIA NASH, 1994

PETER SARNAK, a brash thirty-five-year-old number theorist whose primary interest is the Riemann Hypothesis, joined the Princeton faculty in the fall of 1990. He had just given a seminar. The tall, thin, white-haired man who had been sitting in the back asked for a copy of Sarnak's paper after the crowd had dispersed.

Sarnak, who had been a student of Paul Cohen's at Stanford, knew Nash by reputation as well as by sight, naturally. Having been told many times Nash was completely mad, he wanted to be kind. He promised to send Nash the paper. A few days later, at teatime, Nash approached him again. He had a few questions, he said, avoiding looking Sarnak in the face. At first, Sarnak just listened politely. But within a few minutes, Sarnak found himself having to concentrate quite hard. Later, as he turned the conversation over in his mind, he felt rather astonished. Nash had spotted a real problem in one of Sarnak's arguments. What's more, he also suggested a way around it. "The way he views things is very different from other people," Sarnak said later. "He comes up with instant insights I don't know I'd ever get to. Very, very outstanding insights. Very unusual insights."[1]

They talked from time to time. After each conversation, Nash would disappear for a few days and then return with a sheaf of computer printouts. Nash was obviously very, very good with the computer. He would think up some miniature problem, usually very ingeniously, and then play with it. If something worked on a small scale, in his head, Sarnak realized, Nash would go to the computer to try to find out if it was "also true the next few hundred thousand times."

What really bowled Sarnak over, though, was that Nash seemed perfectly rational, a far cry from the supposedly demented man he had heard other mathematicians describe. Sarnak was more than a little outraged. Here was this giant and he had been all but forgotten by the mathematics profession. And the justification for the neglect was obviously no longer valid, if it had ever been.

· · ·

That was 1990. In retrospect, it is impossible to say exactly when Nash's miraculous remission, which began to be noted by mathematicians around Princeton roughly at the beginning of this decade, really began. But, in contrast to the onset of his illness, which became full-blown in a matter of months, the remission took place over a period of years. It was, by his own account, a slow evolution, "a gradual tapering off in the 1970s and 1980s."[2]

Hale Trotter, who saw Nash nearly every day in the computer center during those years, confirms this: "My impression was of a very gradual sort of improvement. In the early stages he was making up numbers out of names and being worried by what he found. Gradually, that went away. Then it was more mathematical numerology. Playing with formulas and factoring. It wasn't coherent math research, but it had lost its bizarre quality. Later it was real research."[3]

As early as 1983, Nash was beginning to come out of his shell and making friends with students. Marc Dudey, a graduate student in economics, sought Nash out in 1983. "I felt bold enough at the time to want to meet this legend."[4] He discovered that he and Nash shared an interest in the stock market. "We'd be walking along Nassau Street and we'd be talking about the market," Dudey recalled. Nash struck Dudey as a "stock picker" and on occasion Dudey followed his advice (with less than stellar results, it must be said). The following year, when Dudey was working on his thesis and was unable to solve the model he wanted to use, Nash helped to bail him out. "The calculation of an infinite product was involved," Dudey recalled. "I was unable to do it, so I showed it to Nash. He suggested I use Stirling's formula to compute the product and then he wrote down a few lines of equations to indicate how this should be done." All during this time, Nash struck Dudey as no odder than other mathematicians he had encountered.

By 1985, Daniel Feenberg, who had helped Nash factor a number derived from Rockefeller's name a decade earlier and was now a visiting professor at Princeton, had lunch with Nash. He was deeply struck by the change he saw in Nash. "He seemed so much better. He described his work in the theory of prime numbers. I'm not competent to judge it, but it seemed like real mathematics, like real research. That was very gratifying."[5]

The changes were for the most part visible only to a few. Edward G. Nilges, a programmer who worked in Princeton University's computer center from 1987 to 1992, recalled that Nash "acted frightened and silent" at first.[6] In Nilges's last year or two in Princeton, however, Nash was asking him questions about the Internet and about programs he was working on. Nilges was impressed: "Nash's computer programs were startlingly elegant."

And in 1992, when Shapley visited Princeton, he and Nash had lunch and were able, for the first time in many, many years, to have quite an enjoyable conversation. "Nash was quite sharp then," Shapley recalled. "He was free of this distraction. He'd learned how to use the computer. He was working on the Big Bang. I was very pleased."[7]

• • •

That Nash, after so many years of severe illness, was now "within the normal range for the 'mathematical personality' " raises a great many questions. Had Nash really recovered? How rare is such a recovery? Did the "recovery" indicate he had never really had schizophrenia, which, as everyone knows, is incurable? Were his psychotic episodes in the late 1950s through the 1970s really symptoms of bipolar illness, which is generally less debilitating and carries better odds of recovery?

Absent a re-diagnosis based on Nash's psychiatric records, no absolutely definitive answer is possible. Psychotic symptoms alone, psychiatrists now agree, "do not a schizophrenic make," and distinguishing between schizophrenia and bipolar illness when symptoms first appear remains difficult even with today's more precise diagnostic criteria.[8] Nonetheless, there are strong reasons for believing that Nash's initial diagnosis was, in fact, correct and that he is one of a very small number of individuals who suffered a long and severe course of schizophrenia to experience a dramatic remission.

The fact that Nash's younger son has also been diagnosed with paranoid schizophrenia and schizoaffective disorder is strong evidence that Nash himself had schizophrenia.[9] In contrast to the Freudian theories popular in the 1950s, when Nash was first diagnosed, schizophrenia is now thought to have a strong genetic component.[10]

The duration and severity of Nash's symptoms — his inability to do work that was, prior to and since his illness, the principal passion of his life, and his withdrawal from most human contact — is also powerful evidence. Moreover, Nash has described his illness not in terms of highs and lows, bouts of mania followed by disabling depression, but rather in terms of a persistent dreamlike state and bizarre beliefs in terms not dissimilar to those used by other people with schizophrenia.[11] He has spoken of being preoccupied by delusions, of being unable to work, and of withdrawing from the people around him. Mostly, however, he has defined it as an inability to reason.[12] Indeed, he has told Harold Kuhn and others that he is still plagued by paranoid thoughts, even voices, although, in comparison to the past, the noise level has been turned way down.[13] Nash has compared rationality to dieting, implying a constant, conscious struggle. It is a matter of policing one's thoughts, he has said, trying to recognize paranoid ideas and rejecting them, just the way somebody who wants to lose weight has to decide consciously to avoid fats or sweets.[14]

While psychiatry has made progress in defining disease, definitions of recovery remain controversial. The absence of obvious symptoms, as George Winokur and Min Tsuang wrote, "does not necessarily mean that [individuals] are well, since they still may be suffering from a defect state that is stabilized and with which they have now learned to cope." But such an assessment, possibly appropriate to Nash's state in the late 1970s and early 1980s, seems overly pessimistic now. Both the perceptions of those who know Nash and his own indicate a more expansive, far-reaching change. "John has definitely recovered," said Kenneth Fields of Rider College, who has known Nash since the late 1970s and has had a great deal of firsthand experience with people who suffer from schizophrenia.

It would be more accurate to describe Nash's recovery as a "remission." And, it turns out, the remission, though miraculous, is not unique. Until a few years ago, nobody knew much about the life history of people with schizophrenia. The only studies dated to the 1970s and were done by psychiatrists who worked at state hospitals. Since the only older people who were still there to be studied were still sick enough to require constant hospitalization, schizophrenia was viewed as a degenerative disease. Its assault on the brain was thought to continue, more or less evenly, until death.

Manfred Bleuler, a German psychiatrist, was the first researcher to systematically challenge this view.[15] In a twenty-year follow-up of more than two hundred patients, he found 20 percent "fully recovered." Moreover, he concluded that long-lasting recoveries did not result from treatment and hence appeared to be spontaneous.

Then a German team at the University of Bonn did a long-term follow-up of patients who had been admitted to one of the city's psychiatric hospitals during the late 1940s and early 1950s.[16] Going back to the records, they reviewed the diagnosis of schizophrenia and chose only patients whose histories and symptoms were consistent with modern definitions of the disease. There were about five hundred. Then they located the people or their families and, through interviews with the patients and people who knew them, created detailed portraits of what had happened to them.

Many — about a quarter — had died, mostly suicides. Some were still institutionalized, apparently unresponsive to any drugs or to electroshock treatment, which was used far more extensively than in the United States. Another group was living with their families, but still had symptoms, especially the negative symptoms of lethargy, lack of drive, and lack of interest and pleasure in life. But a surprisingly large group — perhaps a quarter — seemed to be symptom-free, living independently, with a circle of friends and jobs in the professions for which they had been trained or had held before they got sick. Most of these had not been under the care of a physician for years.

The researchers were extremely surprised. As news of the study results spread through the small global community of schizophrenia researchers, a team in the United States at the University of Vermont decided to undertake a similar long-term study. Despite their initial skepticism, their results were remarkably similar.[17] Ten years after the disease struck, most patients were still extremely sick. Thirty years later, however, a significant minority were leading fairly normal lives. Only about 5 percent conformed completely to the backward image. Most of those who committed suicide, it turned out, did so in the first ten years of the disease. These appeared to be people who got well enough between acute episodes to appreciate the awfulness of what lay ahead of them and succumbed to despair. And most of the damage to thinking and emotion from the disease seemed to occur in those years as well. After that, symptoms seemed to level out.

Subsequent research has somewhat tempered these optimistic conclusions.[18] All long-term studies are plagued by uncertainties about diagnoses and by differences over what constitutes "recovery." A study by Winokur and Tsuang of 170 patients, perhaps the most rigorous, found that thirty years after the onset of the illness, just 8 percent could be considered well.[19]

Thus, while Nash's dramatic recovery is not unique, it is relatively rare.

While none of the studies was able to pinpoint factors that favored recovery, they suggest that someone with Nash's history prior to the onset of his illness—high social class, high IQ, high achievement, with no schizophrenic relatives, who gets the disease relatively late in the third decade, who experiences very acute symptoms early and gets sick at the time of some great life change—has the best chance of remission.[20] On the other hand, young men like Nash for whom the contrast between early achievement and the state to which they are reduced by the disease is greatest are also most likely to commit suicide. Since suicides are relatively rare for hospitalized patients, Martha may have saved Nash's life by insisting, during the 1960s, that he be hospitalized. Whether or not insulin shock and antipsychotic drugs, which apparently produced the temporary remissions Nash experienced in the first half of the 1960s, increased the odds of a remission later in life is unclear. While a larger number of patients who got sick during the 1950s, when antipsychotic drugs became available on a wide scale, were among those who were symptom free in late middle age, early treatment with drugs wasn't a particularly accurate indicator of what would happen later.[21] At the same time, Nash's refusal to take the antipsychotic drugs after 1970, and indeed during most of the periods when he wasn't in the hospital during the 1960s, may have been fortunate. Taken regularly, such drugs, in a high percentage of cases, produce horrible, persistent symptoms like tardive dyskinesia—stiffening of head and neck muscles and involuntary movements, including of the tongue—and a mental fog, all of which would have made his gentle reentry into the world of mathematics a near impossibility.[22]

Nash's remission did not come about, as many people later assumed, because of some new treatment. "I emerged from irrational thinking," he said in 1996, "ultimately, without medicine other than the natural hormonal changes of aging."[23]

He described the process as one that involved both a growing awareness of the sterility of his delusional state and a growing capacity for rejecting delusional thought. He wrote in 1995:

> Gradually I began to intellectually reject some of the delusionally influenced lines of thinking which had been characteristic of my orientation. This began, most recognizably, with the rejection of politically-oriented thinking as essentially a hopeless waste of intellectual effort.[24]

He believes, rightly or wrongly, that he willed his own recovery:

Actually, it can be analogous to the role of willpower in effectively dieting: if one makes an effort to "rationalize" one's thinking then one can simply recognize and reject the irrational hypotheses of delusional thinking.[25]

"A key step was a resolution not to concern myself in politics relative to my secret world because it was ineffectual," he wrote in his Nobel autobiography. "This in turn led me to renounce anything relative to religious issues, or teaching or intending to teach.

"I began to study mathematical problems and to learn the computer as it existed at the time. I was helped (by mathematicians who got me computer time)."[26]

By the late 1980s, Nash's name was appearing in the titles of dozens of articles in leading economics journals.[27] But Nash himself remained in obscurity. Many younger researchers, of course, simply assumed he was dead. Others thought that he was languishing in a mental hospital or had heard that he had a lobotomy.[28] Even the best-informed saw him, for the most part, as a sort of ghost. In particular, with the exception of the 1978 von Neumann Prize — the result of Lloyd Shapley's efforts — the recognition and honors routinely accorded scholars of his stature simply failed to materialize.[29] One particularly egregious episode in the academic year 1987–88 illustrated just how powerfully the perceptions of Nash's mental illness worked to reinforce his marginalized status, even in the field, economics, that he had helped to revolutionize.

Being elected a Fellow in the Econometric Society is, as one former president of the society put it, tantamount to getting one's membership card in the club of bona-fide economic theorists.[30] By 1987, there were some 350 living Fellows, including every past and future Nobel Laureate to date but Douglass North (presumably excluded because he is an economic historian, not a mathematical economist), as well as every leading contributor to game theory — Kuhn, Shapley, Shubik, Aumann, Harsanyi, Selten, and so forth — but not Nash.[31] In late 1988, Ariel Rubinstein, a recently elected Fellow, was surprised to discover this "historic mistake" and promptly nominated Nash.[32]

The nomination came too late for the November 1989 election. Further, the society's bylaws required any candidate proposed by a sole sponsor to pass muster with the society's five-member nominating committee — one of whose main tasks was, in any case, to "determine whether previous nominating committees had overlooked people" and to correct such oversights.[33] As a result, the nomination was forwarded to the committee, which took it up in the spring of 1989. By then, Rubinstein, a game theorist who holds professorships at the University of Tel Aviv and Princeton University, was a member of the committee. The other members, all professors of economics, were Mervyn King at the London School of Economics (also a vice-chairman of the Bank of England), Beth Allen at the University of Minnesota, Gary Chamberlain at Harvard, and Truman Bewley at Yale.[34]

The proposal to put Nash on the ballot sparked an intense controversy between Rubinstein and the rest of the committee, one that dragged on for months. From the start, the issue was Nash's mental illness. Mervyn King said in 1996: "People felt in some vague sense this was relevant."[35] Other committee members pointed out that Nash had no recent publications, was not even a member of the society, and was unlikely to participate actively, if elected.[36] At one point Truman Bewley, the committee's chairman, wrote to Rubinstein, "I doubt [Nash] would be elected, since he is well known to have been crazy for years," dismissing the nomination as "frivolous."[37] When Rubinstein refused to back down, Bewley asked him to find out more about "the current status of Nash's health." After Rubinstein objected that no other candidates were being similarly investigated, Bewley made his own inquiries, calling, among others, his colleague at Yale Martin Shubik, who had known Nash in graduate school and had received some of Nash's "mad" letters. Bewley reported back to the committee: "Regarding Nash, I inquired and learned that he is still crazy. Fellowship is an activity more than a reward for past work. The fellows are the ultimate governing body of the Econometric Society."[38]

In June, the committee voted four to one to keep Nash off the November 1989 ballot. Rubinstein was the sole dissenter. Beth Allen recalled, "People were asked to give a rank ordering. Nash didn't make it. Ariel had a fit. He insisted Nash be put on the ballot anyway." Bewley made it clear that the matter was closed, a decision he later regretted. "It was the wrong decision," he said in 1996.[39] The episode is reminiscent of the Institute for Advanced Study's refusal, for many years, to grant a mathematics professorship to the world-renowned logician Kurt Gödel.[40] But, in that case, there was considerably more justification, since the Institute's tiny mathematics faculty feared that Gödel's well-known paranoia and terror of decision-making would hamstring its ability to conduct business, which included the selection of each year's visiting scholars.[41]

The crowning irony of this affair is that when Nash did get on the ballot, in the election for 1990 (because Rubinstein circumvented the nominating committee by submitting a joint nomination with Kenneth Binmore, at the University of Michigan, and Roger Myerson, at Northwestern University),[42] he received, according to the Secretary of the society, Julie Gordon, "the overwhelming majority of the votes."[43]

48

The Prize

You will have to wait to find out [the story of Nash's prize] in fifty years. We will never reveal it. — CARL-OLOF JACOBSON, secretary general, Royal Swedish Academy of Sciences, February 1997

IT IS TUESDAY, October 12, 1994. Jörgen Weibull, a personable young professor of economics, looks at his watch for perhaps the fiftieth time.[1] He is standing near the front of the massive Sessions Hall of the Royal Swedish Academy of Sciences — a jewelbox of a room with a heavily ornamented ceiling and portrait-lined walls — which, at the moment, is crowded with reporters and camera crews, jammed in narrow aisles between the U-shaped tables. Near-pandemonium reigns. Everybody is milling around, speculating in loud voices about the delay.

Weibull had been so elated when he left his office at the University of Stockholm that midmorning that he half walked, half ran through the highway underpass and up the hill to the academy half a mile away. Assar Lindbeck, the chairman of the prize committee, had asked him if he wouldn't mind being on hand to answer questions at the press conference — quite an honor. But now Weibull's mouth feels dry, his shoulders ache, and he can feel the first twinges of a headache as he tries to imagine what has gone wrong.

The Nobel press conference had, as usual, been called for eleven-thirty. These staid, heavily scripted events are always held right after the final, ceremonial vote and *always* start on time. But it is one o'clock and there is no sign of any academy officials and no word either. All the reporters are saying that nothing like this has ever happened before.

Suddenly, the enormous doors to his left swing open and a small knot of academy officials burst into the hall, all wearing slightly dazed expressions, like moviegoers stepping out of a theater into daylight. They hurry past the milling, shouting throng, ignoring the questions, brushing aside the demands for explanations. But Weibull, who is standing near the table with the microphones, manages to catch Lindbeck's eye for a fraction of a second. The relief is overwhelming. "Lindbeck didn't signal or anything like that," he said later, "but I saw right away that everything had turned out all right."[2] And the relief turns into something like joy when he listens to Carl-Olof Jacobson, the academy's handsome, silver-haired secretary general, read the first few words of the press release: "John Forbes Nash, Jr., of Princeton, New Jersey . . ."[3]

The behind-the-scenes saga of John Nash's Nobel Prize is almost as extraordinary as the fact that the mathematician became a Laureate at all. For years after the idea of a prize for game theory was first considered, even Nash's most ardent admirers considered the likelihood of his winning impossibly remote.[4] But much later, when the prize was virtually his, after he had been told that he had won it, and within an hour of the official notification, the *ne plus ultra* of honors very nearly eluded him — with far-reaching consequences for the future of the economics prize itself.

This previously untold story is one that the Royal Swedish Academy of Sciences and the Nobel Foundation — intent on preserving the Olympian aura that surrounds the prizes — have tried very hard to keep under wraps. The academy is one of the most secretive of societies, and all details — the nominations, inquiries, deliberations, and votes — of the lengthy selection process are among the most closely guarded secrets in the world. The very statutes of the prize demand it:

> Proposals received for the award of a prize, and investigations and opinions concerning the award of a prize may not be divulged. Should divergent opinions have been expressed in connection with the decision of the prize-winning body concerning the award of the prize, these may not be included in the record or otherwise divulged. A prize-winning body may, however, after due consideration in each individual case, permit access to material which formed the basis for evaluation and decision concerning a prize, for purposes of historical research. Such permission may not be granted until at least 50 years have elapsed after the date on which the decision in question was taken.[5]

There have been breaches, of course. In the 1960s and 1970s, advance rumors of the literature Laureates used to trickle out of the Academy of Arts and Letters with notorious regularity.[6] In 1994, a member of the Norewegian Nobel Committee quit over the impending peace prize to the Palestinian leader Yasir Arafat, and took his protest to the media. Michael Sohlman, the executive director of the Nobel Foundation, still sounds furious when he recounts the incident.[7]

But, few, if any, cracks have appeared, figuratively or otherwise, in the gray Beaux Arts walls of the Royal Swedish Academy of Sciences, guardian of the physics, chemistry, and economics prizes. If not for the mysterious one-and-a-half-hour delay on the day that the Nash prize was announced, the academy might well have succeeded in protecting the secrecy of the process. As it was, academy officials not only refused to explain the delay but denied that it was in any way significant. Indeed, they very quickly began to assert that it had never happened. Recently,

Karl-Göran Mäler, a member of the economics prize committee in 1994 and privy to all of the events that transpired, said, "I do not recall *any* delay."[8]

The prize in economics is something of a stepchild.[9] Alfred Nobel, the Swedish industrialist and inventor, did not have the dismal science in mind when he wrote his famous 1894 will creating Nobel Prizes in physics, chemistry, medicine, literature, and peace. The economics prize was not created until nearly seventy years later, the brainchild of the then head of the Swedish central bank. The prize is financed by the bank and administered by the Royal Swedish Academy of Sciences and the Nobel Foundation. It is not, in fact, a Nobel Prize, but rather "The Central Bank of Sweden Prize in Economic Science in Memory of Alfred Nobel." To the public, that is a distinction without much of a difference. The early winners of the economics prize—among them Paul Samuelson, Kenneth Arrow, and Gunnar Myrdal—were generally acknowledged to be intellectual giants and lent their distinction to the prize. And, so far at least, it has become "the ultimate symbol of excellence for scientists and laymen alike" and does in fact make economics Nobelists "life peers in the world community of scholars."[10]

The criteria, rules, and procedures for the economics prize are patterned after those that apply to the science prizes.[11] Candidates must be living. No more than three can share a prize, which is less of a problem in economics than in physical science, where teamwork is more the norm. Though many people, even those who participate in the nominating process, have failed to appreciate it, the Nobel is not a prize for outstanding individuals nor is it a lifetime achievement award. The prize is awarded for specific achievements, inventions, and discoveries. These can be theories, analytical methods, or purely empirical results. As in physics, in which mathematics plays as big a role as in economics, there is a strong bias against prizes for only mathematics.[12] (Nobel himself is said to have hated mathematicians, though some of the best stories about why—revolving around sexual and professional jealousy—turn out to have been apocryphal).[13]

The prize selection process is also virtually identical to the cycles for the science prizes.[14] A five-member prize committee, composed of senior Swedish economists, gathers nominations and referees reports from elite academics around the world. The committee makes its choice every spring, usually in April. The so-called Social Sciences Class—all academy members in economics and other social sciences—endorses the candidate or candidates in early fall, usually late August or early September. And the academy votes on the nominees in early October, on the day that the winner or winners are announced.

On paper, at least, all the members of the prize committee are as distinguished as the candidates, and the selection of winners is a detached, disinterested, and, ultimately, democratic exercise in scientific judgment—as divorced from personal likes and dislikes, prejudices, or political and pecuniary considerations as the business of determining the winners in a sports tournament. There is some, even a good deal, of truth in this idealized description of what actually goes on, but it is not anything like the whole story.

Assar Lindbeck, who joined the prize committee in 1969 and became its chairman in 1980, has dominated the economics selections for the entire history of the Nobel Prize.[15] Tall, red-haired, powerfully built, he looks like the boss of a machine tool shop or a mine. He is from the far north of Sweden, a little crude, a little uptight, more than a little brusque. He has opinions, strong ones, about nearly all topics that engage his lively mind, and as a result is quite unpopular in the academy. But he is not without a certain earthy charm. His sense of humor is sly and dry. He is a Sunday painter — showing up at prize committee meetings with paint spatters on his horn-rimmed glasses. A large — and extremely graphic — erotic painting hangs in his office at the university.

Lindbeck is Sweden's most important economist. Top academic economists in Sweden, where academia, government, and industry have long been closely entwined, have traditionally wielded a great deal more political power than their American counterparts.[16] Bertil Ohlin, the committee's first chairman, was for years the leader of Sweden's opposition. Gunnar Myrdal, who won the prize in 1974, was a minister in the Social Democratic government. Lindbeck himself was a protégé of Prime Minister Olof Palme, has held many political advisory posts, and has been involved in most public policy debates since the 1960s.

Unlike Ohlin and Myrdal, Lindbeck never abandoned his research career to become a full-time politician. Indeed, he is generally considered a likely contender for a Nobel himself. Even today, at age sixty-eight, there is a small assembly line on the shelves behind his desk at the University of Stockholm: impressively large piles of paper marked "Articles Under Preparation," "Articles Submitted," and "Articles Accepted." And he has used his political savvy and prestige to build up economics departments and research institutes. "He's kind of a mafia leader, a fixer," said Karl-Gustaf Löfgren, an adjunct member of the economics prize committee and a professor of resource economics at the University of Umea.[17] He adds:

> I never did any resource economics, but I became a professor of resource economics. [Lindbeck] has good ideas about who to move here and there. He listens. He has his own opinions. I like him. He's a very sound guy. Very smart.

Lindbeck has a reputation for getting his way. His style is that of a central banker rather than a chief executive officer. As his longtime friend Mäler put it, "Assar never controlled with commands."[18] In an article Lindbeck wrote on the economic prize in the mid-1980s, he bragged: "So far the proposals of the prize committee to the Academy have been unanimous. A consensus has in fact developed quite 'automatically' within the committee, as if by some kind of invisible hand, after intensive discussions." [19] The invisible hand, of course, was his own. "You *could* put it that way," said Löfgren, laughing. "You can *say* it's unanimous. . . . But he's a dominating person. We don't vote officially. You agree." [20]

Kerstin Fredga, the president of the Swedish Academy of Sciences, said at one point, "Very few people have ever dared say no to Assar." [21] Ironically, by December 1994, when Fredga made the remark, it was no longer true.

John Nash's name first appeared as a candidate for a Nobel in the mid-1980s. [22] The Nobel selection process is like a giant funnel. At any given time, the economics prize committee has a dozen "investigations" running of fields and clusters of possible candidates. But, fairly quickly, the focus shifts to the hottest fields and candidates. By 1984, the "obvious" Nobels had been handed out to the likes of Samuelson, Arrow, and James Tobin. The committee was looking further afield among newer branches of economics, and nothing was newer or hotter at that particular moment than game theory. [23]

In 1984, the prize committee contacted a young researcher at Hebrew University in Jerusalem. A combat veteran and an activist in Israel's peace movement, Ariel Rubinstein took months to write a painstaking ten-page report on potential candidates for a prize in game theory. He placed Nash at the top of the list. [24]

The 1982 paper that established Rubinstein as one of the leading researchers in game theory was an extension of Nash's 1950 bargaining paper. [25] Rubinstein's sense of indebtedness to Nash and his appreciation for Nash's original achievement were thus very vivid. Having encountered Nash on a visit to Princeton, Rubinstein also could not help but be struck by the stark contrast between Nash's past contributions and his current circumstances. His outrage was fueled partly by a firsthand encounter with the stigma of mental illness: his mother was once hospitalized for depression, and Rubinstein never forgot the lack of basic human respect accorded her by doctors and relatives. [26]

The Nobel Prize committee did not take up the matter again until 1987, when it commissioned a second report, this time from Weibull. [27] After he submitted it, Lindbeck told him that the committee wanted to ask him some questions and asked him to attend a couple of committee meetings at the Royal Academy. Weibull was, of course, pledged to complete secrecy.

When Weibull walked into the paneled room, introductions were hardly necessary. As a member of Sweden's small academic elite, Weibull already knew the five men, mostly academics, sitting around the enormous table. He was nonetheless slightly awed, realizing from the committee's questions that he was being given the opportunity to participate at the earliest stage of a historic decision. "My impression . . . [was] that it was the first time that the committee had met to consider this." [28]

Weibull presented a verbal summary of his report, telling the committee about the central ideas in game theory, their importance for economic research, and the key contributors. He, too, had placed Nash at the top of his list of half a dozen seminal thinkers.

The committee's questions were carefully phrased to hide the members' own opinions, and focused, in the first session, on whether game theory was just a fad or really an important tool for investigating a wide range of interesting economic problems. By the second meeting, however, Lindbeck, the committee chairman, zeroed in on Nash. Was what Nash did merely mathematics? Lindbeck asked. Did he simply formalize ideas that economists had formulated at least a hundred years earlier? Was it true that Nash had stopped doing research in game theory in the early 1950s? That question was the closest anyone came to mentioning the subject of Nash's mental illness.[29]

When Weibull left the meeting, he thought that there was a good chance that the committee would eventually agree to award a prize in game theory, but he had no reason, given Nash's illness and the decades that had passed since his early papers, to believe that Nash would make the cut.

Eric Fisher, a visitor at Stockholm University's Institute for International Economics that year, recalled being quizzed by Assar Lindbeck about Nash's mental state. Fisher had been an undergraduate at Princeton, where he used to see Nash hanging out in the foyer of Firestone Library. Lindbeck wanted to know whether Nash was "competent enough to handle the publicity that winning [a Nobel] might entail."[30]

It was two years later, the fall of 1989, that Weibull hurried across the Princeton University campus to meet Nash for the first time.[31] After weeks of delicate negotiation, with the chairman of the mathematics department acting as a go-between, the elusive mathematician had finally agreed to have lunch. Weibull had a specific motive for the meeting. Lindbeck had pulled him aside shortly before his departure from Sweden and asked him to report back to him on Nash's mental state. There was some talk, Lindbeck said, that Nash had some sort of remission and was behaving quite reasonably. Was it true? Weibull was about to find out.

Weibull knew instantly that the tall, white-haired, frail-looking man standing in the driveway in front of Prospect House, Princeton's Florentine faculty club, was Nash. He was standing there rather awkwardly, smoking, looking down at the ground, obviously dressed up for the occasion, wearing white tennis shoes but also a long-sleeved dress shirt and long pants. As Weibull drew nearer, he could see that Nash was deathly nervous. When Weibull gave him his ready, friendly smile and extended his hand, Nash was unable to meet his eye and, after the briefest of handshakes, instantly put his hand back into his pocket.

They ate, not in the main, formal restaurant, but downstairs in a small cafeteria. Weibull, a gentle, soft-spoken man, asked Nash questions about his work. Sometimes the conversation took odd turns. When Weibull asked Nash about refining the Nash equilibrium concept by, perhaps, taking into account irrational moves by players, Nash answered him by talking, not about irrationality, but about

immortality. But on the whole, Nash struck Weibull as no more eccentric, irrational, or paranoid than many other academics. Weibull learned interesting details
about Nash's game theory papers that he hadn't known. Nash had gotten his idea
for the bargaining solution as an undergraduate at Carnegie Tech by thinking
about trade agreements between nations. While he had used both Brouwer's and
Kakutani's fixed-point theorems to prove his equilibrium result, he still thought
that the proof relying on Brouwer was both more beautiful and more apt. He said
that von Neumann had opposed his idea of equilibrium, but that Tucker had
supported him.

Afterward, though, what stood out for Weibull about the meeting, and the
thing that transformed him that day from a detached observer and objective infor-
mant into an ardent advocate, was something Nash said before they walked into
the club. "Can I go in?" Nash had asked uncertainly. "I'm not faculty." That this
great, great man did not feel that he had a right to eat in the faculty club struck
Weibull as an injustice that demanded remedy.

By the summer of 1993, rumors about a possible prize in game theory were
rampant.[32] A very small, very select symposium on game theory had taken place in
mid-June, at what used to be Alfred Nobel's old dynamite factory in Bjorkborn, a
few hundred kilometers north of Stockholm.[33] Such symposia, sponsored by the
prize committee, are invariably seen as Nobel beauty contests. This one was organized by Karl-Göran Mäler with the help of Jörgen Weibull and a Cambridge
economist, Partha Dasgupta. Lindbeck, who was spending the spring term in Cambridge, oversaw the preparations by telephone. The dozen or so invited speakers
represented two generations of leading game-theory researchers, mostly theorists
and experimentalists, among them John Harsanyi, Reinhard Selten, Robert Aumann, David Kreps, Ariel Rubinstein, Al Roth, Paul Milgrom, and Eric Maskin.
The topic? Rationality and Equilibrium in Strategic Interaction.

Most of the participants took it for granted that they were performing for
the benefit of the prize committee and assumed that the three graybeards in the
group, Harsanyi, Selten, and Aumann, were the likely Laureates.[34] Aumann, the
white-bearded Israeli dean of game theory, was strutting around "as if he had
already won." Much was made of the choice of topic, which was theoretical and
focused on noncooperative as opposed to cooperative games, and those who hadn't
been invited — Nash most obviously, of course.

As it turned out, the prize committee was far from committing itself to a
candidate.[35] Protestations that the main motivation for the symposium was to
create an opportunity for the committee "to educate itself," as Torsten Persson
of the prize committee put it later, were accurate. Only one other prize
committee member besides Mäler was even there — and that was Ingemar Stahl.
His brother, Ingolf, was one of the speakers, and Ingemar intimated that he had
come to hear him. But everyone assumed that he was there to act as a spy for
the committee.[36]

• • •

A few weeks later, Harold Kuhn, the professor of mathematics and economics at Princeton University, got an urgent fax from Stockholm. It was from Weibull, who wanted Kuhn to send a number of documents, among them Nash's Ph.D. thesis and a RAND memorandum — "no later than mid-August please." [37] Weibull also asked Kuhn to get him a transcript of an interview with Nash conducted by Robert Leonard, the historian. Leonard, who had not taped the interview, wrote Kuhn a note in which he said that the request "sent my mind reeling in the Swedish direction." [38]

In Stockholm, meanwhile, the prize committee was about to report to the so-called Ninth Class of the academy — all the academy members in the social sciences. [39] The bulk of the report, of course, was devoted to the proposed candidates for 1993, two economics historians, Robert Fogel of the University of Chicago and Douglass North of Washington University in St. Louis. But the committee also updated the class on two or three other proposals that constituted the top choices for subsequent prizes. One of them was a prize in game theory; Nash was on the short list of half a dozen candidates. [40]

Nearly the only point the prize committee had agreed on was that it wanted to go ahead with a prize in game theory in 1994, the fiftieth anniversary of John von Neumann and Oskar Morgenstern's great opus.

Lindbeck and the others were still toying with "every possible configuration" of two and three winners. [41] The short list — the candidates that the committee had focused most of its attention on — had scarcely changed since the prize was first conceived. [42] Apart from Nash it included Lloyd Shapley, whom Nash had known as a graduate student at Princeton. Shapley was the most direct intellectual descendant of von Neumann and Morgenstern and the clear leader of the field in the 1950s and 1960s when most of the work was in cooperative theory. Reinhard Selten and John Harsanyi, who had elaborated the theory of noncooperative games, were also on it. Harsanyi's breakthroughs permitted analysis of games of incomplete information while Selten developed a way to discriminate between reasonable and unreasonable outcomes in games. Aumann, who developed the role of common knowledge in games, was also on the list. And Thomas Schelling, who invented the notion of the strategic value of brinkmanship, was being considered because of his broad vision for the application of game theory to the social sciences.

The prize decision is made in stages. [43] Each year the committee starts meeting soon after the January 31 deadline for the two hundred or so nominations that the committee solicits from prominent economists around the world. By April, the committee decides on a particular candidate or candidates. In late August, it submits the proposal — along with a document several inches thick that includes the referee reports, publications, and other supporting material — to the Ninth Class for endorsement. The academy then votes on the candidates in early October. But, as everyone involved was well aware, the power truly resides in the committee and, until recently, in one man, Assar Lindbeck. Löfgren said, "The prize committee meets for a whole year. It's technically impossible for the higher body to make the decision."[44]

Debate in the committee was unusually contentious from the first meeting, attended by Lindbeck, Mäler, Stahl, Persson and Lars Svenson.[45] Lindbeck had come to the conclusion that the prize should be for contributions to noncoopera- tive theory alone. These were the ideas that had proved fruitful for economics, "the most important so far," as Lindbeck later said, adding "cooperative theory has a few interesting applications in economics, but perhaps more in political science."[46] Although Mäler sided with Lindbeck from the start, convincing the rest of the committee was harder than the latter anticipated. "It seemed self-evident afterward. But it took a long time to come to this conclusion. And to convince others."[47] Of course, he later admitted, narrowing the prize down in this way would immediately knock out some of the obvious contenders, namely Shapley and Schelling.[48] And here was the real bone of contention: Focusing on noncooperative theory also meant that it would be difficult to deny Nash the prize. "Once we decided to limit the prize to noncooperative theory then it was very easy to decide who were the . . . [key contributors]. Then it was obvious that Nash is [part of the] Nobel."[49] Lindbeck proposed a three-way prize for the definition of equilibria in non- cooperative games: Nash, Harsanyi, and Selten.[50]

This was where the debate got nasty.

The person on the committee least intimidated by Lindbeck and best equipped intellectually to challenge him was Ingemar Stahl, a sixty-year-old professor at Lund with a joint appointment in economics and law.[51] Stahl is a quick study and a brilliant debater, a man who delights in taking contrarian, often extreme posi- tions, in any debate. He had long been one of the most active committee members and had written many of the committee's prize proposals since the early 1980s.

Stahl is short, with a large head and a big belly. His detractors call him Zwergel or "little dwarf" behind his back. A onetime wunderkind who never quite lived up to his early promise, Stahl owes the prestigious chair at Lund, his academy membership, and his longtime position on the prize committee more to his politi- cal connections and his high-profile posture in public policy debates than to his research output. Like Lindbeck, Stahl began his upward climb early, while he was still in high school, as a protégé of various Social Democratic politicians, including Palme, but he had gone over to the conservative opposition in the late 1960s.

Stahl was deeply and adamantly opposed to awarding the prize to Nash. From the start, he was highly skeptical of game theory — as indeed he is of all pure theory. He is an institutionalist, likes intuitive rather than formal reasoning, and is leery of mathematics and "technicians." He was, for example, a main mover behind the prizes for James Buchanan in 1986 and Ronald Coase in 1991 — economists whose theories focus on the way governments and legal structures affect the workings of markets. He also prides himself on grasping Nobel politics. The more he learned about Nash, the less he liked the idea of giving Nash a prize. In particular, he considered giving the prize to Nash the kind of ill-considered gesture that was likely to result in embarrassment and, more important, make the committee look bad.

"I knew he had been ill," he said later. "I didn't think many people knew about it. I guess I heard Hörmander's version." [52]

Stahl had done quite a bit of digging. In the early fall, he had made a call to Lars Hörmander, Sweden's most eminent mathematician and winner of the 1962 Fields Medal. [53] Hörmander had just retired from the University of Lund. Stahl identified himself as a member of the Nobel Prize committee. He'd heard that Hörmander had known Nash quite well in the 1950s and 1960s, he said. The committee was thinking of giving Nash a Nobel Prize. Could Hörmander give him the lowdown on Nash?

Hörmander was surprised. Like most other pure mathematicians, he didn't think much of Nash's work in game theory. And the last time Hörmander had laid eyes on Nash was in the academic year 1977–78. Hörmander had been in Princeton and he had seen Nash hanging around Fine Hall. Nash was "a ghost." Hörmander didn't think Nash had recognized him or had even been aware of his presence. Hörmander hadn't even tried to speak with him. To give such a man a prize seemed to him "absurd, risky." [54]

Hörmander was precise and frank. His memories of Nash were extremely distasteful. He recalled Nash's decision to give up his citizenship; his deportation, first from Switzerland, then from France; Nash's bizarre behavior at the 1962 conference in Paris; the stream of anonymous cards, with their hints of envy and hostility, that came after Hörmander won the Fields in 1962.

Stahl had also made inquiries among several psychiatrists he knew who, he says, described the illness as unlike depression or mania, where the self remains intermittently at least recognizable. "I knew this type of illness," he said later. "I know some psychiatrists here. Some of the best head shrinkers. When I talked to them I found out that with this disease there is a complete change of personality. He is not the man who did the thing." [55]

Lindbeck, relying on reports from Weibull and Kuhn, was telling committee members that Nash was much improved, that he had, in fact, recovered his sanity. [56] About this, too, Stahl was deeply skeptical. The psychiatrists he spoke to told him that schizophrenia is a chronic, unremitting, degenerative disease. "It's a very tragic illness. It gets calmed down but actually recovering is another matter." [57]

Stahl knew that there was great sympathy for Nash. And he could see that Lindbeck had made up his mind. So he didn't attack frontally, but simply raised question after question. "He'd throw out an argument and somebody would shoot it down," said another member of the committee. "Then he'd shift to another argument. He tried to irritate and confuse us . . . to raise doubts." [58]

Stahl would say, "He's sick. . . . You can't have a person like that." [59]

He asked what would happen at the ceremony. "Would he come? Could he handle it? It's a big show." [60]

He quoted Hörmander and others who had known Nash in the 1950s and 1960s. He read them what he considered a particularly damning quotation from a book by Martin Shubik, who had known Nash as a graduate student.

"The most damning thing," Stahl repeated later, was something Martin Shubik wrote in one of his books: that "you can only understand the Nash equilibrium if you have met Nash. It's a game and it's played alone." [61]

He brought up Nash's work for RAND: "These guys worked with the atom bomb during the cold war. It would be a shameful thing for the prize." [62]

He brought up Nash's lack of interest in game theory after graduate school. As Lindbeck, Jacobson, the academy's secretary general, and others later hinted, Stahl was not the first member of a Nobel Prize committee who was motivated by deep animus toward a particular candidate or who embraced a wide range of intellectual objections in an effort to derail the candidate. [63] But as the spring wore on, Stahl made a great many midnight phone calls. He seemed, Weibull later recalled, to be trying out any and all arguments against Nash's candidacy. [64]

What was certainly the case throughout those months, a member of the Swedish academy said, was a growing feeling on Stahl's and others' part that "a few bad choices would sink the prize. Nash was of course a very weak prize. People were afraid that the thing would blow up. A big scandal." [65] And David Warsh, a syndicated columnist in whom Stahl evidently confided, subsequently wrote, "The whole intellectual world is watching to see what the Swedish Academy of Sciences is going to do about Nash. The Swedes are known to be worried about what Nash might say." [66] Christer Kiselman, head of the mathematics class of the academy at the time and a member of the academy's governing council, remembers talking to Stahl. He recalls that Stahl told him that Nash's work was done too long ago and was too mathematical to warrant a prize. [67] Kiselman, whose son Ola has suffered from schizophrenia since age sixteen, had a different interpretation: "[Stahl] was afraid of schizophrenia. So he had some prejudices. So he thought other people would think the same way. He was afraid of some scandal that would reflect on the committee." [68]

One by one Lindbeck knocked down Stahl's objections. [69] Lindbeck has a reputation for courage. He has never been afraid to take unpopular positions, even at the risk of alienating his political allies. In the late 1970s, for example, he had publicly opposed a favorite Social Democrat proposal to promote worker ownership of manufacturing concerns that had become trendy. [70]

Now Lindbeck took the position that Stahl's objections—that Nash was a mathematician, that Nash had stopped being interested in game theory forty years earlier, that Nash was mentally ill—were irrelevant. He too was worried that Nash would do something peculiar at the ceremony, but he was sure that could be managed. In any case, it was no basis for denying the prize to someone who was, on intellectual grounds, obviously worthy.

Besides, he found that his emotions were involved.[71] Most Laureates were already famous and much honored. The Nobel was only a crowning glory. But in Nash's case it was quite different. Lindbeck thought a great deal about the "misery of his life" and that Nash had been, for all intents and purposes, forgotten. Later, he was to say, "Nash was different. He had gotten no recognition and was living in real misery. We helped lift him into daylight. We resurrected him in a way. It was emotionally satisfying."[72] The only other time Lindbeck had felt similarly was when a Viennese libertarian and critic of Keynes, Friedrich von Hayek, won. "Hayek was so hated, so despised. . . . He'd been in a very deep depression, he told me. It was terribly satisfying to indicate his greatness."[73]

The committee listened to Stahl, but it soon became clear that he wasn't going to win allies. The younger men, Svenson and Persson, were keen on a game-theory prize, and the older ones weren't inclined to pick a fight with Lindbeck.

The normal procedure when there are unresolved disagreements is to append a formal reservation — a minority opinion — to the committee report.[74] Such reservations, which are duly reported to the full academy at the voting session, are not unheard of in physics or chemistry.[75] And, although they are not reported in the announcements at the time of the decision, they become part of the official record and may be made public after fifty years. Things were different in the economics committee. Lindbeck was extremely proud of its record and apparently regarded unanimity as necessary in maintaining the prize's credibility.[76]

As the report to the Ninth Class was being readied, Stahl threatened to register a formal reservation.[77] In the end — whether because of pressure from Lindbeck, advice from his old friend Mäler, or simply a reluctance to go down in history as first to break the former pattern of unanimity — he did not. The Class, which is used to going along with committee proposals, endorsed the proposal.

To Lindbeck, this was the end of the matter. He had prevailed, as he usually did. He felt, however, that extraordinary measures were necessary to make sure that everything would go smoothly once the media furor broke. He took an unprecedented step. He telephoned Kuhn in Princeton and told him that "it's ninety-nine percent certain now" that Nash would get the prize. "The votes were unanimous," he told Kuhn, not giving any hint of the controversy.[78] He gave Kuhn permission to inform the president of Princeton University of the impending award so that the university could make arrangements. As it turns out, Kuhn had to wait until after Labor Day to pass along his exciting news.[79] Harold Shapiro, president of Princeton, was away on vacation.

For once, Lindbeck, for all his political savvy, was wrong. It was not just that Stahl, who was far angrier than Lindbeck appreciated at the time, was a powder keg waiting to explode. Rather, Lindbeck's own long reign, and, indeed, the economics prize itself, were on shakier ground than he imagined. Powerful critics of both within the academy, including a former secretary general of the academy and a number of prominent physicists, were itching to do something. This prize had become an issue for them.

Few people outside Sweden, indeed, few outside the Royal Swedish Academy of Sciences, realize how controversial, even vulnerable, the economics prize has been since its creation in 1968 and continues to be to the present.

The economics prize has never been especially popular within the academy. "Many people question the Nobel Prize [in economics] here," said one longtime member.[80] Oldtimers still thought it had been a grave mistake to add a new Nobel to the original prizes. They thought it cheapened the currency and had, after the "mistake" of accepting the economics prize, successfully fought off efforts to establish other prizes that used the Nobel name. Erik Dahmen, an economist who was a close adviser to one of the richest families in Sweden, the Wallenbergs, calls it "the so-called Nobel Prize in economics."[81] He adds:

> This is not *really* a Nobel Prize. It should never be spoken of together with the other prizes. The academy should never have accepted the prize in economics. I have been against the prize since I became a member of the academy.

One physicist said: "The economics prize was just a way of jumping on the Nobel bandwagon, piggybacking on the Nobel."[82]

Economics was not held in high regard by many of the natural scientists who dominated the academy. It is not, they said, a sufficiently scientific field to deserve equal footing with hard sciences like physics and chemistry. Ideas, they said, slipped in and out of fashion, but one could not point to scientific progress, a body of theories and empirical facts about which there was certainty and near-universal agreement. Anders Karlquist, a physicist, said, "It's not as solid and big an enterprise as chemistry and physics."[83] Lars Gårding, a mathematician at the academy, for example, said later that Nash's prize was for "a very small thing."[84]

Finally, there is a widespread feeling, particularly on the part of natural scientists and mathematicians, that the shallowness of the field was leading to a sharp and rapid decline in the quality of Laureates — a decline that would necessarily worsen with time. Bengt Nagel, secretary of the physics Nobel Prize committee, jokingly quotes an economist who is supposed to have said in the early 1980s, "All the mighty firs have fallen. Now there are only bushes left."[85]

There are occasional calls to abolish the prize. After Myrdal won the prize, he is supposed to have suggested abolishing the prize because there were no longer any prizeworthy candidates.[86] As recently as 1994, Kjell Olof Feldt, the former minister of finance and soon-to-be chairman of the board of the Bank of Sweden — which finances the prize — suggested in a lengthy article in a political monthly that the prize be done away with.[87]

But although many academy members regret that the prize was established in the first place, said Karlquist, they "realize that it's a fact of life." [88] By 1994, in fact, the critics' objective was to wrest control of the prize from the economists. Lindbeck was personally unpopular. It was particularly galling that membership in the economics prize committee seemed to be a lifetime sinecure and that its members could choose winners without any real accountability to the academy.

In February, an academy committee had "suggested" that the economics prize committee be forced to operate by the same rules that apply to the physics and chemistry committees. [89] The suggestion was not binding, but it was a warning note, the first concrete sign that critics of the prize were gaining momentum, and it carried with it the promise that the academy council would, when it got around to it, appoint another group specifically mandated to deal with the matter of the economics prize. The imposition, as for other standing committees, of term limits would, of course, have a drastic and immediate effect on the economics committee. It would eliminate Lindbeck, Mäler, and Stahl, the three longtime members, and virtually end their reign. The other, and more drastic, suggestion was to widen the membership to include non-economists and, most radically, to transform the economics Nobel into, in effect, "the Nobel Prize in social sciences," a notion that appealed not only to natural scientists, but also to the psychologists, sociologists, and other non-economists in the academy's Ninth Class. [90]

Thus the debate between Lindbeck and Stahl over whether Nash was a suitable candidate for the prize, a debate that really turned on whether the choice of Nash would embarrass the committee, took place in an unusually hostile atmosphere and under intense scrutiny. The future of the prize committee and the prize looked more vulnerable than they had in times past. All of these behind-the-scenes opinions and maneuvers explain why, between early September and early October, Stahl acquired a powerful set of allies who joined him for reasons quite apart from Nash's candidacy. [91] The stage was set.

In the end, Nash and the two other candidates for the 1994 economics prize passed by a mere handful of votes — the first in the history to skirt so close to defeat. [92] It is a peculiarity, indeed a major administrative and logistical headache, of the Nobel Prize process that no award can really be said to exist until the members of the full body of the Royal Swedish Academy of Sciences have had their say. They have "the sole right to decide," as a Nobel Foundation booklet puts it: "Even a unanimous committee recommendation may be overruled." [93] Only when the plenary session has cast ballots and the votes are counted and the results announced do the secretary general and members of the prize committee march off to telephone the winners. They then proceed to the Sessions Hall to announce the winners' names to the world press. Other prizes, like the Fields Medal for mathematics or the John Bates Clark medal for economics, by contrast, are settled months ahead of time, their winners notified after a leisurely interval and carefully instructed to sit on the secret until the awarding institutions get around to issuing their press releases or holding their festivities. Presumably, the inconvenience of the last-minute Nobel vote is outweighed by the benefit of being able to avoid leaks before the official announcement.

The Nobel vote, moreover, is traditionally a mostly ceremonial affair, the final flourish after a lengthy selection procedure that is more or less completely dominated by the senior members of the prize committees. In the case of the economics prize, a few dozen random academicians—a fraction of the number who turn out for the physics or chemistry prizes, the other two Nobel awards administered by the academy—assemble in the second week of October largely for the pleasure of hearing a distinguished lecture on the proposed candidates' contributions to scientific progress. As one academy member put it, "Members attend less for the vote itself than for a chance to hear the presentations."[94] In some recent years, the modest quorum of forty academy members has proved difficult to achieve.[95] According to the rules, academy members have three options. They may vote for the candidate or candidates proposed by the committee and endorsed by the Social Sciences Class. They may vote for an alternative candidate of their own choosing. Or they may vote not to give a prize that year. The winner or winners must obtain a simple majority of votes. Until 1994, no candidates proposed by the committee had ever failed to gain a wide majority of votes.

The academy meeting that began promptly at 10:00 A.M. on Tuesday, October 12, in a rather small, poorly lit auditorium tucked in a far corner of the academy's ground floor,[96] promised to be no more or less interesting than previous years' meetings. Fewer than sixty members were scattered around the room, but, as the officials present noted with satisfaction, there was no question of not getting a quorum. (A couple of years earlier, thirty-nine members had sat in that room waiting for a fortieth—who did finally show up.)[97] Kerstin Fredga, the astrophysicist who was the academy's president, and Carl-Olof Jacobson were sitting side by side on the stage. The ballot box was perched at the end of the platform. The five members of the prize committee who belonged to the academy were sitting near the front of the room.

Lindbeck was at the podium in a few long strides. Wearing his thick black-rimmed glasses and usual frown of concentration, Lindbeck dove right into his subject, an overview of the entire process by which the committee had arrived at its recommendation for a prize in game theory. Always intense, Lindbeck stuttered with excitement, waved his long arms, and made a good many very dry jokes.[98] He was followed by Jacobson, low-keyed by contrast, who gave the official endorsement of the Social Sciences Class. Both men claimed that the decisions by the committee as well as the Class were, as always, unanimous. Lindbeck added that unanimity had come about "as if by an Invisible Hand," his standing joke. Finally, Mäler got up and launched the main presentation, a lecture on the contributions of the three candidates.

The lecture was quite disappointing. Mäler, never a brilliant speaker, was more nervous and unsure of himself than usual.[99] He quickly became mired in technicalities and jargon. He read most of it. His wife had left him a few weeks earlier, he was agitated and depressed, and he had had a terrible time preparing the talk.

All this took something like an hour. Had things proceeded as usual there would have been a few rather perfunctory and mostly polite questions from the floor, perhaps a standard monologue by one of the oldtimers about the dubiousness of the economics prize in the first place, before a general silence, a passing-out of plain squares of white paper and number two pencils, quick scribbles, folding, and the drifting down of academicians to the stage to stuff their ballots in the box.

Instead, all hell broke loose. Later the president of the Nobel Foundation remarked wryly that "Troy could only have been destroyed by someone inside the walls. And that's what happened here."[100] No one recalls whether Stahl launched the first verbal grenade, but it was soon obvious to Lindbeck and Mäler that they were in the midst of an ambush. Stahl challenged Mäler to give a single major example showing that the theory had any empirical validity whatsoever. Mäler, who was in particularly poor shape to answer questions, fumbled. Stahl did not — contrary to a story six weeks later in *Dagens Nyheter,* one of Sweden's two dailies — do anything as crass, or risky, as to urge the academy to withhold the prize to Nash because of the mathematician's mental illness.[101] Instead, he argued, forcefully and brilliantly, that a prize for non-cooperative game theory was too narrow, too insubstantial, too technical. He reminded the audience that Nash's contribution had been made nearly half a century earlier and that it was more mathematics than economics. He derided Harsanyi and Selten for being "boring," "mere technicians." Other members of the audience soon chimed in.

Stahl did not make the mistake of merely criticizing the committee's proposal, which, after all, he had signed. He had an alternative, he said.[102] In light of the members' unhappiness, in light of unanswered questions, in light of Mäler's clearly unsatisfactory report, might it not be more prudent to postpone the prize in game theory? Why not vote instead to give the prize to Robert Lucas, the University of Chicago professor whom the committee had virtually decided to propose for the following year.[103] Everybody, he reminded them, was enthusiastic about Lucas, who had invented a theory to explain why governments' efforts to manage the business cycle were doomed to failure — "rational expectations" — and was clearly one of the most important economists of the century. It was an unassailable choice.

Lindbeck, who had at first seemed stunned by the audacity of Stahl's surprise attack, told the members in no uncertain terms what Stahl was implying. He reminded the members that Stahl had signed on to the game-theory prize and accused Stahl of wishing to scuttle the prize because of Nash's illness. He told the membership that it would be a grave injustice to withhold the prize. He did not tell them that, in an absolute breach of the Nobel rules, he had already informed Princeton University's president, Alicia Nash, and Nash himself that he was getting the prize. But those facts were very much in his mind as he appealed to the members.[104]

By the time Carl-Olof Jacobson called for the vote, the atmosphere in the room was tense and bitter. An unusually large number of academicians stayed to hear the vote count. Two members of the academy chosen by the president and Jacobson removed the ballots in front of the audience and tallied the votes. The paper was handed to Jacobson, and Jacobson read the votes one name at a time. For Lindbeck it was, as he later said, a moment of unbearable suspense. Mr. Nash . . . Mr. Harsanyi . . . Mr. Selten . . . Mr. Lucas . . . no prize. . . .

A few moments later, Fredga, Jacobson, Lindbeck, and Mäler, very much shaken, were the only ones left in the room. Their candidates had gotten all that they needed: a slim majority of the votes.

Later, in public, these individuals would all deny that anything extraordinary had happened. They would pretend that Mäler's report had been unusually long, that there had been a great many questions, that the Laureates had been difficult to reach, or simply state baldly that the delay had never occurred. But behind closed doors, within the academy, there would be shock, consternation, and finger-pointing. "It was a unique event. It had never happened before," said one member of the academy. "It's not good for the academy to have close votes," said Kisel-man.[105] The very next day the council hastily appointed an ad hoc committee "to study the future of the economics prize."[106]

Afterward, a committee member friendly to Stahl would say that Stahl had been "used by the physicists."[107] Stahl's double-cross had backfired. Instead of being regarded as the man who saved the prize committee from an embarrassing mistake, he had set into motion the very consequences he feared. Like players in So Long Sucker, the game that Nash and his friends at Princeton had invented forty years earlier, Lindbeck and Mäler formed a temporary coalition with the critics of the economics prize. They threw themselves behind the rules changes. They were determined to punish Stahl and get him off the committee — even if the new rules meant that they had to step aside as well. One prize committee member called their strategy "elegant."[108] Had Nash known about it, he would have appreciated it as a textbook execution of McCarthy's Revenge Rule, especially because Lindbeck could reasonably expect to get elected to the committee again after a three-year interlude, but Stahl, who had provoked the scandal and compounded his sin by talking to a reporter, was out for good.

The consequences did not end there. According to several members of the academy, the ad hoc committee went on record to recommend changing the very nature of the economics prize. In its report, issued a few months later, in February 1995, the committee issued an instruction that essentially redefined the economics prize as a prize in social sciences, open to great contributions in fields like political science, psychology, and sociology.[109] It also ordered the committee membership to be opened to two non-economists. No public announcement of these far-reaching changes was made. But within a year, Lindbeck, Mäler, and Stahl were gone; two social scientists who weren't economists — a statistician and a sociologist — were members of the prize committee; and among the top candidates for the prize was Amos Tversky, an Israeli psychologist who works on irrationality in decision making.[110]

In the auditorium on October 12, the three men rushed over to a small committee room.[111] Jacobson was armed with a page of telephone numbers for the Laureates. It was he who would inform the Laureates of the honor that was about to grace them.

They tried to reach Selten first since Selten was in Germany and, unlike Nash or Harsanyi, would not necessarily be asleep. It was early in the morning for Nash in New Jersey and the middle of the night for Harsanyi in California. As it turns out, Selten was out grocery shopping. Jacobson then tried Harsanyi and, when he got him, quickly put Mäler, who knew Harsanyi well, on the line to quickly assure him, with much joviality, that Jacobson was not some student or, worse, reporter playing a trick on him.[112]

Nash was the last to be called. Jacobson waited expectantly as the telephone rang. Unbeknownst to most of Jacobson's colleagues at the academy, he had a brother who, like Nash, had been diagnosed with schizophrenia as a young man in the 1950s and had been institutionalized ever since.[113] It was a moment of incredible poignancy for Jacobson, "the greatest moment," he later said, of his twenty-year tenure at the academy.

"He was unusually calm," he said afterward. "That was my thought. 'He is taking this very calmly.'"[114]

49

The Greatest Auction Ever

Washington, D.C., December 1994

O<small>N THE AFTERNOON</small> of December 5, 1994, John Nash was riding in a taxi headed to Newark Airport on his way to Stockholm, where he would, in a few days' time, receive from the King of Sweden the gold medal engraved with the portrait of Alfred Nobel.[1] At around the same time, a few hundred miles to the south, in downtown Washington, D.C., Vice-President Al Gore was announcing with great fanfare the opening of "the greatest auction ever."[2]

There was, as *The New York Times* would later report, no fast-talking auctioneer, no banging gavel, no Old Masters.[3] On the auction block was thin air — airwaves that could be used for the new wireless gadgets like telephones, pagers, faxes — worth billions and billions of dollars, enough licenses for every major American city to have three competing cellular phone services. In the secret war rooms and bidding booths were CEOs of the world's biggest communications conglomerates — and an unlikely group of blue-sky economic theoreticians who were advising them. When the auction finally closed the following March, the winning bids totaled more than $7 billion, making it the biggest sale in American history of public assets and one of the most successful (and lucrative) applications of economic theory to public policy ever.[4] Michael Rothschild, dean of Princeton's Woodrow Wilson School, later called it "a demonstration that people thinking hard about a problem can make the world work better . . . a triumph of pure thought."[5]

The juxtaposition of Gore and Nash, the high-tech auction and the medieval pomp of the Nobel ceremony, was hardly an accident. The FCC auction was designed by young economists who were using tools created by John Nash, John Harsanyi, and Reinhard Selten. Their ideas were specifically designed for analyzing rivalry and cooperation among a small number of rational players with a mix of conflicting and similar interests: people, governments, and corporations — and even animal species.[6]

The prize itself was a long-overdue acknowledgment by the Nobel committee that a sea change in economics, one that had been under way for more than a decade, had taken place. As a discipline, economics had long been dominated by Adam Smith's brilliant metaphor of the Invisible Hand. Smith's concept of perfect competition envisions so many buyers and sellers that no single buyer or seller has to worry about the reactions of others. It is a powerful idea, one that predicted how free-market economies would evolve and gave policymakers a guide for encouraging growth and dividing the economic pie fairly. But in the world of megamergers, big government, massive foreign direct investment, and wholesale privatization, where the game is played by a handful of players, each taking into account the others' actions, each pursuing his own best strategies, game theory has come to the fore.[7]

After decades of resistance — Paul Samuelson used to joke about "the swamp of *n*-person game theory"[8] — a younger generation of theorists began using game theory in areas from trade to industrial organization to public finance in the late 1970s and early 1980s.[9] Game theory opened up "terrain for systematic thinking that was previously closed." Indeed, as game theory and information economics have become increasingly entwined, markets traditionally seen as fitting the purely competitive mold have increasingly been studied using game-theory assumptions. The latest generation of texts used in top graduate schools today all recast the basic theories of the firm and the consumer, the foundation of economics, in terms of strategic games.[10]

"Concepts, terminology and models from game theory have come to dominate many areas of economics," said Avinash Dixit, an economist at Princeton who uses game theory in work on international trade and is the author of *Thinking Strategically*. "At last we are seeing the realization of the true potential of the revolution launched by von Neumann and Morgenstern."[11] And because most economic applications of game theory use the Nash equilibrium concept, "Nash is the point of departure."[12]

The revolution has gone far beyond research journals, experimental laboratories at Caltech and the University of Pittsburgh, and classrooms of elite business schools and universities. The current generation of economic policymakers — including Lawrence Summers, undersecretary of the treasury, Joseph Stiglitz, chairman of the Council of Economic Advisers, and Vice-President Al Gore — are steeped in the stuff, which, they say, is useful for thinking about everything from budget proposals to Federal Reserve policy to pollution cleanups.

The most dramatic use of game theory is by governments from Australia to Mexico to sell scarce public resources to buyers best able to develop them. The radio spectrum, T-bills, oil leases, timber, and pollution rights are now sold in auctions designed by game theorists — with far greater success than that of earlier policies.[13]

Economists like Nobel Laureate Ronald Coase have advocated the use of auctions by government since the 1950s.[14] Auctions have long been used in markets where sellers of unusual items — from vintage wines to movie rights — have no idea what bidders are willing to pay. Their basic purpose is to make bidders reveal how much they value the item. But the arguments of Coase and others were stated in abstract, entirely theoretical terms, and little thought was given to how such auctions would actually be conducted. Congress remained skeptical.

Before 1994, Washington simply gave away licenses for free. Until 1982, it had been up to regulators to decide which companies deserved the licenses. Needless to say, the process was dominated by political pressures, outrageously expensive paperwork, and long delays. The pace of licensing lagged hopelessly behind market shifts and new technologies. After 1982, Washington awarded licenses using lotteries, with the winners free to resell licenses. Although the reform did speed up the granting of licenses, the process was still hugely inefficient — and unfair. Bidders with no intention of operating an actual telephone business spent millions to get into the game for the purpose of reaping a windfall. Further, although telephone companies were forced to pay the costs of obtaining licenses, Washington (and taxpayers) did not get the benefit of any revenues. There had to be a better way.

A young generation of game theorists, including Paul Milgrom, John Roberts, and Robert Wilson at the Stanford B-school, came up with that better way.[15] Their chief contribution consisted of recognizing, as Milgrom said, that "the mere design of *some* auction wasn't enough. . . . [G]etting the auction design right was also critically important"[16] In particular, they concluded that the most obvious auction designs — auctioning licenses one by one in sequence using simultaneous sealed bids — was the way least likely to succeed in getting licenses into the hands of corporations that could use them best — Washington's stated objective.

Game theorists treat an auction like a game with rules and try to evaluate how a given set of rules, taken together, is apt to affect the bidders' behavior. They take stock of the options the rules allow, the payoffs to the bidders associated with the options, and bidders' expectations about their competitors' likely choices.

Why did these economists conclude that traditional auction formats would not work? Mainly because the value of each individual license to a user depends — as is the case with a Rembrandt or a Picasso — on what other licenses the user is able to obtain. Some licenses are perfect substitutes for one another. That would be the case for similar spectrum bands to provide a given service. But others are complements. That would be the case for licenses to provide paging services in different parts of the country.

"To permit the efficient license assignment, an auction must allow bidders to consider various packages of licenses, combining complements and switching among substitutes during the course of the auction. Designing an auction to allow this is quite difficult," writes Paul Milgrom, one of the economists who designed the FCC auction of which Gore was speaking.[17]

A second source of complexity, Milgrom says, is that the purpose of the licenses is to create businesses for new services with unknown technology and unknown consumer demand. Since bidders' opinions are bound to be wildly divergent, it is possible that license assignment would depend more on bidders' optimism than on their ability to create a desired service.[18] Ideally, an auction design can minimize that problem.

As Congress and the FCC inched closer to the notion of auctioning off spectrum rights, Australia and New Zealand both conducted spectrum auctions.[19] That they proved to be costly flops and political disasters illustrated that the devil really was in the details. In New Zealand, the government ran a so-called second price auction, and newspapers were full of stories about winners who paid far below their bids. In one case, the high bid was NZ$7 million, the second bid NZ$5,000, and the winner paid the lower price. In another, an Otago University student bid NZ$1 for a television license in a small city. Nobody else bid, so he got it for one dollar. The government expected the cellular licenses to fetch NZ$240 million. The actual revenue was NZ$36 million, one-seventh of the advance estimate. In Australia, a botched auction, in which parvenu bidders pulled the wool over the government's eyes, delayed the introduction of pay television by almost a year.

The FCC's chief economist was an advocate of auctions, but no game theorists were involved in the first stage of the FCC auction design. The theorists' phones started ringing only by accident after the FCC issued a tentative proposal for an auction format with dozens of footnotes to the theoretical literature on auctions.[20] That was how Milgrom and his colleague Robert Wilson, leading auction theorists, got into the game.

Milgrom and Wilson proposed that the FCC adopt a simultaneous, multiple-round auction.[21] In a simultaneous auction, a bunch of licenses are sold at the same time. Multiple rounds mean that, after the first round of bidding, prices are announced, and bidders have a chance to withdraw or raise one another's bids. This is repeated round after round until the auction is over. The chief advantage of this format is that it allows bidders to take account of interdependencies among licenses. Just as sequential, closed-bid auctions let sellers discover what bidders are willing to pay for individual items, the simultaneous, ascending-bid auction lets them discover the market value of different groupings of items.

This early proposal — which the FCC eventually adopted — did not cover seemingly small but critical details.[22] Should there be deposits? Minimum bid increments? Time limits? Should the bidding system be wholly computerized or executed by hand? and so forth. Milgrom, Roberts, and another game theorist, Preston McAfee, an adviser to AirTouch, provided proposals on these issues. The FCC hired another game theorist, John McMillan, of the University of California at San Diego, to help evaluate the effect of every proposed rule. According to Milgrom, "Game theory played a central role in the analysis of the rules. Ideas of Nash equilibrium, rationalizability, backward induction, and incomplete information, though rarely named explicitly, were the real basis of daily decisions about the details of the auction process."[23]

By late spring 1995, Washington had raised more than $10 billion from spectrum auctions. The press and the politicians were ecstatic. Corporate bidders were largely able to protect themselves from predatory bidding and were able to assemble an economically sensible set of licenses. It was, as John McMillan said, "a triumph for game theory."[24]

50

Reawakening

Princeton, 1995–97

Mathematics is a young man's game. Yet it is not bearable to contemplate a brief distinction and burgeoning of activity . . . followed by a lifetime of boredom. — NORBERT WIENER

ON THE AFTERNOON of the Nobel announcement, after the press conference, a small champagne party was in progress in Fine Hall. Nash made a short speech.[1] He was not inclined to give speeches, he said, but he had three things to say. First, he hoped that getting the Nobel would improve his credit rating because he really wanted a credit card. Second, he said that one is supposed to say that one is really glad he is sharing the prize, but he wished he had won the whole thing because he really needed the money badly. Third, Nash said that he had won for game theory and that he felt that game theory was like string theory, a subject of great intrinsic intellectual interest that the world wishes to imagine can be of some utility. He said it with enough skepticism in his voice to make it funny.

All the Swedes' fears — not to mention Harold Kuhn's own private worries — about how Nash would cope with the pomp in Stockholm proved groundless. Everything went swimmingly. The receptions. The press briefings. The Nobel award ceremony itself. The lecture in Uppsala afterward. Indeed, in the weeks between the announcement of the prize and the ceremony, Nash did and felt things that had lain beyond his grasp for decades. When he first arrived in Stockholm, Jörgen Weibull recalled, he behaved pretty much as Weibull had remembered from Princeton a few years before: "He didn't look you in the eye. He mumbled. Socially he was very tentative, very uncertain. But his mood went up from day to day. He got less and less unhappy."[2]

Harold Kuhn, who was to lead a Nobel seminar honoring Nash's work, and his wife Estelle accompanied Nash and Alicia to Stockholm.[3] It was exhilarating. The nicest moment of the week, so full of grand scenes and ceremonies, came when Nash had his much-dreaded private audience with the King. By tradition, the King spends a couple of minutes alone with each Laureate. When Nash's turn came, he grimaced and frowned so much that Harold was afraid he might refuse to go into the King's chambers at the last minute, but finally he followed the aide inside.

Five minutes passed, then seven. Finally, after a full ten minutes, Nash emerged, looking relaxed, even amused. "What did you talk about?" everybody asked at once. Quite a bit, it turned out. In 1958, John told Harold and Estelle, he and Alicia had taken a grand tour of Europe and had driven up into the south of Sweden in their new Mercedes 180. The King had been a student in Uppsala then, addicted to fast sports cars. Around that time, the Swedes were shifting from driving on the left to driving on the right. Nash and the King had spent ten minutes chatting about the pitfalls of driving fast on the lefthand side of the road.

At dusk, Nash and Weibull were riding in a limousine through the countryside north of Stockholm. The farmhouses were lighting up one at a time, the sky was beginning to glimmer. Nash reached over to Weibull and said, "Look, Jörgen. It's so beautiful."[4]

They were on their way back from Uppsala where Nash had given a talk — his first in three decades.[5] Nash hadn't been asked to give the customary hour-long Nobel lecture in Stockholm. The lecture at the University of Uppsala was arranged by Christer Kiselman.[6] Nash's chosen topic was a problem that had interested him before his illness and that he had taken up again since his remission: developing a mathematically correct theory of a non-expanding universe that is consistent with known physical observations. The conventional view, of course, is that the universe is expanding, and attempting to overturn the consensus is exactly the kind of contrarian intellectual bet that Nash has always enjoyed.

Nash's talk on "the possibility that the universe isn't expanding" began with tensor calculus and general relativity — stuff so difficult that Einstein used to say he understood it only in moments of exceptional mental clarity. Though he later confessed to nervousness, he spoke without notes, clearly and convincingly, according to Weibull, who has a doctorate in physics.[7] Physicists and mathematicians in the audience said afterward that Nash's ideas were interesting, made sense, and were expressed with the appropriate degree of skepticism.

It is a quiet life, despite the fairytale of Stockholm and the lofty status of Laureate. The Nashes still live in the Insulbrick house with the hydrangeas out front, next to the alley and across from the Princeton train station. There is a new boiler, a new roof, a few new items of furniture, but that's about it. (Nash was also able to pay down his half of the mortgage.) The few friends they see regularly, among them Jim Manganaro, Felix and Eva Browder, and of course Armand and Gaby Borel, are pretty much the people they have been seeing for some years. Their daily routines have changed less than one might think, dominated as they are by the twin needs of earning a living and caring for Johnny. Alicia takes the train to

Newark every day. Nash, who no longer drives, rides the "Dinky" into town, eats lunch at the Institute, and spends the afternoons in the library or, on rare occasions, in his new office. Very often, when Johnny is not in the hospital or on the road, he takes Johnny with him.

It is a life resumed, but time did not stand still while Nash was dreaming. Like Rip Van Winkle, Odysseus, and countless fictional space travelers, he wakes to find that the world he left behind has moved on in his absence. The brilliant young men that were are retiring or dying. The children are middle-aged. The slender beauty, his wife, is now a mature woman in her sixties. And there is his own seventieth birthday fast approaching.

There are days when he feels that he has escaped the ravages of time, when he believes he can pick up where he left off, when he feels "like a person who wants to do the research he might have done in his 30s and 40s at the delayed time of his 60s and 70s!" In his Nobel autobiography, he writes:

> Statistically, it would seem improbable that any mathematician or scientist, at the age of 66, would be able through continued research efforts to add to his or her previous achievements. However, I am still making the effort, and it is conceivable that with the gap period of 25 years of partially deluded thinking providing a sort of vacation, my situation may be atypical. Thus I have hopes of being able to achieve something of value through my current studies or with any new ideas that come in the future.[8]

But many days he is not able to work. As Nash once told Harold Kuhn, "The Phantom was not in until very late, after 6:00 P.M. because even a Phantom can have ordinary human problems and need to go to a doctor."[9] And there are other days when he discovers an error in his calculations or learns that a promising idea has already been mined by someone else, or when he hears of new experimental data that seem to make certain speculations of his seem less interesting.

On such days, he is full of regrets. The Nobel cannot restore what has been lost. For Nash, the primary pleasure in life had always come from creative work rather than from emotional closeness to other people. Thus, recognition for his past achievements, while a balm, has also cast a harsh light on the vexing issue of what he is capable of doing now. As Nash put it in 1995, getting a Nobel after a long period of mental illness was not impressive; what would be impressive is "persons who AFTER a time of mental illness achieve a high level of mental functioning (and not just a high level of social respectability.)"[10]

Nash gave the starkest assessment of his own situation in front of an audience of psychiatrists to whom he had been introduced as "a symbol of hope." In answer to a question at the end of his 1996 Madrid lecture, he said, "To recover rationality after being irrational, to recover a normal life, is a great thing!" But then he paused, stepped back, and said in a far stronger, more assertive voice: "But maybe it is not such a great thing. Suppose you have an artist. He's rational. But suppose he cannot paint. He can function normally. Is it really a cure? Is it really a salvation? . . . I feel I am not a good example of a person who recovered unless I can do some good work," adding in a wistful, barely audible whisper, "although I am rather old." [11]

These thoughts were much in Nash's mind when he turned down an offer of thirty thousand dollars from the Princeton University Press in 1995 to publish his collected works. "Psychologically I have a problem since I have been, unfortunately, a long time without any publications," he said to Harold Kuhn. He was saying, in short, that he doesn't want to close the door on future work by acknowledging that his lifetime oeuvre is complete.

As Nash says, "I did not want to publish a collected works simply because I wanted to think of myself as, and assume the posture of, a mathematician, still actively engaged in research and not just resting on his laurels (as they say). And of course I knew that if a collected works was not published at this time, then it could be published later when, hopefully, I would have nice new things to add to it." [12] In these feelings, however, he is not so different from his brilliant contemporaries. They, too, are having to face, or have already faced, the prospect that they are likely never to match their past achievements. Some have remained more active than others. But aging is a fact of life, and an especially stringent one for a mathematician. It is, for most of them, a young man's game.

It takes extraordinary courage to return to research after a hiatus of nearly thirty years. But this is exactly what Nash did. As he told the Madrid audience, "I am again engaged in scientific study. I am avoiding routine problems and instead I am 'dabbling.' "

Nash had been thinking about a mathematical theory of the universe since before his meeting with Einstein. Since the lecture in Uppsala, he has suffered various setbacks. In August 1995, he said, "I got results that indicated I had made a fundamental error a long time ago and that I must reformulate . . . [the] theory." Apparently "there was stuff being lost in a singular integration and when I considered distributed matter instead of a point particle, I found the lost stuff which had been erroneously ignored" — adding, with characteristic objectivity, that "this is good since I have avoided publishing a version based on errors."

He went on to describe the specific error:

There was a discrepancy in the field . . . which spoiled things. Recalculation revealed . . . there had been errors in the calculation. Now I must finish up the calculation for a distributed mass of gravitating matter, at least to the first order level of approximation. This level itself could bring an interesting (distinctive result). [13]

· · ·

This evaluation of the difficulties encountered in his research gives a good idea that the problems Nash is working on are ambitious, that he has lost none of his taste for making high-risk bets (whether on ideas or stocks!), and that his thinking is still sharp. And even if his chances of achieving a new breakthrough are statistically small, as he says, the pleasures of thinking about problems are once again his.

The truth, however, is that the research has not been the main thing in his present life. The important theme has been reconnecting to family, friends, and community. This has become the urgent undertaking. The old fear that he depended on others and that they depended on him has faded. The wish to reconcile, to care for those who need him, is uppermost. He and his sister Martha, estranged for nearly twenty-five years, now talk on the telephone once a week. Johnny, of course, is the main thing, the constant.

It was Nash who had told the women to call the police.[14] Johnny had been living at home. He had been all right for a while, but then he began to wear a paper crown. One afternoon, he wanted some money. Because he believed he was a sovereign, he thought that he should be able to get money from Sovereign Bank. But the ATM in front of the bank would not spit out any cash. In fact, it would not return his bank card. Agitated and unhappy, Johnny called his mother, who has an account at Sovereign, and demanded she meet him at the ATM and get his card out of the machine. Alicia told John, who insisted on going with her. The couple tried, vainly, to extract Johnny's card. They also tried, unsuccessfully, to soothe Johnny. At that point, their son became enraged, picked up a big stick, and started to poke first his mother, then his father. Some bystanders across the street stopped when they saw the young man threatening the two elderly people. Nash shouted for one of them to call the police. A squad car pulled up. The police took Johnny, whom they knew well, back to Trenton State.

Johnny was in the hospital when his parents got the news from Stockholm informing them of Nash's Nobel. Nash and Alicia called him first. He thought that they were pulling his leg, that it was a joke, and hung up on them. Later he saw his father's face on CNN.[15]

The subject of Johnny's future is extremely painful. Nash had spoken matter-of-factly about it. Alicia, looking miserable, said nothing and instead sank deep into her seat and closed her eyes. She finally interjected, "He just wants to get on with his life."[16]

The hopeful path that Johnny seemed to be on in his early twenties had long ago petered out. Whether because of the stress of teaching, the social isolation, or because the remission had simply run its course, the year at Marshall University was a disaster. He had come home and has not worked since. "Of course I've been a bad example," Nash admits.[17]

Johnny wanted to get a job, Nash said, but he seemed to think he would be able to get one in a college mathematics department. He had been writing letters introducing himself as the son of a Nobel Laureate and asking for a position. Now Nash was telling the Kuhns that Johnny would not take his medicine when he was not in the hospital. Alicia adds, "He goes to the hospital, he gets better, but when he gets home he doesn't like to take his medication." Then he would get sick again, hearing voices and having delusions. He would be hospitalized again and get better. Then it would start all over again. Watching over Johnny is now Nash's main task in life. Except when Johnny is "on the road" wandering around the country on Greyhound buses, Nash is his caretaker. Nash takes it for granted that his son is his responsibility. As Nash said on one occasion, "My time of delusional thinking is, presumably, in the past, but my son's time of it is right now."[18] They get up in the morning together after Alicia has gone to work. They eat breakfast together. Nash takes him to the library, to the institute, to Fine Hall. On Monday evenings they all attend family therapy together. Nash has tried to get his son interested in the computer and plays computer chess with him. He has said: "Ultimately computers could be a good sort of occupational therapy (as perhaps I was benefited in an OT [occupational therapy] fashion by [Hale] Trotter's help in letting me get familiar with computer use.)[19]

Johnny is thirty-eight years old. He is tall and handsome like his father, and he and his father share an interest in mathematics and chess. But Johnny's illness has dragged on for more than half his life, a quarter of a century. He has been treated with the newest generation of drugs, including Clozaril, Risperadol, and, most recently, Zyprexa. These drugs, which have enabled him, for the most part, to stay out of the hospital, have not given him a life. Time hardly passes for him. He no longer competes in chess tournaments — once his greatest joy. He no longer reads, saying that he has not been able to for a long time. He is often angry and occasionally violent.[20]

Life with Johnny is a tremendous strain on Nash and Alicia. Nash calls it being "perturbed," "tyrannized," and he is often preoccupied with the "drift and danger of degradation."[21] It is a constant disruption even when, as is often the case, Johnny is roaming around the country on Greyhound buses. For instance, Alicia and John go to the Olive Garden to celebrate Nash's birthday, and Johnny calls to say that he has lost his ATM card and has no money. The evening is spent wiring him funds. "We're at our wits' end," Alicia said recently. "You work so hard . . . and then he's out of it. The Nobel hasn't helped Johnny at all."[22]

Johnny draws Nash and Alicia together and tears them apart. There are deep conflicts. They blame each other for Johnny's misbehavior — when he destroys things in the house, attacks them, acts inappropriately in public. Nash feels that Alicia expects him to be the bad cop, a role he's not happy with, while she is the soft one. But they rely on each other. They agree every day on what one or the other should do. They also agree when it is time to hospitalize him. Nash is more judgmental and apt to hold Johnny responsible for his illness. He's sometimes quite cruel, telling Harold Kuhn and others at times that people like Johnny ought to be jailed or that he has chosen to be as he is: "I don't think of my son . . . as entirely a sufferer. In part, he is simply *choosing* to escape from 'the world.' "[23]

Despite such moments of insensitivity, the truth is that Nash expresses hope and pleasure when there is the prospect of a new medication, a new therapy, or when he gets an idea — like teaching Johnny how to play chess on the computer — that he thinks will help him. When his friend Avinash Dixit invites him for dinner, he immediately asks if he might bring Johnny along.[24]

At Dixit's, Johnny takes out a chess set, and father and son sit down to play. Nash is "less than mediocre." At one point, he says he wants to take back a bad move. Johnny lets him. Then Nash wants to take back another.

"Dad, if you keep doing that, you'll win," says Johnny.

"But when I play against the computer, I'm allowed to take back moves," Nash says.

"But, Dad," protests Johnny, "I'm not a computer! I'm a *human being!*"

When it is time to go to the pharmacy for Johnny's "meds," Nash accompanies Alicia.[25] When it is time to attend an open house at the outpatient program where Johnny is sometimes enrolled, Nash is there and on time.[26] Alicia sees this and feels supported by him. She feels that she couldn't do without him.

Marriage is easily the most mysterious of human relationships. Attachments that seem superficial can become surprisingly deep and lasting. Such is the bond between Nash and Alicia. In retrospect, one feels that this is not an accidental pairing, that these two people needed each other. Strong-minded, pragmatic, and independent as she is, Alicia's girlish infatuation has survived the disillusionments, hardships, and disappointments. She takes Nash clothes shopping. She frets, when he travels, that he'll be kidnaped by terrorists or killed in a plane crash or merely worn out. When his ankle swells from a sprain, she leaves a dinner party and sits with him for four hours in the emergency room. More telling, she looks at an old photograph of him in bathing trunks at a poolside in California and says with a giggle, "Aren't his legs beautiful?"[27]

He, meanwhile, sets his clock by her. Stubborn, reserved, self-centered, and jealous of his time (and money) as he is, Nash does nothing without consulting Alicia first, defers to her wishes, and tries to help her, whether it is by washing the dishes, straightening out a problem at the bank, or going with her to family therapy every Monday night. She is the one to whom he faithfully reports the day's events, whom he ran into, what the lecture was about, what he ate for lunch. They argue about money, the housework, Johnny, social engagements, but he has committed himself to making her life easier and more joyful.

Nash is trying to be more sensitive and accommodating. He said, self-critically, "I know I have my social faults and I make Alicia very angry when she is saying something that I can anticipate before she's finished and then I start saying some-thing as if what she's saying is not of an importance."[28] He accepts, with some humor, that his genius does not make him the authority on all matters. When it comes to refinancing their mortgage or choosing between gas and oil heat, he complains humorously that Alicia does not take him seriously as an "economics sage . . . notwithstanding the Nobel."[29]

He does, of course, often wound her. But he catches himself, too, and makes amends. A typical exchange: at Gaby and Armand Borel's dinner party,[30] Alicia announces to the assembled company that their son has received a tentative offer to teach mathematics at a small college in Mexico. Nash engages in an act of cruelty. "Yes," he says, "my son is in a mental hospital in Arkansas but he got a job offer!" He is laughing at the absurdity of this juxtaposition. This is too much for Alicia. "You have to be fair to Johnny," she returns. Nash says nothing. But later in the evening he goes to some lengths to make amends. He brings an offering, maps of Mexico, that he found in books on the Borels' shelves, to Alicia. He takes the opportunity — during a conversation about Andrew Wiles's successful proof of Fermat's Last Theorem — to point out that Johnny had done some "classical" num-ber theory in graduate school. Johnny had published "one correct result, one incorrect, but the correct one was a breakthrough of sorts," he tells the other guests. Alicia responds by paying attention, by taking in what he means.

Much of the renewal of their marriage has taken place since the Nobel. There is now a sense of reciprocity. It is as if regaining the respect of his peers has made Nash feel that he has more to offer the people in his life, and has made those close to him, especially Alicia, feel that he has more to give. This has become self-reinforcing. At one time, before the Nobel, Alicia referred to Nash as her "boarder" and they lived essentially like two distantly related individuals under the same roof. Now there is even some discussion of remarrying, although in what was perhaps an assertion of Nash's old insistence on "rationality," they gave the idea up as impractical, as so many older couples have in light of the attendant tax and Social Security penalties. However, a certificate is not of real importance. They are a real couple again.

John Stier took the first step in ending his twenty-year estrangement from his father, mailing him a copy of the June 1993 *Boston Globe* column that speculated on Nash's chances of winning a Nobel.[31] He sent the clipping anonymously, but Nash immediately guessed its source. He was unsure whether to interpret John Stier's gesture as a taunt or a friendly overture. He told Harold Kuhn that something in the way the letter was addressed to him hinted at mockery. But the following February, two months after his triumph in Stockholm, Nash boarded a shuttle bound for Boston to spend a weekend getting reacquainted with his older son.

Such an encounter, inspired by hopes of putting their sad history behind them, was bound to be bittersweet, an occasion that revived as many painful memories, disappointments, and misunderstandings as it unlocked happier feelings.[32] When the two men finally met face to face, John Stier was no longer the nineteen-year-old Amherst College history major Nash remembered from their last encounter, but a man of forty-four — nearly as old as Nash had been in 1972, when they had last seen each other. Physically, he resembled his father to a striking degree. The impressive stature, broad shoulders, luminous eyes, English complexion, and finely modeled nose were all Nash's. But in his life's choices — and in his ability to derive great satisfaction from helping others — he was his mother's son. John Stier had stayed in Boston, remaining single and pursuing a career as a registered nurse. At the time, he was thinking of returning to graduate school to obtain an advanced degree in nursing.

In the two days they spent in each other's company — the most time they had ever been together at one stretch — they touched on personal topics only occasionally. Indeed, they were mostly with other people; it was important for Nash to have others confirm the reconciliation. They sat looking at old photographs with Eleanor, had a meal with Arthur Mattuck, the closest friend of Nash's "first family," and visited Marvin Minsky in his artificial intelligence laboratory at MIT. At one point, Nash telephoned Martha from John Stier's apartment and put his son on the phone.[33]

When father and son did venture into personal territory, Nash was, as usual, full of the best intentions. He wished to show his son how vitally important he was to him, he wanted to share with him some of his own recent good fortune, he wanted to give him the benefit of paternal advice. He was motivated by love and by a sense of responsibility. He told John that he would divide his estate equally between him and his brother and he invited him to accompany him to a conference in Berlin. All this was to the good. But, as in so many other relationships in his life, Nash's intentions weren't always matched by the emotional means to carry them out satisfactorily. Even as he tried to draw his son closer, he said and did things that could only be called insensitive and alienating.[34] He did not try to hide his own feelings of disappointment. He criticized his son's appearance, calling him fat (which he is not). He criticized his son's choice of profession, suggesting that nursing was beneath a son of his and urging him to go to medical school instead of pursuing a master's in nursing. He hinted strongly that he hoped John would help care for his younger brother, but then angered him by saying it would do Johnny good to be around a "less intelligent older brother."[35] Finally, he said he wanted John to change his name to Nash, a suggestion meant to be magnanimous, but which actually proved hurtful since it implied that he meant for John to renounce all that he was and had been. Eleanor, of course, felt injured.

A few months later, Nash did take John Stier to Berlin with him. The tensions of their first reunion surfaced again.[36] Nash remorselessly needled his son about trifles, making him turn out the light when he wanted to read, not letting him order dessert, telling him not to eat butter or bread. Yet even so, John Stier felt great pride when Nash gave his lectures.[37] And Nash was able to write to Harold Kuhn, "Berlin was a great experience . . . my son enjoyed the trip."[38]

• • •

A Nobel award has a finality about it. Yet despite the unique honor, life continues beyond the fairytale celebration in Stockholm. More so than for other Laureates, Nash's immediate future is uncertain. Nobody knows whether his remission is permanent. People have relapsed after many years of being symptom-free. The present is precious.

Unlike a game of Hex, outcomes in real life aren't predetermined by the first or even the fiftieth move. The extraordinary journey of this American genius, this man who surprises people, continues. The self-deprecating humor suggests greater self-awareness. The straight-from-the-heart talk with friends about sadness, pleasure, and attachment suggests a wider range of emotional experiences. The daily effort to give others their due, and to recognize their right to ask this of him, bespeaks a very different man from the often cold and arrogant youth. And the disjunction of thought and emotion that characterized Nash's personality, not just when he was ill, but even before are much less evident today. In deed, if not always in word, Nash has come to a life in which thought and emotion are more closely entwined, where getting and giving are central, and relationships are more symmetrical. He may be less than he was intellectually, he may never achieve another breakthrough, but he has become a great deal more than he ever was — "a very fine person," as Alicia put it once.

As we leave him now he is perhaps just hurrying under the Eisenhart gate on his way to Fine Hall . . . or sitting next to Alicia on the living-room sofa watching *Dr. Who* on the big television . . . or losing a game of chess to Johnny . . . or spending 105 minutes on the telephone comforting Lloyd Shapley after his wife's death . . . or giving Harold Kuhn a look like a naughty boy's when Harold asks whether the lecture notes for Pisa are ready . . . or sitting at the institute math table with his lunch tray, nodding while Enrico Bombieri, who has just read the love letters of Carrington, bemoans the lost art of letter writing . . . or, after listening to an astronomy lecture, gazing through a telescope at some distant star glimmering in the night sky. . . .

Epilogue

THE FESTIVE SCENE at the turn-of-the-century frame house opposite the train station might have been that of a golden wedding anniversary: the handsome older couple posing for pictures with family and friends, the basket of pale yellow roses, the 1950s photo of the bride and groom on display for the occasion.

In fact, John and Alicia Nash were about to say "I do" for the second time, after a nearly forty-year gap in their marriage. For them it was yet another step — "a big step," according to John — in piecing together lives cruelly shattered by schizophrenia. "The divorce shouldn't have happened," he told me. "We saw this as a kind of retraction of that." Alicia said simply, "We thought it would be a good idea. After all, we've been together most of our lives."

After Mayor Carole Carson pronounced them man and wife, John was asked to kiss his bride again for the camera. "A second take?" he quipped. "Just like a movie."

A few moments before the ceremony Alicia's cousin spoke to me about "the amazing metamorphosis" he had witnessed in John's life since the Nobel. It's not just the many other honors and speaking invitations from around the world that have followed, or the much wider audience that now appreciates the full range of exciting intellectual contributions made during his brief but brilliant career, or even the glamour of having his remarkable story told by Hollywood.

At seventy-three, John looks and sounds wonderfully well. He feels increasingly certain that he won't suffer a relapse. "It's like a continous process rather than just waking up from a dream," he told a *New York Times* reporter recently. "When I dream . . . it sometimes happens that I go back to the system of delusions that's typical of how I was . . . and then I wake and then I'm rational again." Growing self-confidence may be one reason that he is less embarrassed by talking about his past, and now speaks to groups that see his experience as "something that helps to reduce the stigma against people with mental illness."

For the first time since resigning from MIT in 1959, he now enjoys a modicum of personal security for himself and his family. Little things that the rest of us take for granted — having a driver's license again, or getting a credit card — mean a lot. "I feel I can go into a coffee place and spend a few dollars," Nash told me last year when I was working on a story about how economics Laureates spend their prize checks. "Lots of other academics do that," he said. "If I was really poor, I couldn't do that. I was like that."

Once threatened by homelessness, John values his home and personal belongings as few of us can. Back at the house after the ceremony, he was looking at a 1950 Parker Brothers version of Hex, the game he'd invented as a Princeton graduate student. He once owned a copy, he said. "I lost so many of my possessions due to my mental illness."

He has been able to return to mathematics. "I am working," he told the *Times* reporter. He no longer dreams of picking up where he left off, but is glad to be able to do serious work and make a contribution. John is once more a fixture at the math table at the Institute for Advanced Study and at tea in the Fine Hall common room. He now has a grant from the National Science Foundation. The other day he gave a seminar at the Institute about his new research on the theory of bargaining. "It actually wouldn't have been possible in those earlier days because I'm using computational facilities that didn't exist in the '50s and '60s," he said. "I'm ready to do a publication now."

Even more important, his remission and the Nobel have enabled him to renew broken ties. He has reconnected with old acquaintances from Bluefield, Carnegie, Princeton, and MIT. After today's ceremony, he gossiped happily with a mathematician and an engineer he first met in his twenties. He and Alicia were going to spend their second honeymoon among friends in Switzerland, where John will be giving a talk at a memorial celebration for Jürgen Moser, who died last year.

John has been able to share his good fortune with those closest to him. He's been in touch with John David, the older son who was once lost to him. He spends much of his time with his younger son, John Charles. On his wedding day, he proudly described a mathematical result that Johnny has lately been trying to publish. He and his sister, Martha, still talk on the phone every week. And, as today's scene suggests, he has come to acknowledge Alicia's central role in his life.

As for his biographer, John's attitude has changed dramatically. While this book was being written, he said to a *New York Times* reporter, "I adopted a position of Swiss neutrality." Since its publication, however, "A lot of my friends, family, and relations persuaded me it was a good thing." Besides, there is so much in the book that he had forgotten or never even knew. At this point in life, he made it clear, retrieving some of the past has been something of a solace.

When John met Russell Crowe, who plays him in the movie inspired by his life, he told me that his first words to the Australian actor were, "You're going to have to go through all these transformations!" Even in the three years since the publication of this book, the transformations in Nash's life have been as remarkable as any that will be portrayed on screen.

Princeton Junction, New Jersey, June 1, 2001

Notes

Prologue

1. George W. Mackey, professor of mathematics, Harvard University, interview, Cambridge, Mass., 12.14.95.
2. See, for example, David Halberstam, *The Fifties* (New York: Fawcett Columbine, 1993).
3. Mikhail Gromov, professor of mathematics, Institut des Hautes-Études, Bures-sur-Yvette, France, and Courant Institute, interview, 12.16.97. The claim that Nash ranks among the greatest mathematicians of the postwar era is based on judgment of fellow mathematicians. The topologist John Milnor expressed a nearly universal opinion among mathematicians when he wrote: "To some, the brief paper, written at age 21, for which he has won a Nobel prize in economics, may seem like the least of his achievements." In "A Celebration of John F. Nash, Jr.," a special volume, *Duke Mathematical Journal*, vol. 81, no. 1 (Durham, N.C.: Duke University Press, 1995), the game theorist Harold W. Kuhn calls Nash "one of the most original mathematical minds of this century."
4. Paul R. Halmos, "The Legend of John von Neumann," *American Mathematical Monthly*, vol. 80 (1973), pp. 382–94.
5. Donald J. Newman, professor of mathematics, Temple University, interview, Philadelphia, 3.2.96.
6. Harold W. Kuhn, professor of mathematics, Princeton University, interview, 7.26.95.
7. John Forbes Nash, Jr., remarks at the American Economics Association Nobel luncheon, San Francisco, 1.5.96; plenary lecture, World Congress of Psychiatry, Madrid, 8.26.96.
8. John Nash, "Parallel Control," RAND Memorandum no. 1361, 8.7.54; plenary lecture, Madrid, 8.26.96, op. cit.
9. Interviews with Newman, 3.2.96; Eleanor Stier, 3.13.96.
10. John Nash, plenary lecture, Madrid, 8.26.96, op. cit.
11. Jürgen Moser, professor of mathematics, ETH, Zurich, interview, New York City, 3.21.96.
12. Interviews with Paul Zweifel, professor of physics, Virginia Polytechnic Institute, 10.94; Solomon Leader, professor of mathematics, Rutgers University, 7.9.95; David Gale, professor of mathematics, University of California at Berkeley, 9.20.95; Martin Shubik, professor of economics, Yale University, 9.27.95; Felix Browder, president, American Mathematical Society, 11.2.95; Melvin Hausner, professor of mathematics, Courant Institute, 1.26.96; Hartley Rogers, professor of mathematics, MIT, Cambridge, 2.16.96; Martin Davis, professor of mathematics, Courant Institute, 2.20.96; Eugenio Calabi, 3.2.96.
13. Atle Selberg, professor of mathematics, Institute of Advanced Study, interview, Princeton, 8.16.95.
14. George W. Boehm, "The New Uses of the Abstract," *Fortune* (July 1958), p. 127: "Just turned thirty, Nash has already made a reputation as a brilliant mathematician who is eager to tackle the most difficult problems." Boehm goes on to say that Nash is working on quantum theory and that he invests in the stock market as a hobby.
15. John von Neumann, "Zur Theorie der Gesellschaftsspiele," *Math. Ann.*, vol. 100 (1928), pp. 295–320. See also Robert J. Leonard, "From Parlor Games to Social Science: Von Neumann, Morgenstern and the Creation of Game Theory, 1928–1944," *Journal of Economic Literature* (1995).
16. See, for example, Harold Kuhn, ed., *Classics in Game Theory* (Princeton: Princeton University Press, 1997); John Eatwell, Murray Milgate, and Peter Newman, *The New Palgrave: Game Theory* (New York: Norton, 1987); Avinash K. Dixit and Barry J. Nalebuff, *Thinking Strategically* (New York: Norton, 1991).
17. Robert J. Leonard, "Reading Cournot, Reading Nash: The Creation and Stabilization of the Nash Equilibrium," *The Economic Journal* (May 1994), pp. 492–511; Martin Shubik, "Antoine Augustin Cournot," in Eatwell, Milgate, and Newman, op. cit., pp. 117–28.
18. Joseph Baratta, historian, interview, 6.12.97.
19. John Nash, "Non-Cooperative Games," Ph.D. thesis, Princeton University Press (May 1950). Nash's thesis results were first published as "Equilibrium Points in N-Person Games," *Proceedings of the National Academy of Sciences, USA* (1950), pp. 48–49, and later as "Non-Cooperative Games," *Annals of Mathematics* (1951), pp. 286–95. See also "Nobel Seminar: The Work of John Nash in Game Theory," in *Les Prix Nobel 1994* (Stockholm: Norstedts Tryckeri, 1995). For a reader-friendly exposition of the Nash equilibrium, see Avinash Dixit and Susan Skeath, *Games of Strategy* (New York: Norton, 1997).
20. See, for example, Anthony Storr, *Solitude: A Return to the Self* (New York: Ballantine Books, 1988);

Robert Heilbroner, *The Worldly Philosophers* (New York: Simon & Schuster, 1992); E. T. Bell, *Men of Mathematics* (New York: Simon & Schuster, 1986); Stuart Hollingdale, *Makers of Mathematics* (New York: Penguin, 1989); Ray Monk, *Ludwig Wittgenstein: The Duty of Genius* (New York: Penguin, 1990); John Dawson, *Logical Dilemmas: The Life and Work of Kurt Gödel* (Wellesley, Mass.: A. K. Peters, 1997); Roger Highfield and Paul Carter, *The Private Lives of Albert Einstein* (New York: St. Martin's Press, 1994); Andrew Hodges, *Alan Turing: The Enigma* (New York: Simon & Schuster, 1983).

21. Anthony Storr, *The Dynamics of Creation* (New York: Atheneum, 1972).

22. Ibid.

23. John G. Gunderson, "Personality Disorders," *The New Harvard Guide to Psychiatry* (Cambridge: The Belknap Press of Harvard University, 1988), pp. 343–44.

24. Ibid.

25. Ibid.

26. Havelock Ellis, *A Study of British Genius* (Boston: Houghton Mifflin, 1926).

27. Rogers, interview, 2.16.96.

28. Zipporah Levinson, interview, Cambridge, 9.11.95.

29. Irving I. Gottesman, *Schizophrenia Genesis: The Origins of Madness* (New York: W. H. Freeman, 1991). For a contrary view, which states that cases of schizophrenia have been documented as long as 3,400 years ago, see Ming T. Tsuang, Stephen V. Faraone, and Max Day, "Schizophrenic Disorders," *New Harvard Guide to Psychiatry,* op. cit.

30. Tsuang, Faraone, and Day, op. cit., p. 259.

31. Gottesman, op. cit.; Tsuang, Faraone, and Day, op. cit.; Richard S. E. Keefe and Philip D. Harvey, *Understanding Schizophrenia: A Guide to the New Research on Causes and Treatment* (New York: Free Press, 1994); E. Fuller Torrey, *Surviving Schizophrenia: A Family Manual* (New York: Harper & Row, 1988).

32. Gottesman, op. cit.

33. For an excellent summary see Michael R. Trimble, *Biological Psychiatry* (New York: John Wiley & Sons, 1996), p. 224.

34. Eugen Bleuler, quoted in Louis A. Sass, *Madness and Modernism* (New York: Basic Books, 1992), p. 14.

35. Emil Kraepelin, quoted in ibid., pp. 13–14.

36. Torrey, op. cit.

37. Gottesman, op. cit.

38. Ibid.

39. See, for example, Tsuang, Faraone, and Day, op. cit.

40. See, for example, Gottesman, op. cit.

41. Ibid.

42. See, for example, Storr, *Solitude,* op. cit.; Gale Christianson, *In the Presence of the Creator* (New York: Free Press, 1984); Richard S. Westfall, *The Life of Isaac Newton* (Cambridge, U.K.: Cambridge University Press, 1993).

43. George Winokur and Ming Tsuang, *The Natural History of Mania, Depression and Schizophrenia* (Washington, D.C.: American Psychiatric Press, 1996), pp. 253–68; Manfred Bleuler, *The Schizophrenia Disorders: Long-Term Patient and Family Studies* (New Haven: Yale University Press, 1978).

44. M. Bleuler, op. cit., quoted in Sass, op. cit., p. 14.

45. Storr, *The Dynamics of Creation,* op. cit.

46. See, for example, Gottesman, op. cit. For discussions of differences between manic depressive illness and schizophrenia, see Torrey, op. cit.; Kay Redfield Jamison, *Touched with Fire: Manic-Depressive Illness and the Artistic Temperament* (New York: Free Press, 1993).

47. Sass, op. cit., prologue.

48. Emil Kraepelin, *Dementia Praecox and Paraphrenia* (Huntington, N.Y.: R. E. Krieger, 1971), quoted in Sass, op. cit., pp. 13–14.

49. Sass, op. cit., p. 4.

50. Letter from John Nash to Emil Artin, written in Geneva, undated (1959).

51. Letter from John Nash to Alex Mood, 11.94.

52. R. Nash, interview, 1.7.96.

53. Confidential source.

54. See, for example, Mikhail Gromov, *Partial Differential Relations* (New York: Springer-Verlag, 1986); Heisuke Hironaka, "On Nash Blowing Up," *Arithmetic and Geometry II* (Boston: Birkauser, 1983), pp. 103–11; P. Ordehook, *Game Theory and Political Theory: An Introduction* (Cambridge, U.K.: Cambridge University Press, 1986); Richard Dawkins, *The Selfish Gene* (Oxford: Oxford University Press, 1976); John Maynard Smith, *Did Darwin Get It Right?* (New York: Chapman and Hall, 1989); as well as *Math Reviews* and *Social Science Citation Index,* various dates.

55. Eatwell, Milgate, Newman, op. cit., p. xii.

56. Ariel Rubinstein, professor of economics, Princeton University and University of Tel Aviv, interview, 10.18.95.

57. Eatwell, Milgate, Newman, op. cit.

58. Member, School of Historical Studies, Institute for Advanced Study, interview, 1995.

59. Freeman Dyson, professor of physics, Institute for Advanced Study, interview, Princeton, 12.5.96.

60. Enrico Bombieri, professor of mathematics, Institute for Advanced Study, interview, 12.6.96.

61. See, for example, Winokur and Tsuang, op cit., p. 268.

62. Kuhn, interview, 10.94.

Part One: A BEAUTIFUL MIND

1: Bluefield

1. John Forbes Nash, Jr., autobiographical essay, *Les Prix Nobel 1994,* op. cit.
2. "Nash-Martin," *Appalachian Power & Light Searchlight,* vol. 3, no. 9 (September 1924), p. 14.
3. Ibid.
4. Martha Nash Legg, interview, Roanoke, 7.31.95.
5. The history of the Nashes is based on genealogical materials, regional histories, and newspaper clippings supplied by Martha Legg and Richard Nash, including *The History of Grayson County, Texas,* vol. 2 (Grayson County Frontier Village, 1981) and Graham Landrum and Allan Smith, *Grayson County: An Illustrated History* (Fort Worth, Tex.: Historical Publishers). The facts of John Forbes Nash, Sr.'s early life are based on interviews with Martha Nash Legg as well as his obituary.
6. Obituaries of Martha Nash, *Baptist Standard* (1944); M. Legg, interview, 8.1.95; R. Nash, interview, San Francisco, 1.7.97.
7. M. Legg, interview, 7.31.95.
8. The history of the Martins and the facts of Virginia Martin's early life are based on interviews with Martha Legg as well as obituaries of Emma Martin and Virginia Martin in the *Bluefield Daily Telegraph.*
9. Letter from John Forbes Nash, Jr., to Martha Legg, undated (1969).
10. For a short history of the marriage bar, see Claudia Goldin, "Career and Family: College Women Look to the Past," Working Paper No. 5188 (Cambridge, Mass.: National Bureau of Economic Research, July 1995).
11. C. Stuart McGehee, *The City of Bluefield: Centennial History 1889–1989* (Bluefield Historical Society).
12. Ibid.; John E. Williams, professor of psychology, Wake Forest University, interview, 8.95.
13. John Nash, *Les Prix Nobel 1994,* op. cit.
14. Williams, interview, 10.24.95; William Lewis, McKinsey & Partners, interview, 10.94.
15. John Nash, *Les Prix Nobel 1994,* op. cit.
16. M. Legg, interview, 8.3.95.
17. Ibid.
18. John G. Gunderson, "Personality Disorders," op. cit., pp. 343–44; also Nikki Erlenmeyer-Kimling, professor of genetics and development, Columbia University, interview, 1.17.98.
19. M. Legg, interview.
20. George Thornhill, quoted in William Archer, *Bluefield Daily Telegraph,* 10.94.
21. Report cards, various years, supplied by Martha Legg.
22. John Nash, *Les Prix Nobel 1994,* op. cit.
23. M. Legg, interview, 8.1.95.
24. Eddie Steele, quoted in William Archer, *Bluefield Daily Telegraph,* 10.13.94.
25. Donald V. Reynolds, interview, 6.29.97.
26. Ibid.
27. Ibid.
28. M. Legg, interview, 8.2.95.
29. Ibid.
30. E. T. Bell, *Men of Mathematics,* op. cit.; Betty Umberger, quoted in William Archer, *Bluefield Daily Telegraph,* 10.13.94.
31. Janice Thresher Frazier, personal communication, 9.97.
32. The origin of this quotation is unknown.
33. M. Legg, interview, 10.94.
34. Kuhn, interview, 3.97.
35. John Nash, *Les Prix Nobel 1994,* op. cit.
36. Bell, op. cit.
37. Ibid.
38. Ibid.
39. Denis Brian, *Einstein: A Life* (New York: John Wiley & Sons, 1996).
40. Bell, op. cit.; also Kuhn, interview, 10.21.97.
41. Bell, op. cit.
42. M. Legg, interview, 8.1.95.
43. Williams, interview.
44. Donald V. Reynolds, interview.
45. Interviews with Peggy Wharton, 12.96; Robert Holland, 6.9.97; John Louthan, 6.21.97; John Williams; Reynolds.
46. Reynolds, interview.
47. Ibid.
48. Felix Browder, president, American Mathematics Society, interview, 11.2.95.
49. M. Legg, interview, 11.94.
50. Nelson Walker, quoted in William Archer, *Bluefield Daily Telegraph,* 10.94.
51. Edwin Elliot, quoted in William Archer, *Bluefield Daily Telegraph,* 11.14.94.
52. M. Legg, interview, 8.2.95.

53. Reynolds, interview; see also William Archer, "Boys Will Be Boys," *Bluefield Daily Telegraph,* 11.14.94.

54. Julia Robinson, in Donald Albers, Gerald L. Alexanderson, and Constance Reid, *More Mathematical People* (New York: Harcourt Brace Jovanovich, 1990), p. 271.

55. Anthony Storr, *The Dynamics of Creation,* op. cit.

56. M. Legg, interview, 11.94.

57. Vernon Dunn, quoted in William Archer, *Bluefield Daily Telegraph,* 11.94.

58. Beaver High School Yearbook, 1945.

59. Interviews with Williams and Louthan.

60. M. Legg, interview, 8.1.95.

61. John Nash, *Les Prix Nobel 1994,* op. cit.

62. John F. Nash and John F. Nash, Jr., "Sag and Tension Calculations for Cable and Wire Spans Using Catenary Formulas," *Electrical Engineering,* 1945.

63. *Uncle App's News,* 7.45.

2: Carnegie Institute of Technology

1. Nash's interest in number theory, topology, and other branches of pure mathematics was recalled by Robert Siegel, professor of physics, College of William and Mary, interview, 10.30.97; Hans F. Weinberger, professor of mathematics, University of Minnesota, interviews, 9.6.95, 10.28.95, and 10.29.95; Paul F. Zweifel, professor of mathematics, Virginia Polytechnic Institute, interviews, 10.94 and 9.6.95; Richard J. Duffin (deceased), emeritus professor of mathematics, Carnegie-Mellon University, interviews, 10.94, 8.95, and 10.26.96.

2. See, for example, Stephan Lorant, *Pittsburgh: The Story of an American City* (Lenox, Mass.: author's edition, 1980) and interviews with Nash's contemporaries.

3. Richard Cyert, former president, Carnegie-Mellon University, interview, 10.26.95. Also Herbert Simon, Nobel Laureate, Carnegie-Mellon University, interview, 10.26.95.

4. Duffin, interview, 10.26.96; Robert E. Gleeson, professor of history, Carnegie-Mellon University, interview, 10.27.95; Glen U. Cleeton, *The Story of Carnegie Tech, II: The Doherty Administration, 1936–1950* (Pittsburgh: Carnegie Press, 1965); Robert E. Gleeson and Steven Schlossman, *George Leland Bach and the Rebirth of Graduate Management Education in the United States, 1945–1975* (Graduate Management Admission Council, Spring 1995); Robert E. Gleeson and Steven Schlossman, *The Many Faces of the New Look: The University of Virginia, Carnegie Tech and the Reform of American Management Education in the Postwar Era* (Graduate Management Admission Council, Spring 1992).

5. Interviews with Weinberger, 10.28.95; Zweifel, 10.94; George W. Hinman, professor of physics, Washington State University, 10.30.97; David R. Lide, editor, *CRC Handbook of Chemistry and Physics,* 10.30.97; Edward Kaplan, professor of statistics, Oregon State University, 5.21.97.

6. Interviews with Martha Nash Legg, 8.2.95; Weinberger, 10.28.95; Zweifel, 10.94.

7. Interviews with Siegel, 10.30.97; Hinman, 10.30.97.

8. John Nash, autobiographical essay, *Les Prix Nobel 1994,* op. cit.

9. Lide, interview, 10.30.97.

10. Hinman, interview, 10.30.97.

11. Lide, interview, 10.30.97.

12. John Nash, *Les Prix Nobel 1994,* op. cit.

13. Interviews with Raoul Bott, professor of mathematics, Harvard University, 11.5.95; Hinman, 10.30.97; Cathleen S. Morawetz, professor of mathematics, Courant Institute, and daughter of J. Synge, 2.29.96.

14. Duffin, interview, 10.26.95.

15. Duffin, interview, 10.94.

16. Morawetz, interview.

17. Ibid.

18. Interviews with Lide, 10.30.97, and Duffin, 10.26.95.

19. Weinberger, interview, 9.6.95.

20. Siegel, interview, 10.30.97. Siegel may have been mistaken; Bluefield had both a symphony and concert series before the war.

21. Bott, interview, 11.5.95.

22. Patsy Winter, Williamsburg, Virginia, interview, 10.30.97.

23. Weinberger, interview, 10.28.96.

24. Lide, interview, 10.30.97.

25. Interviews with Zweifel, 10.94, and Lide, 10.30.97.

26. Weinberger, interview, 10.28.95.

27. Siegel, interview, 10.30.97.

28. Hinman, interview, 10.30.97.

29. Zweifel, interview, 10.94.

30. Zweifel, interview, 1.21.98.

31. Ibid.; also interviews with Hinman, 10.30.97, and Siegel, 10.30.97.

32. Siegel, interview, 10.30.97.

33. Weinberger, interview, 10.28.95.

34. Zweifel, interview, 10.94.
35. Fletcher Osterle, professor of mechanical engineering, Carnegie-Mellon University, interview, 5.21.97.
36. *Mathematical Monthly* (September 1947), p. 400.
37. Leonard F. Klosinski, director, the William Lowell Putnam Mathematical Competition, interview, 10.96; Gerald L. Alexanderson, associate director, the William Lowell Putnam Mathematical Competition, interview, 10.96; Garrett Birkhoff, "The William Lowell Putnam Mathematical Competition: Early History," and L. E. Bush, "The William Lowell Putnam Mathematical Competition: Later History and Summary of Results," reprinted from *American Mathematical Monthly,* vol. 72 (1965), pp. 469–83.
38. Hinman, interview.
39. Harold Kuhn, interview, 7.97.
40. John Nash, *Les Prix Nobel 1994,* op. cit.
41. This scene is based on recollections of Duffin, interview, 10.94 and 10.26.95; Bott, interview, 10.94; and Weinberger, interviews, 9.6.95 and 10.28.95.
42. Duffin, interview, 10.94.
43. Bott, interview, 10.94.
44. Martin Burrow, professor of mathematics, Courant Institute, interview, 2.4.96.
45. Duffin, interviews, 10.94 and 10.26.95.
46. Duffin, interview, 10.94.
47. Bott, interview, 11.5.95.
48. Weinberger, interview, 10.28.95.
49. Siegel, interview, 10.30.97.
50. Weinberger, interview.
51. John Nash, *Les Prix Nobel 1994,* op. cit.
52. See Chapter 9.
53. *The Carnegie Tartan,* 4.20.48.
54. Interviews with Kuhn, 10.97, and M. Legg, 8.3.95.
55. John Nash, *Les Prix Nobel 1994,* op. cit.
56. The perception of Harvard's relative decline and Princeton's ascendancy by the late 1940s was widespread among Nash's contemporaries.
57. Duffin, interview, 10.26.95.
58. Letter from Solomon Lefschetz to Nash, 4.8.48.
59. Details about the JSK Fellowship, named after John S. Kennedy, a Princeton alumnus, are based on a memorandum from Sandra Mawhinney to Harold Kuhn, 10.27.97.
60. Graduate Catalog, Princeton University, various years; Report to the Dean of Faculty, Princeton University, various years.
61. John Nash, *Les Prix Nobel 1994,* op. cit.
62. Letter from S. Lefschetz to J. Nash.
63. Letter from John Nash to Solomon Lefschetz, undated, mid-April 1948.
64. Clifford Ambrose Truesdell, interview, 8.14.96.
65. Letter from J. Nash to S. Lefschetz. For the events transpiring then, see *Chronicle of the Twentieth Century* (Mount Kisco, N.Y.: Chronicle Publications, 1987).
66. Interviews with Charlotte Truesdell, 8.14.96, and Kaplan, 5.21.97.
67. Letter from J. Nash to S. Lefschetz, 4.26.48.
68. Clifford Truesdell, interview, 8.14.96.
69. Charlotte Truesdell, interview, 8.14.96.

3: The Center of the Universe

1. Martha Nash Legg, interview, 8.3.95.
2. See, for example, Rebecca Goldstein, *The Mind-Body Problem* (New York: Penguin, 1993); Ed Regis, *Who Got Einstein's Office?* (Reading, Mass.: Addison Wesley, 1987); and recollections of Nash's contemporaries, including interviews with Harold Kuhn and Harley Rogers and letter from George Mowbry, 4.5.95.
3. F. Scott Fitzgerald, *This Side of Paradise* (New York: Scribner, 1920).
4. Albert Einstein, quoted in Goldstein, op. cit.
5. As recalled by her niece Gillian Richardson, interview, 12.14.95.
6. Donald Spencer, professor of mathematics, Princeton University, interview, Durango, Colorado, 11.18.95.
7. Leopold Infeld, *Quest* (New York: Chelsea Publishing Company, 1980).
8. Virginia Chaplin, "Princeton and Mathematics," *Princeton Alumni Weekly* (May 9, 1958).
9. John D. Davies, "The Curious History of Physics at Princeton," *Princeton Alumni Weekly* (October 2, 1973).
10. Harold W. Kuhn, interview, 1.97.
11. Eugene Wigner, *Recollections of Eugene Paul Wigner as Told to Andrew Szanton* (New York: Plenum Press, 1992).
12. Regis, op. cit.
13. Infeld, op. cit.

14. Chaplin, op. cit.; William Aspray, "The Emergence of Princeton as a World Center for Mathematical Research, 1896–1939," in *A Century of Mathematics in America, Part II* (Providence, R.I.: American Mathematical Society, 1989); Gian-Carlo Rota, "Fine Hall in Its Golden Age," in *Indiscrete Thoughts* (Washington, D.C.: Mathematical Association of America, 1996), pp. 3–20.

15. Davies, op. cit.

16. Solomon Lefschetz, "A Self Portrait," typewritten, 1.54, Princeton University Archives.

17. Davies, op. cit.

18. Ibid.

19. Ibid.

20. Robert J. Leonard, "From Parlor Games to Social Science," op. cit.

21. Davies, op. cit.

22. Woodrow Wilson, quoted in ibid.

23. George Gray, Confidential Monthly Trustees Report, Rockefeller Foundation Archives (November 1945).

24. Wigner, op. cit.

25. The account of the Institute's history is based on Regis, op. cit.; Bernice M. Stern, *A History of the Institute for Advanced Study 1930–1950*, unpublished two-volume manuscript (1964).

26. Garrett Birkhoff, "Mathematics at Harvard 1836–1944," in *A Century of Mathematics in America, Part II*, op. cit., pp. 3–58; William Aspray, "The Emergence of Princeton as a World Center for Mathematical Research, 1896–1939," in *A Century of Mathematics in America, Part II*, op. cit., pp. 195–216; Gian-Carlo Rota, "Fine Hall in Its Golden Age," in *A Century of Mathematics in America, Part II*, op. cit., pp. 223–36.

27. Robin E. Rider, "Alarm and Opportunity: Emigration of Mathematicians and Physicists to Britain and the United States, 1933–1945," *Historical Studies in the Physical and Biological Sciences*, vol. 15, no. 1 (1984), pp. 108–71.

28. Paul Samuelson, "Some Memories of Norbert Wiener," provided by author, undated.

29. William James, "Great Men, Great Thoughts and Environment," *Atlantic Monthly*, vol. 46 (1880), pp. 441–59, quoted in Silvano Arieti, *Creativity: The Magic Synthesis* (New York: Basic Books, 1976), p. 299.

30. See, for example, Davies, op. cit.; Chaplin, op. cit.; Nathan Rheingold, "Refugee Mathematicians in the United States of America, 1933–1941: Reception and Reaction," *Annals of Science*, vol. 38 (1981), pp. 313–38; Rider, op. cit.; Lipman Bers, "The European Mathematician's Migration to America," in *A Century of Mathematics in America, Part I* (Providence, R.I.: American Mathematical Society, 1988).

31. See, for example, Mina Rees, "The Mathematical Sciences and World War II," in *A Century of Mathematics in America, Part I*; op. cit., Peter Lax, "The Flowering of Applied Mathematics in America," in *A Century of Mathematics in America, Part II*, op. cit., pp. 455–66; Fred Kaplan, *The Wizards of Armageddon* (New York: Simon & Schuster, 1983).

32. Chaplin, op. cit.

33. Andrew Hodges, *Alan Turing: The Enigma* (New York: Simon & Schuster, 1983).

34. Chaplin, op. cit.

35. Ibid.

36. See Kaplan, op. cit.; William Poundstone, *Prisoner's Dilemma* (New York: Doubleday, 1992); David Halberstam, *The Fifties*, op. cit.

37. Rees, "The Mathematical Sciences and World War II," op. cit.; Lax, "The Flowering of Applied Mathematics in America," op. cit., pp. 455–66.

38. Herman H. Goldstine, "A Brief History of the Computer," in *A Century of Mathematics in America, Part I*, op. cit., pp. 311–22; Poundstone, op. cit., pp. 76–78, on von Neumann's role in the development of the computer; Halberstam, op. cit., pp. 93–97, on von Neumann and the computer.

39. Hartley Rogers, professor of mathematics, MIT, interview, 1.26.96.

4: School of Genius

1. Solomon Leader, professor of mathematics, Rutgers University, interview, 6.9.95.

2. The portrait of Solomon Lefschetz is based on interviews with Harold W. Kuhn, 11.97; William Baumol, 1.95; Donald Spencer, 11.18.95; Eugenio Calabi, 3.2.96; Martin Davis, 2.20.96; Melvin Hausner, 2.6.96; Solomon Leader, 6.9.95; and other contemporaries of Nash's at Princeton. Also consulted were several memoirs, including Solomon Lefschetz, "Reminiscences of a Mathematical Immigrant in the United States," *American Mathematical Monthly*, vol. 77 (1970); A. W. Tucker, *Solomon Lefschetz: A Reminiscence;* Sir William Hodge, *Solomon Lefschetz, 1884–1972;* Phillip Griffiths, Donald Spencer, and George Whitehead, *Solomon Lefschetz: Biographical Memoirs* (Washington, D.C.: National Academy of Sciences, 1992); Gian-Carlo Rota, *Indiscrete Thoughts*, op. cit.

3. Lefschetz's obituary in *The New York Times* (October 7, 1972) credits him for "develop[ing] [the *Annals of Mathematics*] into one of the world's foremost mathematical journals."

4. "It should be noted that although Lefschetz was Jewish, he was not above engaging in a mild form of anti-semitism. He told Henry Wallman that he was the last Jewish graduate student that would be admitted to Princeton because Jews could not get a job anyway and so why bother," Ralph Phillips, "Reminiscences of the 1930s," *The Mathematical Intelligencer*, vol. 16, no. 3 (1994). Lefschetz's attitude toward Jewish students was well known. Phillips's impressions were confirmed by Leader, interview, 6.9.95; Kuhn, interview, 11.97; Davis, interview, 2.20.96; and Hausner, interview, 2.6.96.

5. Baumol, interview, 1.95.
6. See, for example, Gian-Carlo Rota, "Fine Hall in Its Golden Age," op. cit. DOD personnel security application, 3.10.56, Princeton University Archives.
7. Solomon Lefschetz, "A Self Portrait," typewritten, 1.54, Princeton University Archives.
8. Ibid., p. iii.
9. Donald Spencer, interviews, 11.28.95; 11.29.95; 11.30.95.
10. Rota, op. cit.
11. Ibid.
12. Ibid.
13. Leader, interview, 6.9.95.
14. Davis, interview, 2.6.96.
15. Hausner, interview, 2.6.96.
16. Leader, interview, 6.9.95.
17. Spencer, interviews.
18. Virginia Chaplin, "Princeton and Mathematics," op. cit.; Davis, interview, 2.20.96; Hartley Rogers, interview, 1.26.96.
19. Ibid.
20. Hausner, interview.
21. Ibid.
22. Ibid.
23. Joseph Kohn, interview, 7.25.96.
24. Robert Kanigel, *The Man Who Knew Infinity* (New York: Pocket Books, 1991); G. H. Hardy, "The Indian Mathematician Ramanujan," lecture delivered at the Harvard Tercentenary Conference of Arts and Sciences, August 31, 1936, reprinted in *A Century of Mathematics* (Washington, D.C.: Mathematical Association of America, 1994), p. 110.
25. Hardy, op. cit.
26. J. Davies, op. cit.; Gerard Washnitzer, professor of mathematics, Princeton University, interview, 9.25.96.
27. Graduate Catalog, Princeton University, various years; Report to the President, Princeton University, various years.
28. Letter from John Nash Forbes, Jr., to Solomon Lefschetz referring to request for private room, 4.46; Calabi, interview.
29. Interviews with Kuhn, 11.97; Washnitzer, 9.25.96; Felix Browder, 11.2.96; Calabi, 3.12.96; John Tukey, professor of mathematics, Princeton University, 9.30.97; John Isbell, professor of mathematics, State University of New York at Buffalo, 8.97; Leader, 6.9.95; Davis, 2.6.96.
30. Kuhn, interview.
31. Davis, interview.
32. Interviews with Washnitzer and Kuhn.
33. Washnitzer, interview.
34. Tukey, interview.
35. Kuhn, interview.
36. Calabi, interview.
37. Martin Shubik, "Game Theory at Princeton: A Personal Reminiscence," Cowles Foundation Preliminary Paper 901019, undated.
38. Interviews with Hausner; Davis; Kuhn; Spencer; Leader; Rogers; Calabi; and John McCarthy, professor of computer science, Stanford University, 2.4.96.
39. Hausner, interview, 2.6.96.
40. Interviews with Davis, Leader, Spencer; Rota, op. cit.
41. Rota, op. cit.
42. Isbell, interview.
43. Tukey, interview.
44. David Yarmush, interview, 2.6.96.
45. Princeton Alumni Directory 1997.
46. John W. Milnor, professor of mathematics and director, Institute for Mathematical Sciences, State University of New York at Stony Brook, interviews, 10.28.94 and 7.95.
47. Interviews with Kuhn, Hausner, John McCarthy.
48. Interviews with Hausner and Davis.

5: Genius

1. Kai Lai Chung, professor of mathematics, Stanford University, interview, 1.96; letter, 2.6.96.
2. Abraham Pais, *Subtle Is the Lord: The Science and Life of Albert Einstein* (New York: Oxford University Press, 1982).
3. Interviews with Charlotte Truesdell, 8.14.96; Martin Davis, 2.20.96; Hartley Rogers, 2.16.96; and John McCarthy, 2.4.96; John Forbes Nash, Jr., Personnel Security Questionnaire, 5.26.50, Princeton University Archives.

4. "Trivial," Melvin Hausner, interview; "burbling," Patrick Billingsley, professor of statistics, University of Chicago, interview, 8.12.97; "hacker," Hausner, interview.

5. Rogers, interview.

6. Davis, interview.

7. Peggy Murray, former secretary, department of mathematics, Princeton University, interview, 8.25.97.

8. Davis, interview.

9. John Milnor, interview, 9.26.95.

10. John Nash, autobiographical essay, *Les Prix Nobel 1994,* op. cit.

11. Mentioned by many of his contemporaries, this was confirmed by Nash in a conversation with Harold Kuhn.

12. Harold Kuhn, personal communication, 8.96.

13. Eugenio Calabi, interview.

14. Ibid.

15. Interviews with Solomon Leader and Calabi.

16. Letter from John Nash to Solomon Lefschetz, 4.48.

17. Calabi, interview.

18. John Milnor, "A Nobel Prize for John Nash," *The Mathematical Intelligencer,* vol. 17, no. 3 (1995), p. 5.

19. Leader, interview, 6.9.96.

20. Ibid.

21. David Gale, interview, 9.20.95.

22. Davis, interview.

23. Kuhn, interview, 9.96.

24. Hausner, interview.

25. Milner, interview, 9.26.95.

26. Norman Steenrod, letter, 1950, quoted by Harold Kuhn, introduction, "A Celebration of John F. Nash, Jr.," *Duke Mathematical Journal,* vol. 81, no. 2 (1996).

27. E. T. Bell, *Men of Mathematics,* op. cit.

28. Steenrod, letter, 2.5.53.

29. For this assessment, I relied on Hale Trotter and Harold Kuhn.

30. Milnor, interview.

31. Kuhn, interview, 8.97.

32. Ed Regis, *Who Got Einstein's Office?* op. cit.; Denis Brian, *Einstein: A Life,* op. cit.

33. John Forbes Nash, Jr., plenary lecture, World Congress of Psychiatry, Madrid, 8.26.96, op. cit.

34. Ibid.

35. Regis, op. cit.

36. Ibid.; also Brian, op. cit.

37. Brian, op. cit.

38. Ibid.

39. Nash, as told to Harold Kuhn; see also Brian, op. cit., for description of Kemeny's assistantship under Einstein in 1948–49.

40. Brian, op. cit.

41. John Nash, as told to Kuhn, November 1997.

42. Ibid.

43. Ibid.

44. Ibid.

45. Calabi, interview.

46. William Browder, professor of mathematics, Princeton University, interview, 12.6.96.

47. Steenrod, letter, 2.5.53.

48. Milnor, interview, 9.26.95.

49. Interviews with Leader and Kuhn.

50. Princeton University Archives.

51. Ibid.

52. Melvin Peisakoff, interview, 6.3.97.

53. Rogers, interview.

54. Calabi, interview.

55. Hausner, interview.

56. Rogers, interview.

57. Hausner, interview.

58. Felix Browder, interview, 11.2.95.

59. Leader, interview.

60. Harold Kuhn witnessed the scene, and Mel Peisakoff confirmed that it took place.

61. Donald Spencer, interview.

62. Letter from Al Tucker to Alfred Koerner, 10.8.56.

63. The portrait of Artin is based on Gian-Carlo Rota, *Indiscrete Thoughts,* op. cit., as well as recollection of John Tate; Spencer, interview, 11.18.96; Hauser, interview; and materials from the Princeton University Archives.

64. Spencer, interview.

65. Kuhn, interview.

6: Games

1. Albert W. Tucker, as told to Harold Kuhn, interview.
2. Interviews with Marvin Minsky, professor of science, MIT, 2.13.96; John Tukey, 9.30.97; David Gale, 9.20.96; Melvin Hausner, 1.26.96 and 2.20.96; and John Conway, professor of mathematics, Princeton University, 10.94; John Isbell, e-mails, 1.25.96, 1.26.97, 1.27.97.
3. Isbell, e-mails.
4. Letter from John Nash to Martin Shubik, undated (1950 or 1951); Hausner, interviews and e-mails.
5. William Poundstone, *Prisoner's Dilemma,* op. cit.; John Williams, *The Compleat Strategyst* (New York: McGraw Hill, 1954).
6. Poundstone, op. cit.
7. Solomon Leader, interview, 6.9.95.
8. Martha Nash Legg, interview, 8.1.95.
9. Isbell, e-mails.
10. Hartley Rogers, interview, 1.26.96.
11. Ibid.
12. Ibid.
13. Nash may have had the idea while he was at Carnegie. This, in any case, is Hans Weinberger's recollection, interview, 10.28.95.
14. Martin Gardner, *Mathematical Puzzles and Diversions* (New York: Simon & Schuster, 1959), pp. 65–70.
15. Gardner's comment, in 1959, was that Hex "may well become one of the most widely played and thoughtfully analyzed new mathematical games of the century."
16. Gale, interview, 9.20.95.
17. Dinner at which John Nash, David Gale, and the author were present, January 5, 1996, San Francisco.
18. Gale, interview.
19. Ibid.
20. Phillip Wolfe, mathematician, IBM, interview, 9.9.96.
21. John Milnor, "A Nobel Prize for John Nash," op. cit.
22. Ibid.; Gardner, op. cit.
23. Gale, interview.
24. Ibid.
25. Ibid.
26. Kuhn, interview.
27. Ibid.
28. Milnor, interview, 9.26.95.

7: John von Neumann

1. See, for example, Stanislaw Ulam, "John von Neumann, 1903–1957," *Bulletin of the American Mathematical Society,* vol. 64, no. 3, part 2 (May 1958); Stanislaw Ulam, *Adventures of a Mathematician* (New York: Scribner's, 1983); Paul R. Halmos, "The Legend of John von Neumann," *American Mathematical Monthly,* vol. 80 (1973); William Poundstone, *Prisoner's Dilemma,* op. cit.; Ed Regis, *Who Got Einstein's Office?,* op. cit.
2. Poundstone, op. cit.
3. Ulam, "John von Neumann," op. cit.; Poundstone, op. cit., pp. 94–96.
4. Harold Kuhn, interview, 1.10.96.
5. In remarks at a Nobel luncheon at the American Economics Association meeting on 1.5.96, Nash traced a lineage from Newton to von Neumann to himself. Nash shared von Neumann's interest in game theory, quantum mechanics, real algebraic variables, hydrodynamic turbulence, and computer architecture.
6. See, for example, Ulam, "John von Neumann," op. cit.
7. Norman McRae, *John von Neumann* (New York: Pantheon Books, 1992), pp. 350–56.
8. John von Neumann, *The Computer and the Brain* (New Haven: Yale University Press, 1959).
9. See, for example, G. H. Hardy, *A Mathematician's Apology* (Cambridge, U.K.: Cambridge University Press, 1967), with a foreword by C. P. Snow.
10. Ulam, "John von Neumann," op. cit.
11. Poundstone, op. cit.
12. Poundstone, *Prisoner's Dilemma,* p. 190.
13. Clay Blair, Jr., "Passing of a Great Mind," *Life* (February 1957), pp. 89–90, as quoted by Poundstone, op. cit., p. 143.
14. Poundstone, op. cit.
15. Ulam, "John von Neumann," op. cit.
16. Harold Kuhn, interview, 3.97.
17. Paul R. Halmos, "The Legend of John von Neumann," op. cit.
18. Ibid.
19. Poundstone, op. cit.
20. Halmos, op. cit.
21. Ibid.

22. Poundstone, op. cit.
23. Ulam, *Adventures of a Mathematician,* op. cit.
24. Ulam, "John von Neumann," op. cit.
25. Ibid.
26. Ibid., p. 10; Robert J. Leonard, "From Parlor Games to Social Science," op. cit.
27. Richard Duffin, interview, 10.94.
28. Halmos, op. cit.
29. Ulam, "John von Neumann," op. cit., pp. 35–39.
30. Interviews with Donald Spencer, 11.18.95; David Gale, 9.20.95; and Harold Kuhn, 9.23.95.
31. Poundstone, op. cit.
32. Herman H. Goldstine, "A Brief History of the Computer," *A Century of Mathematics in America, Part I,* op. cit.
33. John von Neumann, as quoted in ibid.

8: The Theory of Games

1. John von Neumann and Oskar Morgenstern, *The Theory of Games and Economic Behavior* (Princeton: Princeton University Press, 1944, 1947, 1953).
2. Both von Neumann and Morgenstern came to the seminar. Albert W. Tucker, interview, 10.94. See also Martin Shubik, "Game Theory and Princeton, 1940–1955: A Personal Reminiscence," Cowles Foundation Preliminary Paper, undated, p. 3; David Gale, interview, 9.20.95; and Harold Kuhn, interview, 9.20.95.
3. A. W. Tucker, "Combinatorial Problems Related to Mathematical Aspects of Logistics: Final Summary Report" (U.S. Department of the Navy, Office of Naval Research, Logistics Branch, February 28, 1957), p. 1.
4. Melvin Hausner, interview, 2.6.96.
5. Interviews with David Yarmush, 2.6.96, and John Mayberry, 4.15.96.
6. David Gale, interview.
7. Kuhn, interview.
8. Ibid.; Hausner, interview.
9. Robert J. Leonard, "From Parlor Games to Social Science," op. cit.
10. See, for example, H. W. Kuhn and A. W. Tucker, "John von Neumann's Work in the Theory of Games and Mathematical Economics," *Bulletin of the American Mathematical Society* (May 1958).
11. Leonard, "From Parlor Games to Social Science," op. cit.
12. Ibid.
13. Ibid.
14. Dorothy Morgenstern Thomas, interview, 1.25.96. Morgenstern kept a portrait of the kaiser hanging in his home.
15. Letter from George Mowbry to author, 4.5.95.
16. Leonard, "From Parlor Games to Social Science," op. cit.
17. As quoted in ibid.
18. Ibid.
19. Ibid.
20. Ibid.
21. Ibid.
22. Ibid.
23. Ibid.
24. Ibid.
25. A. W. Tucker, who knew both men well, said, "If he hadn't been forced to write a book, it wouldn't have gotten written," interview, 10.94. Von Neumann was interested in economics before he met Morgenstern.
26. Leonard, "From Parlor Games to Social Science," op. cit.
27. Ibid.
28. Von Neumann and Morgenstern, op. cit., p. 6.
29. Leonid Hurwicz, "The Theory of Economic Behavior," *The American Economic Review* (1945), pp. 909–25.
30. Von Neumann and Morgenstern, op. cit., p. 7.
31. Ibid., p. 3.
32. Ibid.
33. Ibid., p. 4.
34. Ibid., p. 7.
35. Ibid., p. 2.
36. Ibid.
37. Ibid., p. 6.
38. *New York Times,* 3.46.
39. See, for example, Herbert Simon, *The American Journal of Sociology,* no. 50 (1945), pp. 558–60. Hurwicz, op. cit.; Jacob Marschak, "Neumann's and Morgenstern's New Approach to Static Economics," *Journal of Political Economy,* no. 54 (1946), pp. 97–115; John McDonald, "A Theory of Strategy," *Fortune* (June 1949), pp. 100–110.
40. Leonard, "From Parlor Games to Social Science," op. cit.
41. Ibid.

42. Ibid.
43. Shubik, "Game Theory and Princeton," op. cit., p. 2.
44. Von Neumann and Morgenstern, op. cit. See also Eatwell, Milgate, and Newman, op. cit.
45. Von Neumann and Morgenstern, op. cit.
46. Ibid.
47. See, for example, John C. Harsanyi, "Nobel Seminar," in *Les Prix Nobel 1994.*
48. Von Neumann and Morgenstern, op. cit.
49. Ibid.
50. Ibid.
51. Harsanyi, op. cit.

9: The Bargaining Problem

1. John Forbes Nash, Jr., "The Bargaining Problem," *Econometrica,* vol. 18 (1950), pp. 155–62.
2. Nash's bargaining solution was "virtually unanticipated in the literature," according to Roger B. Myerson, "John Nash's Contribution to Economics," *Games and Economic Behavior,* no. 14 (1996), p. 291. See also Ariel Rubinstein, "John Nash: The Master of Economic Modeling," *The Scandinavian Journal of Economics,* vol. 97, no. 1 (1995), pp. 11–12; John C. Harsanyi, "Bargaining," in Eatwell, Milgate, and Newman, op. cit., pp. 56–60; Andrew Schotter, interview, 10.25.96; Ariel Rubinstein, interview, 11.25.96; James W. Friedman, professor of economics, University of North Carolina, interview, 10.2.96.
3. "This is the classical problem of exchange and, more specifically, of bilateral monopoly as treated by Cournot, Bowley, Tintner, Fellner and others," Nash, "The Bargaining Problem," p. 155. As Harold Kuhn points out, Nash's delineation of the history of the problem was undoubtedly supplied by Oskar Morgenstern, "It is now clear that Nash had not read those writers," Harold Kuhn, "Nobel Seminar," *Les Prix Nobel 1994.* For a delightful short history of exchange, including the references to pharaohs and kings, see Robert L. Heilbroner, *The Worldly Philosophers,* 6th edition (New York: Touchstone, 1992), p. 27.
4. John C. Harsanyi, "Approaches to the Bargaining Problem Before and After the Theory of Games: A Critical Discussion of Zeuthen's, Hick's and Nash's Theories," *Econometrica,* vol. 24 (1956), pp. 144–57.
5. In his now-classic reformulation of the Nash bargaining model, Ariel Rubinstein traces the bargaining problem to Edgeworth, "Mathematical Psychics: An Essay on the Application of Mathematics to the Moral Sciences" (London: C. Kegan Paul, 1881), reprinted in *Mathematical Psychics and Other Essays* (Mountain Center, Calif.: James & Gordon, 1995). Martin Shubik writes, "Even as a graduate student I was struck by the contrast between cooperative game theory, the seeds of which I regarded as already present in Edgeworth and noncooperative theory which was present in Cournot," Martin Shubik, *Collected Works,* forthcoming, p. 6. For lively accounts of Edgeworth's life and contributions, see Heilbroner, op. cit., pp. 174–76, and John Maynard Keynes, "Obituary of Francis Isidro Edgeworth, March 26, 1926," reprinted in Edgeworth, op. cit.
6. Heilbroner, op. cit., p. 173.
7. Ibid., p. 174.
8. Edgeworth, op. cit.
9. Ibid.
10. Ibid.
11. Harsanyi, op. cit.
12. John von Neumann and Oskar Morgenstern, *The Theory of Games and Economic Behavior,* op. cit., p. 9. "It may also be regarded as a nonzero-sum two-person game," Nash, "The Bargaining Problem," op. cit., p. 155; "even though von Neumann and Morgenstern's theory of games was an essential step toward a strong bargaining theory, their own analysis of two-person bargaining games did not go significantly beyond the weak bargaining theory of neoclassical economics," Harsanyi, "Bargaining," op. cit., pp. 56–57.
13. See, for example, Robert J. Leonard, "From Parlor Games to Social Science," op. cit., for a history of the axiomatic approach, and a superb interpretive discussion of "axiomatics" in Robert J. Aumann, "Game Theory," in John Eatwell, Murray Milgate, and Peter Newman, *The New Palgrave,* op. cit., pp. 26–28.
14. Von Neumann and Morgenstern used the axiomatic method to derive their theory of expected or von Neumann–Morgenstern utilities in the second, 1947, edition of *Theory of Games and Economic Behavior.* The first application to a problem in social sciences, I believe, was Kenneth J. Arrow's Ph.D. thesis *Social Choice and Individual Values* (New York: John Wiley & Sons, 1951). Lloyd S. Shapley's "A Value of N-Person Games," *Contributions to the Theory of Games II* (Princeton: Princeton University Press, 1953), pp. 307–17, is another stellar example.
15. John Nash, "The Bargaining Problem," op. cit., p. 155.
16. John Nash, *Les Prix Nobel 1994,* op. cit., pp. 276–77.
17. The sketch of Bart Hoselitz is based on an interview with his friend Sherman Robinson, professor of economics, University of Chicago, 7.95, and questionnaires, letters, and a curriculum vitae from Carnegie-Mellon University archives.
18. This bit of history about international trade theory after World War II was supplied by Kenneth Rogoff, professor of economics, Princeton University, interview.
19. John Nash, *Les Prix Nobel 1994,* op. cit., pp. 176–77.
20. Nash told Myerson that he was inspired by a problem posed by Hoselitz. Roger Myerson, professor of economics, Northwestern University, interview, 8.7.97.
21. Myerson, e-mail, 8.11.97.
22. Letter from John Nash to Martin Shubik, undated (written in 1950 or 1951).

23. Harold Kuhn was for many years convinced that Nash had mailed a copy of his first draft to von Neumann while he was still at Carnegie. Also interviews with David Gale, 9.20.95, and William Browder, 12.6.96.

24. After historian Robert Leonard published the established version of the origins of the paper in "Reading Cournot, Reading Nash: The Creation and Stabilisation of the Nash Equilibrium," *The Economic Journal,* no. 164 (May 1994), p. 497, Nash corrected the record at a lunch with Harold Kuhn and Roger Myerson, 5.96, Kuhn, personal communication, 5.96.

25. John Nash, "The Bargaining Problem," op. cit., p. 155.

26. John Nash, *Les Prix Nobel 1994,* op. cit., p. 277.

10: Nash's Rival Idea

1. Harold Kuhn, interview, 4.14.97.

2. Albert William Tucker, interview, 10.94.

3. The beer party scene was reconstructed from the recollections of Melvin Hausner, 2.6.96, Martin Davis, 2.20.96, and Hartley Rogers, 1.16.96, who attended several such parties in the course of their graduate school careers.

4. Davis, interview.

5. Ibid. Amazingly, Davis was able, forty years later, to recall the entire song, a few lines of which are given here, interview.

6. Kuhn, interview, 4.16.97.

7. Ibid.

8. Henri Poincaré, quoted in E. T. Bell, *Men of Mathematics,* op. cit., p. 551.

9. John Nash to Robert Leonard, e-mail, 2.20.93. Further details supplied by Harold Kuhn, interview, 4.17.97.

10. "All the graduate students were afraid of him," according to Donald Spencer, interview, 11.8.95.

11. Von Neumann's dress and manner are described by George Mowbry in a letter, 4.5.95. Harold Kuhn, interview, 5.2.97.

12. See, for example, Norman McRae, *John von Neumann,* op. cit., pp. 350–56.

13. As told to Harold Kuhn, 4.17.97.

14. John Nash, *Les Prix Nobel 1994,* op. cit.

15. Silvano Arieti, *Creativity,* op. cit., p. 294.

16. J. Nash to R. Leonard, e-mail.

17. Ibid.

18. The conversation between Nash and Gale was recounted by Gale in an interview, 9.20.95. Gale also suggested that Nash use Kakutani's fixed point theorem instead of Brouwer's to simplify the proof, a suggestion that Nash followed in the note in the National Academy of Sciences *Proceedings*.

19. John F. Nash, Jr., "Equilibrium Points in N-Person Games," communicated by S. Lefschetz, 11.16.49, pp. 48–49.

20. Gale, interview.

21. Tucker, interview, 10.94.

22. Gian-Carlo Rota, interview, 12.12.95.

23. Tucker's account of Minsky's thesis on computers and the brain, "Neural Networks and the Brain Problem," is given in an interview with Stephen B. Maurer published in the *Two Year College Mathematics Journal,* vol. 14, no. 3 (June 1983).

24. Tucker, interview.

25. Harold Kuhn, "Nobel Seminar," *Les Prix Nobel 1994,* op. cit., p. 283.

26. Tucker, interview, 10.94.

27. Ibid.

28. Ibid.

29. John Nash, *Les Prix Nobel 1994,* op. cit.

30. Tucker, interview.

31. Letter from Albert W. Tucker to Solomon Lefschetz, 5.10.50.

32. Ibid.

33. See, for example, introduction, John Eatwell, Murray Milgate, and Peter Newman, *The New Palgrave,* op. cit.

34. "It so happens that the concept of the two-person zero-sum games has *very few* real life applications," John C. Harsanyi, "Nobel Seminar," *Les Prix Nobel 1994,* op. cit., p. 285.

35. Ibid.

36. Nobel citation.

37. Avinash Dixit and Barry Nalebuff, *Thinking Strategically,* op. cit.

38. Ibid.

39. "Nowadays it almost seems to be obvious that the correct application of Darwinism to problems of social interaction among animals requires the use of non-cooperative game theory," according to Reinhard Selten, "Nobel Seminar," *Les Prix Nobel 1994,* op. cit., p. 288.

40. "Game Theory," in Eatwell, Milgate, and Newman, op. cit., p. xiii.

41. Michael Intriligator, personal communication, 6.27.95.

42. Selten, op. cit., p. 297.

43. Von Neumann, as Nash always acknowledged, nonetheless helped to gain attention for Nash's ideas. For example, the preface to the third edition (1953) of *Theory of Games and Economic Behavior* directs readers to Nash's work on noncooperative games, p. vii.

11: Lloyd

1. T. S. Ferguson, "Biographical Note on Lloyd Shapley," in *Stochastic Games and Related Topics in Honor of Professor L. S. Shapley,* edited by T. E. S. Raghavan, T. S. Ferguson, T. Parthasarathy, and O. J. Vrieze (Boston: Kluwer Academic Publishers, 1989).
2. See, for example, Carl Sagan, *Broca's Brain* (New York: Random House, 1979).
3. David Halberstam, *The Fifties,* op. cit.
4. The description of Shapley's experiences during the war, at Princeton, and at RAND draw on the recollections of Harold Kuhn, 11.18.96; Norman Shapiro, 2.9.96; Martin Shubik, 9.27.95 and 12.13.96; Melvin Hausner, 2.6.96; Eugenio Calabi, 3.2.96; John Danskin, 10.19.96; William Lucas, 6.27.95; Hartley Rogers, 1.26.96; John McCarthy, 2.4.96; Marvin Minsky, 2.13.96; Robert Wilson, 3.7.96; Michael Intriligator, 6.27.95.
5. Letter from John von Neumann, 1.54.
6. Solomon Leader, interview, 6.9.95.
7. Rogers, interview, 1.26.96.
8. "It was like ESP. Shapley seemed to know where all of the pieces were all of the time," Minsky, interview.
9. Hausner, interview, 2.6.96.
10. Danskin, interview, 10.19.95.
11. Letter from Lloyd Shapley to Solomon Lefschetz, 4.4.49.
12. Interviews with Nancy Nimitz, 5.21.96, and Kuhn, 4.4.96.
13. Shapiro, interview, 12.13.96.
14. Intriligator, interview, 6.27.95.
15. Shubik, interview, 12.13.96.
16. Lloyd S. Shapley, interview, 10.94.
17. Ibid.
18. Shubik, interview, 12.13.96.
19. Interviews with Shapley, Shubik, McCarthy, Calabi.
20. Calabi, interview.
21. Ibid.
22. Ibid.
23. Shubik, interview, 9.27.95.
24. Shubik, interview, 9.27.95.
25. Letter from Nash to Martin Shubik, undated (1950 or 1951).
26. McCarthy, interview.
27. McCarthy, interview.
28. Hausner, interview, 2.6.96; M. Hausner, J. Nash, L. Shapley, and M. Shubik, "So Long Sucker—A Four-Person Game," mimeo provided by Hausner.
29. Interviews with Shubik and McCarthy.
30. John Nash and Lloyd Shapley, "A Simple Three-Person Poker Game," *Annals of Mathematics,* no. 24 (1950).
31. "To some extent there was a competition between Nash, Shapley, and me," Shubik, interview, 12.13.96.
32. Shapley, interview.
33. Shapley, *Additive and Non-Additive Set Functions,* Ph.D. thesis, Princeton University, 1953. Shapley published his famous result—the so-called Shapley value—a value for n-person games, in 1953.
34. Martin Shubik, "Game Theory at Princeton," op. cit., p. 6: "We all believed that a problem of importance was the characterization of the concept of threat in a two person game and the incorporation of the use of threat in determining the influence of the employment of threat in a bargaining situation. [Nash, Shapley, and I] worked on this problem, but Nash managed to formulate a good model of the two person bargain utilizing threat moves to start with." Shubik is referring here to Nash's "Two-Person Cooperative Games," published in *Econometrica* in 1953 but actually written in August 1950 during Nash's first summer at RAND.
35. Letter from Albert W. Tucker, 1953.
36. Ibid.
37. Letter from Frederick Bohnenblust, spring 1953.
38. Letter from John von Neumann, 1.54.
39. Kuhn, interview, 11.18.96.
40. Shapley, interview, 10.94.

12: The War of Wits

1. John McDonald, "The War of Wits," *Fortune* (March 1951).

2. William Poundstone, *Prisoner's Dilemma*, op. cit.; Fred Kaplan, *The Wizards of Armageddon*, op. cit.; *The RAND Corporation: The First Fifteen Years* (Santa Monica, Calif.: RAND, November 1963) and *40th Year Anniversary* (Santa Monica: RAND, 1963); John D. Williams, An Address, 6.21.50; Bruce L. R. Smith, *The RAND Corporation* (Cambridge: Harvard University Press, 1966); Bruno W. Augenstein, *A Brief History of RAND's Mathematics Department and Some of Its Accomplishments* (Santa Monica, Calif.: RAND, March 1993); Alexander M. Mood, "Miscellaneous Reminiscences," *Statistical Science,* vol. 5, no. 1 (1990), pp. 40–41.

3. Herman Kahn, *On Thermonuclear War* (Princeton: Princeton University Press, 1960), as quoted in Poundstone, op. cit., p. 90.

4. Isaac Asimov, *Foundation* (New York: Bantam Books, 1991).

5. Poundstone, op. cit.

6. Kaplan, op. cit., p. 52.

7. Ibid., p. 10.

8. Oskar Morgenstern, *The Question of National Defense* (New York: Random House, 1959), as quoted in Poundstone, op. cit., pp. 84–85.

9. McDonald, "The War of Wits," op. cit.

10. The account of RAND's beginnings is based on Poundstone, op. cit.

11. Ibid., p. 93.

12. See, for example, Stanislaw Ulam, *Adventures of a Mathematician,* op. cit.; Richard Rhodes, *The Making of the Atomic Bomb* (New York: Simon & Schuster, 1986); Hodges, *Alan Turing: The Enigma,* op. cit.

13. Mina Rees, "The Mathematical Sciences and World War II," op. cit.

14. The sketch of RAND's mathematics, economics, and computer groups is based largely on interviews with RAND staff and consultants from the early Cold War period, including Kenneth Arrow, 6.26.95; Bruno Augenstein, 6.13.96; Richard Best, 5.22.96; Bernice Brown, 5.22.96; John Danskin, 10.19.95; Martha Dresher, 5.21.96; Theodore Harris, 5.24.96; Mario Juncosa, 5.21.96 and 5.24.96; William Karush, 5.96; William F. Lucas, 6.26.95; John W. Milnor, 9.95; John McCarthy, 2.4.96; Alexander M. Mood, 5.23.96; Evar Nering, 6.18.96; Nancy Nimitz, 5.21.96; Melvin Peisakoff, 6.3.96; Harold N. Shapiro, 2.20.96; Norman Shapiro, 2.29.96; Lloyd S. Shapley, 11.94; Herbert Simon, 10.16.95; Robert Specht, 2.96; Albert W. Tucker, 12.94; Willis H. Ware, 5.24.96; Robert W. Wilson, 8.96; Charles Wolf, Jr., 5.22.96.

15. Augenstein, interview, 6.13.96.

16. R. Duncan Luce, interview, 1996.

17. The descriptions of Arrow's contributions are taken from Mark Blaug, *Great Economists Since Keynes* (Totowa, N.J.: Barnes & Noble, 1985), pp. 6–9.

18. Kenneth Arrow, professor of economics, Stanford University, interview, 6.26.95.

19. McDonald, interview.

20. Richard Best, former manager of security, RAND Corporation, interview, 5.22.96.

21. Interviews with Alexander M. Mood, professor of mathematics, University of California at Irvine, former deputy director, mathematics department, RAND Corporation, 5.23.96, and Mario L. Juncosa, mathematician, RAND, 5.21.96 and 5.24.96.

22. Kaplan, op. cit., p. 51.

23. Bernice Brown, retired statistician, RAND, interview, 5.22.96.

24. Augenstein, interview.

25. Arrow, interview.

26. *Chronicle of the Twentieth Century,* op. cit., p. 667.

27. David Halberstam, *The Fifties,* op. cit.

28. Ibid.

29. Ibid., p. 46.

30. Kaplan, op. cit.

31. Martha Dresher, interview.

32. Best, interview.

33. Halberstam, *The Fifties,* op. cit., p. 45; *Chronicle of the Twentieth Century,* op. cit., p. 677.

34. Halberstam, op. cit., p. 49.

35. *Chronicle of the Twentieth Century,* op. cit., p. 750.

36. Best, interview.

37. Ibid.

38. Letter from Col. Walter Hardie, U.S. Air Force, to RAND, 10.25.50.

39. As told to Harold Kuhn, interview, 8.97.

40. Letter from John Nash to John and Virginia Nash, 11.10.51.

41. Best, interview.

42. The Eisenhower guidelines refer to DOD directive 52206, 1953 and Executive Order 10450, 1953.

43. Danskin, interview.

44. Robert Specht, interview, 10.96.

45. John Williams, *The Compleat Strategyst,* op. cit.

46. The account of mathematicians' work habits is based on interviews with Brown, Mood, Juncosa, Danskin, and Shapiro.

47. Interviews with Mood and Juncosa.

48. Juncosa, interview.
49. Mood, interview.
50. The description of Williams is based on interviews with Best, Brown, Mood, and Juncosa; Poundstone, op. cit.; and Kaplan, op. cit.
51. Mood, interview.
52. As quoted in Poundstone, op. cit., p. 95.
53. Mood, interview.
54. Danskin, interview.
55. Arrow, interview.
56. Mood, interview.
57. Best, interview.
58. Harold Shapiro, interview.
59. Mood, interview.
60. Danskin, interview.
61. Ibid.
62. Best, interview.

13: Game Theory at RAND

1. Kenneth Arrow, interview, 6.26.95.
2. M. Dresher and L. S. Shapley, *Summary of RAND Research in the Mathematical Theory of Games (RM-293)* (Santa Monica, Calif.: RAND, 7.13.49).
3. Arrow, interview.
4. Fred Kaplan, *The Wizards of Armageddon*, op. cit.
5. Thomas C. Schelling, *The Strategy of Conflict* (Cambridge: Harvard University Press, 1960).
6. Ibid.
7. Arrow, interview.
8. See, for example, Martin Shubik, "Game Theory and Princeton," op. cit.; William Lucas, "The Fiftieth Anniversary of TGEB," *Games and Economic Behavior*, vol. 8. (1995), pp. 264–68; Carl Kaysen, interview, 2.15.96.
9. John McDonald, "The War of Wits," op. cit.
10. For a humorous account of Prussian military's romance with probability theory see John Williams, *The Compleat Strategyst*, op. cit.
11. McDonald, op. cit.
12. Bernice Brown, interview, 5.22.96.
13. Rosters, RAND Department of Mathematics.
14. Dresher and Shapley, op. cit. For a lucid description of game theoretic analyses of duels, see Dixit and Skeath, op. cit.
15. Dresher and Shapley, op. cit.
16. For von Neumann's views, see Clay Blair, Jr., "Passing of a Great Mind," *Life* (February 1957), pp. 88–90, as quoted in William Poundstone, *Prisoner's Dilemma*, op. cit., p. 143.
17. Arrow, interview.
18. See Poundstone, op. cit.; Joseph Baratta, interview, 8.12.97.
19. Arrow, interview.
20. John H. Kagel and Alvin E. Roth, *The Handbook of Experimental Economics* (Princeton: Princeton University Press, 1995), pp. 8–9.
21. Albert W. Tucker, interview, 12.94.
22. See, for example, Avinash Dixit and Barry Nalebuff, *Thinking Strategically*, op. cit.
23. See, for example, Anatole Rappaport, "Prisoner's Dilemma," in John Eatwell, Murray Milgate, and Peter Newman, *The New Palgrave*, op. cit., pp. 199–204.
24. Dixit and Nalebuff, op. cit.
25. Harold Kuhn, interview, 7.96.
26. Poundstone, op. cit.; also Kagel and Roth, op. cit.
27. John F. Nash, Jr., as quoted in Kagel and Roth, op. cit.
28. Martin Shubik, "Game Theory at Princeton, 1949–1955: A Personal Reminiscence," in *Toward a History of Game Theory*, edited by E. Roy Weintraub (Durham, N.C.: Duke University Press, 1992).
29. The first version of Nash's analysis of the role of threats in bargaining was published as a RAND memorandum, "Two-Person Cooperative Games, P-172" (Santa Monica, Calif.: RAND, 8.31.50). A final version appeared under the same title in *Econometrica* (January 1953), pp. 128–40. Also "Rational Non-Linear Utility," RAND Memorandum, D-0793, 8.8.50.
30. Kaplan, op. cit.
31. Ibid.
32. Ibid.
33. Ibid., pp. 91–92.
34. Ibid.
35. Bruno Augenstein, interview.
36. R. Duncan Luce and Howard Raiffa as quoted in Poundstone, op. cit., p. 168.
37. Thomas Schelling, *The Strategy of Conflict* (Cambridge, Mass.: Harvard University Press, 1960).

14: The Draft

1. Department of Mathematics, Princeton University.
2. Recommendations of 5.11.50 by Solomon Lefschetz, chairman, mathematics department, to president, Princeton University, that John Forbes Nash, Jr., be appointed research assistant, three-quarters time, on A. W. Tucker's ONR Contract A-727.
3. See, for example, David Halberstam, *The Fifties,* op. cit.
4. Proceedings of the International Congress of Mathematicians, August 30–September 6, 1950, vol. 1, p. 516.
5. Letter from John Nash to Albert W. Tucker, 9.10.50. Letter from John Nash to Solomon Lefschetz, undated (probably written between April 10 and April 26, 1948), gives the clearest statement of why Nash wanted to avoid the draft: "Should there come a war involving the U.S. I think I should be more useful, and better off, working on some research project than going, say, into the infantry."
6. Letter from Fred D. Rigby, Office of Naval Research, Washington, D.C., to Albert W. Tucker, 9.15.50.
7. Letter from J. Nash to A. W. Tucker, 9.10.50.
8. Letters from A. W. Tucker to Local Board No. 12, 9.13.50; Raymond J. Woodrow to Local Board No. 12, 9.15.50 and 9.18.50; Raymond J. Woodrow, Committee on Project Research and Inventions, Princeton University, to Local Board No. 12, Bluefield, W.Va., re occupational deferment for John F. Nash, Jr. (with reference to RAND consultancy).
9. Letter from F. D. Rigby to A. W. Tucker, 9.10.50.
10. Ibid.
11. Halberstam, op. cit.
12. Hans Weinberger, interview, 10.28.95.
13. Harold Kuhn, interview, 9.6.96.
14. Gottesman, *Schizophrenia Genesis,* op. cit., pp. 152–55; also Bruce Dohrenwind, professor of social psychology, Columbia University, interview, 1.16.98.
15. H. Steinberg and J. Durrel, "A Stressful Situation as a Precipitant of Schizophrenic Symptoms," *British Journal of Psychiatry,* vol. 111 (1968), pp. 1097–1106, as quoted in Gottesman, *Schizophrenia Genesis,* op. cit.
16. Notes of telephone call from Alice Henry, secretary, department of mathematics, Princeton University, re I-A classification of John Nash and request that Dean Douglas Brown write a letter to ONR to be forwarded to the Bluefield draft board, 9.15.50.
17. "Information Needed in National Emergency," form filled out 9.50 by John F. Nash, Jr., refers to I-A status, pending application for II-A, ONR and RAND research roles.
18. Letter from Raymond J. Woodrow, Committee on Project Research and Inventions, Princeton University, to commanding officer, Office of Naval Research, New York Branch, re deferment for John F. Nash, Jr., 9.18.50.
19. Letter from W. S. Keller, Office of Naval Research, New York Branch, to Selective Service Board No. 12, Bluefield, W.Va., re deferment for John F. Nash, Jr., 9.28.50.
20. Richard Best, interview, 5.96.
21. Melvin Peisakoff, interview, 5.96.
22. Best, interview.
23. Letter from Raymond J. Woodrow to John Nash, 10.6.50.
24. Ibid.; letter from L. L. Vivian, ONR, New York Branch, to commanding officer, ONR, New York Branch Office, re notification of Nash by draft board that active service postponed until June 30, 1951, and continued I-A status, 11.22.50.

15: A Beautiful Theorem

1. Richard J. Duffin, interview, 10.26.95.
2. "He can hold his own in pure mathematics, but his real strength seems to lie on the frontier between mathematics and the biological and social sciences," letter from Albert W. Tucker to Marshall Stone, 12.14.51.
3. John Nash, "Algebraic Approximations of Manifolds," *Proceedings of the International Congress of Mathematicians,* vol. 1 (1950), p. 516, and "Real Algebraic Manifolds," *Annals of Mathematics,* vol. 56, no. 3 (November 1952; received October 8, 1951). For expositions of Nash's result, see John Milnor, "A Nobel Prize for John Nash," op. cit., pp. 14–15, and Harold W. Kuhn, introduction, "A Celebration of John F. Nash, Jr.," *Duke Mathematical Journal,* vol. 81, no. 1 (1995), p. iii.
4. Harold Kuhn, interview, 11.30.97.
5. See, for example, June Barrow-Green, *Poincaré and the Three-Body Problem* (Providence, R.I.: American Mathematical Society, 1977); also Kuhn, interview.
6. George Hinman, interview, 10.30.97.
7. John F. Nash, Jr., *Les Prix Nobel 1994,* op. cit.
8. See, for example, E. T. Bell, *Men of Mathematics,* op. cit., and Norman Levinson, "Wiener's Life," in "Norbert Wiener 1894–1964," *Bulletin of the American Mathematical Society,* vol. 72, no. 1, part II, p. 8.
9. Martin Davis, interview, 2.6.96.
10. Norman Steenrod, letter of recommendation, 2.51, as quoted by Kuhn, introduction, "A Celebration of John F. Nash, Jr.," op. cit.
11. John Nash, "Algebraic Approximations of Manifolds," op. cit., p. 516.

12. Solomon Lefschetz, President's Report, Princeton University Archives, 7.18.80.
13. Solomon Lefschetz, memorandum, 3.9.49, on Spencer's appointment as visiting professor at Princeton in academic year 1948–49; Donald Spencer, interviews, 11.28.95 and 11.29.95.
14. Lefschetz, memorandum, 3.9.49.
15. Donald Clayton Spencer, Biography, 10.61, Princeton University Archives.
16. See, for example, "Analysis, Complex," *Encyclopaedia Britannica* (1962).
17. Kodaira won the Fields in 1954; David C. Spencer, "Kunihiko Kodaira (1915–1997)," *American Mathematical Monthly,* 2.98.
18. Spencer won the Bôcher in 1947, Biography, op. cit.
19. Lefschetz, memorandum, 3.9.49.
20. Joseph Kohn, professor of mathematics, Princeton University, interview, 7.19.95.
21. Ibid. Also Phillip Griffiths, director, Institute for Advanced Study, interview, 5.26.95.
22. In his recommendation for Spencer's appointment as visiting professor in 1949, Lefschetz remarks on his "warm and sympathetic personality." Spencer had an unusual willingness to reach out to colleagues in trouble. He became deeply involved in helping Max Shiffman, a bright young mathematician at Stanford who was diagnosed with schizophrenia; John Moore, a mathematician who suffered a severe depression; and John Nash after Nash returned to Princeton in the early 1960s. See Spencer, op. cit.
23. Spencer, op. cit.
24. As slightly restated by Milnor, "A Nobel Prize for John Nash," op. cit., p. 14.
25. Intersectional Nomination: Class Five; 1996 Election, John F. Nash, Jr.
26. Michael Artin, professor of mathematics, MIT, interview, 12.2.97.
27. See, for example, Michael Artin and Barry Mazur, "On Periodic Points," *Annals of Mathematics,* no. 81 (1965), pp. 82–99. Milnor calls this an "important" application.
28. Barry Mazur, professor of mathematics, Harvard University, interview, 12.3.97.
29. Nash cites, for example, H. Seifert, "Algebraische Approximation von Mannigfaltigkeiten," *Math. Zeit.,* vol. 41 (1936), pp. 1–17.
30. Ibid.
31. Steenrod, letter, 2.51, as quoted by Kuhn, introduction, "A Celebration of John F. Nash, Jr.," op. cit.
32. Spencer, op. cit.
33. Nash, as told to Harold Kuhn, private communication, 12.2.97. The subsequent Nash–Moser theorem has even more profound implications for celestial mechanics. See Chapter 30.
34. Albert W. Tucker, interview, 11.94. Nash still dabbled in game theory, perhaps partly to maintain his RAND connection. For example, he wrote "N-Person Games: An Example and a Proof," RAND Memorandum, RM-615, June 4, 1951, as well as, with graduate students Martin Shubik and John Mayberry, "A Comparison of Treatments of a Duopoly Situation," RAND Memorandum P-222, July 10, 1951.
35. Kuhn, interview.
36. Letter from Albert W. Tucker to Hassler Whitney, 4.5.55.
37. Artin supervised the honors calculus program, which, according to John Tate (interview, 6.29.97), he took very seriously. Later documents refer to Nash's having been a poor teacher; the comments undoubtedly stem from his experiences in 1950–51.
38. "There is no doubt that the department should look towards keeping Milnor permanently as a member of our faculty," Solomon Lefschetz, President's Report, Princeton University Archives, 9.51.
39. Letter from A. W. Tucker to H. Whitney, op. cit.
40. William Ted Martin, professor of mathematics, MIT, interview, 9.7.95.
41. Letter from Albert W. Tucker to Marshall Stone, 2.26.51.
42. Nash told Kuhn that his desire to live in Boston played a role in his accepting the MIT position, Kuhn, personal communication, 7.97.

16: MIT

1. Lindsay Russell, interview, 1.14.96.
2. Patrick Corcoran, retired captain, Cambridge City Police, interview, 8.12.97.
3. Felix Browder, interview, 11.14.95.
4. Gian-Carlo Rota, professor of mathematics, MIT, interview, 10.29.94.
5. Paul A. Samuelson, professor of economics, MIT, interview, 11.94.
6. Harvey Burstein, former FBI agent who set up the campus police at MIT, interview, 7.3.97.
7. Samuelson, interview.
8. William Ted Martin, professor of mathematics, MIT, interview, 9.7.95.
9. Samuelson, interview.
10. Department of Physics, MIT, communication, 1.98.
11. Course catalog, MIT, various years.
12. Samuelson, interview.
13. Ibid.
14. Arthur Mattuck, professor of mathematics, MIT, e-mail, 6.23.97.
15. Joseph Kohn, professor of mathematics, Princeton University, interview, 7.25.95.
16. Samuelson, interview. See also Report to the President, MIT, various years.
17. Jerome Lettvin, professor of electrical engineering and bioengineering, MIT, interview, 7.25.97; Emma Duchane, interview, 6.26.97.

18. Samuelson, interview.
19. Gian-Carlo Rota, interview.
20. Hearing before Committee on Un-American Activities (HUAC), House of Representatives, Eighty-third Congress, First Session, Washington, D.C., April 22 and 23, 1953.
21. Samuelson, interview.
22. Martin, interview.
23. Ibid.
24. See, for example, Wiener's obituary, *New York Times,* 3.19.64; Paul Samuelson, "Some Memories of Norbert Wiener," 1964, Xerox provided by Samuelson; and Norbert Wiener, *Ex-Prodigy* (New York: Simon & Schuster, 1953) and *I Am a Mathematician* (New York: Simon & Schuster, 1956).
25. Samuelson, "Some Memories of Norbert Wiener," op. cit.
26. Ibid.
27. Zipporah Levinson, interview, 9.11.95.
28. Samuelson, "Some Memories of Norbert Weiner," op. cit.
29. Z. Levinson, interview.
30. Ibid.
31. Ibid.
32. Ibid.
33. Note from John Nash to N. Wiener, 11.17.52.
34. Letter from John Nash to Albert W. Tucker, 10.58.
35. Jerome Neuwirth, professor of mathematics, University of Connecticut at Storrs, interview, 5.21.97.
36. The sketch of Levinson is based on recollections of his widow, Zipporah Levinson; Arthur Mattuck; F. Browder, 11.2.95; Gian-Carlo Rota, 11.94; and many others. Also Kenneth Hoffman, Memorandum to President J. B. Wiesner, 3.14.74; William Ted Martin et al., obituary of Norman Levinson, 12.17.75.
37. HUAC, op. cit. See also Chapter 19.
38. Arthur Mattuck, "Norman Levinson and the Distribution of Primes," address to MIT shareholders, 10.6.78.

17: Bad Boys

1. Donald J. Newman, professor of mathematics, Temple University, interview, 12.28.95; Leopold Flatto, Bell Laboratories, interview, 4.25.96.
2. Sigurdur Helgason, professor of mathematics, MIT, interview, 2.13.96.
3. Course catalog, MIT, various years.
4. Arthur Mattuck, interview, 11.7.95.
5. Robert Aumann, professor of mathematics, Hebrew University, interview, 6.25.95.
6. Joseph Kohn, interview, 7.19.95.
7. Ibid.
8. Aumann, interview.
9. Seymour Haber, professor of mathematics, Temple University, interviews, 3.14.95 and 3.19.95.
10. George Whitehead, professor of mathematics, MIT, interview, 12.12.95.
11. Eva Browder, interview, 9.6.97.
12. Barry Mazur, interview, 12.3.97.
13. Harold Kuhn quotes Nash taking credit for introducing the tea hour at MIT in his introduction to the special volume in honor of Nash, "A Celebration of John F. Nash, Jr.," op. cit.
14. Isadore M. Singer, professor of mathematics, MIT, interview, 12.13.95.
15. Kohn, interview.
16. Singer, interview.
17. Jerome Neuwirth, interview, 5.21.97.
18. Mattuck, interview, 2.13.96.
19. Descriptions of this legendary crowd are based on interviews with Kohn; Felix Browder, 11.2.95, 11.10.95, 9.6.97; Aumann; Neuwirth; Newman; H. F. Mattson, 10.29.97 and 11.18.97; Larry Wallen, 5.16.97 and 5.20.97; Mattuck; Paul Cohen, 1.5.96; Jacob Bricker, 5.22.97; and others.
20. F. Browder, interview, 9.6.97.
21. Haber, interview.
22. Ibid.
23. Martha Nash Legg, interview, 3.29.96.
24. Neuwirth, interview.
25. Ibid.
26. Mattuck, interview, 2.13.96.
27. Interviews with Neuwirth and F. Browder, 11.2.95.
28. Jürgen Moser, professor of mathematics, Eidgenössische Techische Hochschule, Zurich, interview, 3.23.96.
29. Marvin Minsky, professor of science, MIT, interview, 2.13.96.
30. Herta Newman, interview, 3.2.96.
31. Andrew Browder, professor of mathematics, Brown University, interview, 6.18.97.
32. Haber, interview.
33. Flatto, interview.

34. D. Newman, interview, 2.4.96.
35. Zipporah Levinson, interview, 9.11.95.
36. Neuwirth, interview.
37. D. Newman, interview.
38. Ibid.
39. Lawrence Wallen, professor of mathematics, University of Hawaii, interviews, 5.20.97 and 6.4.97.
40. Kohn, interview.
41. H. F. Mattson, professor of computer science, Syracuse University, interview, 5.16.97; also Wallen, interview.
42. J. C. Lagarias, "The Leo Collection: Anecdote and Stories," AT&T Bell Laboratories, 4.29.95 (Xerox).
43. Mattuck, interview, 5.21.95, and Neuwirth, interview.
44. Neuwirth, interview.
45. The sketch of Donald J. Newman is based on an interview with him and on interviews with Flatto, Kohn, Mattuck, Singer, and Harold S. Shapiro, professor of mathematics, Royal Institute of Technology, Stockholm, Sweden, e-mail, 5.21.97.
46. Singer, interview, 12.13.95.
47. Mattuck, interview, 11.7.95.
48. D. Newman, interview, 3.2.96.
49. Helgason, interview, 12.3.94; also interviews with Mattuck and Singer.
50. Flatto, interview.
51. Ibid.
52. Ibid.
53. Singer, interview.
54. Haber, interview.
55. Ibid.
56. Flatto, interview.
57. Ibid.
58. Ibid.
59. Neuwirth, interview.
60. Ibid.
61. D. Newman, interview, 3.2.96.
62. Ibid.
63. H. Newman, interview.
64. Fred Brauer, professor of mathematics, University of Wisconsin, interview, 5.22.97.

18: Experiments

1. Harold N. Shapiro, professor of mathematics, Courant Institute, interview, 2.20.96.
2. John Milnor, interview, 9.26.95.
3. The account of the cross-country trip is based largely on recollections of Martha Nash Legg, interviews, 8.29.95 and 3.29.96, and Ruth Hincks Morgenson, interview, 6.22.97.
4. John Nash to Harold Kuhn, personal communication, 6.24.97; also Morgenson, interview.
5. M. Legg, interview.
6. Ibid.
7. Ibid.
8. Ibid.; Milnor, interview.
9. John M. Danskin, interview, 10.29.95.
10. M. Legg, interview.
11. Ibid.
12. John Milnor, "Games Against Nature," in *Decision Processes,* edited by R. M. Thrall, C. H. Coombs, and R. L. Davis (New York: John Wiley & Sons, 1954).
13. "Some Games and Machines for Playing Them," RAND Memorandum, D-1164, 2.2.52.
14. John Nash and R. M. Thrall, "Some War Games," RAND Memorandum, D-1379, 9.10.52.
15. G. Kalisch, J. Milnor, J. Nash, and E. Nering, "Some Experimental N-Person Games," RAND Memorandum, RM-948, 8.25.52.
16. M. Legg, interview.
17. The description of the experiment is based on, apart from the original paper, Evar Nering, professor of mathematics, University of Minnesota, interview, 6.18.96; R. Duncan Luce and Howard Raiffa, *Games and Decisions* (New York: John Wiley & Sons, 1957), pp. 259–69; John H. Kagel and Alvin E. Roth, *The Handbook of Experimental Economics,* op. cit., pp. 10–11.
18. Kagel and Roth, op. cit.
19. Milnor, interview, 10.28.94.
20. John Milnor, "A Nobel Prize for John Nash," op. cit.
21. See, for example, Kagel and Roth, op. cit.
22. Milnor, interview, 1.27.98.
23. Letter from John Nash to John Milnor, 12.27.64.

19: Reds

1. Zipporah Levinson, interview, 9.11.95.
2. Hearing before Committee on Un-American Activities, House of Representatives, Washington, D.C., 4.22.53 and 4.23.53. Unless otherwise noted, all references to the hearing are based on this transcript.
3. David Halberstam, *The Fifties,* op. cit.
4. Letter from Harold W. Dodds, president, Princeton University, to Colonel S. R. Gerard, Screening Division, Western Industrial Personnel Security Board, 10.14.54, Princeton University Archives.
5. See, for example, F. David Peat, *Infinite Potential: The Life and Times of David Bohm* (Reading, Mass.: Addison Wesley, 1997).
6. Z. Levinson, interview.
7. Ibid. See also Felix Browder, interview, 11.10.95.
8. Z. Levinson, interview.
9. Ibid.
10. *The Tech,* spring 1953, various issues.
11. Z. Levinson, interview.
12. Ibid.
13. William Ted Martin, interview.
14. Z. Levinson, interview.
15. Fred Brauer, e-mail, 6.23.97; Arthur H. Copeland, professor of mathematics, University of New Hampshire, e-mail, 6.24.97; Arthur Mattuck, e-mail, 6.25.97.
16. John Nash, plenary lecture, World Congress of Psychiatry, Madrid, 8.26.96, op. cit.

20: Geometry

1. Letter from Warren Ambrose to Paul Halmos, undated (written spring 1953).
2. The portrait of Ambrose is based on the recollections of Isadore Singer, 2.13.95; Lawrence Wallen, 6.4.97; Felix Browder, 11.2.95; Zipporah Levinson, 9.11.95; William Ted Martin, 9.7.95; H. F. Mattson, 10.29.97, 11.18.97, 11.28.97; Gian-Carlo Rota, 10.94; George Mackey, 12.14.95.
3. See, for example, I. M. Singer and H. Wu, "A Tribute to Warren Ambrose," *Notices of the AMS* (April 1996).
4. Robert Aumann, interview, 6.28.95.
5. Gabriel Stolzenberg, professor of mathematics, Northeastern University, interview, 4.2.96.
6. Leopold Flatto, interview, 4.15.96. See also "The Leo Collection: Anecdotes and Stories," AT&T Bell Laboratories, 4.29.94.
7. Ibid.
8. George Mackey, interview, 12.14.95.
9. Felix Browder, interview, 11.2.95.
10. Flatto, interview.
11. Despite its apocryphal ring, the story appears to be true and has been confirmed by Nash. Harold Kuhn, personal communication, 8.97.
12. Armand Borel, professor of mathematics, Institute for Advanced Study, interview, 3.1.96.
13. F. Browder, interview.
14. Ibid.
15. Joseph Kohn, interview, 7.19.95. Phrasing the question precisely, Ambrose would have used the adverb "isometrically" — meaning "to preserve distances" — after "embedding."
16. Shlomo Sternberg, professor of mathematics, Harvard University, interview, 3.5.96.
17. Mikhail Gromov, interview, 12.16.97.
18. John Forbes Nash, Jr., *Les Prix Nobel 1994,* op. cit.
19. Gromov, interview.
20. John Conway, professor of mathematics, Princeton University, interview, 10.94.
21. Jürgen Moser, e-mail, 12.24.97.
22. Richard Palais, professor of mathematics, Brandeis University, interview, 11.6.95.
23. Moser, interview.
24. Donald J. Newman, interview, 3.2.96.
25. Jürgen Moser, "A Rapidly Convergent Iteration Method and Non-linear Partial Differential Equations, I, II," *Annali della Scuola Normale Superiore di Pisa,* vol. 20 (1966), pp. 265–315, 499–535.
26. See, for example, Kyosi Ito, ed., *Encyclopedic Dictionary of Mathematics* (Mathematical Society of Japan; Cambridge: MIT Press, 1987), p. 1076; Lars Hörmander, "The Boundary Problems of Physical Geodesy," *Archive for Rational Mechanics and Analysis,* vol. 62, no. 1 (1976), pp. 1–52; and S. Klainerman, *Communications in Pure and Applied Mathematics,* vol. 33 (1980), pp. 43–101.
27. John Nash, "C^1 Isometric Imbeddings," *Annals of Mathematics,* vol. 60, no. 3 (November 1954), pp. 383–96.
28. Kohn, interview.
29. John Forbes Nash, Jr., *Les Prix Nobel 1994,* op. cit.
30. Rota, interview, 11.14.95.
31. Flatto, interview.
32. Jacob Schwartz, professor of computer science, Courant Institute, interview, 1.29.96.

33. Isadore Singer, interview, 12.14.95.
34. Paul J. Cohen, professor of mathematics, Stanford University, interview, 1.6.96.
35. Moser, interview, 3.23.96.
36. The Nash–Federer correspondence wasn't saved, and Federer declined to be interviewed (personal communication, 6.25.96). The account is based on the recollections of several individuals, including Wendell Fleming (interview, 6.97), a longtime collaborater and friend of Federer.
37. Fleming, interview.
38. John Nash, "The Imbedding Problem for Riemannian Manifolds," *Annals of Mathematics,* vol. 63, no. 1 (January 1956, received October 29, 1954, revised August 20, 1955).
39. Borel, interview.
40. Letter from John Forbes Nash, Jr., to Virginia and John Nash, Sr., 4.54.
41. Rota, interview.
42. Stolzenberg, interview, 4.2.96.
43. Ibid.
44. Schwartz, interview.
45. Moser, interview.
46. Ibid.
47. Ibid.
48. Rota, interview, 10.94.
49. George Whitehead, professor of mathematics, MIT, interview, 12.12.95.
50. Flatto, interview.
51. Lawrence Wallen, interview, 6.4.97.

Part Two: SEPARATE LIVES

21: Singularity

1. Postcard from John Nash to Arthur Mattuck, 1968. *B* stood for Jacob Bricker, *T* for Ervin D. Thorson, *F* for Herbert Amasa Forrester, and *R* for Donald V. Reynolds.

22: A Special Friendship

1. Letter from John Forbes Nash, Jr., to Martha Nash Legg, 11.4.65.
2. Ibid.
3. Herta Newman, interview, 3.2.96.
4. D. Newman, interview.
5. Joseph Kohn, interview, 2.15.96.
6. H. Newman, interview.
7. D. Newman, interview.
8. In his 11.4.65 letter, Nash describes Thorson as one of three "special friendships." Thorson was working in Santa Monica, California, at Douglas Aircraft.
9. The references to *T* in Nash's letters continued until at least 1968, usually in conjunction with references to *B* (for Bricker) and *F.*
10. M. Legg, interview, 3.30.96.
11. Douglas Aircraft could supply no biographical or professional information on Thorson (Donald Hanson, personal communication, 6.17.97). Nash did not recall Thorson when asked about him by Harold Kuhn (6.97). What details are known of Thorson are based solely on an obituary in the *Hemet News* and a brief conversation with his surviving sister, Nelda Troutman, 5.28.97.
12. Hanson, interview.
13. Ibid.
14. Troutman, interview, 5.28.97.
15. Ibid.
16. Ibid.
17. Under the Eisenhower guidelines, homosexuals were not permitted to have security clearances.

23: Eleanor

1. The description of Nash's stay at Mrs. Grant's house is based on interviews with Lindsay Russell, 1.14.96, 4.23.96, and 7.97.
2. Postcard from John Nash, Jr., to Virginia and John Nash, Sr., 9.52.
3. Martha Nash Legg, interview, 9.3.95.
4. Eleanor Stier, interview, 2.14.96.
5. Ibid., 3.15.96.
6. Ibid., 2.14.96 and 3.18.96.
7. Arthur Mattuck, interview, 11.7.95.
8. Eleanor's history was taken from interviews with her, 3.15.95, and John David Stier, 9.20.97.

9. E. Stier, interview, 2.14.96.
10. Ibid., 3.15.96.
11. That Nash was interested in, and experimented with, various drugs was recalled by Donald Newman, interview, 3.2.96. Eleanor Stier confirmed this, interview, 3.18.96, although neither witnessed Nash's experiments, if indeed they ever took place. Their possible significance is twofold. First, it suggests Nash's concern with enhancing his mental powers but also his concerns about his own "manliness."
12. E. Stier, interview, 3.13.96.
13. Ibid.
14. M. Legg, interview.
15. E. Stier, interview, 3.15.96. Confirmed by Jacob Bricker, interview, 5.22.97, and Arthur Mattuck, interview.
16. Bricker, interview.
17. E. Stier, interview, 7.95.
18. Ibid.
19. Bricker, interview.
20. E. Stier, interview, 3.15.96.
21. John David Stier, interview, 6.29.96.
22. E. Stier, interview, 3.15.96.
23. J. D. Stier, interview, 9.20.97.
24. E. Stier, interview, 3.15.96.
25. Ibid.
26. Ibid, 3.18.96.
27. Ibid., 3.18.96, and J. D. Stier, interview, 9.20.97.
28. J. D. Stier, interview, 9.20.97.
29. A. Mattuck, interview.
30. E. Stier, interview, 3.18.96.
31. Bricker, interview; Mattuck, interview.
32. E. Stier, interview, 3.18.96.
33. Mattuck, interview.
34. E. Stier, interview, 3.18.96.
35. Ibid., 3.15.96.
36. Mattuck, interview.
37. Best, interview, 5.22.96.
38. Mattuck, interview, 5.21.97.
39. Bricker, interview.
40. E. Stier, interview.
41. Ibid., 3.18.96.
42. Ibid.
43. J. D. Stier, interview, 9.20.97.
44. Ibid.

24: Jack

1. Donald J. Newman, interview, 3.12.96.
2. Arthur Mattuck, interview, 5.21.97.
3. The portrait of Bricker is based on interviews with Mattuck; Newman; Herb Kamowitz; Jerome Neuwirth, 5.23.97 and 6.5.97; Leopold Flatto, 4.25.96; Lawrence Wallen, 5.20.97.
4. Jacob Bricker, interview, 5.22.97.
5. Jack Kotick, interview, 1.21.98.
6. D. Newman, interview, 3.12.96.
7. Ibid., 1.25.98.
8. Eleanor Stier, interview.
9. Letter from John Nash to Martha Nash Legg, 11.4.65.
10. Herta Newman, interview, 3.2.96.
11. Sheldon M. Novick, *Henry James: The Young Master* (New York: Random House, 1996).
12. Letter from J. Nash to M. Legg.
13. Alfred C. Kinsey et al., *Sexual Behavior of the Human Male* (Philadelphia: Saunders, 1948).
14. Letter from J. Nash to M. Legg.
15. Bricker, interview, 5.22.97.
16. Neuwirth, interviews.
17. Mattuck, interviews, 5.20.97 and 5.28.97.
18. Bricker, interview, 5.22.97.
19. Postcard from John Nash to Jacob Bricker, 8.3.67.
20. Letter from John Nash to Arthur Mattuck, 7.10.68. "Mattuckine" seems to be a reference to the Mattachine Society, the first American advocacy group for homosexuals, founded in 1951 (source: Neil Miller, *Out of the Past: Gay and Lesbian History from 1869 to the Present* [New York: Vintage Books, 1995], pp. 334–38).
21. Bricker, interview.
22. Bricker, interview, 1.26.98.

25: The Arrest

1. Nash mostly pursued his growing interest in computers and wrote a paper in which he proposed the idea of parallel control. "Higher Dimensional Core Arrays for Machine Memories," RAND Memorandum, D-2495, 7.22.54; "Parallel Control," RAND Memorandum, RM-1361, 8.27.54. He wrote two other papers as well, including "Continuous Iteration Method for Solution of Differential Games," RAND Memorandum, RM-1326, 8.18.54.

2. *The Evening Outlook* (Santa Monica, California), summer 1954, various dates.

3. Ibid.

4. Melvin P. Peisakoff, interview, 6.3.97.

5. Richard Best, interview, 5.22.96. All direct quotations attributed to Best throughout chapter 25 come from the 5.22.96 interview.

6. Letter from John Nash to Arthur Mattuck, 1.15.73. In a reference to his 1954 arrest, Nash named the arresting officer.

7. Best, interview.

8. Ibid.

9. DOD Directive 52206, 1953; Executive Order 10450, 1953; *Greene* v. *McElroy,* 360 US 474, 1959.

10. Best, interview.

11. "The Consenting Adult Homosexual and the Law: An Empirical Study of Enforcement and Administration in Los Angeles County," *UCLA Law Review,* vol. 13 (1966), pp. 643, 691. "Solicitation" and "police decoys": Thomas E. Lodge, "There May Be Harm in Asking: Homosexual Solicitations and the Fighting Words Doctrine," in *Homosexuality, Criminology and the Law,* edited by Wayne R. Dynes and Steven Donaldson (New York: Garland Publishing, 1992), pp. 461–93. "In 1961 every state in the United States had sodomy laws," from *Lesbians, Gay Men and the Law,* edited by William B. Rubenstein (New York: The New Press, 1993), p. xvi.

12. See, for example, Jerel McCrary and Lewis Gutierrez, "The Homosexual Person in the Military and in National Security Employment," *Journal of Homosexuality,* vol. 5, nos. 1 and 2 (Fall 1979–Winter 1980); Ellen Schrecker, *The Age of McCarthyism: A Brief History with Documents* (New York: St. Martin's Press, 1994).

13. McCrary and Gutierrez, op. cit.

14. Nancy Nimitz, retired economist, RAND Corporation, interview, 5.21.96.

15. Best, interview.

16. Ibid.

17. Ibid.

18. McCrary and Gutierrez, op. cit.

19. Best, interview.

20. Ibid.; "The Consenting Adult Homosexual and the Law," op. cit.

21. Best, interview.

22. Ibid.

23. Ibid.

24. Ibid.

25. Postcard from John Nash to Virginia and John Nash, Sr., 9.54.

26. Alexander M. Mood, interview, 5.22.96.

27. RAND mathematics department roster, 1954, RAND Archives.

28. Letter from J. Nash to A. Mattuck, 1.15.73.

29. John W. Milnor, interview, 1.27.98.

30. Lloyd Shapley retold the story of Nash's arrest at a Thanksgiving dinner in 1994. Norman Shapiro, former RAND employee, interview, 2.29.96.

31. Felix Browder, interview, 9.6.97. Browder's recollection was that "Norman Levinson had to take care of it," and that Levinson later regarded the arrest as a sign of "approaching schizophrenia."

32. As quoted by N. Shapiro, interview. "Lloyd told me it was John."

33. Irving I. Gottesman, professor of psychology, University of Virginia, interview, 1.16.98.

34. Nikki Erlenmeyer-Kimling, professor of genetics and development, Columbia University, interview, 1.17.98.

35. "J. C. C. McKinsey" (obituary), *Proceedings and Addresses of the American Philosophical Association,* vol. 27 (1954).

36. Andrew Hodges, *Alan Turing: The Enigma,* op. cit.

26: Alicia

1. Alicia Nash, interviews, 10.94 and 4.18.97.

2. Peter Munstead, chief librarian, music library, MIT, interview, 9.19.97; also Lawrence Wallen, interview, 6.4.97.

3. The portrait of Alicia at age twenty-one is based largely on interviews with two women who knew her as an undergraduate at MIT: Joyce Davis, 5.17.97 and 6.30.97, and e-mails, various dates; and Emma Duchane, 4.30.96 and 6.26.97. It also draws on interviews with Wallen, 6.5.97; Arthur Mattuck, 11.7.97; Herta Newman, 3.2.96; Jacob Bricker, 5.22.97.

4. Duchane, interviews.

5. Ibid.
6. J. Davis, interview.
7. Ibid.
8. The Larde family history is based on interviews with Alicia Nash, Odette Larde, Enrique L. Larde, and the senior Enrique Larde's self-published history, *The Crown Prince Rudolf: His Mysterious Life After Mayerling* (Pittsburgh: Dorrance Publishing, 1994).
9. E. Larde, *The Crown Prince Rudolf,* op. cit.
10. A. Nash, interview, 5.14.97.
11. O. Larde, interview, 1.7.97.
12. See, for example, Patricia Parkman, *Nonviolent Insurrection in El Salvador* (Tucson: University of Arizona Press, 1988).
13. O. Larde, interview.
14. Tinker Cassell, Veterans Administration, Biloxi, Mississippi, interview, 8.97.
15. The sketch of Marymount is based on interviews with A. Nash, 4.18.97; Elizabeth Keegen, 4.18.97; Sister Kathleen Fagan, Marymount High School, 5.22.97; Sister Raymond, Marymount High School, 5.22.97.
16. Sister Raymond, interview.
17. Fagan, interview.
18. A. Nash, interview.
19. Duchane, interview.
20. A. Nash, interview.
21. O. Larde, interview.
22. J. Davis, interview.
23. Sister Raymond, interview.
24. A. Nash, interview.
25. Sister Raymond, interview.
26. *The Tech*, 9.51.
27. A. Nash, interview, 8.22.95.
28. J. Davis, interview.
29. Ibid.
30. Duchane, interview.
31. J. Davis, interview.
32. Letters from Joyce Davis to her parents, 1951–53.
33. J. Davis, interview.
34. Letter from Alicia Nash to Joyce Davis, June or July 1952.
35. J. Davis, interview.
36. Ibid.
37. H. Newman, interview, 3.2.96.
38. Duchane, interview.
39. A. Nash, interview, 11.94.
40. J. Davis, interview.
41. Letter from J. Davis to her parents, 4.24.54.
42. Letter from A. Nash to J. Davis, June or July 1954.
43. A. Nash, interview, 7.18.96.
44. John Moore, professor of mathematics, Princeton University, interview, 10.6.95.

27: The Courtship

1. Arthur Mattuck, interview, 11.7.95.
2. Letter from Alicia Nash to Joyce Davis, 7.55.
3. Ibid.
4. Emma Duchane, interview, 4.30.96.
5. Jacob Bricker, interview, 5.22.97.
6. Duchane, interview, 6.26.97.
7. Ibid.
8. Ibid., 4.30.96.
9. Ibid., 6.26.97.
10. Mattuck, interview.
11. Eleanor Stier, interview, 2.14.96.
12. Duchane, interview, 4.30.96.
13. "Grant in Aid, Support for Dr. John F. Nash, Jr., as Alfred F. Sloan Research Fellow in Mathematics," 5.15.56; also, Report for 1955–56, Alfred F. Sloan Foundation, New York, New York.
14. "The application is quasi-tentative . . . the draft problem is a complication." Letter from John Nash to Albert W. Tucker, undated (probably written in early fall 1955).
15. Letter from John Nash to Hassler Whitney, 10.55; John Forbes Nash, Jr., membership application, Institute for Advanced Study, 5.23.55. Nash's application was formally approved in January (source: letter from Robert Oppenheimer to John Nash, 1.17.56).
16. Letter from A. Nash to J. Davis, 2.56.

17. Nesmith Ankeny, who joined the MIT faculty in the fall of 1955, witnessed the incident and related the anecdote to Harold and Estelle Kuhn not long after it occurred (source: Harold Kuhn, e-mail, 5.21.97, and interview, 5.22.97).

18. J. Davis, interview, 5.19.97.

28: Seattle

1. The Institute on Differential Geometry took place from mid-June to the end of July 1956 at the University of Washington in Seattle. Dates and participants given in a memorandum from Carl B. Allendoerfer, chairman, department of mathematics, University of Washington, Seattle, 5.23.56.

2. John Milnor, e-mail, 8.97.

3. Eugenio Calabi, interview, 3.2.96; John Isbell, professor of mathematics, State University of New York at Buffalo, interview, 6.14.97; Raoul Bott, professor of mathematics, Harvard University, interview, 11.5.95.

4. E-mail from John Nash to Harold Kuhn, 4.16.96.

5. Letter from John Nash to Martha Nash Legg, 11.4.65.

6. The description of Forrester is based on: Arthur Mattuck, interview, 5.21.97, e-mail, 6.13.97; Isbell, interview, 6.14.97; Calabi, interview, 3.2.96; Albert Nijenhuis, interview, 6.17.97, e-mails, 6.13.97; Victor Klee, e-mails, 6.13.97, 6.14.97, 6.16.97; Kuhn, e-mails, 4.16.96, 4.17.96, 4.18.96; Joseph Kohn, interview, 4.17.96; John Walter, interview, 6.13.97; Robert L. Vaught, interview, 6.13.97; Ramesh Gangolli, interview, 6.16.97. Mary Sheetz provided the dates of Forrester's employment at the University of Washington, e-mail, 6.16.97.

7. Nijenhuis, interview.

8. Mattuck, interview.

9. Isbell, interview.

10. Vaught, interview.

11. Nijenhuis, interview.

12. Vaught, interview.

13. Ibid.

14. Walter, interview.

15. Nash was in Seattle in February of 1967, apparently for a month. Letter from John Nash to Virginia Nash, 2.67.

16. Klee, interview.

17. This scene is reconstructed on the basis of recollections from Martha Nash Legg, interview, 9.2.95.

18. Postcard from John Nash to Virginia and John Nash, Sr., 7.12.56.

19. Jerome Neuwirth, interview, 5.21.97.

20. Jacob Bricker, interview, 5.22.97.

29: Death and Marriage

1. Postcard from John Nash to Virginia and John Nash, Sr., 8.11.56.

2. Ibid., 9.18.56.

3. Elizabeth Hardwick, "Boston: A Lost Ideal," *Harper's*, December 1959, quoted in Paul Mariani, *Lost Puritan: A Life of Robert Lowell* (New York: Norton, 1994), p. 271.

4. Postcards from John Nash to Virginia and John Nash, Sr., 8.53, 9.53, 12.2.53, 1.2.55.

5. Martha Nash Legg, interview, 3.29.96.

6. Harold Kuhn, interview, 8.97.

7. M. Legg, interview.

8. Letter from John Nash to Martha Nash Legg, from Paris, 9.28.59.

9. M. Legg, interview.

10. Letter from J. Nash to H. Kuhn, 8.97.

11. Death certificate of John Nash, Sr., 9.12.56.

12. M. Legg, interview.

13. Eleanor Stier, interview, 3.15.96.

14. Natasha Brunswick, interview, 9.25.95.

15. Leo Goodman, as told to Harold Kuhn, 1.95.

16. Alicia Nash, interview, 5.14.97.

17. Letter from Alicia Nash to Joyce Davis, 10.26.56.

18. Ibid.

19. Sylvia Plath, *The Bell Jar* (New York: Harper & Row, 1971).

20. M. Legg, interview.

21. John Nash, dinner party at Gaby and Armand Borel's, 3.22.96.

22. M. Legg, interview.

23. A. Nash, interview, 10.11.97; also M. Legg, interview.

24. Postcard from J. Nash to V. Nash, 2.57.

25. Enrique Larde, interview, 12.21.95.

Part Three: A SLOW FIRE BURNING

30: Olden Lane and Washington Square

1. Institute for Advanced Study, Directory, 1956–57, Institute for Advanced Study Archive, Princeton, New Jersey.
2. Regis, *Who Got Einstein's Office?*, op. cit., p. 5.
3. John Danskin, interview, 10.19.95.
4. Paul S. Cohen, professor of mathematics, Stanford University, interview, 1.6.96.
5. Peter Lax, professor of mathematics, Courant Institute, interview, 2.29.96.
6. Cathleen Morawetz, professor of mathematics, Courant Institute, interview, 2.29.96.
7. George Boehn, "The New Uses of the Abstract," *Fortune,* July 1958.
8. Constance Reid, *Courant in Göttingen and New York: The Story of an Improbable Mathematician* (New York: Springer Verlag, 1976).
9. Ibid.
10. Ibid.
11. Lax, interview.
12. Boehm, "The New Uses of the Abstract," op. cit.
13. Nash told Harold Kuhn that he kept a car in New York City that year and that parking it caused him innumerable headaches, personal communication, 7.97.
14. Postcard from John Nash to Virginia and John Nash, Sr., 8.11.56.
15. Natasha Brunswick, interview, 9.25.95.
16. Tilla Weinstein, professor of mathematics, Rutgers University, interview, 8.25.97.
17. Morawetz, interview.
18. Lars Hörmander, professor of mathematics, University of Lund, interview, 2.13.97.
19. Lax, interview.
20. Hörmander, interview.
21. John Isbell, e-mail, 3.28.95.
22. Boehm, "The New Uses of the Abstract," op. cit.
23. Stanislaw Ulam, "John von Neumann, 1903–57," *Bulletin of the American Mathematical Society,* vol. 64, no. 3, part ii (May 1958).
24. John Nash, "Continuity of Solutions of Parabolic and Elliptic Equations," *American Journal of Mathematics,* vol. 80 (1958), pp. 931–54.
25. See Chapters 2 and 16.
26. John Nash, "Continuity of Solutions of Parabolic and Elliptic Equations," op. cit.
27. Louis Nirenberg, professor of mathematics, Courant Institute, interview, 10.94. See also Lax, interview.
28. Ibid.
29. Ibid.
30. Lax, interview.
31. Ibid.
32. Nirenberg, interview.
33. Hörmander, interview.
34. Ibid.
35. Lax, interview.
36. Nirenberg, interview.
37. Armand Borel, professor of mathematics, Institute for Advanced Study, interview, 3.1.96.
38. Lax, interview.
39. Morawetz, interview; Gian-Carlo Rota, interview, 10.94.
40. Paul R. Garabedian, professor of mathematics, Courant Institute, interview, 2.20.96.
41. "Ennio De Giorgi, 1928–1996" and "Interview with Ennio De Giorgi," *Notices of the American Mathematical Society,* 10.97.
42. John Nash, Jr., *Les Prix Nobel 1994,* op. cit.
43. Rota, interview.
44. Lax, interview.
45. Letter from John Nash to Robert Oppenheimer, 7.10.57.
46. Ibid.
47. John Nash, plenary lecture, World Congress of Psychiatry, Madrid, 8.26.96, op. cit.
48. Institute for Advanced Study, directories, various years.
49. Letter from J. Nash to R. Oppenheimer.
50. John Nash, plenary lecture, op. cit.

31: The Bomb Factory

1. Richard Emery, attorney, interview, 4.4.96.
2. Ibid.
3. Postcard from John Nash to Virginia Nash, 9.57.
4. Emma Duchane, interview, 6.26.96.

5. Alicia Nash, interview, 7.1.97.
6. Duchane, interview.
7. Hartley Rogers, interview, 2.16.96.
8. Zipporah Levinson, interview, 9.11.95.
9. A. Nash, interview, 10.94.
10. Nash's chief result was initially published in a note—submitted by Marston Morse of the Institute for Advanced Studies on 6.10.57—in the *Proceedings of the National Academy of Sciences,* no. 43 (1957), pp. 754–58. The full paper was submitted to the *American Journal of Mathematics* nearly a year later, on 5.26.58, and published in vol. 80 (1958), pp. 931–58.
11. Elias Stein, professor of mathematics, Princeton University, interview, 12.2.95.
12. Lennart Carleson, professor of mathematics, University of Stockholm, interview, 10.3.95.
13. Ibid.
14. Stein, interview.
15. Ibid.
16. Ibid.
17. Paul R. Garabedian, interview, 2.20.96.
18. George Boehm, "The New Mathematics," two-part series, *Fortune* (June and July 1958).
19. Martha recalled Nash's telling her that he was considering accepting a post at Caltech in order to raise the likelihood of an offer from Harvard, possibly because Harvard and MIT had an informal nonraiding policy. Martha Nash Legg, interview, 3.30.96.
20. Letter from John Nash to Albert W. Tucker, 10.58.
21. At that time, tenure was normally not awarded until the candidate's seventh year. At MIT, unlike some other institutions, tenure was paired with promotion to full, not associate, professor.
22. Gian-Carlo Rota, interview, 10.94.
23. John Forbes Nash, Jr., *Les Prix Nobel 1994,* op. cit.
24. *Awards, Honors and Prizes,* 8th edition, vol. II (Detroit: Gale Research, 1989), p. 129.
25. Lars Hörmander, interview, 2.13.97.
26. Confidential source.
27. *Proceedings, International Congress of Mathematicians, 1958* (Providence, R.I.: American Mathematical Society, 1960).
28. Jürgen Moser, interview, 3.21.96.
29. *Proceedings, International Congress of Mathematicians,* op. cit.
30. Confidential source.
31. Confidential source.
32. Moser, e-mail, 12.24.97.
33. Peter Lax, interview, 2.6.96.
34. Moser, interview, 3.21.96.
35. Ibid.
36. For the history of the Bôcher Prize, see the Web site for the American Mathematical Society.
37. Letter from Lars Hörmander to author, 1.3.96; Hörmander, interview, 2.13.97.
38. Hörmander, e-mail, 12.16.97.
39. Ibid.

32: Secrets

1. John Forbes Nash, Jr., plenary lecture, World Congress of Psychiatry, Madrid, 8.26.96, op. cit.
2. G. H. Hardy, *The Mathematician's Apology* (Cambridge, U.K.: Cambridge University Press, 1967), with a foreword by C. P. Snow.
3. Paul S. Cohen, interview, 1.5.96.
4. Stanislaw Ulam, "John von Neumann, 1903–1957," op. cit., p. 5.
5. Hardy, op. cit.
6. Felix Browder, interview, 11.10.95.
7. Harold Kuhn, interview, 7.95.
8. Ibid.
9. John Nash, plenary lecture, op. cit.
10. Elias Stein, interview, 12.28.95.
11. Cohen, interview.
12. E. T. Bell, *Men of Mathematics,* op. cit.
13. Enrico Bombieri, interview, 12.6.95.
14. Bell, op. cit.
15. Andrew Wiles, professor of mathematics, Princeton University, personal communication, 6.97.
16. Lars Hörmander, interview, 2.13.97.
17. F. Browder, interview.
18. John Forbes Nash, Jr., *Les Prix Nobel 1994,* op. cit.
19. Bell, op. cit.
20. Ibid.
21. Ibid.
22. Jacob Schwartz, professor of computer science, Courant Institute, interview, 1.29.96.

23. Jerome Neuwirth, interview, 5.27.97.
24. Stein, interview.
25. Ibid.
26. Richard Palais, professor of mathematics, Brandeis University, interview, 11.6.95.
27. Bell, op. cit.
28. Atle Selberg, interview.
29. Eugenio Calabi, interview, 3.2.96.
30. Letter from John Nash to Martha Nash Legg, 11.4.65.
31. Stein, interview.
32. Hörmander, interview.
33. Harold Kuhn, e-mail, 7.97.
34. Paul A. Samuelson, interview.
35. William Ted Martin, interview, 9.7.95.
36. Robert Solow, professor of economics, MIT, interview, 1.95.
37. Martin, interview.
38. Cathleen Morawetz, interview, 2.29.96.
39. Alicia Nash, interview, 1.3.97.
40. Ibid.
41. John Nash, personal communication, 3.22.96.
42. Eva Browder, interview, 9.6.97.
43. Ibid.
44. A. Nash, interview
45. F. Browder, interview.
46. John Moore, professor of mathematics, Princeton University, interview, 10.5.95.

33: Schemes

1. Alicia Nash, interview, 7.1.97.
2. Ibid.
3. Letter from John Nash to Albert W. Tucker, early October 1958.
4. George Mackey, interview, 1.21.96.
5. Letter from C. Ralph Buncher, professor of biostatistics and epidemiology, University of Cincinnati Medical Center, to author, 5.20.96.
6. A. Nash, interview.
7. John Nash, letter to A. Tucker, 10.58.
8. Ibid.
9. Martha Nash Legg, interview, 3.29.96.
10. Paul A. Samuelson, interview, 3.13.96.
11. Saunders McLane, former chairman, department of mathematics, University of Chicago, interview, 3.4.96.
12. Shlomo Sternberg, interview, 3.5.96.
13. Ibid. Also membership application, Institute for Advanced Studies, fall 1958.
14. Letter from Albert W. Tucker to John Nash, 10.8.58.
15. Letter from Albert W. Tucker to Sloan Foundation, 10.8.58.
16. Letter from Albert W. Tucker to Guggenheim Foundation, 11.26.58.
17. Gian-Carlo Rota, interview, 11.14.95.
18. Robert Solow, emeritus professor of economics, MIT, interview, 1.95.
19. Letter from John Nash to Virginia Nash, 10.15.58.
20. *New York Times,* 11.14.63.
21. Paul S. Cohen won the Fields in 1966 and the Bôcher in 1964. The sketch of Paul Cohen is based on interviews with Raoul Bott, 11.95 and 11.5.96; Lennart Carleson, 10.18.95; Elias Stein, 12.28.95; Felix Browder, 11.2.95; Adriano Garsia, professor of mathematics, University of California at San Diego, 12.31.95; Lars Hörmander, 2.13.97; Jürgen Moser, 3.21.96; Jerome Neuwirth, 5.27.97.
22. Cohen, interview, 1.5.96.
23. Stein, interview, 12.28.95.
24. Ibid.
25. Garsia, interview, 12.31.95.
26. Cohen, interview.
27. Garsia, interview; Neuwirth, interview, 5.27.97.
28. F. Browder, interview, 11.10.95.
29. Ibid., 11.2.95.

34: The Emperor of Antarctica

1. Richard Emery, interview, 4.4.96. The party scene described by Emery is also based on the recollections of Jürgen and Gertrude Moser, John and Karen Tate, Adriano Garsia, Gian-Carlo Rota, and Alicia Nash.
2. Alicia Nash, interview, 2.7.96.

3. Paul S. Cohen, interview, 1.5.96.

4. Al Vasquez, professor of mathematics, City University of New York, interview, 6.17.97.

5. Raoul Bott, interview, 11.5.95.

6. Emma Duchane, interview, 6.26.97.

7. Letter from C. Ralph Buncher to author, 5.20.96; also letter from Henry Y. Wan, professor of economics, Cornell University, to author, 6.5.96. Tony Phillips, professor of mathematics, State University of New York at Stony Brook, interview, 8.26.97, recalled Nash's question to the class.

8. Ramesh Gangolli, professor of mathematics, University of Washington, interview, 6.12.95. Also, Alberto R. Galmarino, professor of mathematics, Northeastern University, interview, 6.95.

9. Atle Selberg, interviews, 8.16.95 and 1.23.96.

10. Gian-Carlo Rota, interview, 10.29.94; Gangolli, interview; Galmarino, interview. Martha Nash Legg put this episode later, but Gangolli and Galmarino recall that Nash didn't meet his classes for the last couple of weeks of the term which ended 1.21.59 and Rota recalled that Nash stopped by his apartment before "driving south."

11. Jerome Neuwirth, interview, 6.4.97; also Garsia, interview, 12.31.95.

12. Hartley Rogers, interview, 2.16.96.

13. Duchane, interview, 4.30.96.

14. Confidential source.

15. Vasquez, interview.

16. Kate Tate, interview, 8.11.97.

17. John Nash, plenary lecture, op. cit.

18. A. Nash, interview.

19. Cohen, interview.

20. Vasquez, interview.

21. Harold Kuhn, interview, 8.94.

22. Cohen, interview.

23. Neuwirth, interview.

24. Moser, interview, 3.23.96.

25. William Ted Martin, interview, 9.7.95.

26. Felix Browder, interview, 11.2.95; Paul A. Samuelson, interview, 10.94.

27. John Danskin, interview, 10.19.96.

28. The account of this incident is based on interviews with the following sources: Sigurdur Helgason, 2.13.96; F. Browder; Samuelson, 10.94 and 3.15.96; Harold Kuhn, interview, 1.95. Browder, who later became chairman of the Chicago department, recalled seeing the letter in the files. Efforts by the current chairman to locate it proved fruitless.

29. Vasquez, interview.

30. Eugenio Calabi, interview, 3.2.96.

31. Ibid.

32. Selberg, interview.

33. Program, 554th Meeting, Columbia University, New York, February 28, 1959, *Bulletin of the American Mathematical Society*, vol. 65 (1959), p. 149.

34. Harold N. Shapiro, interview, 2.29.96.

35. Peter Lax, interview, 2.6.96.

36. Donald J. Newman, interview, 3.2.96.

37. Cathleen Morawetz, interview, 2.29.96.

38. F. Browder, interview.

35: In the Eye of the Storm

1. Alicia Nash, interview, 7.1.97.

2. Emma Duchane, interview, 6.26.97.

3. A. Nash, interview.

4. Donald V. Reynolds, interview, 6.29.97.

5. A. Nash, interview.

6. Duchane, interview.

7. Martha Nash Legg, interview, 3.29.96.

8. Duchane, interview.

9. A. Nash, interview.

10. Duchane, interview.

11. A. Nash, interview.

12. Duchane, interview.

13. Ibid.

14. William Ted Martin, interview, 9.7.95.

15. Gian-Carlo Rota, interview, 10.29.94.

16. Letter from John Nash to Virginia Nash, 3.12.59.

17. Letter from John Nash to Martha Nash Legg, 3.12.59.

18. A. Nash, interview, 7.1.97.

19. Al Vasquez, interview, 6.17.97.

20. Duchane, interview.
21. Ibid.
22. Paul S. Cohen, interview, 1.5.96.
23. Gertrude Moser, interview, 8.25.95.
24. Kay Whitehead, professor of mathematics, Tufts University interview, 12.12.95.

36: Day Breaks in Bowditch Hall

1. Paul S. Cohen, interview, 1.5.96.
2. Adriano Garsia, interview, 12.31.95.
3. Cohen, interview.
4. My description of how MIT's psychiatric service likely handled Nash's commitment is based on interviews with Benson Rowell Snyder, who was hired by President Julius Stratton to reorganize the service, interview, 7.24.97; Wade Rockwood, interview, 7.26.97; Merton J. Kahne, professor, MIT, interview, 5.15.96; Harvey Burstein, former FBI agent who was brought in by Stratton to expand MIT's campus police, interview, 7.3.97.
5. The description of how Nash was taken to McLean against his will is based on a contemporaneous account by a former dean of Tufts Medical School, A. Warren Stearns, who interviewed Nash shortly after his commitment (letter fron Stearns to Bernard Bradley, 4.14.59), and a further elaboration by Nash (E-mail, 5.15.98).
6. Snyder, interview.
7. For a portrait of McLean as it was in the 1950s, I relied on an official history by S. B. Sutton, *A History of McLean Hospital* (Washington, D.C.: American Psychiatric Press, 1986); annual reports; firsthand accounts by Sylvia Plath, Robert Lowell, and Ray Charles, as well as Suzanna Kaysen's more recent report, *Girl, Interrupted;* and interviews with individuals associated with McLean in that era, including Paul Howard, former associate psychiatrist in chief and director of the clinical service, 2.15.95; Kahne; Joseph Brenner, 7.23.97; Arthur Cain, psychiatrist, 8.20.97; Alfred Pope, senior neuropathologist, McLean Hospital, and professor of neuropathology, Harvard Medical School, 12.13.95 and 2.16.96.
8. Robert Garber, former president, American Psychiatric Association, interview, 5.6.96.
9. Sylvia Plath, *The Bell Jar,* op. cit.; Ray Charles, *Brother Ray* (New York: Da Capo, 1978, 1992).
10. Letter from A. W. Stearns to B. Bradley, 5.14.53.
11. Zipporah Levinson, interview, 9.11.95.
12. Emma Duchane, interview, 6.26.97.
13. Robert Lowell was hospitalized at McLean at the end of April 1959. Lowell was confined to Bowditch, as he had been two years earlier when he wrote "Day Breaks at Bowditch Hall," one of the poems in *To the Union Dead.* Several of Nash's visitors, including Gian-Carlo Rota, Isadore Singer, and Arthur Mattuck, recall encounters with Lowell, and therefore it seems that Nash, too, was confined to Bowditch. Since we have no firsthand reports from Nash, I have made use of Lowell's impressions from 1957 and 1959, augmented by the impressions of some of Lowell's visitors, including his wife, writer Elizabeth Hardwick, letter, 8.8.97; poet Stanley Kunitz, interview, 8.2.97; and Lowell's executor, Frank Bidart, interview, 7.27.97. See also Ian Hamilton, *Robert Lowell: A Biography* (New York: Random House, 1982); Paul Mariani, *The Lost Puritan,* op. cit., and interview, 7.28.97; Peter Davison, *The Fading Smile: Poets in Boston, 1955–1960, from Robert Frost to Robert Lowell to Sylvia Plath* (New York: Knopf, 1994), and interview, 8.11.97.
14. "I've been conditioning here for about a month," letter from Robert Lowell to Edmund Wilson, 5.19.59, from Bowditch House; "In the hospital I spent a mad month or more rewriting everything in my three books," letter from Robert Lowell to Elizabeth Bishop, 7.24.59.
15. Elizabeth Hardwick, personal communication, 9.8.97.
16. Arthur Mattuck, e-mail, 8.8.97.
17. "The house I was in was divided between ex-paranoid boys and senile old men," letter from Robert Lowell to Peter Taylor, 3.15.58.
18. Letter from R. Lowell to E. Bishop, 3.15.58.
19. Ibid.; also "Waking in the Blue," Robert Lowell, *Life Studies and For the Union Dead* (New York: Farrar, Straus and Giroux, 1992). Quotes in this and the following paragraphs are taken from "Waking" unless otherwise noted.
20. From "Waking in the Blue"; also Duchane, interview.
21. Letter from R. Lowell to E. Bishop; also "Waking in the Blue."
22. Seymour Krim, "The Insanity Bit," in *View of a Nearsighted Cannoneer* (New York: E. P. Dutton, 1968).
23. Al Vasquez, interview, 6.17.97.
24. Z. Levinson, interview.
25. Vasquez, interview.
26. Garsia, interview.
27. Jürgen Moser, interview, 3.23.96.
28. Duchane, interview.
29. George Mackey, interview, 12.14.95.
30. Herta Newman, interview, 3.2.96.
31. Felix Browder, interview, 1.2.95.
32. Gian-Carlo Rota, interview, 10.29.94.
33. Garsia, interview.

34. This is Jerome Lettvin's term, Jerome Lettvin, professor of electrical engineering, MIT, interview, 7.25.97.
35. John McCarthy, interview, 2.4.96.
36. Arthur Mattuck, interview, 11.7.95.
37. I am assuming that Nash's treatment was similar to that of other patients and have based my account on the recollections of Paul Howard, clinical director of McLean at the time, as well as other McLean staffers, including Joseph Brenner, psychiatrist, interview, 7.25.97; Cain, interview; Kahne, interview.
38. Letter from A. W. Stearns to B. Bradley, 5.20.59.
39. Kahne, interview.
40. Brenner, interview, 7.23.97.
41. Z. Levinson, interview.
42. Cohen, interview; F. Browder, interview.
43. Francine M. Benes, psychiatrist, McLean Hospital, interview, 2.13.96.
44. See, for example, Mariani, op. cit., and Hamilton, op. cit.
45. Kahne, interview; also Howard, interview.
46. Kahne, interview.
47. Howard, interview.
48. Brenner, interview.
49. Z. Levinson, interview.
50. Isadore Singer, interview, 12.13.95.
51. Letter from A. W. Stearns to B. Bradley, 5.20.59.
52. Duchane, interview.
53. Letter from A. W. Stearns to B. Bradley, 5.20.59.
54. Taffy Griffiths, physician, Princeton, 5.20.59, and interview, 7.95.
55. Notes of a telephone conversation between A. Warren Stearns and Bernard E. Bradley, attorney, 5.13.59. In an interview (8.19.97), Bradley said that he handled many similar cases, but did not recall Nash.
56. The sketch of A. Warren Stearns is based on a biographical essay provided by the Tufts University archives; an interview with his son Charles Stearns, 3.14.96; and an interview with Paul Samuelson, who knew Stearns, 3.15.96.
57. A. W. Stearns and B. Bradley phone conversation, 5.14.59.
58. Letter from A. W. Stearns to B. Bradley, 5.20.59.
59. Ibid.
60. Letter from Robert A. Grimes, attorney, Hardy, Hall & Grimes, to A. Warren Stearns, 6.18.59.
61. Letter from A. W. Stearns to B. Bradley, 5.20.59.
62. Ibid.

37: Mad Hatter's Tea

1. Emma Duchane, interview, 6.26.97. The sketch of Alicia Nash and the final months of her pregnancy are based on this.
2. Confidential source.
3. Confidential source.
4. Michael Artin, interview, 12.12.95.
5. Confidential source.
6. Zipporah Levinson, interview, 9.11.95.
7. Al Vasquez, interview, 6.17.97.
8. Letter from John Nash to Lars Hörmander, undated (arrived around 6.1.59).
9. Gaby Borel, interview, 9.94.
10. John Nash, plenary lecture, World Congress of Psychiatry, Madrid, 8.26.96, op. cit.
11. Paul Samuelson, interview, 3.16.97.
12. Z. Levinson, interview.
13. William Ted Martin, interview, 9.7.95.
14. A. Warren Stearns, note for file, 6.15.59.
15. Samuelson, interview.
16. Letter from Henry Y. Wan, Jr., to author, 6.5.96.
17. Enrique Larde, interview, 12.21.95.
18. John Danskin, interview, 10.19.95.
19. Alicia Nash, interview, 7.1.97.

Part Four: THE LOST YEARS

38: Citoyen du Monde

1. Postcard from John Nash to Virginia Nash, 7.18.59.
2. Ibid., 7.20.59.
3. Janet Flanner, *Paris Journal 1944–1965* (New York: Atheneum, 1965).
4. John Moore, interview, 10.6.97.

5. Alicia Nash, interview, 8.15.97.
6. Odette Larde, interview, 12.8.95.
7. *International Herald Tribune*, 7.10.59, 7.11.59, 7.12.59, 8.7.59.
8. Interviews with Joseph Baratta, historian, 8.12.97; Francis Bourne, 8.12.97; David Gallup, attorney, 8.12.97.
9. *New York Times*, 5.27.48; Garry Davis, World Citizen Foundation, interview, 8.13.97. See also Art Buchwald, *I'll Always Have Paris* (New York: G. P. Putnam & Sons, 1996), and Garry Davis, *My Country Is the World: The Adventures of a World Citizen* (New York: G. P. Putnam & Sons, 1961).
10. *New York Times*, 9.18.48.
11. *International Herald Tribune*, 6.16.49.
12. Buchwald, op. cit.
13. *International Herald Tribune*, 6.16.49.
14. Louis Sass, *Madness and Modernism*, op. cit., pp. 324–25.
15. Postcard from J. Nash to V. Nash, 7.29.59.
16. Section 1481 of the 1941 Immigration and Naturalization Act.
17. Edward A. Betancourt, Overseas Citizens Services, Immigration and Naturalization Service, interview, 8.26.97.
18. 1941 Immigration and Naturalization Act.
19. John Nash, plenary lecture, World Congress of Psychiatry, Madrid, 8.26.96, op. cit.
20. Martha Nash Legg, interview, 3.29.96.
21. Armand Borel, interview, 3.1.96.
22. Postcard from J. Nash to V. Nash, 7.31.59.
23. Ibid.
24. Denis Brian, *Einstein: A Life*, op. cit.
25. *International Herald Tribune*, various issues, August 1959.
26. John Nash, plenary lecture, op. cit.
27. See, for example, Paul Hofmann, *Switzerland* (New York: Henry Holt & Co., 1994).
28. Mary Wollstonecraft Shelley, *Frankenstein or the Modern Prometheus* (New York: Penguin, 1985).
29. Postcard from J. Nash to V. Nash, 8.12.59.
30. As quoted by Sass, op. cit.
31. Letter from John Nash to Lars Hörmander, 2.10.60.
32. Zurbuchen, Le Directeur, Contrôle de l'Habitant, Geneva, 9.29.59, provided by Schweizerisches Bundesarchiv.
33. Franz Kafka, *The Castle* (New York: Scholastic Books, 1992), with an introduction by Irving Howe.
34. Ibid.
35. Ibid.
36. Postcard from J. Nash to V. Nash, 9.28.59.
37. Convention Relating to the Status of Refugees of July 28, 1951, United Nations High Commissioner for Refugees, Geneva.
38. Zurbuchen, op. cit.
39. Ibid.
40. Direktion der Eidg. Militarverwaltung, Berne to Contrôle de l'Habitant, Geneva, 11.21.59.
41. John Nash, plenary lecture, op. cit.
42. Ibid.
43. Harold Kuhn, interview, 1.95.
44. John Haslam, as quoted by Sass, op. cit.
45. Sass, op. cit.
46. Postcard from J. Nash to V. Nash, 9.28.59.
47. Letter from M. Legg to John Nash, 9.59.
48. A. Nash, interview.
49. Telegram from Amory Houghton, U.S. ambassador to France, to Secretary of State Christian A. Herter, 12.15.59.
50. Letter from J. Nash to L. Hörmander, from Paris, 1.18.60.
51. Postcard from J. Nash to V. Nash, 10.11.59.
52. After returning to the U.S., Nash claimed to be a resident of Liechtenstein, which levied no income tax, and refused to sign U.S. tax forms (source: H. Kuhn, interview, 8.92).
53. O. Larde, interview, 12.8.96.
54. Letter from John Nash to Virginia Nash, 11.10.59.
55. The anecdote concerns Paul Erdos and was told by Donald Spencer, interview, 11.28.95.
56. O. Larde, interview, 12.8.95.
57. M. Legg, interview, 3.29.96.
58. Sass, op. cit.
59. Letter from John Nash to Norbert Wiener, 12.9.95.
60. Letter from J. Nash to V. Nash, 12.13.59.
61. Franz Kafka, *The Metamorphosis* (New York: Schocken Books, 1995).
62. Irving Howe introduction, Kafka, *The Castle*, op. cit.
63. James M. Glass, *Delusion* (Chicago: University of Chicago Press, 1985).
64. Telegram from A. Houghton to C. A. Herter.

65. Telegram from Henry S. Villard, U.S. consul to Switzerland, to Secretary of State Christian A. Herter, 12.16.59.
66. Ibid.
67. Theodore Friend, obituary of Edward Hill Cox, 8.4.75, Swarthmore College Archive.
68. A. Nash, interview.
69. Telegram from A. Houghton to C. A. Herter.
70. Telegram from H. S. Villard to C. A. Herter.'
71. Letter from J. Nash to V. Nash, 12.26.59; O. Larde, interview, 12.8.95.
72. O. Larde, interview, 12.8.95.
73. Shiing-shen Chern, professor of mathematics, University of California at Berkeley, interview, 6.17.97.
74. A. Nash, interview.
75. "Alexandre Grothendieck," History of Mathematics Archive, School of Mathematical and Computational Sciences, University of St. Andrews, Scotland; see also interviews with Nick Katz, professor of mathematics, Princeton University, 8.26.97; Arthur Mattuck, 9.19.97; Paulo Ribenboim, professor of mathematics, Queens University, Kingston, Ontario, Canada, 9.28.97; Tony Phillips, 8.26.97.
76. O. Larde, interview, 12.8.85.
77. A. Nash, interview.
78. Felix Browder, interview, 9.6.97. See also Larkin Farinholt's obituary, *New York Times*, 7.17.90, for details of his career.
79. Letter from J. Nash to L. Hörmander, 2.10.60.
80. John Nash, plenary lecture, op. cit.
81. Letter from Lars Hörmander to John Nash, 2.12.60.
82. Postcard from J. Nash to V. Nash, 3.2.60.
83. John Nash, conversation with author, 6.25.95.
84. F. Browder, interview.
85. Ibid.
86. Letter from J. Nash to V. Nash, 3.60.
87. Michael Artin, interview, 12.12.95.
88. Al Vasquez, interview, 6.17.97.
89. Cathleen Morawetz, interview, 2.29.96.
90. John Danskin, interview, 10.19.95.
91. M. Legg, interview.
92. Eleanor Stier, interview, 3.18.96.
93. Letter from J. Nash to V. Nash, 4.9.60.
94. Ibid.
95. Telegram from Allyn C. Donaldson, Department of State, to Virginia Nash, 4.21.60.
96. Emma Duchane, interview, 4.30.95.
97. Vasquez, interview.
98. A. Nash, interview.
99. G. Davis, interview.

39: Absolute Zero

1. Alicia Nash, interview, 8.15.97.
2. Martha Nash Legg, interview, 8.1.95.
3. Interviews with John Danskin, 10.19.95, and Joyce Davis, 5.30.97.
4. Handwritten note from Alicia Nash to Joyce Davis, summer 1960.
5. Odette Larde, interview, 12.7.95.
6. A. Nash, interview.
7. Jean-Pierre Cauvin, professor of French, University of Texas at Austin, interview, 8.25.97; also Agnes Sherman, interview, 8.26.96.
8. O. Larde, interview.
9. Cauvin, interview.
10. Danskin, interview.
11. Ibid.
12. Elvira Leader, interview, 6.9.95.
13. Solomon Leader, interview, 6.9.95.
14. Danskin, interview.
15. Samuel C. Howell, memorandum to file, 11.10.60.
16. Notes of conversations between Oskar Morgenstern and Douglas Brown, Princeton University Archives, 11.2.50.
17. Letter from Raymond J. Woodrow to John F. Nash, Jr., 10.21.60.
18. Letter from Donald Spencer to Jean Leray, 10.31.60.
19. Ibid.
20. Burton Randol, professor of mathematics, City University of New York, interview, 8.26.97.
21. Ibid.
22. Ibid.

23. Ibid.
24. Confidential source.
25. Confidential source.
26. Randol, interview.
27. Danskin, interview.
28. Martin Shubik, interview, 10.94.
29. Paul Zweifel, interview, 9.6.95.
30. Edmond Nelson, professor of mathematics, Princeton University, interview, 8.17.95.
31. Armand Borel, interview, 3.1.96.
32. Danskin, interview. Robert Goheen, president of Princeton University, was unable to confirm these events, which would have been handled by someone on the campus security detail in any case, interview, 9.10.97.
33. A. Nash, interview.
34. O. Larde, interview.
35. Confidential source.

40: Tower of Silence

1. Martha Nash Legg, interview, 8.2.95.
2. Ibid.
3. Gerald N. Grob, *The Mad Among Us* (Cambridge: Harvard University Press, 1994), and "Abuse in American Mental Hospitals in Historical Perspective: Myth and Reality," *International Journal of Law and Psychiatry,* vol. 3 (1980), pp. 295–310. Also interview with Grob, professor of history, Rutgers University, 8.4.97.
4. See biographies of Dorothea Dix, including Rachel Basker, *Angel of Mercy: The Story of Dorothea Dix* (New York: Messner, 1955); also Penny Colman, *Breaking the Chains: The Crusade of Dorothea Lynde Dix* (White Hall, Va.: Shoetree Press, 1992).
5. Descriptions of Trenton State are based on interviews with psychiatrists who were affiliated with the hospital, including Robert Garber, former president, American Psychiatric Association, 5.6.96; Peter Baumecker, 5.1.96, 5.2.96, 5.9.96; Arthur A. Sugarman, 8.25.97.
6. Baumecker, interview.
7. Ibid.
8. Ariel Rubinstein, e-mail, 2.3.97.
9. Baumecker, interview. "B" probably refers to Jacob Bricker (see Chapter 44).
10. John Danskin, interview, 10.19.96. For an account of the hijacking, see *Time* magazine, 2.3.61.
11. M. Legg, interview.
12. Danskin, interview.
13. Robert Winters, interview, 8.9.95.
14. Letter from Robert Winters to Joseph Tobin, 2.2.61.
15. Letter from Robert Winters to Harold Magee, 2.2.59. Also interview with Tobin, 6.10.97.
16. Seymour Krim, "The Insanity Bit," op. cit.
17. Baumecker, interview.
18. Phillip Ehrlich, psychiatrist, Princeton Hospital, interview, 8.24.97.
19. Baumecker, interview.
20. M. Legg, interview.
21. Interviews with Garber and Baumecker.
22. Baumecker, interview.
23. Danskin, interview.
24. Garber, interview.
25. Baumecker, interview.
26. Ibid.
27. Burton Randol, interview, 8.25.97.
28. Lenore McCall, *Between Us and the Dark* (Philadelphia: J. B. Lippincott, 1947).
29. Baumecker, interview.
30. Garber, interview.
31. Jerome Lettvin, interview, 7.25.97.
32. Grob, *The Mad Among Us,* op. cit., p. 185.
33. Garber, interview.
34. Letter from John Nash to Alexander Mood, 12.17.94, one of many references Nash has made to his insulin treatments and memory loss.
35. Richard Nash, interview, 1.6.96.
36. Interviews with Grob and Lettvin.
37. Baumecker, interview.
38. Ibid.
39. Ibid.
40. Postcard from John Nash to Virginia Nash, 7.14.61. Nash says he's due to be released the following day.
41. Baumecker, interview.
42. Postcard from J. Nash to V. Nash, 7.14.61.
43. Baumecker, interview.

41: An Interlude of Enforced Rationality

1. John Forbes Nash, Jr., *Les Prix Nobel 1994,* op. cit.
2. Louis Sass, *Madness and Modernism,* op. cit.
3. A decline in measured intelligence within a short time of the onset of schizophrenia has been documented in a series of studies. Jed Wyatt, personal communication, 6.97.
4. Letter from John Nash to Donald Spencer, undated, spring 1961.
5. Interviews with Armand Borel, 3.1.96, and Atle Selberg, 1.23.96.
6. Letter from Atle Selberg to John Nash, 9.25.61; letter from Robert Oppenheimer to John Nash, 10.3.61.
7. John Nash, membership application, 7.17.61, Institute for Advanced Study Archive.
8. Letter from J. Nash to D. Spencer.
9. Shlomo Sternberg, interview, 3.5.96. Also postcards from John Nash to Virginia Nash, 8.1.61 and 8.3.61.
10. Alicia Nash, interview, 8.15.96.
11. Interviews with John Danskin, 10.19.95, and Odette Larde, 12.7.95.
12. O. Larde, interview.
13. "Recent Advances in Game Theory," Princeton, October 4–6, 1961.
14. Reinhard Selten, professor of economics, University of Bonn, interview, 6.27.95.
15. John Harsanyi, interview, 6.27.95.
16. Harold Kuhn, personal communication, 8.97.
17. John Nash, "Le Problème de Cauchy Pour Les Equations Differentielles d'une Fluide Générale," *Bulletin de la Société Mathématique de France,* vol. 90 (1962), pp. 487–97. Submitted 1.19.62.
18. John Nash, *Les Prix Nobel 1994,* op. cit.
19. According to the *Encyclopedia of Mathematics,* "Mathematical study of [the Cauchy problem for the general Navier-Stokes equation] has become active since J. Nash and N. Itaya proved the existence of unique regular solutions local in time."
20. Selberg, interview.
21. Gillian Richardson, interview, 12.14.97.
22. Karl Uitti, professor of French, Princeton University, interview, 8.22.97.
23. Confidential source.
24. Uitti, interview.
25. Jean-Pierre Cauvin, interview, 8.25.97.
26. Hubert Goldschmidt, Columbia University, interview, 3.20.97.
27. Letter from Robert Oppenheimer to Leon Motchane, Institut des Hautes Études, 4.26.62.
28. Memorandum from Robert Oppenheimer to Atle Selberg, 4.26.62.
29. Stefan A. Burr, professor of computer science, City College of New York, interview, 5.95.
30. A. Borel, interview.
31. Ibid.
32. Gaby Borel, interview, 10.94.
33. Al Vasquez, interview, 6.17.97.
34. Lloyd S. Shapley, interview, 10.94.
35. Ibid.
36. Postcard from J. Nash to V. Nash, 7.62.
37. Ed Nelson, professor of mathematics, Princeton University, interview, 8.17.95.
38. Lars Hörmander, interview, 2.13.97.
39. John Nash, personal communication with Harold Kuhn, 8.97.
40. Hörmander, interview.
41. Ibid.
42. Death certificate of Carlos Larde, State Department of Health, New Jersey, 7.2.62.
43. Postcard from John Nash to Martha Nash Legg, 7.24.63.
44. John Danskin, interview, 10.19.95.
45. Confidential source.
46. Proceedings, International Congress of Mathematicians, Stockholm, 1962.
47. Letter from John Nash to Martha Nash Legg, 9.20.62.
48. Unsigned postcard to mathematics department, Princeton University, 9.1.62.
49. Uitti, interview.
50. Letter from John Nash to M. Legg, 11.19.62.
51. Ibid., 1.26.63.
52. M. Legg, interview, 3.30.96.
53. *Alicia L. Nash* vs. *John Forbes Nash,* Complaint, Superior Court of New Jersey, Mercer County, 12.27.62; Frank L. Scott, attorney, interview, 8.12.97.
54. M. Legg, interview, 8.2.95.
55. *A. Nash* vs. *J. Nash,* op. cit.
56. Judgment Nisi, *Alicia Nash* vs. *John Forbes Nash,* Superior Court of New Jersey, Mercer County, 5.1.63.
57. Final Judgment (Divorce), Alicia L. Nash and John Forbes Nash, 8.2.63.
58. Robert Winters, interview, 8.9.95.

59. Letter from James G. Miller to Albert E. Meder, Jr., treasurer, American Mathematical Society, 4.2.63.

60. Harold Kuhn, interview, 8.95.

61. Letter from William Ted Martin to Albert W. Tucker, 4.1.63.

62. Ibid.

63. Letter from Albert E. Meder to William Ted Martin, 3.28.63.

64. Confidential source.

65. Donald Spencer, interview, 11.28.95.

66. Winters, interview.

67. Letter from Martha Nash Legg to Donald Spencer, 4.24.63.

42: The "Blowing Up" Problem

1. Robert Garber, interview, 5.6.96.

2. Ken Kesey, *One Flew Over the Cuckoo's Nest* (New York: Viking, 1962); Joanne Greenberg, *I Never Promised You a Rose Garden* (New York: Signet, 1964); Thomas S. Szasz, *The Myth of Mental Illness* (New York: Hoeber-Harper, 1961).

3. William Otis, psychiatrist, interview, 5.3.96.

4. Garber, interview.

5. Alicia Nash, interview, 8.15.97.

6. Otis, interview.

7. A. Nash, interview.

8. Martha Nash Legg, interview, 3.30.96.

9. Garber, interview.

10. Ibid.

11. Frank L. Scott, interview, 11.12.97.

12. Garber, interview.

13. Letter from John Nash to Norbert Wiener, 5.1.63.

14. Interviews with A. Nash; Donald Spencer, 11.28.95; Gaby Borel, 3.14.96.

15. Howard Mele declined to be interviewed, 4.9.96.

16. New Jersey Board of Medicine.

17. Interviews with Garber and Otis.

18. Belle Parmet, social worker, interview, 8.24.97.

19. Letter from J. Nash to N. Wiener.

20. Garber, interview.

21. Letter from John Nash to Virginia Nash, 8.10.63.

22. Ibid., 8.22.63.

23. Ibid., 8.29.63.

24. Richard S. E. Keefe and Phillip D. Harvey, *Understanding Schizophrenia* (New York: Free Press, 1994), p. 48.

25. Louisa Cauvin, interview, 8.25.97.

26. Armand Borel, interview, 3.1.96.

27. Ibid.

28. Memorandum from Robert Oppenheimer to Atle Selberg, 9.30.63.

29. Letter from David Gale to Deane Montgomery, 1.3.64.

30. Letter from J. Nash to V. Nash, 10.31.63.

31. Ibid., 3.14.64.

32. Ibid., 10.31.64 and 12.13.64.

33. John Nash, plenary lecture, World Congress of Psychiatry, Madrid, 8.26.96, op. cit.

34. Heisuke Hironaka, "On Nash Blowing Up," in *Arithmetic and Geometry II* (Boston: Birkhauser, 1983).

35. William Browder, interview.

36. Memorandum from John Milnor to Dean of Faculty J. Douglas Brown, 4.8.64.

37. Ibid.

38. Letter from Howard S. Mele to John Milnor, 3.30.64.

39. Garber, interview.

40. Letter from H. S. Mele to J. Milnor.

41. Memorandum from J. Douglas Brown to Robert F. Goheen, 4.6.64.

42. Letter from Ernest J. Johnson to John Nash, 5.1.64.

43. Letter from J. Nash to V. Nash, 2.18.64.

44. Ibid., 3.14.64.

45. Ibid., 3.64.

46. During the spring, Nash wrote to a colleague in Europe saying that he hoped to accept a visiting position at the Institut des Hautes Études near Paris, arranged by Alexandre Grothendieck.

47. M. Legg, interview, 3.29.96.

48. Ibid.

49. Letter from John Nash to Martha Nash Legg, 4.64.

50. Karl Uitti, interview, 8.22.97.

51. Letter from J. Nash to V. Nash, 2.18.64.

52. Letter from John Nash to a colleague, 5.64 or 6.64.

53. Letter from John Nash to Robert Oppenheimer, 5.24.64.

54. The 1964 Summer Research Institute on Algebraic Geometry, American Mathematical Society, *Notices*, October 1963; also John Tate, professor of mathematics, University of Texas, interview, 6.20.97.

55. Letter from J. Nash to V. Nash, 8.31.64.

56. Ibid.

57. John Nash, plenary lecture, op. cit.

58. Ibid.

59. Ibid.

60. Letter from John Nash to Arthur Mattuck, 11.13.71.

61. Harold Kuhn, e-mail, 5.96.

62. Letter from J. Nash to V. Nash, 8.31.64.

63. Postcard from John Nash to Virginia Nash, 9.2.64.

64. Jean Pierre Serre, e-mail, 2.15.96.

65. Postcard from J. Nash to V. Nash, 9.7.64.

66. Memorandum from A. W. Tucker to J. D. Brown, 9.18.64.

67. Postcard from J. Nash to V. Nash, 9.64.

68. Atle Selberg, interview, 1.23.96.

69. Letter from John Nash to John Milnor, 12.27.64.

70. Interviews with John Danskin, 10.9.96; also with William Lucas, professor of mathematics, Claremont Graduate School, 6.27.95, and Herbert Scarf, professor of mathematics, Yale University, 8.97.

71. Danskin, interview.

72. Kuhn, interview.

73. Richard C. Palais, professor of mathematics, Brandeis University, interview, 11.6.95.

74. A. Borel, interview.

75. Palais, interview.

76. Letter from J. Nash to V. Nash, 7.29.65.

43: Solitude

1. Letter from John Nash to Martha Nash Legg, 1.16.66.

2. Martha Nash Legg, interview, 3.29.96.

3. Letter from J. Nash to M. Legg, 7.27.65.

4. Ibid., 8.2.65.

5. John David Stier, interviews, 6.29.96 and 9.20.97.

6. Letter from J. Nash to M. Legg, 10.31.65.

7. Ibid., 5.1.66.

8. Ibid.

9. J. D. Stier, interviews, 6.29.96 and 9.20.97. Except where noted, the facts of John David Stier's childhood are drawn from these interviews.

10. Eleanor Stier, interview, 3.25.96.

11. J. D. Stier, interview, 9.20.97.

12. Letter from J. Nash to M. Legg, 1.16.66.

13. Ibid., 2.22.66.

14. Ibid., 2.27.66.

15. Ibid., 4.24.66.

16. Ibid., 5.8.66.

17. Letter from John Nash to Virginia Nash, 10.31.65.

18. Ibid.

19. Letter from J. Nash to M. Legg, 11.14.65.

20. Letters from J. Nash to V. Nash, 10.31.65 and 1.16.65.

21. Letter from J. Nash to M. Legg, 11.28.65.

22. Ibid.

23. Ibid., 1.9.66.

24. Letters from J. Nash to V. Nash, 1.16.65, and to M. Legg, 2.22.66; also Joan Berkowitz, interview, 8.28.97.

25. Palais, interview.

26. Al Vasquez, interview, 6.17.97.

27. "Analyticity of Solutions of Implicit Function Problems with Analytic Data," *Annals of Mathematics*, vol. 84 (1966), pp. 345–55.

28. Harold Kuhn, interview, 7.17.97.

29. Letter from J. Nash to M. Legg, 9.19.66.

30. Egbert Brieskorn, professor of mathematics, University of Bonn, interview, 1.27.98.

31. Letters from J. Nash to M. Legg, 12.5.65 and 5.1.66.

32. Letter from J. Nash to M. Legg, 2.27.66.

33. Letter from J. Nash to V. Nash, 1.9.66.

34. Kuhn, interview, 5.96. The paper was not rejected, according to Nash, but the editors asked for revisions that he never made.

35. Mikhail Gromov, interview, 12.15.97.

36. This point was raised by Francine M. Benes, psychiatrist, McLean Hospital, interview, 2.13.96.

37. John Nash visited Gian-Carlo Rota in New York City sometime during his first year in Boston. Rota recalled that at lunch Nash traced patterns on his plate and complained that shock treatments had caused him "to forget all my mathematics," interview, 10.29.94.

38. Richard Wyatt, personal communication, 6.97.

39. This was Max Shiffman at Stanford University. Donald Spencer, interview, 11.29.95.

40. Letter from J. Nash to M. Legg, 6.26.96.

41. Zipporah Levinson, interview, 11.15.96.

42. Letter from J. Nash to M. Legg, 5.22.66.

43. Letter from John Nash to Harold Kuhn, 5.17.66.

44. Palais, interview.

45. Vasquez, interview.

46. Letter from J. Nash to M. Legg, 9.1.66.

47. Martha Legg quoting her letter of 9.28.66 to Pattison Esmiol.

48. M. Legg, interview.

49. Letter from Pattison Esmiol to Martha Nash Legg, 10.7.66.

50. Letter from J. Nash to M. Legg, 10.8.66.

51. M. Legg, interview.

52. Letter from J. Nash to M. Legg, 11.66.

53. Ibid., 11.28.66.

54. Vasquez, interview.

55. Joseph Kohn, interview, 1.16.96.

56. Z. Levinson, interview, 11.15.96.

57. Richard Nash, interview, San Francisco, 1.6.96.

58. Letter from J. Nash to M. Legg, 2.67, saying that he had been in Seattle since February.

59. Postcard from John Nash to Martha Nash Legg, 3.11.67, saying that he had been in Santa Monica for about ten days and would be returning to Roanoke by March 22.

60. Jacob Bricker, interview, 5.22.97.

61. Letter from P. Esmiol to M. Legg, 4.19.67.

62. Gilbert Strand, professor of mathematics, MIT, e-mail, 6.5.97.

63. Letter from Armand Borel to Norman Levinson, 5.17.67.

64. Greeting card from John Nash to Arthur Mattuck, 1.15.73.

65. Palais, interview.

66. Letter from John Nash to Jürgen Moser, 5.23.67.

67. Z. Levinson, interview, 11.15.96.

68. Letter from J. Nash to M. Legg, 6.26.67.

69. Z. Levinson, interview.

70. Anna Rosa Kohn, interview, 1.16.96.

71. Letter from Norman Levinson to Martha Nash Legg, 6.30.67.

44: A Man All Alone in a Strange World

1. Letter from John Nash to Arthur Mattuck, 8.5.68.

2. Ibid.

3. Letter from John Nash to a colleague, 1967.

4. Martha Nash Legg, interview, 3.2.96.

5. James Glass, *Delusion* (Chicago: University of Chicago Press, 1985).

6. M. Legg, interview, 10.94.

7. Ibid., 8.31.95.

8. Letter from J. Nash to A. Mattuck, 8.8.67.

9. See, for example, *Diagnostic and Statistical Manual of Mental Disorders* (Washington, D.C.: American Psychiatric Press, 1987). Ming T. Tsuang, Stephen V. Faraone, and Max Day, "Schizophrenic Disorders," op. cit.

10. E. Fuller Torrey, *Surviving Schizophrenia* (New York: Harper & Row, 1988).

11. "... symptoms of clouded consciousness and disorientation in schizophrenia are relatively rare," Richard S. E. Keefe and Phillip D. Harvey, *Understanding Schizophrenia*, op. cit.

12. Letter from J. Nash to A. Mattuck, 3.18.68.

13. See, for example, Torrey, op. cit. Also Glass, op. cit., and James Glass, professor of government and politics, University of Maryland, research affiliate of the Sheppard and Enoch Pratt Hospital, interview, 10.94.

14. Letter from J. Nash to A. Mattuck, 7.24.67.

15. Ibid., 8.8.67.

16. Ibid., 9.9.67.

17. Ibid., 10.7.67.

18. Ibid., 9.9.67.

19. Ibid., 1.10.68.

20. References to the story of Jacob and Esau appear in numerous letters and postcards written by Nash between 1967 and 1969, including 8.8.67, 9.25.67, 10.7.67, 11.8.67, 12.24.67, and 6.16.69.

21. Letter from J. Nash to A. Mattuck, 1.20.68.

22. Ibid., 2.22.68.

23. Ibid., 3.10.68.

24. Ibid., 6.16.69.

25. Letter from John Nash to Eleanor Stier, 8.20.68.

26. Letter from J. Nash to A. Mattuck, 8.11.67.

27. Ibid., 11.8.67.

28. Letter from J. Nash to A. Mattuck, 3.18.68.

29. Ibid., 2.27.68.

30. Ibid., 4.24.69.

31. See, for example, Keefe and Harvey, op. cit., p. 110.

32. Letter from J. Nash to A. Mattuck, 11.11.69.

33. See, for example, Keefe and Harvey, op. cit., pp. 6–7.

34. Peter Newman, interview, 12.12.95.

35. Letter from J. Nash to V. Nash, 8.8.68.

36. The example given combines phrases from two letters to Arthur Mattuck, 9.9.67 and 3.18.68. Nash ended virtually every letter in this period with a variation on this paragraph.

37. M. Legg, interview, 3.2.96. The account of the remainder of Nash's interlude in Roanoke comes from this interview.

45: Phantom of Fine Hall

1. Joseph Kohn, interview, 7.25.95.

2. David Raoul Derbes, University of Chicago, e-mail, 3.27.95; Daniel Rohrlich, University of Tel Aviv, e-mail, 9.3.97.

3. Derbes, e-mail.

4. Sylvain Cappell, professor of mathematics, Courant Institute, 2.29.96.

5. Lee Mosher, professor of mathematics, Rutgers University at Newark, interview, 9.20.97.

6. Derbes, e-mail.

7. Mark Reboul, interview, 8.30.97.

8. Steven Ebstein, e-mail, 3.28.95.

9. Sara Beck, University of Tel Aviv, e-mail, 5.31.95.

10. Ibid.

11. Ibid.

12. Ibid.

13. Frank Wilczek, professor of physics, Institute for Advanced Study, interview, 9.11.97.

14. Letter from Mark B. Schneider, professor of physics, Grinnell College, to author, 9.20.95.

15. Letter from David A. Cox, professor of mathematics, Amherst College, to author, 3.27.95.

16. Letter from M. Schneider to author, 9.28.95.

17. Marc D. Rayman, chief mission engineer, New Millennium Program, NASA, e-mail, 11.24.95.

18. Letter from M. Schneider to author.

19. Wilczek, interview.

20. Ibid.

21. Harold Kuhn, interview, 8.30.97.

22. Margaret Wertheim, "When 1 Plus 1 Makes Neither 2 Nor 11," *New York Times,* 1997.

23. Hale Trotter, professor of mathematics, Princeton University, interview, 11.29.95.

24. Peter Cziffra, librarian, Fine Hall, interview, 8.26.97.

25. William Browder, interview, 12.6.95.

26. James Glass, interview, 10.94.

27. Ibid.

28. Roger Lewin, professor of psychiatry, University of Maryland, interview, 10.94.

29. Steven Bottone, e-mail, 9.2.97.

30. Daniel Feenberg, research associate, National Bureau of Economic Research, interview, 10.94.

31. Trotter, interview, 9.11.97.

32. Reboul, interview.

33. Feenberg, interview.

34. Trotter, interview, 9.30.96.

35. Marc Fisher, reporter, *Washington Post,* e-mail, 3.29.95.

36. Charles Gillespie, professor of history, Princeton University, interview, 7.26.95.

37. Amir H. Assadi, professor of mathematics, University of Wisconsin, interview, 12.13.95.

38. Kohn, interview.

39. Claudia Goldin, professor of economics, Harvard University, interview, 8.30.95.

40. Feenberg, interview.

41. Alicia Nash, interview, 12.6.97.

42. Interviews with Alan Hoffman, 10.94; Lloyd Shapley, 10.94; George Nemhauser, 8.29.97; Albert W. Tucker, 10.94.

43. Shapley, interview.
44. Ibid.
45. Nemhauser, interview.
46. Hoffman, interview.
47. Ibid.

46: A Quiet Life

1. Letter from Alicia Nash to Martha Nash Legg and Virginia Nash, 11.8.68.
2. Ibid.
3. Gillian Richardson, interview, 12.14.95.
4. John Coleman Moore, professor of mathematics, Princeton University, interview, 10.6.95.
5. George Whitehead, interview, 12.12.95.
6. Interviews with Moore, also with Gaby Borel, 10.94 and 3.14.96.
7. Herb Gurk, RCA, interview, 4.23.96.
8. Alicia Nash, private communication, 12.6.97.
9. Martha Nash Legg, interview, 3.30.96; confirmed by Alicia Nash in private communication.
10. Interview with Moore, and with G. Borel, 10.6.95.
11. A. Nash, private communication, and interview, 12.28.95.
12. A. Nash, interview, 12.28.95.
13. Ibid., 1.10.95.
14. Ibid.
15. Odette Larde, interview, 12.8.95.
16. Moore, interview, 10.94.
17. Richard Keefe, interview, 5.95.
18. Richard S. E. Keefe and Phillip D. Harvey, *Understanding Schizophrenia,* op. cit., p. 9.
19. A. Nash, interview, 1.10.95.
20. A. Nash, private communication, 12.6.97.
21. Joyce Davis, interview, 5.30.96.
22. Anna Bailey, interview, 5.29.97.
23. A. Nash, interview, 1.10.95. In addition, interviews with John Charles Martin Nash, Harold Kuhn, Gaby Borel, and others.
24. David Salowitz, "It's Not a Matter of Degrees: John Nash, Shy High School or College Degree, Seeks Ph.D.," *The Princeton Packet,* 7.1.81.
25. A. Nash, interview, 1.10.95.
26. Amir Assadi, interview, 2.4.96.
27. Solomon Leader, interview.
28. A. Nash, interview, 5.16.95.
29. Salowitz, op. cit.
30. Ibid.
31. A. Nash, interview, 5.16.95. Also letter from John Nash to Richard Keefe, 1.14.95.
32. Salowitz, op. cit.
33. Bailey, interview.
34. A. Nash, interview, 5.16.95.
35. Armand Borel, interview, 3.1.96.
36. Moore, interview, 10.5.94.
37. G. Borel, interview, 10.94.
38. John David Stier, interview, 9.20.97.
39. Letter from Alicia Nash to Arthur Mattuck, 11.27.71.
40. J. D. Stier, interview.
41. Norton Starr, professor of mathematics, Amherst College, interviews, 7.95 and 1.20.98.
42. Eleanor Stier, interview, 3.18.96.
43. John Stier, interview, 1.21.98.
44. Letter from John Nash to Arthur Mattuck, 1.15.73.
45. E. Stier, interview, 3.18.96.
46. Irving I. Gottesman, professor of psychology, University of Virginia, interview, 1.16.98.
47. Kenneth L. Fields, professor of mathematics, Rider University (formerly Rider College), interview, 1.30.98.
48. Melvyn B. Nathanson, professor of mathematics, Graduate Center of the City University of New York, interview, 1.31.98.
49. John C. M. Nash (with Melvyn B. Nathanson), "Cofinite Subsets of Asymptotic Bases for the Positive Integers," *Journal of Number Theory,* vol. 20, no. 3 (1985), pp. 363–72; John C. M. Nash, "Results in Bases in Additive Number Theory," Ph.D. thesis, Rutgers University, 1985.
50. John C. M. Nash, "Some Applications of a Theorem of M. Kneser," *Journal of Number Theory,* vol. 44, no. 1 (1993), pp. 1–8.
51. John C. M. Nash, "On B_4 Sequences," *Canadian Mathematical Bulletin,* vol. 32, no. 4 (1989), pp. 446–49.
52. Alicia Nash, interview, 9.97.

Part Five: THE MOST WORTHY

47: Remission

1. Peter Sarnak, professor of mathematics, Princeton University, interview, 8.25.95.

2. E-mail from John Nash to Harold Kuhn, 6.20.96.

3. Hale Trotter, interviews, 11.29.95 and 9.10.97.

4. Mark Dudey, professor of economics, Rice University, interviews, 10.94 and 6.24.95.

5. Daniel Feenberg, interview, 10.94.

6. Letter from Edward G. Nilges to author, 8.19.95.

7. Lloyd S. Shapley, interview, 10.94.

8. George Winokur and Ming T. Tsuang, *The Natural History of Mania, Depression and Schizophrenia* (Washington, D.C.: American Psychiatric Press, 1996), p. 28.

9. Letter from John Nash to Richard Keefe, 1.14.95. Nash gives Johnny's diagnosis as "paranoid schizophrenia" and "schizo-affective disorder."

10. See, for example, Irving I. Gottesman, *Schizophrenia Genesis,* op. cit., p. 18; Michael R. Trimble, *Biographical Psychiatry* (New York: John Wiley & Sons, 1996), pp. 184–85.

11. John Forbes Nash, Jr., *Les Prix Nobel 1994,* op. cit.

12. John Nash, plenary lecture, World Congress of Psychiatry, Madrid, 8.26.96, op. cit.

13. Harold Kuhn, interview, 9.95.

14. Letter from John Nash to Richard Keefe, 1.14.95. Nash has made the same point to many people.

15. Winokur and Tsuang, op. cit., p. 30; also Manfred Bleuler, *The Schizophrenic Disorders: Long-Term Patient and Family Studies* (New Haven: Yale University Press, 1978).

16. Gerd Huber, Gisela Gross, Reinhold Schuttler, and Maria Linz, "Longitudinal Studies of Schizophrenic Patients," *Schizophrenia Bulletin,* vol. 6, no. 4 (1980).

17. C. M. Harding, G. W. Brooks, T. Ashikaga, J. S. Strauss, and A. Brier, "The Vermont Longitudinal Study of Persons with Severe Mental Illness, I and II," *American Journal of Psychiatry,* vol. 144 (1987), pp. 718–26, 727–35. E. Johnstone, D. Owens, A. Gold et al., "Schizophrenic Patients Discharged from Hospital: A Follow-Up Study," *British Journal of Psychiatry,* no. 145 (1984), pp. 586–90, found that 18 percent of the 120 in the study had no significant symptoms and were functioning satisfactorily; 50 percent were still psychotic; and the remainder were somewhere in between. Only two subjects, both of whom had been hospitalized only once, were considered *truly* well.

18. Richard Wyatt, head of neuropsychiatry, National Institute of Mental Health, personal communication, 12.97. See also Winokur and Tsuang, op. cit., pp. 199–217.

19. Winokur and Tsuang, op. cit., pp. 267–68.

20. Huber et al., op. cit.

21. Richard Wyatt, interview, 5.5.96.

22. E. Fuller Torrey, *Surviving Schizophrenia,* op. cit.

23. E-mail from J. Nash to H. Kuhn, 6.1.95.

24. John Forbes Nash, Jr., *Les Prix Nobel 1994,* op. cit.

25. Letter from J. Nash to R. Keefe.

26. John Forbes Nash, Jr., *Les Prix Nobel 1994,* op. cit.

27. Social Science Citation Index, various dates.

28. John Conway, professor of mathematics, Princeton University, interview, 10.94.

29. Nash's work on Riemannian embeddings and partial differential equations would likely have made him a strong candidate for a Fields in the 1960s and his contributions to game theory might easily have been honored with a Nobel as early as 1983, when Gerard Debreu won for his work on general equilibrium theory. He would certainly have garnered lesser honors such as membership in the National Academy of Sciences and the American Academy of Arts and Sciences.

30. Amartya Sen, professor of economics, Harvard University, interview, 12.92.

31. Fellows of the Econometric Society as of January 1988, *Econometrica,* vol. 56, n. 3 (May 1988).

32. Ariel Rubinstein, professor of economics, University of Tel Aviv and Princeton University, interviews, 1.96 and 2.96.

33. Mervyn King, professor of economics, London School of Economics, and vice-chairman, Bank of England, interview, 2.28.96.

34. Letter from Julie Gordon, executive director, The Econometric Society, to author, 2.2.96.

35. King, interview.

36. Interviews with Gary Chamberlain, professor of economics, Harvard University, 2.28.96; Beth E. Allen, professor of economics, University of Minnesota, 2.26.96.

37. Letter from Truman Bewley, professor of economics, Yale University, to Ariel Rubinstein, undated (spring 1989).

38. Ibid., 6.4.89.

39. Truman Bewley, interview, 2.20.96.

40. John Dawson, *Logical Dilemmas: The Life and Work of Kurt Gödel,* op. cit.

41. Ibid.

42. Ken Binmore, Roger Myerson, Ariel Rubinstein, "Nomination of Candidates as a Fellow," 1990.

43. Letter from J. Gordon to author, 1.31.96.

48: The Prize

1. Jörgen W. Weibull, Stockholm School of Economics and member economics prize committee, interview, 11.14.96.
2. Ibid.
3. Carl-Olof Jacobson, secretary-general of the Royal Swedish Academy of Sciences, interview, 2.12.97.
4. Kenneth Birnum, game theorist at the London School of Economics, for example, recently wrote to Harold Kuhn (e-mail, 1.7.98) that he had nominated Nash for the Nobel once in the 1980s. "I didn't persist in nominating him because nobody seemed to take the idea seriously."
5. Statutes of the Nobel Foundation, 4.27.95.
6. Michael Sohlman, executive director, Nobel Foundation, interview, 2.11.97.
7. Ibid.
8. Karl-Göran Mäler, executive director, Beijer Institute of the Royal Swedish Academy of Sciences, interview, 2.12.97.
9. Assar Lindbeck, "The Prize in Economic Science in Memory of Alfred Nobel," *Journal of Economic Literature*, vol. 23 (March 1985), pp. 37–56.
10. Harriet Zuckerman, *Scientific Elite: Nobel Laureates in the United States* (London: Free Press, 1977).
11. Lindbeck, op. cit.
12. See, for example, John E. Morrill, "A Nobel Prize in Mathematics," *The American Mathematical Monthly*, vol. 102, no. 10 (December 1995).
13. Lars Gårding and Lars Hörmander, "Why Is There No Nobel Prize in Mathematics?" *The Mathematical Intelligencer* (July 1985), pp. 73–74.
14. Jacobson, interview.
15. The sketch of Lindbeck is based on the author's interview with him in Stockholm on 2.12.97, two autobiographical essays, and the impressions of members of the prize committee and the Academy of Sciences, including Carl-Olof Jacobson, 2.12.97; Karl-Gustaf Löfgren, professor of economics, University of Umea, 2.12.97; Karl-Göran Mäler, 2.12.97; Jörgen Weibull and Torsten Persson, visiting professor, Harvard University, 10.4.94 and 3.7.97.
16. Persson, interview, 3.7.97.
17. Löfgren, interview.
18. Mäler, interview.
19. Lindbeck, "The Prize in Economic Science," op. cit.
20. Löfgren, interview.
21. Kerstin Fredga, as told to Harold Kuhn at the 12.94 Nobel ceremony in Stockholm, 1.95.
22. By the late 1980s, Harold Kuhn and other game theorists were nominating Nash. Others, however, saw no point in doing so. "I did not nominate him," Shubik later recalled. "He was better than several of the people I nominated, but it seemed that they'd throw him out because he's nuts. The other reason was that I thought the bargaining work was better than the stuff on noncooperative equilibrium," interview, 12.13.96.
23. Lindbeck, interview, 2.12.97.
24. Ariel Rubinstein, interview, 6.26.95.
25. Ariel Rubinstein, "Perfect Equilibrium in a Bargaining Model," *Econometrica*, no. 50 (1982), pp. 97–109.
26. Rubinstein, interview, 6.95.
27. Weibull, interview, 1.14.96.
28. Ibid.
29. Ibid.
30. E-mail from Eric Fisher, assistant professor of economics, Ohio State University, to author, 7.25.95.
31. Weibull, interview, 11.6.96.
32. Gene Grossman, professor of economics, Princeton University, interview, 9.93. Grossman was the first to point out to the author, a reporter at *The New York Times*, that Nash might share a Nobel.
33. Nobel Symposium on Game Theory: Rationality and Equilibrium in Strategic Interaction, Bjorkborn, Sweden, June 18–20, 1993.
34. Confidential source who attended the conference.
35. Persson, interview.
36. Confidential source who attended the conference.
37. Fax from Jörgen Weibull to Harold Kuhn, 7.14.93.
38. Letter from Robert J. Leonard to Harold Kuhn, 7.27.93.
39. Jacobson, interview.
40. Lindbeck, interview.
41. Ibid.
42. Confidential source.
43. Jacobson, interview.
44. Löfgren, interview.
45. Lindbeck, interview.
46. Ibid.
47. Ibid.
48. Shapley's most important work is in cooperative game theory while Schelling's work is in applications of game theory.

49. Lindbeck, interview.
50. Ibid.
51. The sketch of Stahl is based on interviews with his brother Ingolf Stahl, 2.12.97; Mäler; Lindbeck; Löfgren; Weibull; David Warsh, columnist, *Boston Globe,* 2.5.97; and others.
52. Ingemar Stahl, professor of law, Lund University, interview, 2.4.97.
53. Letter from Lars Hörmander to Ingemar Stahl, 9.10.93, with Nash bibliography.
54. Ibid.
55. Ingemar Stahl, interview.
56. Ibid.
57. Ibid.
58. Confidential source present at the discussion.
59. Ibid.
60. Ingemar Stahl, interview.
61. Confidential source.
62. Ibid.
63. Interviews with Lindbeck and Jacobson.
64. Weibull, interview.
65. Confidential source.
66. David Warsh, "Game Theory Plays Strategic Role in Economics' Most Interesting Problems," *Chicago Tribune,* 7.24.94.
67. Christer Kiselman, professor of mathematics, University of Uppsala, interview, 3.5.97.
68. Ibid.
69. Confidential source.
70. Weibull, interview, 11.6.96.
71. Lindbeck, interview.
72. Ibid.
73. Ibid.
74. Jacobson, interview.
75. Confidential source.
76. Lindbeck, interview.
77. Ibid.; also confidential source.
78. As quoted by Harold Kuhn, interview, 1.95.
79. E-mail from Harold Kuhn to Harold Shapiro, president, Princeton University, 9.1.94.
80. Confidential source.
81. Erik Dahmen, professor of economics, Stockholm Institute of Economics, and member, Royal Swedish Academy of Sciences, interview, 2.12.97.
82. Confidential source.
83. Anders Karlquist, interview, 3.17.97.
84. Lars Gårding, professor of mathematics, Lund University, personal communication, 2.10.97.
85. Bengt Nagel, personal communication, 2.10.97.
86. Confidential source.
87. Kjell Olof Feldt, "I Nationalekonomns Atervandsgrand," *Moderna Tider* (March 1994).
88. Karlquist, interview.
89. Confidential source.
90. Lindbeck, interview.
91. Confidential source.
92. Ibid.
93. Statutes of the Nobel Foundation.
94. Confidential source.
95. Ibid.
96. Jacobson, interview.
97. Confidential source.
98. Jacobson, interview.
99. Ingemar Stahl, interview.
100. Sohlman, interview.
101. Johann Schuck, reporter, article in *Dagens Nyheter,* 12.10.94. Schuck broke the story of the behind-the-scenes fight between Stahl and Lindbeck that delayed the announcement of the prize. A translation was provided by Hans Carlsson, professor of economics, Lund University, 12.4.95.
102. Confidential source.
103. Ibid.
104. Harold Kuhn informed Alicia Nash on Friday, October 7, and Nash himself on October 10, the day before the official announcement.
105. Kiselman, interview.
106. Confidential source with access to the report.
107. Confidential source.
108. Ibid.
109. Confidential source with access to the report.
110. Confidential source.
111. Jacobson, interview.

112. Mäler, interview.
113. Jacobson, interview.
114. Ibid.

49: The Greatest Auction Ever

1. Harold Kuhn, interview, 1.95.
2. William Safire, "The Greatest Auction Ever," *New York Times,* 3.16.95, as quoted by Paul Milgrom, *Auction Theory for Privatization* (New York: Cambridge University Press, forthcoming).
3. Edmund Andrews, "Wireless Bidders Jostle for Position," *New York Times,* 12.5.94.
4. Milgrom, *Auction Theory for Privatization,* op. cit.
5. Michael Rothschild, dean of the Woodrow Wilson School, remarks at conference, "Market Design: Spectrum Auctions and Beyond," Princeton University, 11.9.95.
6. Peter C. Cramton, "Dealing with Rivals? Allocating Scarce Resources? You Need Game Theory" (Xerox, 1994). Nash provided the fundamental theory used to analyze and predict behavior in simple games in which rational players have complete knowledge of each other's preferences and abilities. Harsanyi, in papers published in 1967 and 1968, analyzed games in which some parties had private information. Selten, in 1976, extended the theory to dynamic games, games that take place over time. Cramton gives the offers and counteroffers during a merger negotiation as an example of a dynamic game.
7. Peter Passell, "Game Theory Captures a Nobel," *New York Times,* 10.12.94.
8. Paul Samuelson as quoted by Vincent P. Crawford, "Theory and Experiment in the Analysis of Strategic Interaction," Symposium on Experimental Economics, Econometric Society, Seventh World Congress, August 1995 (draft: September 1994).
9. See, for example, Robert Gibbons, "An Introduction to Applicable Game Theory," *Journal of Economic Perspectives,* vol. 11, no. 1 (Winter 1997), pp. 127–49.
10. Avinash Dixit, interview, 7.97.
11. Avinash Dixit, as quoted by Passell, op. cit.
12. Ibid.
13. John McMillan, *Games, Strategies and Managers* (New York: Oxford University Press, 1992).
14. R. H. Coase, "The Federal Communications Commission," *Journal of Law and Economics* (October 1959), pp. 1–40, quoted by John McMillan, "Selling Spectrum Rights," *Journal of Economic Perspectives,* vol. 8, no. 3 (Summer 1994).
15. Peter C. Cramton, "The PCS Spectrum Auction: An Early Assessment," *The Economist* (August 25, 1995).
16. Milgrom, *Auction Theory for Privatization,* op. cit.
17. Ibid. See also McMillan, "Selling Spectrum Rights," op. cit., pp. 153–55.
18. Ibid.
19. See, for example, McMillan, "Selling Spectrum Rights," op. cit.; Paul Milgrom, "Game Theory and Its Use in the PCS Spectrum Auction," Games '95, conference, Jerusalem, 9.29.95.
20. Milgrom, *Auction Theory for Privatization,* op. cit.
21. Ibid.
22. Ibid.
23. Ibid.
24. McMillan, "Selling Spectrum Rights," op. cit.

50: Reawakening

1. Sylvain Cappell, interview, 2.29.96.
2. Jörgen Weibull, interview, 11.14.96.
3. Harold and Estelle Kuhn, interviews, 1.95.
4. Weibull, interview.
5. Lena Koster, "For the First Time in 30 Years: Economy Prize Winner Lectured in Uppsala," *Uppsala Nya Tidning,* 12.94.
6. Christer Kiselman, interview, 3.4.97.
7. Weibull, interview.
8. John Forbes Nash, Jr., *Les Prix Nobel 1994,* op. cit.
9. As quoted by Harold Kuhn, interview, 7.24.96.
10. E-mail from John Nash to Harold Kuhn, 3.26.96.
11. John Nash, plenary lecture, World Congress of Psychiatry, Madrid, 8.26.96, op. cit.
12. E-mail from J. Nash to H. Kuhn, 11.94.
13. Ibid., 8.6.95 and 8.26.95.
14. Harold Kuhn, interview, 1.95.
15. Armand Borel, interview, 3.1.96.
16. This conversation took place in a taxi on the way to Newark Airport on 12.5.94 and was recounted by Harold Kuhn, interview, 1.95.
17. As quoted by H. Kuhn, interview, 1.95.
18. E-mail from John Nash to Herbert Meltzer, 7.8.97.

19. E-mail from J. Nash to H. Kuhn, 7.16.95.
20. Confidential source.
21. E-mail from J. Nash to H. Kuhn, 5.12.95.
22. Alicia Nash, interview, 5.16.95.
23. H. Kuhn, interview, 7.26.95.
24. Avinash Dixit, personal communication, 1.31.96.
25. E-mail from J. Nash to H. Kuhn, 8.6.95.
26. Ibid.
27. Alicia Nash, personal communication, 11.29.97.
28. E-mail from J. Nash to H. Kuhn, 6.6.96.
29. Ibid., 9.94.
30. John Nash, personal communication, 3.22.96.
31. H. Kuhn, interview, 8.95.
32. Interviews with John David Stier, 9.20.97; Eleanor Stier, 7.95; Arthur Mattuck, 11.7.95.
33. Martha Nash Legg, interview, 3.1.96.
34. J. D. Stier, interview.
35. Ibid.
36. E. Stier, interview.
37. J. D. Stier, interview.
38. E-mail from J. Nash to H. Kuhn, 9.26.95.

Select Bibliography

Bell, E. T. *Men of Mathematics*. New York: Simon & Schuster, 1986.

Blaug, Mark. *Great Economists Since Keynes*. Totowa, N.J.: Barnes & Noble Books, 1985.

Bleuler, Manfred. *The Schizophrenic Disorders: Long-Term Patient and Family Studies*. New Haven: Yale University Press, 1978.

Boehm, George W. "The New Uses of the Abstract." *Fortune* (July 1958).

Brian, Denis. *Einstein: A Life*. New York: John Wiley & Sons, 1996.

Buchwald, Art. *I'll Always Have Paris*. New York: G. P. Putnam & Sons, 1996.

A Century of Mathematics in America. Providence, R.I.: American Mathematical Society, 1988.

Chaplin, Virginia. "Princeton and Mathematics." *Princeton Alumni Weekly* (May 9, 1958).

Chronicle of the Twentieth Century. Mt. Kisco, N.Y.: Chronicle Publications, 1987.

Community of Scholars: Institute for Advanced Study Faculty and Members, 1930–1980, A. Princeton: Institute for Advanced Study, 1980.

Davies, John D. "The Curious History of Physics at Princeton." *Princeton Alumni Weekly* (October 2, 1973).

Davison, Peter. *The Fading Smile: Poets in Boston from Robert Frost to Robert Lowell to Sylvia Plath, 1955–1960*. New York: Knopf, 1994.

Diagnostic and Statistical Manual for Mental Disorders, 3rd ed. Washington, D.C., American Psychiatric Association, 1987.

Dixit, Avinash K., and Barry J. Nalebuff. *Thinking Strategically*. New York: W. W. Norton, 1991.

Dixit, Avinash, and Susan Skeath. *Games of Strategy*. New York: W. W. Norton, 1997.

Eatwell, John, Murray Milgate, and Peter Newman, eds. *The New Palgrave: Game Theory*. New York: W. W. Norton, 1989.

Ewing, John H., ed. *A Century of Mathematics*. Washington, D.C.: The Mathematical Association of America, 1994.

Gardner, Howard. *Creating Minds*. New York: Basic Books, 1993.

Gardner, Martin. *Mathematical Puzzles and Diversions*. New York: Simon & Schuster, 1959.

Glass, James M. *Delusion*. Chicago: University of Chicago Press, 1985.

Goldstein, Rebecca. *The Mind-Body Problem*. New York: Penguin, 1993.

Gottesman, Irving I. *Schizophrenia Genesis: The Origins of Madness*. New York: W. H. Freeman & Co., 1991.

Grob, Gerald N. *The Mad Among Us*. Cambridge: Harvard University Press, 1994.

Halberstam, David. *The Fifties*. New York: Fawcett Columbine, 1993.

Hale, Nathan G., Jr. *The Rise and Crisis of Psychoanalysis in the United States*. New York: Oxford University Press, 1995.

Halmos, Paul R. "The Legend of John von Neumann." *American Mathematical Monthly*, vol. 80 (1973), pp. 382–94.

Hardy, G. H. *A Mathematician's Apology*, with foreword by C. P. Snow. Cambridge, U.K.: Cambridge University Press, 1967.

Heilbroner, Robert. *The Worldly Philosophers*. New York: Simon & Schuster, 1992.

Hironaka, Heisuke. "On Nash Blowing Up," *Arithmetic and Geometry II*. Boston: Birkhauser, 1983.

Hollingdale, Stuart. *Makers of Mathematics*. New York: Penguin, 1989.

Ito, Kyosi, ed. *Encyclopedic Dictionary of Mathematics*, vols. I, II, and III, 3rd ed. Mathematical Society of Japan; Cambridge: MIT Press, 1987.

Jamison, Kay Redfield. *Touched with Fire: Manic-Depressive Illness and the Artistic Temperament*. New York: The Free Press, 1993.

"John von Neumann 1903–1957." *Bulletin of the American Mathematical Society* (May 1958).

Kafka, Franz. *The Castle,* with introduction by Irving Howe. New York: Scholastic Books, 1992.
————. *The Metamorphosis.* New York: Shocken Books, 1995.
Kagel, John H., and Alvin E. Roth. *The Handbook of Experimental Economics.* Princeton: Princeton University Press, 1995.
Kanigel, Robert. *The Man Who Knew Infinity: A Life of the Genius Ramanujan.* New York: Pocket Books, 1992.
Kaplan, Fred. *The Wizards of Armageddon.* Stanford: Stanford University Press, 1983.
Keefe, Richard S. E., and Philip D. Harvey. *Understanding Schizophrenia: A Guide to the New Research on Causes and Treatment.* New York: The Free Press, 1994.
Kuhn, Harold W. Introduction, "A Celebration of John F. Nash, Jr.," *Duke Mathematical Journal,* vol. 81, no. 1 (1995), pp. i–v.
————. "Nobel Seminar: The Work of John Nash in Game Theory, December 8, 1994," *Les Prix Nobel 1994.* Stockholm: Norstedts Tryckeri, 1995.
Larde, Enrique. *The Crown Prince Rudolf: His Mysterious Life After Mayerling.* Pittsburgh: Dorrance, 1994.
Leonard, Robert J. "From Parlor Games to Social Science: Von Neumann, Morgenstern and the Creation of Game Theory, 1928–1944." *Journal of Economic Literature* (1995).
————. "Reading Cournot, Reading Nash: The Creation and Stabilization of the Nash Equilibrium." *The Economic Journal* (May 1994), pp. 492–511.
Lindbeck, Assar. "The Prize in Economic Science in Memory of Alfred Nobel." *Journal of Economic Literature,* vol. 23 (March 1985), pp. 37–56.
Lowell, Robert. "Waking in the Blue." *Life Studies and For the Union Dead.* New York: Farrar Straus and Giroux, 1992.
Luce, R. Duncan, and Howard Raiffa. *Games and Decisions.* New York: John Wiley & Sons, 1957.
McDonald, John. "The War of Wits." *Fortune* (March 1951).
Milnor, John. "A Nobel Prize for John Nash." *The Mathematical Intelligencer,* vol. 17, no. 3 (1995), pp. 14–15.
Nash, John Forbes, Jr. "Sag and Tension Calculations for Cable and Wire Spans Using Catenary Formulas" (with John F. Nash, Sr.). *Electrical Engineering* (1945).
————. "Equilibrium Points in N-Person Games." *Proceedings of the National Academy of Sciences, USA,* vol. 36 (1950), pp. 48–49.
————. *Non-Cooperative Games,* Ph.D. thesis, Princeton University, May 1950.
————. "A Simple Three-Person Poker Game" (with Lloyd S. Shapley). *Annals of Mathematics Study,* vol. 24 (1950).
————. "The Bargaining Problem." *Econometrica,* vol. 18 (1950), pp. 155–62.
————. "Non-Cooperative Games." *Annals of Mathematics,* vol. 54 (1951), pp. 286–95.
————. "Real Algebraic Manifolds." *Annals of Mathematics,* vol. 56, no. 3 (November 1952), pp. 405–21.
————. "Some Experimental N-Person Games" (with G. Kalisch, J. W. Milnor, and E. D. Nering). *Decision Processes,* ed. R. M. Thrall, C. H. Coombs, and R. L. Davis. New York: John Wiley & Sons, 1954.
————. "Two-Person Cooperative Games." *Econometrica,* vol. 21 (1953), pp. 405–21.
————. "A Comparison of Treatments of a Duopoly Situation" (with J. P. Mayberry and M. Shubik). *Econometrica,* vol. 21 (1953), pp. 141–54.
————. "Higher Dimensional Core Arrays for Machine Memories." RAND Memorandum, D-2495, 7.22.54.
————. "LODAR." RAND Memorandum, D-2349, 7.23.54.
————. "Continuous Iteration Method for Solution of Differential Games." RAND Memorandum, RM-1326, 8.18.54.
————. "Parallel Control." RAND Memorandum, RM-1361, 8.27.54.
————. "C^1 Isometric Imbeddings." *Annals of Mathematics,* vol. 60, no. 3 (November 1954), pp. 382–96.
————. "Results on Continuation and Uniqueness of Fluid Flow." *Bulletin of the American Mathematical Society,* vol. 60 (1954), pp. 165–66.
————. "A Path Space and the Stiefel-Whitney Classes." *Proceedings of the National Academy of Sciences USA,* vol. 41 (1955), pp. 320–21.
————. "The Imbedding Problem for Riemannian Manifolds." *Annals of Mathematics,* vol. 63, no. 1 (January 1956), pp. 20–63.
————. "Parabolic Equations." *Proceedings of the National Academy of Sciences USA,* vol. 43 (1957), pp. 754–58.

————. "Continuity of Solutions of Parabolic and Elliptic Equations." *American Journal of Mathematics,* vol. 80 (1958), pp. 931–58.

————. "Le probleme de Cauchy pour les equations differentielles d'un fluide general." *Bull. Soc. Math., France,* vol. 90 (1962), pp. 487–97.

————. "Analyticity of Solutions of Implicit Function Problems with Analytic Data." *Annals of Mathematics,* vol. 84 (1966), pp. 345–55.

————. "Arc Structure of Singularities." *Duke Mathematical Journal,* vol. 81, no. 1 (1996), pp. 31–38.

————. Autobiographical essay, *Les Prix Nobel 1994.* Stockholm: Norstedts Tryckeri, 1995.

————. Plenary lecture, World Congress of Psychiatry, Madrid, 8.26.96 (unpublished).

Nicholi, Armand M., Jr. *The New Harvard Guide to Psychiatry.* Cambridge: The Belknap Press of Harvard University, 1988.

"Norbert Wiener 1894–1964." *Bulletin of the American Mathematical Society,* vol. 72, no. 1, part ii (1964).

Poundstone, William. *Prisoners' Dilemma.* New York: Doubleday, 1992.

Regis, Ed. *Who Got Einstein's Office?* Reading, Mass.: Addison-Wesley, 1987.

Reid, Constance. *Courant in Göttingen and New York.* New York: Springer Verlag, 1976.

Rota, Gian-Carlo. *Indiscrete Thoughts.* Boston: Birkhauser, 1997.

Sass, Louis A. *Madness and Modernism.* New York: Basic Books, 1992.

Schelling, Thomas C. *The Strategy of Conflict.* Cambridge: Harvard University Press, 1960.

Storr, Anthony. *Solitude: A Return to the Self.* New York: Ballantine Books, 1988.

————. *The Dynamics of Creation.* New York: Atheneum, 1972.

Torrey, E. Fuller. *Surviving Schizophrenia: A Family Manual.* New York: Harper & Row, 1988.

Trimble, Michael R. *Biological Psychiatry.* New York: John Wiley & Sons, 1996.

Ulam, Stanislaw. *Adventures of a Mathematician.* New York: Scribner, 1983.

U.S. House of Representatives. *Hearings.* Committee on Un-American Activities, April 22 and 23, 1953.

von Neumann, John, and Oskar Morgenstern. *Theory of Games and Economic Behavior.* Princeton: Princeton University Press, 1944, 1947, 1953.

Williams, John. *The Compleat Strategyst.* New York: McGraw Hill, 1954.

Winokur, George, and Ming Tsuang. *The Natural History of Mania, Depression and Schizophrenia.* Washington, D.C.: American Psychiatric Press, 1996.

Zuckerman, Harriet. *Scientific Elite: Nobel Laureates in the United States.* London: The Free Press, 1977.

Acknowledgments

MANY PEOPLE contributed to this book, two above all: my friend of twenty-five years, Ellen Tremper, who cheered me on and rendered invaluable assistance every step of the way, and Harold W. Kuhn, whose enthusiasm for the enterprise and intimate knowledge of John Nash and the mathematics community was a constant source of guidance and inspiration. No one could have done more.

I am deeply indebted to Alicia Larde Nash and Martha Nash Legg, without whose support I could not have embarked on this biography, much less completed it. I am also grateful to John David Stier, Eleanor Stier, and John Charles Martin Nash for their cooperation, and appreciate John Nash's benign "attitude of Swiss neutrality" toward the undertaking.

No author was ever in better hands than those of Alice Mayhew, my editor, and Kathy Robbins, my agent — not to mention those of Simon & Schuster's terrific publishing team, especially Robert Labrie, Victoria Meyer, Elizabeth Hayes, and Nira Weisel.

I am thankful to Amartya Sen and Phillip Griffiths for enabling me to spend a vital year as a Director's Visitor at the Institute for Advanced Study in Princeton; Gian-Carlo Rota for a shorter but equally critical interlude at the MIT mathematics department; and Vivien Arterberry for a productive week at the RAND Corporation.

Joseph Lelyveld, Soma Golden Behr, and Glenn Kramon of *The New York Times* granted me a generous leave of absence and enthusiastic support.

My colleagues Doug Frantz at *The New York Times* and Rob Norton at *Fortune* gave much-appreciated advice and encouragement at every stage.

Avinash Dixit, Harold Kuhn, Roger Myerson, Ariel Rubinstein, and Robert Wilson patiently shared their insights about game theory and served as valuable sounding boards.

Donald Spencer, Harold Kuhn, Lars Hörmander, Michael Artin, Joseph Kohn, John Milnor, Louis Nirenberg, and Jürgen Moser worked hard to help me convey the originality of Nash's contributions to pure mathematics clearly and accurately.

Superb histories by John McDonald, William Poundstone, Fred Kaplan, and David Halberstam provided much of the context for Nash's tenure at RAND. Ed Regis's lively history of the Institute for Advanced Study and Rebecca Goldstein's delightful novel *The Mind-Body Problem* were also invaluable.

Richard Jed Wyatt guided me through the vast and fascinating literature on schizophrenia. The extraordinary work of Louis Sass, Anthony Storr, John Gunderson, Kenneth Kendler, Irving Gottesman, Richard Keefe, James Glass, Kay Redfield Jamison, and E. Fuller Torrey provided inspiration as well as important information. Special thanks to Connie and Steve Lieber, the founders of the National Alliance for Research on Schizophrenia and Depression, for their interest in this project.

Psychiatrists Paul Howard, Joseph Brenner, Robert Garber, and Peter Baumecker provided firsthand descriptions of the institutions where Nash was treated and glimpses into the mysteries of clinical psychiatry.

Jörgen Weibull and other members of the economics prize committee and the Swedish Academy of Sciences were wonderfully hospitable during my visit to Stockholm and helped me decipher the seemingly inscrutable process by which the *ne plus ultra* of honors is bestowed. Sociologist Harriet Zuckerman's landmark study of Nobel Laureates served as an excellent road map.

Lloyd Shapley's loving and lovely phrase "a beautiful mind" became, at Kathy Robbins's suggestion, the title of the book.

I am infinitely grateful to the hundreds of individuals — mathematicians, economists, psychiatrists, and others who knew John Nash — who supplied the memories from which I've woven together his remarkable story. Every fragment, however tiny, added to the vividness of the whole, and each was gratefully received and treasured. In addition to those already cited, I am particularly indebted to Paul Samuelson, Arthur Mattuck, Paul Cohen, Odette Larde, Dorothy Thomas, Peter Lax, Cathleen Morawetz, Donald Newman, Al Vasquez, Richard Best, John Moore, Armand and Gaby Borel, Zipporah Levinson, Jerome Neuwirth, Felix and Eva Browder, Leopold Flatto, John Danskin, Emma Duchane, and Joyce Davis.

Archivists and librarians at Carnegie Mellon University, Princeton University, MIT, Harvard University, the Institute for Advanced Study, the Rockefeller Archive Center, McLean Hospital, the Swiss National Archives, and the National Archive provided important material and expert guidance. Special thanks to Arlen Hastings, Momota Ganguli, and Elise Hansen at the Institute for Advanced Study for making my year at the institute so productive, and to Richard Wolfe for sharing his knowledge of the Cambridge intellectual community.

Ellen Tremper, Geoffrey O'Brien, Harold Kuhn, Avinash Dixit, Lars Hörmander, Jürgen Moser, Michael Artin, Donald Spencer, Richard Wyatt, and Rob Norton read and commented on various drafts. Their painstaking efforts eliminated mistakes, improved expositions, and added important new insights. All errors that remain are, of course, mine.

My husband, Darryl McLeod, and children, Clara, Lily, and Jack, not only lived with this book and its harried author for three years, but pitched in — on the computer, in the library, around the house — when deadlines were looming and the sky seemed about to fall. For their love and patience I am most indebted.

Index

Photo Credits

1–7: Courtesy of Martha Nash Legg.
8–12, 21, 22: Courtesy of John D. Stier.
13–16, 18, 19, 23: Courtesy of Alicia Nash.
17: Adriano Garsia; courtesy of Alicia Nash.
20: Courtesy of Richard Nash.
24, 25: Pressens Bild.
26: Dick Pettersson, *Upsala Nya Tidning.*